# PACIFIC SYMPOSIUM ON
# BIOCOMPUTING 2005

# PACIFIC SYMPOSIUM ON
# BIOCOMPUTING 2005

Hawaii, USA
4–8 January 2005

*Edited by*

## Russ B. Altman
Stanford University, USA

## A. Keith Dunker
Indiana University, USA

## Lawrence Hunter
University of Colorado Health Sciences Center, USA

## Tiffany A. Jung
Stanford University, USA

## Teri E. Klein
Stanford University, USA

**World Scientific**

NEW JERSEY • LONDON • SINGAPORE • SHANGHAI • HONG KONG • TAIPEI • BANGALORE

*Published by*

World Scientific Publishing Co. Pte. Ltd.

5 Toh Tuck Link, Singapore 596224

*USA office:* 27 Warren Street, Suite 401-402, Hackensack, NJ 07601

*UK office:* 57 Shelton Street, Covent Garden, London WC2H 9HE

**British Library Cataloguing-in-Publication Data**

A catalogue record for this book is available from the British Library.

ISBN 981-256-046-7

Printed in Singapore.

# PACIFIC SYMPOSIUM ON BIOCOMPUTING 2005

2005 marks the tenth year of the Pacific Symposium on Biocomputing (PSB). The first meeting was in 1996 and was organized by Teri Klein and Larry Hunter. They brought Keith Dunker and Russ Altman on the team for 1997, and the organizing group has been in place and stable since then. PSB is a unique conference among those in bioinformatics and computational biology: the session topics are proposed each year and selected in a competitive process. They are chosen based on the opportunity to bring together a critical mass of scientists with shared interests in hot or emerging areas. Grass-roots session organizers who make successful proposals then solicit papers and moderate their review. The PSB organizers take pride in the fact that PSB is often the first venue to cluster papers related to a particular theme. With only a few exceptions, all PSB published manuscripts are available online at http://psb.stanford.edu/psb-online/. In fact, a look at the sessions over the last ten years is a useful way to understand the evolution of our field. A complete listing of the sessions is presented at the end of this preface. With the publication of this issue of the proceedings, PSB has now published more than 500 manuscripts. They are all in PubMed and can be retrieved using the NIH abbreviation for the proceedings "Pac Symp Biocomput."

What has been the impact of PSB papers? PSB is not indexed by the Thomson ISI service, and so overall impact factor is not available (we are working on this!). However, there are a few ways to gauge our impact. Informally, we have frequently noticed publications in other major scientific journals with references to PSB papers. This is particularly true in emerging areas that PSB "covered" before they hit mainstream, and includes gene expression array analysis, network reconstruction, and text analysis. The Citeseer program (http://citeseer.ist.psu.edu/) principally tracks references in computer science, and even then has only partial coverage of all citations. Nonetheless, it provides some measure of impact. The coverage of citations in biology is not good (the *Journal of Molecular Biology* has only 244 citations!), but in the computer science literature, PSB proceedings have been cited 567 times, and there are 71 papers with two or more citations. The top 10 papers in terms of impact in Citeseer are also provided at the end of this preface. We also have provided a histogram showing the relationship of citation frequency to date of publication.

v

In the biology literature, one measure of impact of PSB is the number of citations in the online Highwire Press collection of journals (http://highwire.stanford.edu/). A search for "Pac Symp Biocomput" in Highwire shows 390 Highwire journal manuscripts referring to PSB papers. Highwire provides an automated indexing of these papers, showing which keyword/topics appear most frequently. The list of topics with greater than five papers is also provided below. The three most common topics of papers citing PSB are "gene expression," "networks," and "alignment." Again, the information from the Highwire analysis is limited because it only contains a fraction of all relevant biological publications.

In any case, it is clear that PSB papers are getting attention and are being cited in the archival literature. The organizers are working on linking the online proceedings to PubMed, a task which is straightforward but requires resources that we have so far been unable to muster. In addition, we are working on convincing the Thomson ISI folks to include PSB in their impact factor ratings.

The PSB organizers would like to thank the National Library of Medicine/National Institutes of Health and the U.S. Department of Energy for support. Applied Biosystems continues to sponsor PSB, and as a result, we are able to provide travel grants to meeting participants. PSB values its relationship with the International Society for Computational Biology (http://www.iscb.org/), and is pleased to offer registration discounts to members.

We look forward to the keynote addresses by David Eisenberg of UCLA and Arti Rai of Duke University. Tiffany Jung has expertly managed the process of assembling the proceedings, in addition to many other organization tasks. Of course, the meeting exists because of the dedicated efforts of the session organizers. These busy scientists moderate a high quality review process that must be completed in less than two months, with an overall accept rate of less than 35%. They are:

**Carlos Bustamante, Shamil Sunyaev and Matt Dimmic**
*Inferring SNP Function Using Evolutionary, Structural and Computational Methods*

**Alexander Tropsha and Herbert Edelsbrunner**
*Biogeometry: Applications of Computational Geometry to Molecular Structure*

**Patricia Babbitt, Philip Bourne, and Sean Mooney**
*Inferring Function from Structural Genomics Targets*

**Marylyn Ritchie, Michelle Carrillo, and Russell Wilke**
*Computational Approaches for Pharmacogenomics*

**Alex Hartemink and Eran Segal**
*Joint Learning from Multiple Types of Genomic Data*

**Olivier Bodenreider, Joyce Mitchell, and Alexa McCray**
*Biomedical Ontologies*

PSB 2005 will also feature four tutorials: *Use of Multi-locus Data in Gene Mapping Studies* by Jo Knight; *Association Mapping: Design Issues and Data Analysis Approaches* by Jotun Hein and Leif Schauser; *Data Analysis and Sharing with Web Services* by Michael Jensen and Timothy Patrick; and *Function Prediction: From High Throughput to Individual Proteins* by Yanay Ofran, Marco Punta, and Burkard Rost.

We would like to acknowledge the efforts of those who reviewed the submitted manuscripts on a very tight schedule. The partial list that is included after this preface does not include those who wished to remain anonymous. We apologize to others who we may have left off inadvertently.

We look forward to another exciting meeting, and encourage you to consider proposing a new session topic, teaching a new tutorial, or submitting your work for the conference as it enters its second decade.

Aloha!

*Pacific Symposium on Biocomputing Co-Chairs*　　　　　　*October 1, 2004*

*Russ B. Altman*
*Department of Genetics, Stanford University*

*A. Keith Dunker*
*Center for Computational Biology & Bioinformatics, Indiana University School of Medicine*

*Lawrence Hunter*
*Department of Pharmacology, University of Colorado Health Sciences Center*

*Teri E. Klein*
*Department of Genetics, Stanford University*

# PSB Session Topics Over Last Ten Years (1996-2005)

1996
- The Evolution of Biomolecular Structures
- Discovering, Learning, Analyzing and Predicting Protein Structure.
- Stochastic Models, Formal Systems and Algorithmic Discovery for Genome Informatics.
- Interactive Molecular Visualization.
- Educational Issues in Biocomputing.
- Internet Tools for Computational Biology.
- Population Modelling.
- Hybrid Quantum and Classical Mechanical Methods for Studying Biopolymers in Solution.
- Control in Biological Systems.

1997
- Distributed and Intelligent Databases
- Modern Concepts in Molecular Modeling,
- Extracting Biological Knowledge from DNA Sequences
- Understanding and Predicting Protein Structure
- Biopolymer Structures: Where Do They Come From? Where Are They Going? Evolutionary Perspectives on Biopolymer Structure and Function
- Computing with Biomolecules
- Computation in Biological Pathways
- Biocomputing Education: Further Challenges

1998
- Gene expression and genetic networks
- Molecules to maps: tools for visualization and interaction
- Gene structure identification in large-scale genomic sequence
- Molecular modeling in drug design and biotechnology
- Protein structure prediction
- The relationships between protein structure and function
- Computing with biomolecules
- Complexity and information theoretic approaches to biology
- Distributed and intelligent databases
- Building bioinformation infrastructure in the Pacific Rim

1999
- Gene expression and genetic networks
- Data mining and knowledge discovery in molecular databases
- Computer modeling in physiology: from cell to tissue
- Information theoretic approaches to biology
- Molecules to maps: tools for visualization and interaction
- Computer-aided drug design
- Protein structure prediction
- Disorder in protein structure and function

2000
- Protein evolution and structural genomics
- Protein structure prediction in biology and medicine
- Molecules to maps: tools for visualization and interaction
- Molecular network modeling and data analysis
- Data mining and knowledge discovery in molecular databases
- Identification of coordinated gene expression and regulatory sequences
- Natural language processing for biology
- Computer-aided combinatorial chemistry and cheminformatics
- Applications of information theory in biology
- Human genome variation:  analysis of SNP data

2001
- Human Genome Variation: Linking Genotypes to Clinical Phenotypes
- Disorder and Flexibility in Protein Structure and Function
- DNA Structure, Protein-DNA Interactions, and DNA-Protein Expression
- Structures, Phylogenies, and Genomes
- High Performance Computing for Computational Biology
- Natural Language Processing for Biology
- Genome, Pathway and Interaction Bioinformatics
- Phylogenetics in the Post-Genomic Era
- Bioethics, Fiction Science, and the Future of Mankind

2002
- Human Genome Variation: Disease, Drug Response, and Clinical Phenotypes
- Genome-Wide Analysis and Comparative Genomics
- Expanding Proteomics to Glycobiology
- Literature Data Mining for Biology
- Genome, Pathway and Interaction Bioinformatics
- Phylogenetic Genomics and Genomic Phylogenetics
- Proteins: Structure, Function and Evolution

2003
- Gene Regulation
- Genome, Pathway, and Interaction Bioinformatics
- Informatics Approaches in Structural Genomics
- Genome-wide Analysis and Comparative Genomics
- Linking Biomedical Language, Information and Knowledge
- Human Genome Variation: Haplotypes, Linkage Disequilibrium, and Populations
- Biomedical Ontologies

2004
- Alternative Splicing
- Computational Tools for Complex Trait Gene Mapping
- Biomedical Ontologies
- Joint Learning from Multiple Types of Genomic Data
- Informatics Approaches in Structural Genomics
- Computational and Symbolic Systems Biology

2005

- Inferring SNP Function Using Evolutionary, Structural and Computational Methods
- Biogeometry: Applications of Computational Geometry to Molecular Structure
- Inferring Function from Structural Genomics Targets
- Computational Approaches for Pharmacogenomics
- Joint Learning from Multiple Types of Genomic Data
- Biomedical Ontologies

## CiteSeer's Most Commonly Cited PSB Papers in Computer Science Literature

K. Fukuda, T. Tsunoda, A. Tamura, and T. Takagi. *Toward information extraction: Identifying protein names from biological papers.* 1998.

Liang S., Fuhrman S. and Somogyi R. REVEAL, *A General Reverse Engineering Algorithm for Inference of Genetic Network Architectures.* 1998.

A. Regev, W. Silverman, and E. Shapiro. *Representation and simulation of biochemical processes using the pi-calculus process algebra.* 2001

Matsuno, H., Doi, A., Nagasaki, M., and Miyano, S. *Hybrid Petri net representation of gene regulatory network.* 2000.

C. Leslie, E. Eskin, and W. S. Noble. *The spectrum kernel: A string kernel for SVM protein classification.* 2002

S. Eker, M. Knapp, K. Laderoute, P. Lincoln, J. Meseguer, and K. Sonmez. *Pathway logic: Symbolic analysis of biological signaling.* 2002.

A. J. Hartemink, D. K. Gifford, T. S. Jaakkola, and R. A. Young, *Using graphical models and genomic expression data to statistically validate models of genetic regulatory networks.* 2001

Raychaudhuri, S., Stuart, J. M. and Altman, R. B. *Principal components analysis to summarize microarray experiments: application to sporulation time series.* 2000.

James Thomas, David Milward, Christos Ouzounis, Stephen Pulman, and Mark Carroll. *Automatic extraction of protein interactions from scientific abstracts.* 2000.

Wroe CJ, Stevens RD, Goble CA, Ashburner M. *A methodology to migrate the Gene Ontology to a description logic environment using DAML+ OIL.* 2003.

**CiteSeer number of citations** in computer science literature plotted based on year of publication of PSB papers. There is a 4-5 year lag between having papers published and having them cited.

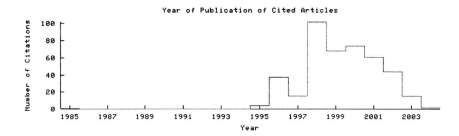

**Highwire Press automated index of topics in papers referring to PSB proceedings. Number of citations discussing the topic in parentheses.**

Gene expression (188)
Network (110)
Alignment (88)
Extracted (56)
Ontology (35)
Side Chain (30)
Disordered; Regions (27)
Splicing (24)
Metabolic pathways (16)
NMR (16)
Support vector machines (14)
Haplotype; Markers (12)
Free energy (12)
HMMs; Models (11)
Protein kinase (10)
Genome maps (10)

Sample sizes (9)
Duplication; Evolution (9)
Precision And Recall (8)
Controlling Gene (7)
Simulation system (7)
Molecular structure (7)
Cross-Linking (7)
Known 3D structure (7)
Pain; Modulation (6)
Prediction Meta (6)
Promoter prediction (6)
Protein Evolution (6)
Hormone; Temporal (6)
Knowledge discovery (6)
Combinatorial Library (6)

**Thanks to the reviewers...**

Finally, we wish to thank the scores of reviewers. PSB requires that every paper in this volume be reviewed by at least three independent referees. Since there is a large volume of submitted papers, paper reviews require a great deal of work from many people. We are grateful to all of you listed below and to anyone whose name we may have accidentally omitted or who wished to remain anonymous.

Constantin Aliferis
Bill Amadio
Dominik Aronsky
Lan Aronson
Vladimir Babenko
Ziv Bar-Joseph
Khemissa Bejaoui
Philip Benfey
Mukesh Bickol
Judy Blake
Erik Boczko
Olivier Bodenreider
Dan Brown
Celeste Brown
Martha Bulyk
Anita Burgun
Carlos Bustamante
Jeff Canter
Gal Chechik
Tin Chen
Cheng Cheng
Christopher Coffey
Greg Cooper
Ronald Cornet
Ann Daly
David Danks
Diego di Bernardo
Matt Dimmic
Adrian Dobra
Dave Edwards
Eleazar Eskin
Brendan Frey

Aldo Gangemi
Zerues
 Gebreegziabhere
Georg Gerber
Nader Ghebranious
Catherine Grasso
Mike Gruninger
Udo Hahn
Lance Hahn
Daniel Hahs
Jonathan Haines
Midori Harris
Frank Hartel
Elizabeth Hauser
Monica M. Horvath
Dimitar Hristovski
Fang-Chi Hsu
Toni Kazic
Mikko Koivisto
Jill Kolesar
Stephen Lake
Guiseppe Lancia
Chris Langmead
Ross Lazarus
Christina Leslie
Shawn Levy
Chun Li
Jane Lomax
Phillip Lord
Yves Lussier
Dimitris Margaritis
Eden Martin

Dan McCarty
Alexa McCray
Robin McEntire
Brett McKinney
Leonid Mirny
Joyce Mitchell
Quaid Morris
Mark Musen
Pauline Ng
Rasmus Nielsen
Uwe Ohler
Grier Page
Tim Patrick
Bret Payseur
Leon Peshkin
Predrag Radivojac
Marco F. Ramoni
Sarah Rayan
Aviv Regev
Rachel Richesson
Tom Rindflesch
Igor Rogozin
Pedro R. Romero
Stefan Schulz
Steffen Schulze-
 Kremer
Jeff Shrager
Saurabh Sinha
Barry Smith
Chris Sorkness
Padmini Srinivasan
Robert Stevens

Chris Stoeckert
Josh Stuart
Shamil Sunyaev
Tricia Thornton
Ioannis Tsamardinos
Vladimir N. Uversky
Leszek Vincent
Nathan Walsh
Wendell Weber
Anja Wille
Jen Williams
Jennifer Williams
Chris Wroe
Gerald Wyckoff
Eric Xing
Chen-Hsiang Yeang
Songmao Zhang
Pierre Zweigenbaum

# CONTENTS

## COMPUTATIONAL APPROACHES FOR PHARMACOGENOMICS

## INFORMATICS APPROACHES IN STRUCTURAL GENOMICS: MODELING AND REPRESENTATION OF FUNCTION FROM MACROMOLECULAR STRUCTURE

## INFERRING SNP FUNCTION USING EVOLUTIONARY, STRUCTURAL, AND COMPUTATIONAL METHODS

## JOINT LEARNING FROM MULTIPLE TYPES OF GENOMIC DATA

# BIOGEOMETRY: APPLICATIONS OF COMPUTATIONAL GEOMETRY TO MOLECULAR STRUCTURE

ALEXANDER TROPSHA

*School of Pharmacy, University of North Carolina,*
*Chapel Hill, NC, USA (alex_tropsha@unc.edu)*

HERBERT EDELSBRUNNER

*Duke University, Computer Science Department,*
*Box 90129, Durham, NC 27708, USA (edels@cs.duke.edu)*

The analysis of structure-function relationships has traditionally been an area of interest of the Pacific Symposium on Biocomputing. Because (macro)molecular shape frequently defines the function, it seems evident that geometric methods should be an essential component of any attempt to understand and simulate biological systems. Existing techniques in computational structural biology and bioinformatics, however, rely primarily on sequence and, in some cases, structure information (in the context of 3D contacts or patterns of contacts) and use statistical and/or energy based methods to analyze the relationship between biological structure and function. They have been developed over three decades and have their roots in methods first applied by computational chemists to much smaller molecular systems. Although there have been significant advancements in the field, a systematic solution of many of the most important biological problems is still elusive, including *ab initio* protein structure prediction, the protein folding process, and ligand to protein docking.

Biogeometry is an emerging scientific discipline at the interface between computational geometry, biochemistry and biophysics, statistics, and chemistry that brings together specialists in the above disciplines to develop new computational techniques and paradigms for representing, storing, searching, simulating, analyzing, and visualizing biological structures. Biogeometry embraces ideas from a wide range of areas of computer science and mathematics, including algorithms, geometry, topology, graphics, robotics, and databases to address some of the most fundamental biological problems such as structure-function relationships for biological molecules.

Although a new discipline, Biogeometry has been a subject of intensive research in several groups for a number of years. The "Computational Geometry

---

*This work is supported by the National Science Foundation grant CCR-0086013.

1

for Structural Biology and Bioinformatics" project has been funded by NSF since 2001. It has involved researchers and students from Duke University, Stanford University, University of North Carolina Chapel Hill, and North Carolina A&T University (see http://biogeometry.duke.edu/ for additional information). Collectively, the collaborating researchers have published many dozens of papers and made numerous presentations at various national and international meetings. The Biogeometry session as part of PSB'05 is the first specialized session with such title ever included in a major computational biology conference, and the papers included in these proceedings present novel methods and developing ideas in a broad range of topics covered by Biogeometry.

The first two papers of the session apply computational geometry approaches to the problem of protein-protein recognition. The paper by Wang *et al* describes a coarse alignment algorithm for efficient protein-protein docking. This algorithm detects protrusions and cavities as local maxima of the novel elevation function, aligns them, and employs a simple scoring function to produce a reliable set of potential docking positions. Using a test set of 25 protein complexes, the authors demonstrate that their algorithm is able to generate near native conformations in all but one case. The paper by Li and Liang presents a novel method for designing peptide libraries to modulate protein-protein interactions. Based on the alpha shapes of antibody-antigen complexes, they develop an empirical pair potential for antigen-antibody interactions that depends on local packing. They demonstrate that this potential successfully discriminates the native interface peptides from a simulated library of 10,000 random peptides for 34 antigen-antibody complexes.

Three papers explore various aspects of protein folding and design problems. For many practical tasks associated with the protein folding problem such as energy functions for folding simulations or fold recognition approaches to structure prediction, it is important to have a set of structure decoys. To this end, Singh and Berger describe their CHAINTWEAK algorithm for rapid generation of near native decoys starting from the native protein conformation. Russell and Guibas present the first application of so-called geometric spanners (geometric graphs with a sparse set of edges which approximate the $n(n-1)/2$ interatom distances with paths) to the segmentation of folding trajectories. They show that this representation affords easy visualization of the protein conformations over the entire folding trajectory of a protein and easy detection of the formation of secondary and tertiary structures as the protein folds. Leaver-Fay *et al* describe the novel application of a dynamic programming algorithm to a side chain placing problem, which facilitates the task of rational protein design.

Although most of the studies in the area of macromolecular structure and biocomputing have been done on proteins, there is a growing interest among computational biologists to study nucleic acids. The contribution from Karklin *et al* applies graph representation of non-coding RNA secondary structure to develop a structure classification method. They show that the combination of labeled dual graph representations and kernel machine learning methods (such as support vector machines) has potential for use in automated classification of uncharacterized RNA molecules or efficient genome-wide screens for RNA molecules from existing families.

As the Biogeometry session chairs, we are convinced that such interdisciplinary topic will continue to attract attention of leading specialists in computational, statistical, and biochemical/biophysical sciences who are interested in the role of shape in such fundamental computational problems as ligand-to-protein docking, *ab initio* and knowledge-based structure prediction, and visualization. Because of their fundamental role in structural biology, methods and applications to be discussed in this session will be of a great value for all participants of the PSB'05 conference.

# CLASSIFICATION OF NON-CODING RNA USING GRAPH REPRESENTATIONS OF SECONDARY STRUCTURE

YAN KARKLIN

*Department of Computer Science*
*Carnegie Melon University*
*Pittsburgh, PA USA*
*E-mail: yan+@cs.cmu.edu*

RICHARD F. MERAZ* AND STEPHEN R. HOLBROOK

*Physical Biosciences Division*
*Lawrence Berkeley National Laboratory*
*Berkeley, CA USA*
*E-mail: rfmeraz@gmail.com, srholbrook@lbl.gov*

Some genes produce transcripts that function directly in regulatory, catalytic, or structural roles in the cell. These non-coding RNAs are prevalent in all living organisms, and methods that aid the understanding of their functional roles are essential. RNA secondary structure, the pattern of base-pairing, contains the critical information for determining the three dimensional structure and function of the molecule. In this work we examine whether the basic geometric and topological properties of secondary structure are sufficient to distinguish between RNA families in a learning framework. First, we develop a *labeled dual graph* representation of RNA secondary structure by adding biologically meaningful labels to the dual graphs proposed by Gan *et al* [1]. Next, we define a similarity measure directly on the labeled dual graphs using the recently developed marginalized kernels [2]. Using this similarity measure, we were able to train Support Vector Machine classifiers to distinguish RNAs of known families from random RNAs with similar statistics. For 22 of the 25 families tested, the classifier achieved better than 70% accuracy, with much higher accuracy rates for some families. Training a set of classifiers to automatically assign family labels to RNAs using a *one vs. all* multi-class scheme also yielded encouraging results. From these initial learning experiments, we suggest that the *labeled dual graph* representation, together with kernel machine methods, has potential for use in automated analysis and classification of uncharacterized RNA molecules or efficient genome-wide screens for RNA molecules from existing families.

---

*to whom correspondence should be addressed.

## 1. Introduction

Non-coding RNA (ncRNA) molecules are those RNAs that do not encode proteins, but instead serve some other function in the cell [3]. They play a variety of critical roles and are ubiquitous in all kingdoms of life [4]. The function of non-coding RNAs is uniquely determined by the three dimensional structure of the molecule. To reach its functional form, a single stranded RNA molecule undergoes folding – driven by GC/AU/GU base-pairing and stacking interactions – to form short helices and various single stranded loop regions that define its secondary structure [5]. Some RNAs require metals or proteins to chaperone the folding process, but for the most part, the final three dimensional structure, and hence the functional role, is fully determined by the secondary structure [6]. This suggests that development of computational tools based on RNA secondary structure is essential for discovery of new non-coding RNAs and classification of their functional roles.

A variety of computational methods have used the secondary structure of RNA molecules to search and categorize ncRNAs, but many of these methods are limited in their use of secondary structure. Regular-expression-like pattern matching algorithms have been used to scan genome sequences for regions that fold into the canonical structures of specific families [7]. However, they are designed to match stringent configurations of secondary structure elements, and therefore perform poorly on families with variations in folding. Pair Stochastic Context Free Grammars (P-SCFG) look for evidence of secondary structure conservation by modeling covariance of mutations from related genomes [8] – but determining an appropriate grammar is a non-trivial problem [9]. Some discriminative classifiers use secondary structure stability as an input feature to distinguish non-coding RNAs from intergenic sequence [10], but they ignore important topological information. On the other hand, methods that use computable representations of secondary structure, such as trees and graphs, have been restricted to categorization and enumeration of gross topological features [11, 1].

Here we present a kernel-based machine learning method for classifying RNA families that avoids some of these limitations by learning directly from a graphical representation of secondary structure. This discriminative method does not require the estimation of any parameters or training of cumbersome generative models, yet it captures some of the topological relationships of RNA secondary structures. First, we define an appropriate representation of RNA secondary structure by extending the RNA dual graph representation [1] with a biologically relevant labeling scheme. Second, we define a similarity measure between RNA secondary structures by applying the recently developed marginalized kernel [2] to compare RNA molecules represented as labeled dual graphs. We tested the ability of this method to learn non-coding RNA structure by training Support Vector Machine [12] classifiers to distinguish known ncRNAs from random RNA sequences with similar nucleotide statistics. We also tested whether this approach can pick up on and generalize from structural features that distinguish non-coding RNA families.

## 2. An Algorithm for Classification Based on Secondary Structure Topology

Classification of RNA secondary structures with Support Vector Machines (SVMs) requires both a representation that captures the secondary structure and a kernel function that provides a reasonable similarity measure for the chosen representation. Below we present a graph representation of RNA secondary structure, the *labeled dual graph*, and show how it captures the basic structural features of the molecule. We then describe a method for applying kernel functions to the labeled dual graphs.

### 2.1. *Labeled Dual Graphs*

Given a secondary structure of an RNA molecule (see Figure 1A for examples), we want to construct a graph that captures essential properties of the structure. The dual graph [1] is a concise representation that captures basic topological properties of the folded RNA molecule, such as the number and relative position of the helical regions. In this representation, helical regions of the RNA are represented as vertices of a graph, while single RNA strands that connect the helical regions are edges. Thus, internal loops, bulges, and multi-loops become edges that connect vertices (helices adjacent to the loops), and external loops become edges from a vertex to itself. The result is a multigraph – up to two edges may connect a pair of vertices when a bulge or an internal loop separates two helical regions – that excludes the free 5' and 3' ends and ignores the directionality of the molecule, but captures its basic topology.

We augment the graph representation by adding labels that correspond to the length and type of secondary structure elements. The resulting *labeled dual graphs* (LDGs) are comprised of vertices labeled according to the number of nucleotide-pairs in the helical region they represent, and edges labeled according to the length (in number of nucleotides) and type (internal/external) of the loop they represent. See Figure 1B for an illustration of labeled dual graphs.

## 2.2. *Marginalized Kernels for Labeled Dual Graphs*

In order to use an SVM classifier on graph objects, we need a kernel function to define a similarity between two labeled dual graphs. Several kernels for graph objects have been proposed [13, 2]; here we use the recently developed *marginalized kernel* for labeled graphs [2] because it is relatively simple to implement, computationally efficient, and yielded promising results. Intuitively, this kernel function computes a similarity measure between two arbitrary labeled graphs by comparing the label sequences produced by taking random walks on each of the two graphs; the more similar the sets of label sequences, the higher the similarity score for the pair of graphs.

The computation of the kernel function between two graphs $G$ and $G'$ proceeds as follows. First, generate a random walk $h$ on graph $G$ and a walk $h'$ on graph $G'$, according to some defined probability of transitioning from vertex to vertex. Each walk produces a sequence of vertex and edge labels, $z = \{v_1, e_{12}, v_2, e_{23}, v_3, \ldots\}$ and $z' = \{v'_1, e'_{12}, v'_2, e'_{23}, v'_3, \ldots\}$ (see Figure 1C for an example). Next, define the label sequence kernel $K_z(z, z')$ as the product of the vertex label kernels $K_v(v, v')$ and the edge label kernels $K_e(e, e')$ over the sequence of labels,

$$K_z(z, z') = K_v(v_1, v'_1)K_e(e_{12}, e'_{12})K_v(v_2, v'_2)\ldots . \tag{1}$$

If the two walks are of different lengths, we define the label sequence kernel to be 0. Now that a similarity measure $K_z(z, z')$ is defined for each pair of walks, the value of the full graph kernel $K(G, G')$ is computed as the expected value of $K_z(z, z')$ over all possible walks $h$ and $h'$, weighted by the probability of generating the walks,

$$K(G, G') = \langle K_z(z, z')\rangle_{h,h'} . \tag{2}$$

The probability of taking a random walk on a graph, $p(h, h')$ depends on the probability of starting at a particular vertex and transitioning to subsequent vertices. We assumed a uniform starting probability over all vertices, a uniform probability of transitioning from a vertex to one of its neighbors, and a constant probability (0.1) of terminating the walk after any step.

Finally, we need to specify the edge and the vertex kernel functions, $K_e(.,.)$, $K_v(.,.)$. These should reflect the similarities in RNA structural motifs – similar helices should produce high similarity scores, as should

8

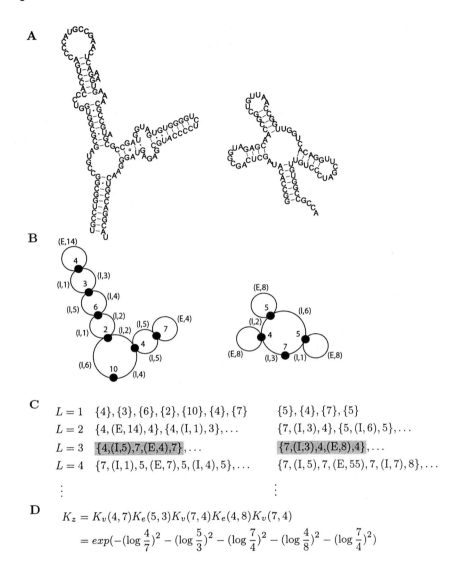

A

B

(E,14)

4

(I,1)    3    (I,3)

(I,4)

(I,5)    6    (I,2)

(I,1)    2    (I,2)    (I,5)    7    (E,4)

4

(I,6)    10    (I,5)

(I,4)

(E,8)

5    (I,6)

(I,2)    4    5

(E,8)    (I,3)    7    (I,1)    (E,8)

C

$L = 1$   $\{4\}, \{3\}, \{6\}, \{2\}, \{10\}, \{4\}, \{7\}$   $\{5\}, \{4\}, \{7\}, \{5\}$

$L = 2$   $\{4, (E, 14), 4\}, \{4, (I, 1), 3\}, \ldots$   $\{7, (I, 3), 4\}, \{5, (I, 6), 5\}, \ldots$

$L = 3$   $\{4, (I, 5), 7, (E, 4), 7\}, \ldots$   $\{7, (I, 3), 4, (E, 8), 4\}, \ldots$

$L = 4$   $\{7, (I, 1), 5, (E, 7), 5, (I, 4), 5\}, \ldots$   $\{7, (I, 5), 7, (E, 55), 7, (I, 7), 8\}, \ldots$

$\vdots$   $\vdots$

D

$$K_z = K_v(4, 7) K_e(5, 3) K_v(7, 4) K_e(4, 8) K_v(7, 4)$$

$$= exp(-(\log \frac{4}{7})^2 - (\log \frac{5}{3})^2 - (\log \frac{7}{4})^2 - (\log \frac{4}{8})^2 - (\log \frac{7}{4})^2)$$

Figure 1. (A) Secondary structure diagram and (B) *labeled dual graph* (LDG) representation of 5S rRNA (left) and tRNA (right) molecules. In the LDG, the numbers and
ordered pairs are the vertex (helix) and edge (loop) labels, respectively. The labels $E$ and
$I$ are used to distinguish external from internal loops. (C) A subset of label sequences
generated by taking random walks on the two graphs. Here $L$ refers to the length of the
path. (D) An example of the label sequence kernel and its output, as it is applied to the
highlighted pair of paths in (C). The full kernel between the two graphs is computed as
the expected value of the path kernels over all possible pairs of walks.

comparable loops. The choice of biophysical parameters that can serve as the basis for similarity comparisons is large – base composition, sequence or structural alignment, feature lengths, among others. As a first step we chose edge and vertex kernels that reflect the most basic structural parameters: the number of nucleotides that comprise a secondary structure motif. The vertices and edges of the dual graphs are labeled with these distances, and the vertex and edge kernels are defined as the Gaussian distance on the log-ratio of the two labels (lengths). This choice of kernel means that, compared to a particular structural element, elements twice its length or half its length score similarly, and that the similarity measure drops off smoothly as the ratio of the lengths deviates from 1. The vertex kernel is thus defined as

$$K_v(v_i, v_j) = exp(-\lambda_{ij}^2) \,, \tag{3}$$

where $\lambda_{ij} = log(v_i/v_j)$. For two edges of the same type of loop (internal or external), the edge kernel is similarly defined,

$$K_e(e_{ij}, e_{kl}) = exp(-\lambda_{ij,kl}^2) \,, \tag{4}$$

and for edges of different types, the edge kernel is 0. See Figure 1D for an illustrative example.

Effectively, two labeled dual graphs are considered similar when the two sets of all possible walks on each graph are similar. The similarity between individual walks is calculated as a product of simple functions defined on their constituent labels. Thus, if all the vertex and edge labels in the two walks match up, the output of the kernel function on the two walks will be high; and if many of the walks on the two graphs are similar, the kernel function will return a high value (with $\max K(G, G') = 1$) for the two graph objects. Hence this computation captures some topological relationships between structural elements of RNA secondary structure.

## 3. Methods

We performed two sets of experiments to test the ability of the classifier to learn RNA secondary structure and predict RNA family labels. First we trained SVM classifiers to distinguish non-coding RNAs from random RNAs with similar di-nucleotide composition. We also trained a system of multi-class SVMs to determine the family labels of RNA sequences.

Single family classification was tested on a number of RNA families from the RFAM database [14] (see the Results section for the list of tested RFAM families). When possible, we trained and tested the classifier on 500

RNA sequences, randomly selected from all RNAs in the family. However, some RFAM families contained fewer sequences, in which case all were used for classification. The negative data set was constructed by shuffling the nucleotide sequences of the positive data set while preserving the dinucleotide frequencies (see [15] for methods), which destroys characteristic secondary structure but produces random RNAs with sequence statistics similar to real RNA.

RNA sequences were converted to secondary structures with the Vienna RNA [16] folding prediction package, then converted to labeled dual graphs as described above. We implemented the kernel computation using an iterative method described in [2]; one thousand kernel computations took between 2 and 40 seconds on a desktop machine (2GHz Athlon), depending on the average complexity of the secondary structure. SVM classification was performed with 10 fold cross validation, with the precision parameter set to 10000. We assessed classifier performance with sensitivity and specificity measures and by computing the area under the Receiver Operating Characteristic (ROC) curve, a general measure of the discrimination ability [17].

We also trained a multiclass classifier on nine large RFAM families using the *one vs. all* method, a simple and frequently used approach to multiclass classification [18]. In this method, a separate classifier is trained to distinguish each class from the remaining ones. During classification, a test sample (in this case an RNA sequence) is tested against each of the trained classifiers, and a label assigned according to the classifier that produced the highest decision value.

In this experiment, we grouped together several related RNA families in order to have a sufficient number of sequences in each class for training and testing (see Results for details). We assessed performance with the generalized *class sensitivity* and *class specificity* measures [19]. For each classifier, the class sensitivity $(Q^D)$ represents the percentage of samples correctly predicted relative to the total number of samples in that family, while the class specificity $(Q^M)$ captures the number of samples correctly predicted relative to the total number of samples predicted to be in that family.

## 4. Results

### 4.1. *Single Family SVM*

Figure 2 shows the results of SVM classifiers trained to identify individual RFAM families. For sufficient training data we used only families with 50 or more sequences. The generation of negative training data is described in the previous section. The classifiers showed good performance for a large number of families, with $A_{ROC} > 0.7$ for 22 of 25 families tested. This suggests that the learning method is useful for learning a variety of secondary structure topologies. A notable result is the good classifier performance on several riboswitch and microRNA families, two particularly exciting non-coding RNA classes that have recently been shown to be involved in novel mechanisms for regulating gene expression.

### 4.2. *Multi-class SVM*

Table 1 shows the cross validation results of the *one vs. all* multi-class SVM trained on nine RFAM families. The MICRO and RNASE groups represent aggregates of functionally related individual RFAM families (see the caption for details). Again, classifier sensitivity and specificity were good over a range of families, although specificity clearly degraded for RNA families with larger molecules and possibly more complicated secondary structures. In these instances, it is possible that shorter walks pick up spurious similarities.

## 5. Discussion

The method presented here was able to learn to distinguish a number of non-coding RNA families; however, it is worth highlighting a few factors that may have adversely impacted its performance. First and most important is the reliance of the algorithm on accurate secondary structure prediction. Because the classifier uses solely secondary structure as input, it is sensitive to incorrectly predicted structures. As an example, training and testing a classifier on tRNAs for which correct folding was manually verified increased the accuracy from 89% to 98% ($A_{ROC}$). Nevertheless, because the kernel computation considers local paths over the entire structures, parts of the molecule that are correctly folded will still contribute to the correct computation of the kernel, even if some parts of the molecule are mis-folded. More accurate folding algorithms will likely improve the performance of this classifier. Alternatively, we can incorporate the con-

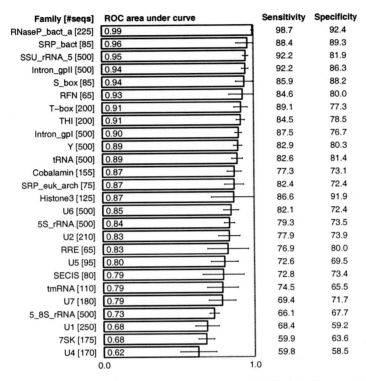

| Family [#seqs] | ROC area under curve | Sensitivity | Specificity |
|---|---|---|---|
| RNaseP_bact_a [225] | 0.99 | 98.7 | 92.4 |
| SRP_bact [85] | 0.96 | 88.4 | 89.3 |
| SSU_rRNA_5 [500] | 0.95 | 92.2 | 81.9 |
| Intron_gpII [500] | 0.94 | 92.2 | 86.3 |
| S_box [85] | 0.94 | 85.9 | 88.2 |
| RFN [65] | 0.93 | 84.6 | 80.0 |
| T–box [200] | 0.91 | 89.1 | 77.3 |
| THI [200] | 0.91 | 84.5 | 78.5 |
| Intron_gpI [500] | 0.90 | 87.5 | 76.7 |
| Y [500] | 0.89 | 82.9 | 80.3 |
| tRNA [500] | 0.89 | 82.6 | 81.4 |
| Cobalamin [155] | 0.87 | 77.3 | 73.1 |
| SRP_euk_arch [75] | 0.87 | 82.4 | 72.4 |
| Histone3 [125] | 0.87 | 86.6 | 91.9 |
| U6 [500] | 0.85 | 82.1 | 72.4 |
| 5S_rRNA [500] | 0.84 | 79.3 | 73.5 |
| U2 [210] | 0.83 | 77.9 | 73.9 |
| RRE [65] | 0.83 | 76.9 | 80.0 |
| U5 [95] | 0.80 | 72.6 | 69.5 |
| SECIS [80] | 0.79 | 72.8 | 73.4 |
| tmRNA [110] | 0.79 | 74.5 | 65.5 |
| U7 [180] | 0.79 | 69.4 | 71.7 |
| 5_8S_rRNA [500] | 0.73 | 66.1 | 67.7 |
| U1 [250] | 0.68 | 68.4 | 59.2 |
| 7SK [175] | 0.68 | 59.9 | 63.6 |
| U4 [170] | 0.62 | 59.8 | 58.5 |

0.0                  1.0

Figure 2. Performance of SVM classifiers trained on single RFAM families vs. shuffled sequences with the same di-nucleotide composition. Area under the ROC curve ($A_{ROC}$) is computed as the mean of the areas for each ROC curve of the 10 cross validation trials; error bars are standard deviation of $A_{ROC}$.

fidence in the secondary structure prediction into the learning algorithm, or even use the set of predicted suboptimal structures provided by folding algorithms as input to the classifier.

As a representation of RNA secondary structure, the labeled dual graph captures the basic features of the molecule: the number and length of helical regions and their relative position. However, some of the structural information is not represented. For example, the natural 5'-3' directionality of the molecule, the lengths of free 5' and 3' strands, as well as more complex topological features such as chirality. Much of this information could be included with natural extensions to the labeling scheme.

The computation of similarity between graphs (implemented with the

Table 1. Contingency table showing results for 10 fold cross validation of *one vs. all* multi-class SVM. For each RNA family (table row), the number of RNAs classified as a certain family appears in the respective column. $Q^D$ and $Q^M$ refer to generalized sensitivity and specificity, respectively. If $z_{ij}$ is an element in the contingency table, then $Q_i^D = \frac{z_{ii}}{\sum_j z_{ij}}$ and $Q_j^M = \frac{z_{jj}}{\sum_i z_{ij}}$. Several functionally related small RFAM families were grouped together to form aggregate families, **MICRO**: *let-7, lin-4, mir-1, mir-10, mir-101, mir-103, mir-124, mir-130, mir-135, mir-148, mir-156, mir-16, mir-160, mir-166, mir-17, mir-181, mir-19, mir-192, mir-194, mir-196, mir-199, mir-2, mir-218, mir-219, mir-24, mir-26, mir-29, mir-30, mir-46, mir-6, mir-7, mir-8, mir-9*; and **RNASE**: *RNaseP_bact_a, RNaseP_bact_b, RNaseP_nuc, RNase_MRP*, and these were trained and tested as single classes.

| | Histone3 | Intron_gpI | Intron_gpII | MICRO | RNASE | SSU_rRNA | tRNA | U6 | Y | $Q^D$ |
|---|---|---|---|---|---|---|---|---|---|---|
| Histone3 | 123 | 0 | 0 | 0 | 0 | 0 | 1 | 3 | 0 | .97 |
| Intron_gpI | 0 | 355 | 82 | 1 | 18 | 33 | 1 | 5 | 5 | 0.71 |
| Intron_gpII | 0 | 27 | 443 | 0 | 5 | 7 | 6 | 9 | 3 | 0.89 |
| MICRO | 0 | 3 | 2 | 165 | 0 | 0 | 1 | 1 | 8 | 0.92 |
| RNASE | 0 | 26 | 5 | 1 | 251 | 52 | 0 | 3 | 3 | 0.74 |
| SSU_rRNA | 0 | 17 | 0 | 0 | 5 | 474 | 0 | 1 | 3 | 0.95 |
| tRNA | 0 | 16 | 8 | 4 | 8 | 27 | 370 | 33 | 26 | 0.75 |
| U6 | 0 | 17 | 4 | 0 | 4 | 25 | 10 | 409 | 31 | 0.82 |
| Y | 0 | 32 | 3 | 4 | 11 | 31 | 14 | 30 | 375 | 0.75 |
| $Q^M$ | 1.0 | 0.72 | 0.81 | 0.94 | 0.83 | 0.73 | 0.92 | 0.83 | 0.83 | |

marginalized kernels) is also an imperfect measure. It does not account for relative position of helical regions, it is sensitive to bulges in helical regions, and it ignores global features such as the number of helices and the size of the molecule. Some of these might not be critical for discriminating ncRNAs – we tried a variant of LDGs that ignores bulges and observed no improvement in performance – but others should be incorporated into the representation and the kernel computation. Finally, the parameters used for computing the marginalized kernels also have an impact on the kernel output. For larger walks the random walk transition probabilities affect the relative contributions of local or global structural features to the similarity measure. Instead of adapting these parameters for optimal performance, we simply chose a set of sensible values, and it is possible that performance can be improved by adjusting these parameters. In order to address these concerns, it will be essential to look at exactly what aspects of the representation and the kernel allow the algorithm to learn to distinguish ncRNAs and to generalize to new structures. This will help us understand where and why it succeeds, and which aspects require improvement, and would also suggest areas of application for which this method is particularly suited.

## 6. Conclusion

We have presented a novel, simple, and computationally efficient approach for learning RNA secondary structures that requires no tuning of parameters and can be applied to a wide range of learning problems. It uses graph representations of folded RNA structures and kernels defined on graph objects to train SVM classifiers. Applied to non-coding RNAs from the RFAM database, the method gave promising results. It could distinguish many families from random RNA sequences with identical di-nucleotide composition, and showed some ability to differentiate one family from another. Because this conceptually simple approach produced relatively accurate classifiers, and because no other automated discriminative method for classification or discovery of ncRNA families exists, we believe there is great potential for extending this method or combining it with other techniques. Specific applications could include automated class-discovery of uncharacterized RNA molecules and computationally efficient heuristic filters in conjunction with other methods for RNA family prediction.

## Acknowledgments

We thank the anonymous reviewers for their insightful comments. This work was supported by grant 5RO1H6002665-02 from the NHGRI. Yan Karklin is supported by a Department of Energy Computational Science Graduate Fellowship (DOE-CSGF).

## References

1. H. H. Gan, S. Pasquali, and T. Schlick. Exploring the repertoire of RNA secondary motifs using graph theory; implications for RNA design. *Nucleic Acids Res*, 31(11):2926–43, 2003.
2. H. Kashima, K. Tsuda, and A. Inokuchi. Marginalized kernels between labeled graphs. In *International Conference on Machine Learning*, volume 20, pages 321–328. AAAI Press, 2003.
3. S. R. Eddy. Noncoding RNA genes. *Curr Opin Genet Dev*, 9(6):695–9, 1999.
4. G. Storz. An expanding universe of noncoding RNAs. *Science*, 296(5571):1260–3, 2002.
5. M. Zuker. Calculating nucleic acid secondary structure. *Curr Opin Struct Biol*, 10(3):303–10, 2000.
6. I. Tinoco and C. Bustamante. How RNA folds. *J. Mol Biol*, 293(2):271–281, 1999.

7. V. Tsui, T. Macke, and D. A. Case. A novel method for finding tRNA genes. *RNA*, 9(5):507–17, 2003.

8. E. Rivas, R. J. Klein, T. A. Jones, and S. R. Eddy. Computational identification of noncoding RNAs in E. coli by comparative genomics. *Curr Biol*, 11(17):1369–73, 2001.

9. R. D. Dowell and S. R. Eddy. Evaluation of several lightweight stochastic context-free grammars for RNA secondary structure prediction. *BMC Bioinformatics*, 5(1):71, 2004.

10. R. J. Carter, I. Dubchak, and S. R. Holbrook. A computational approach to identify genes for functional RNAs in genomic sequences. *Nucleic Acids Res*, 29(19):3928–38, 2001.

11. G. Benedetti and S. Morosetti. A graph-topological approach to recognition of pattern and similarity in RNA secondary structures. *Biophys Chem*, 59(1-2):179–84, 1996.

12. C. J. Burges. A tutorial on support vector machines for pattern recognition. *Data Mining and Knowledge Discovery*, 2(2):121–167, 1998.

13. T. Gartner. A survey of kernels for structured data. *ACM Special Interest Group on Knowledge Discovery Explorations*, 5(1):49–58, 2003.

14. S. Griffiths-Jones, A. Bateman, M. Marshall, A. Khanna, and S. R. Eddy. Rfam: an RNA family database. *Nucleic Acids Res*, 31(1):439–41, 2003.

15. C. Workman and A. Krogh. No evidence that mRNAs have lower folding free energies than random sequences with the same dinucleotide distribution. *Nucleic Acids Res*, 27(24):4816–22, 1999.

16. S. Wuchty, W. Fontana, I. L. Hofacker, and P. Schuster. Complete suboptimal folding of RNA and the stability of secondary structures. *Biopolymers*, 49(2):145–65, 1999.

17. T. Fawcett. ROC graphs: Notes and practical considerations for data mining researchers. Technical report, HP Laboratories, 1/17/2003 2003.

18. R. Rifkin and A. Klautau. In defense of one-vs-all classification. *Journal of Machine Learning Research*, 5:101–141, 2004.

19. P. Baldi, S. Brunak, Y. Chauvin, C. A. Andersen, and H. Nielsen. Assessing the accuracy of prediction algorithms for classification: an overview. *Bioinformatics*, 16(5):412–24, 2000.

# AN ADAPTIVE DYNAMIC PROGRAMMING ALGORITHM FOR THE SIDE CHAIN PLACEMENT PROBLEM

ANDREW LEAVER-FAY     BRIAN KUHLMAN     JACK SNOEYINK

*UNC-Chapel Hill*
*Chapel Hill, NC 27514, USA*
*leaverfa@email.unc.edu*
*bkuhlman@email.unc.edu*
*snoeyink@email.unc.edu*

Larger rotamer libraries, which provide a fine grained discretization of side chain conformation space by sampling near the canonical rotamers, allow protein designers to find better conformations, but slow down the algorithms that search for them. We present a dynamic programming solution to the side chain placement problem which treats rotamers at high or low resolution only as necessary. Dynamic programming is an exact technique; we turn it into an approximation, but can still analyze the error that can be introduced. We have used our algorithm to redesign the surface residues of ubiquitin's beta sheet.

## 1. Introduction

Current techniques for protein redesign fit new amino acids onto the rigid backbone scaffold of a known protein. Side chain conformations are almost always modeled by a rotamer library that discretely samples the conformation space. These libraries are obtained from observed side chain conformations in solved structures and/or by quantum calculations with small molecules.[5,15] The energy of a placement depends on the rotamer (or state, $S_v$) assigned to each residue $v$, and can be expressed as a sum of internal/background energies for each rotamer and interaction energies between pairs of residues:

$$\sum_v \mathcal{E}_{self}(S_v) + \sum_{v_1<v_2} \mathcal{E}_{pair}(S_{v_1}, S_{v_2}). \tag{1}$$

Minimizing energy expression (1) is the *side chain placement problem*. There are two natural subproblems: the *rotamer relaxation problem*, where the search is only over conformational space, and the *redesign problem* that explicitly involves the additional search over sequence space.

Many computational methods have been applied to side chain placement, including simulated annealing (SA),[2,8,16] dead-end elimination (DEE),[4,14] branch-and-bound,[7] potentials of mean force,[10] genetic algorithms,[3] and integer linear programming.[6] We are not aware of a viable dynamic programming[1] (DP) solution that scales to the size of practical rotamer libraries. DP, unlike all other methods except DEE and branch-and-bound, is guaranteed to find the global optimal state assignment. To make DP viable, we capture all pairs with non-negligible interactions in an *interaction graph*, which guides the dynamic programming as we describe in the next section. This is an extension of our previous work on the hydrogen placement problem.[13] For side chain placement, we make the algorithm adaptively explore the rotamers at either high or low resolution.

## 2. Methods

### 2.1. *Interaction Graph Formulation*

A *graph* $G = \{V, E\}$ consists of a set $V$ of *vertices* and $E \subset V \times V$ of *edges*. We say that vertices $u$ and $v$ are *adjacent* if the pair $(u, v) \in E$. A *hypergraph* is a generalization of a graph in which an edge (sometimes called a *hyperedge*) can contain any number of vertices of $V$. The *degree* of a hyperedge is the number of vertices it is incident upon. In Fig. 1b, we draw a hypergraph with the vertices as points and the hyperedges as curves encircling them.

We can initially create an *interaction graph*, a hypergraph, $G = \{V, E\}$, that captures energy expression (1). Each residue is represented by a vertex, $v \in V$, and each vertex carries a state space $\mathcal{S}(v)$, which is the set of rotamers that can be placed at $v$.

We initially create a (hyper)edge for each vertex and each pair. We assign each (hyper)edge $e \in E$ a *scoring function* $f_e(S)$ that maps the states chosen for its vertices to their interaction potentials: $f_e \colon \prod_{v \in e} \mathcal{S}(v) \to \Re$. Specifically, each vertex, $v \in V$, has a corresponding degree-1 hyperedge, $\{v\} \in E$, with a hyperedge scoring function $f_{\{v\}}(S_v) = \mathcal{E}_{self}(S_v)$. Each pair of vertices, $v_1, v_2 \in V$, has a corresponding degree-2 hyperedge, $\{v_1, v_2\} \in E$, with a hyperedge scoring function $f_{\{v_1, v_2\}}(S_{v_1}, S_{v_2}) = \mathcal{E}_{pair}(S_{v_1}, S_{v_2})$.

We can omit an edge $e$ when $f_e$ is zero under all possible state assignments to the vertices of $e$. To ignore small contributions, we can also omit edges for which $|f_e|$ is always less than a chosen magnitude threshold, $\mu$. Figure 1a shows an interaction graph for ubiquitin design.

We denote a state assignment to all vertices by $S_V$, and the induced assignment to any subset (particularly hyperedges) $e$ by $S_e$. Any assignment induces a score for the interaction graph, $\sum_{e \in E} f_e(S_e)$, and the optimal assignment minimizes the score.

## 2.2. *Dynamic Programming*

A dynamic programming (DP) algorithm can optimize an interaction graph by *eliminating* vertex $v$ from the graph by solving for the optimal state of $v$ for all possible states of its neighbors, then replacing $v$ with a hyperedge containing these neighbors.

Specifically, let $E_v$ be the set of hyperedges that contain $v$, and $N_v = \bigcup_{e \in E_v} e \setminus \{v\}$ be the neighbors of $v$. If $N_v$ is not already a hyperedge, create $N_v$ with an initial scoring function $f_{N_v} = 0$. Next, add to $f_{N_v}$ the scoring functions of $E_v$ with the best assignment to $v$, and eliminate $v$ and edges $E_v$ from the hypergraph. Let $\hat{f}_{e,v=s}$ denote the function whose domain is the states of $N_v$ obtained from $f_e$ by restricting the state of $v$ to be $s \in \mathcal{S}(v)$. Then we can write $f_{N_v} = \min_{s \in \mathcal{S}(v)} \sum_{e \in E_v} \hat{f}_{e,v=s}$. Also record $v$'s optimal state as a function of its neighbors' states for later retrieval.

The scoring function for $N_v$ now represents the simultaneous interaction of the vertices of $N_v$ with the optimal state of $v$. The minimum score for this reduced interaction graph is the same as the score of the original. Each function is represented in a multi-dimensional table; the table dimensionality is the degree of its corresponding hyperedge. Computing $f_{N_v}$ amounts to filling all cells in the table representing it.

We can reduce a graph to a single vertex by repeated elimination; Figure 1 illustrates the first two vertex eliminations. Once we have the optimal state of this remaining vertex, we can trace back the optimal states for eliminated vertices by reading off their optimal states in the reverse order of their elimination.

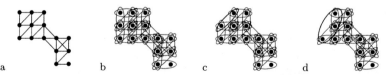

Figure 1. Interaction graph reduction. *The edges in (a) are drawn as curves encircling the vertices they contain in (b). The DP algorithm eliminates the upper-left vertex of (b) and updates the existing degree-2 hyperedge (c). It creates a degree-3 hyperedge as it eliminates the next vertex (d).*

In the remainder of this subsection we give upper bounds on the running time and memory for an interaction graph with $n$ vertices. If we assume at most $s$ states per vertex, and at most $w$ neighbors at each elimination, then DP runs in $O(ns^{w+1})$ time and uses $O(ns^w)$ space.

The parameter $w$ is known as the *treewidth*[9] of the interaction graph. Our algorithm shows that side chain placement is another instance of an NP-Hard problem with a polynomial time solution for graphs of bounded treewidth.

To define treewidth, we first define the *tree decomposition* of a hypergraph $G = \{V, E\}$ as a tree with $T$ whose vertices $\{X_1, X_2, \ldots, X_m\}$ represent subsets of $V$ that satisfy the following three properties:

(1) The union of sets $\bigcup_{1 \leq i \leq m} X_i = V$.
(2) Each edge $e \in E$ is in some set: $e \subseteq X_i$ for some $1 \leq i \leq m$.
(3) Each vertex $v \in V$ occupies a connected part of tree $T$: for any $X_j$ on a path in $T$ from $X_i$ to $X_k$, the intersection $X_i \cap X_k \subseteq X_j$.

The treewidth of a tree decomposition is $\max_i |X_i| - 1$. The treewidth of a graph $G$ is the minimum treewidth over all possible tree decompositions of $G$. Figure 2b illustrates a treewidth-3 tree decomposition of the interaction graph in Fig. 2a.

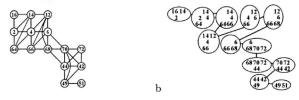

Figure 2. Tree decomposition. *The tree in (b) is a tree-decomposition of the labeled graph in (a). Each tree node contains the vertex label set, $X_i$.*

To show that bounded-treewidth interaction graphs yield polynomial time optimization, we convert a tree decomposition into a canonical form, which will then specify the elimination order for the vertices of the graph.

**Lemma 2.1.** *A tree decomposition $T$ with treewidth $w$ of the graph $G$ can be converted into a rooted tree decomposition with treewidth $w$ in which the vertex sets $X_i$ satisfy $|X_i \setminus X_j| = 1$ whenever $X_j$ is the parent of $X_i$.*

**Proof.** We give a constructive proof. Begin by choosing an arbitrary node of $T$ as the root. Now, consider any node $X_i$ and parent $X_j$ for which $|X_i \setminus X_j| \neq 1$. If $|X_i \setminus X_j| = 0$, then delete $X_i$ and connect any children of $X_i$ to the parent $X_j$. Otherwise take one vertex from the set difference, $v \in X_i \setminus X_j$, and create a new node, $Y = (X_i \cap X_j) \cup \{v\}$ between $X_i$ and $X_j$.

Node $X_i$ now has the desired property. Tree $T$ is still a tree decomposition: the first two properties for tree decompositions hold trivially, and the third property holds because $(X_i \cap X_j) = (X_i \cap Y \cap X_j)$. The treewidth of the new decomposition is still $w$ since $|Y| < |X_i| \leq (w+1)$. Recurse on $Y$ until the desired property holds. $\square$

Now, a depth-first traversal of a rooted, canonical tree decomposition gives an order for vertex elimination: Reaching a node, write down the one vertex in the set difference between it and its parent. At the root, $r$, write down the vertices of $X_r$ in any order. Because each vertex appears in a connected part of the tree, each vertex is written down once, and the recorded list of vertices provides the elimination order.

**Theorem 2.1.** *For an interaction graph $G$ on $n$ vertices with at most $s$ states per vertex and a tree decomposition $T$ of treewidth $w$, we can compute the optimum state assignment in $O(ns^{w+1} + wn)$ time and $O(ns^w + wn)$ space by dynamic programming.*

**Proof.** The $O(wn)$ terms come from the time and space to construct and use the canonical tree decomposition. To bound the running time of DP, we use induction on the number of nodes eliminated so far. Before eliminating node $v$, we assert that $G$ is the interaction graph that results from having eliminated the vertices up to vertex $v$, and that the canonical tree decomposition $T$ that we maintain is a tree decomposition of treewidth $w$ for $G$. If no vertices before $v$ have been eliminated, then our assertion is trivially true. Vertex $v$ was chosen for elimination because it was contained in a leaf node $X_i$ of $T$, but not in the parent of $X_i$. If $X_i$ was the root, then $X_i$ has fewer than $w + 1$ nodes, and we may apply brute force optimization in $O(s^{w+1})$ time. Otherwise, let $X_j$ be $X_i$'s parent. By definition of the canonical tree decomposition, $\{v\} = X_i \setminus X_j$. Therefore $v$ is adjacent to at most $w$ vertices, and we can record the best state for $v$ for each assignment to these vertices in $O(s^{w+1})$ time and $O(s^w)$ space. The hyperedge that $v$'s elimination produces is a subset of $X_j$, so we can delete $X_i$ from $T$ to obtain a tree decomposition of $G$ after the elimination of $v$. Since each vertex is eliminated once, the theorem is established. $\square$

Although computing the treewidth of a graph is NP-hard, the interaction graphs we have observed are small enough that we can make low-treewidth tree decompositions by hand or by heuristics and feed them as input to our algorithm.

## 2.3. *The Move Towards Large Rotamer Libraries*

The analysis of the DP algorithm shows that it is principally limited by two parameters: the number of states per node, $s$, which can be in the tens to the thousands, and the treewidth, $w$, which is fixed by the interaction graph and where we have only tried problems when it was between three to five. Since the number of vertices $n$ is in the tens for a redesign problem, we seek to reduce the impact of increasing $s$. How and why does $s$ increase?

Large rotamer libraries are typically cousins of the canonical rotamer libraries,[5,15] obtained by sampling around the canonical rotamers. Specifically, rotamers are usually defined as particular values of $\chi$ dihedral angles for side chains of amino acids. A large rotamer library might expand the canonical leucine rotamer represented by $\chi$ angles $(60, 180)$ into many samples inside the box $(60 \pm 5, 180 \pm 10)$. The canonical rotamers represent the bottoms of shallow energy wells for side chain conformations; the large rotamer libraries sample these wells. We can therefore organize our rotamer libraries as a set of base-rotamers (boxes) which contain many sub-rotamers (points): we denote the sub-rotamers of a base-rotamer $b$ by subscripts $b_1, b_2, \ldots$

Protein designers have found that large rotamer libraries produce better designs. The improvement can be attributed to the high penalty assigned to colliding atoms. The Lennard-Jones "6-12" potential suggests a larger rotamer library: the $\frac{1}{d^{12}}$ repulsive term is sensitive to small changes when $d$ is small, and slight flexes of the $\chi$ dihedrals resolve some high penalty collisions. The remaining terms of the energy functions are less sensitive: electrostatic interaction, for instance, is a function of $\frac{1}{d}$. Thus, for rotamers that are not near each other, we hope to represent many sub-rotamer interactions by a single base-rotamer interaction. We formalize this idea by defining the concept of stiff rotamer interactions.

## 2.4. *Stiff Interactions*

An assignment of base-rotamers, $a, b$, to the adjacent vertices $u$ and $v$ is *stiff in* $\{u, v\} \in E$ if any sub-rotamers $a_i, a_j$ and $b_k, b_l$ satisfy

$$|f_{\{u,v\}}(a_i, b_k) - f_{\{u,v\}}(a_j, b_l)| > \epsilon.$$

If a base-rotamer interaction fails to exceed the stiffness threshold, then we will approximate the energy between pairs of sub-rotamers by the energy between the canonical rotamers and $\epsilon$ will bound the error in this approximation.

Our analogy is with stiff differential equations: the stiff interactions are those where the energy is rapidly changing over a collection of sub-rotamers. As adaptive schemes for numerical integration increase their temporal resolution when their input ODEs become stiff, and decrease their resolution when the ODEs behave smoothly, our algorithm increases its spatial resolution when it encounters stiff base-rotamer combinations, examining all possible sub-rotamer combinations, and decreases its resolution for non-stiff base-rotamer combinations.

## 2.5. *Adaptive Dynamic Programming*

In adaptive dynamic programming, each vertex carries two levels of state spaces. At the top level is the base-state state space, $\mathcal{L}(v)$, and each base-state $b \in \mathcal{L}(v)$ carries a sub-state state space, $\mathcal{H}(b)$. The hyperedge scoring functions are defined as a mapping $f_e : \prod_{v \in e} \mathcal{L}(v) \rightarrow (\prod_{b \in B} \mathcal{H}(b) \rightarrow \Re)$ where $B$, a base state assignment to the vertices in $e$, is the input argument to the outer function. For each edge $e$ and each base-state assignment $B$, we maintain a *stiffness descriptor*, $C_e^B$ which is a family of subsets of B. Let $C_{e,k}^B \subseteq B$ represent the $k^{th}$ element of $C_e^B$ where $1 \leq k \leq |C_e^B|$. Alongside the stiffness descriptor, we define the set of all stiff base-states in an assignment, $U_e^B = \bigcup_{k=1}^{|C_e^B|} C_{e,k}^B$.

If $e$ is an input hyperedge $\{u, v\}$ under the base-state assignment $\{a, b\}$, the stiffness descriptor is either $C_{\{u,v\}}^{\{a,b\}} = \{\{a, b\}\}$ if $a$ and $b$'s interaction is stiff or $\{\{\}\}$ otherwise. A hyperedge created over the course of the reduction corresponds to an eliminated vertex; the stiffness descriptors for such a hyperedge correspond to the stiffnesses that existed with the optimal states of the eliminated vertex.

The elimination of vertex $v$ with incident edges $E_v$ leaves behind a hyperedge $N_v$ containing $v$'s former neighbors. We will describe the stiffness descriptor for a single base-state assignment, $B$, to the vertices of $N_v$. Define the stiffly-interacting base-states of a single base-state of $v$, $b \in \mathcal{L}(v)$ to be $T_{E_v}^{(B,b)} = \bigcup_{e \in E_v} U_e^{(B,b)_e} - b$ where $(B, b)_e$ represents the base-states assigned to the vertices $e$ contains. Let $P^B$ represent the power set of $B$, where if $|N_v| = d$ then $|P^B| = 2^d$, and let $P_i^B$ be the $i^{th}$ member of $P_i^B$. Let $\mathcal{L}_{P_i^B}(v)$ be $\{b \mid T_{E_v}^{(B,b)} = P_i^B\}$, that is, the set of all base-states of $v$ which have the same stiffly-interacting base-states of $P_i^B$.

Let $\hat{f}_e((B, b)_e)_{v=s}$ denote the function whose domain is the sub-states of $N_v$ under the base-state assignment $B$ obtained from $f_e((B, b)_e)$ by restricting $v$ to sub-state $s$. Then for each $i$ with a non-empty $\mathcal{L}_{P_i^B}(v)$, compute the function $f_{P_i^B} : \prod_{b \in P_i^B} \mathcal{H}(b) \to \Re$ so that

$$f_{P_i^B} = \min_{b \in \mathcal{L}_{P_i^B}(v)} \min_{s \in \mathcal{H}(b)} \sum_{e \in E_v} \hat{f}_e((B, b)_e)_{v=s}$$

We represent the range of $f_{P_i^B}$ by $[\text{best}_{P_i^B} \dots \text{worst}_{P_i^B}]$. We define the *best-worst* score as $\min_i \text{worst}_{P_i^B}$ and define the set of competing stiffnesses as $I = \{i \mid \text{best}_{P_i^B} < \textit{best-worst}\}$.

Now, there is a natural partial order based on the subset property for power sets. We use this partial order to define a *maximal* family of sets to be any where no element in the family is a subset of any other. Given a family of sets, $P$ we define maximal($P$) to be the function which returns the maximal family produced by throwing out any set if it is a subset of any other. The stiffness descriptor $C_{N_v}^B$ will be assigned maximal($\{P_i^B \mid i \in I\}$). Let the set $I_k$, corresponding to $C_{N_v,k}^B$, represent $\{i \mid P_i^B \subseteq C_{N_v,k}^B\}$. Then we define a set of functions, $f_{C_{N_v,k}^B} = \min_{i \in I_k} f_{P_i^B}$. Then the hyperedge scoring function with the state assignment $B$ can be defined as $f_{N_v}(B) = \min_k f_{C_{N_v,k}^B}$.

Where we represented each hyperedge scoring function before as a multi-dimensional table, we now represent each function as multi-resolution multi-dimensional table: a table of tables. The top level table has an entry for each base-state assignment to the vertices the edge contains. In each entry resides a set of tables holding the $f_{C_{N_v,k}^B}$ functions. Each table requires $\prod_{b \in C_{N_v,k}^B} |\mathcal{H}(b)|$ space. The smaller tables reduce the memory required by the adaptive algorithm compared to standard dynamic programming.

### 2.6. *Irresolvable Collisions*

With a final parameter, $\tau$, we define a pair of base-rotamers $b$ and $c$ to be in an irresolvable collision if

$$\neg \exists_{i,j} \mathcal{E}_{pair}(b_i, c_j) < \tau$$

While some pairs of base-rotamers can resolve their collisions by slight dihedral flexes, others cannot. As long as we have hope that a collision-free placement of side chains exists, we need not examine the colliding pairs.

Within the DP framework, this means we may avoid calculating the best state of a vertex for any combination of neighbors's base-states that put them in an irresolvable collision. Since the $\frac{1}{d^{12}}$ collision term is fluctuating wildly, irresolvable collisions will meet our stiff interaction threshold, $\epsilon$, and we would wastefully treat them at high resolution if we did not ignore them.

## 2.7. Error Analysis

**Theorem 2.2.** *The score induced by the adaptive algorithm's state assignment $S_V$ is within $2\epsilon|E_2|$ of the global optimum score where $|E_2|$ is the number of degree-2 hyperedges in the input interaction graph.*

**Proof.** Let $\sigma_{S_V}$ be the energy of the adaptive algorithm's state assignment and $\sigma_{opt}$ be the global minimum energy. Now, in selecting a sub-state of some base-state and representing it's interaction energy with another base-state non-stiffly, the algorithm can misrepresent the energy by at most $\epsilon$. On one hand, this may decrease the apparent score of the state assignment $S_V$ while on the other increasing the apparent score of the global optimal state. Over the course of the optimization, the algorithm may misrepresent at most $|E_2|$ non-stiff interactions, so the apparent score of $S_V$ may decrease to $\sigma_{S_V} - \epsilon|E_2|$ and the apparent score of the optimal state may increase to $\sigma_{opt} + \epsilon|E_2|$. Because the algorithm chose $S_V$, we know

$$\sigma_{S_V} - \epsilon|E_2| \leq \sigma_{opt} + \epsilon|E_2|$$

and thus $\sigma_{S_V} - \sigma_{opt} \leq 2\epsilon|E_2|$. □

## 3. Results

We tested our algorithm at the rotamer relaxation task and the redesign task. For both tasks, we selected 15 surface residues from ubiquitin's $\beta$-sheet, pictured in Fig. 3a. We excluded the following amino acids to keep the treewidth of our interaction graphs low: arginine, lysine, and methionine.

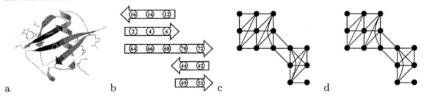

Figure 3. Ubiquitin's $\beta$-sheet. *The $\beta$-sheet in (a) is flattened in (b) with it's 15 surface residues shown. We observed the treewidth-4 interaction graph in (c) by including edges between residues if any pair of rotamers ever interacted with an energy magnitude at least $\mu = 0.2$ kcal/mol. We artificially created the treewidth-3 interaction graph in (d) by dropping a single edge.*

For the rotamer relaxation task, we first created 100 sequences for the ubiquitin backbone, asking the design module of the Rosetta molecular modeling program[11] to stochastically redesign these 15 surface residues. We then evaluated Rosetta's experimentally validated energy function[12] between all pairs of sub-rotamers, and included hyperedges that met our interaction magnitude threshold $\mu = 0.2$ kcal/mol. This produced a treewidth-4 interaction graph, shown in Fig. 3c. We set our irresolvable collision cutoff to $\tau = 1$ kcal/mol. We compared the standard DP algorithm against the adaptive algorithm with $\epsilon$ values of 0, $10^{-4}$, $10^{-3}$, $10^{-2}$, 0.1, and 1.0. Against DP, we compared the time a single Rosetta SA design required and scores it produced.

In the relaxation problem, the average residue had 32 total rotamers, breaking down into 3 base-states and 10 sub-states per base-state. The median state space size was $\sim 10^{18}$. We measured performance on a dual 2 GHz AMD Athlon with 2 GB RAM. In Fig. 4 we plot the relative running time of the adaptive and standard DP algorithms against the actual error observed. In table 1, we present the actual running times. Except for three instances, SA produced the optimal answer when run for as long as standard DP.

Table 1. Average running time comparison, in milliseconds, at the rotamer relaxation task.

| Run Time | DP | $\epsilon = 0$ | $10^{-4}$ | $10^{-3}$ | $10^{-2}$ | 0.1 | 1.0 | SA |
|---|---|---|---|---|---|---|---|---|
| Mean | 206.2 | 63.7 | 62.9 | 63.0 | 61.2 | 17.5 | 6.4 | 65.1 |
| Median | 117.3 | 53.7 | 53.2 | 53.7 | 50.7 | 11.3 | 4.8 | 65.0 |
| Std Dev | 399.3 | 49.1 | 48.6 | 48.9 | 48.8 | 38.2 | 7.6 | 6.5 |

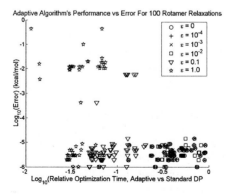

Figure 4. Rotamer relaxation task. *Increasing $\epsilon$ to as high as 1.0 kcal/mol gives a theoretical error bound of $\pm$ 64 kcal/mol but actually preserves accuracy and greatly decreases running time.*

For the redesign task, we artificially imposed a treewidth-3 interaction graph on the problem, pictured in figure 3d. This interaction graph differs by a single edge from the graph in 3c. The absence of this edge decreases the quality of the design. We none the less include the task as it pushes the DP algorithm's limits.

Each residue in the design problem averaged 680 total rotamers. This broke down into about 57 base-rotamers per residue and 12 sub-rotamers per base-rotamer. The size of the state space was $\sim 10^{42}$. We measured the performance of both the standard and the adaptive DP algorithms on a dual 900 MHz 64-bit Itanium-2 with 10 GB of RAM. We compared against a single SA run on the 2 GHz Athlon. In table 2 we present the results.

Table 2.   Redesign task performance comparison.

|  | DP | $\epsilon = 0.1$ | $\epsilon = 1.0$ | SA |
|---|---|---|---|---|
| Run Time | 15.99 hrs. | 5.07 hrs. | 1.52 hrs. | 3.42 seconds |
| Memory Usage | 3.7 GB | 3.4 GB | 1.5 GB | 0.2 GB |
| Score (kcal / mol) | -42.5893 | -42.5893 | -42.5579 | -42.5692 |
| Error (kcal / mol) |  | 0.0000 | 0.0314 | 0.0201 |

## 4. Discussion

We have presented a novel application of our DP algorithm, and an improvement upon it that begins to make it competitive. DP offers an alternative to DEE and branch-and-bound for finding the global optimal solution in the side chain placement problem. The other algorithms are designed to solve a very difficult problem; the generic interaction graph implied by energy expression (1) is fully connected. Distant amino acids, however, have interaction energies of zero, so the interaction graphs we encounter are sparsely connected. Actual instances of the side chain placement problem may not require algorithms as generic as DEE and branch-and-bound.

We were surprised to observe the small error for high values of $\epsilon$. None of our experiments with $\epsilon > 10^{-2}$ produced an error near the theoretical bound we proved in Sec. 2.7. When we set $\epsilon$ as high as 1 kcal/mol, we limit our high resolution focus to only those sometimes-colliding pairs of base-rotamers. It is possible that in general the interactions in the global optimal solution induces are either very stiff or almost totally insensitive to small change; rotamers will either pack tightly or are far apart. It is also possible that the non-canonical rotamers' internal strain prevents their selection except in the instances of collision resolution. By representing base-state interactions with the canonical rotamer interaction energies, we would introduce error only for the non-stiff interactions induced alongside a resolved collision—the resolved collision itself would be modeled stiffly.

## 5. Future Work

We have allowed our adaptive algorithm only two spatial resolutions: high or low. We want to heirarchically group rotamers with varying resolutions dictated by the sub-rotamer energy range. This would let us make tighter theoretical bounds while hopefully preserving our performance gains. Moreover, we want to treat some states at even lower resolution. If two base-rotamers for a residue reached away from the vertex $v$ being eliminated so that their interaction energies with the rotamers of $v$ were the same, we could treat these two base-rotamers as one and reduce the redundant computations.

We would also like to incorporate partial dynamic programming into a simulated annealing algorithm. We can decrease the problem complexity if we allow DP to eliminate all degree-1 and -2 vertices. This fixes the eliminated vertices in their optimal states. We expect this adaptation will produce better designs.

## Acknowledgements

This research was partially funded by NSF grant 0076984.

## References

1.  R. Bellman. *Dynamic Programming*. Princeton Univ. Press, 1957.
2.  B. I. Dahiyat and S. L. Mayo. *Science*, **278** 82, (1997).
3.  J. R. Desjarlais and T. M. Handel. *Prot. Sci.*, **4** 2006, (1995).
4.  J. Desmet, M. D. Maeyer, B. Hazes, and I. Lasters. *Nature*, **356** 539, (1992).
5.  R. L. Dunbrack. Jr. *Curr. Opin. Struct. Biol.*, **12** 431, (2002).
6.  O. Eriksson, Y. Zhou, and A. Elofsson. In *WABI'01*, 128, (2001).
7.  D. B. Gordon and S. L. Mayo. *Structure* **7** 1089, (1999).
8.  L. Holm and C. Sander. *Proteins*, **14**(2):213, (1992).
9.  T. Kloks. *Treewidth: computations and approximations*. Springer, (1994).
10. P. Koehl and M. Delarue. *J Mol Biol*, **239**(2) 249, (1994).
11. B. Kuhlman, and D. Baker. *Proc Natl Acad Sci USA*, **97** 10383, (2000).
12. B. Kuhlman, G. Dantas, G. C. Ireton, G. Varani, B. L. Stoddard, and D. Baker. *Science*, **302** 1364, (2003).
13. A. Leaver-Fay, Y. Liu, and J. Snoeyink. In *ALENEX'04*, (2004).
14. L. L. Looger and H. W. Hellinga. *J Mol Biol*, **307**(1) 429, (2001).
15. S. C. Lovell, J. M. Word, J. S. Richardson, and D. C. Richardson. *Proteins*, **40** 389, (2000).
16. J. G. Saven and P. G. Wolynes. *J. Phys. Chem. B*, **101** 8375, (1997).

# COMPUTATIONAL DESIGN OF COMBINATORIAL PEPTIDE LIBRARY FOR MODULATING PROTEIN-PROTEIN INTERACTIONS [a]

XIANG LI[1,2] and JIE LIANG[2]

[1] *Graduate Program in Bioinformatics and* [2]*Department of Bioengineering, SEO,*
*MC-063*
*University of Illinois at Chicago*
*851 S. Morgan Street, Room 218*
*Chicago, IL 60607–7052 U.S.A.*

Screening phage-displayed combinatorial peptide library is an effective approach for discovery of peptide modulators for protein-protein interactions. However, as peptide length increases, the chance of finding active peptides in a finite size library diminishes. To increase the likelihood of finding peptides that bind to a protein, we develop statistical potential for computational construction of biased combinatorial antibody-like peptide libraries. Based on the alpha shapes of antibody-antigen complexes, we developed an empirical pair potential for antigen-antibody interactions that depends on local packing. We validate this potential and show that it can successfully discriminate the native interface peptides from a simulated library of 10,000 random peptides for 34 antigen-antibody complexes. In addition, we show that it can successfully recognize the native binding surface patch among all possible surface patches taken from either the antibody or the antigen for seven antibody-antigen protein complexes contained in the CAPRI (Critical Assessment of Predicted Interactions) dataset. We then develop a Weighted Amino Acid Residue sequence Generator (WAARG) for design of biased peptide library. When compared with a random peptide library, WAARG libraries contain more native-like binding peptides at a significantly smaller size. Our method can be used to construct peptide library for screening of antibody variants with improved specificity and affinity to a target antigen. It can also be used for screening of antibody-like antagonist peptides modulating other protein-protein interactions.

## 1 Introduction

Modulating protein-protein interactions has the promise of obtaining many novel therapeutic agents. An effective approach for discovery of such modulators is through screening of combinatorial libraries of peptides. To identify peptides that interact with an antigen, the technique of phage display is effective [1,2], because it can produce synthetic peptides that may have the target-recognition qualities of natural antibodies. By fusing the DNA sequence encoding a particular peptide to the gene of a coat protein of a bacterial virus

---

[a]This work is supported by grants from NSF(CAREER DBI0133856 and DBI0078270), NIH(GM68958), and ONR(N000140310329). We thank Dr. Brian Kay for many stimulating discussions.

(called phage), the peptide is displayed on the virus coat. The coding sequence for the selected peptide inside the phage can be readily retrieved for further analysis and amplification. A phage library is formed by a collection of a large number of antibody-covered phages. The new technique of trinucleotide-phosphoramidite-based synthesis enables the design of peptide library at individual residue level [3].

Random phage-display peptide libraries have been applied to identify binding peptides of a specific target. They can also be used to predict binding sites on a 3D structure [4]. However, random combinatorial libraries meet their limitations because of the huge sequence space. The entire mutated fragment of each peptide phage libraries can contain up to a billion different peptides, a size comparable to that of the repertoire of human immune system. It is still too small to cover the space of $3 \times 10^{19}$ possible 15-mers. When a typical random phage library containing $10^9$ unique peptide sequences is screened, there is only a minuscule chance that a peptide of length 15 with reasonably high affinity for an intended protein target is actually contained in the library. It is necessary to generate biased libraries that are enriched with active peptides.

In this study, we develop a method for computational design of phage display libraries, with the goal to improve the likelihood of finding effective peptide modulators. Our method requires a known protein target structure of antigen to which a modulating peptide will bind. A critical ingredient is a potential function that can be computed efficiently to guide the generation of promising candidate peptides. We develop such an empirical potential function based on statistical analysis of contact interactions across protein-protein interfaces in protein database. Specifically, we select a set of protein-protein complexes from Protein Data Bank and compute their alpha shapes. Statistical models are then developed for estimating propensity for two residues on two proteins to interact.

An important consideration of our model for contact interactions is the local environment. It is well known that protein-protein interface is not evenly packed: some regions are tightly packed, but others contain voids and pockets [5]. Both energetically important interfacial residues, termed as "hot spots" [6], and structurally conserved residues are more likely to be located at tightly packed regions [7]. The importance of hot spots in such tightly packed region are due to not only the numerous contact interactions with the binding partner, but also the dehydrated environment where H-bonding interactions are enhanced because of reduced dialectic constant [8]. A parameter related to local packing environment is explicitly included in our model.

We organize this paper as follows. First, we describe the geometric model for contact interactions and for the coordination shell surrounding a contacting residue pair, and how they can be computed using alpha shape. We then discuss the probabilistic model of packing-dependent empirical interface contact potential. This empirical potential is then validated by testing its ability to discriminate native antibody interfaces from random peptides, and its ability to recognize native binding surface patches from other surface patches for antibody and antigen complexes used in the CAPRI competition. We then describe how this scoring function can be used to generate biased peptide library. We conclude with discussion.

## 2   Model and Methods

**Empirical potential of residue interactions.**   Due to its simplicity and fast evaluation, empirical potential based on statistics of protein database is well-suited for rapid generation of peptide library of tens of thousands peptides. Molecular mechanics and other methods based on potential functions derived from physical modes are difficult to use for this purpose.

Empirical potential can be derived based on either description of protein structure at residue level[9−11] or at atomic level[12, 13]. It has been applied with success in fold recognition and structure prediction. Because the atomic details of protein-protein interactions are difficult to obtain, we develop potential function based on residue level representation for designing peptide library.

We use a simplified residue model for protein structure. We follow the union of ball model and represent the $i$-th amino acid residue as a ball $b_i$[14], whose center $x_i \in \mathbb{R}^3$ coincides with the geometric center of its side-chain. Because Gly residue has no side chain, the position of $C_\alpha$ atom is taken as the center of the ball $x_i$. The radius $r_i$ of each ball depends on the size of side chain, and is taken from values by Levitt listed in[15], with an added 0.5 Å increment to account for uncertainty due to side-chain flexibility. This is necessary to reduce spurious contacts.

**Alpha contacts and coordination shell.**   We are interested in identifying contacting residues that are spatial nearest neighbors. We use the dual complex calculated by the alpha shape software to identify such residues[14, 16−20]. Briefly, the Voronoi diagram decomposes the space and the union of residue balls $\bigcup B = \bigcup b_i$ into convex regions $V_B$, and the dual complex $\mathcal{K}$ or the alpha shape of the molecule records the overlap pattern among these regions[14]: $\mathcal{K} = \{\sigma = \mathrm{conv} x_B \mid \bigcap V_B \cap \bigcap B \neq \emptyset\}$, where $x_B$ is the set of residue centers $\{x_i\}$ of a set of balls $B$, $V_B$ is the set of Voronoi cells of balls $B$, whose intersection $\bigcap V_B$ overlap with the intersection $\bigcap B$ of the balls. $\mathrm{conv} x_B$ is the convex hull of residue centers $x_B$, which forms a simplex $\sigma$. In this study, we only make use of a subset of 1-simplices $\sigma_{ij}$ (or alpha edges), such that the corresponding two residues $i$ and $j$ are from two proteins. Denote the name of the protein of residue $i$ as $\mathbb{I}(i)$, we have $\mathbb{I}(i) \neq \mathbb{I}(j)$.

For a pair of such balls $B = \{b_i, b_j\}$ located on the protein interface, we examine the set of residues $S_{ij}$ that are connected by an alpha edge to either residue $b_i$ or residue $b_j$: $S_{ij} = \{ b_k | \sigma_{ki,k\neq j} \in \mathcal{K} \text{ or } \sigma_{kj,k\neq i} \in \mathcal{K}\}$. We call this set of residues the *coordination shell* of the interacting residue pair $i$ and $j$. The number of such residues $z_{ij} = |S_{ij}|$ is termed the coordination number of contacting residue pair $ij$.

The Delaunay triangulation is computed using the DELCX program, and the alpha shapes computed using the MKALF program [17,19]. Both can be downloaded from the web-site at (http://www.alphashape.org).

**Probabilistic model.** The propensity $p(k, l, z)$ for residue of type $k$ interacting with residue of type $l$ with coordination number $z$ is modeled as an odds ratio. We first estimate the probability $q(k, l, z)$ of residues of type $k$ and type $l$ interact across protein-protein interface with a coordination number $z$. The random probability $q_R(k, l, z)$ of a pairwise contact involving both residue $k$ and $l$ with coordination number $z$ is calculated from a null model (or reference state). Specifically, we have: $p(k, l, z) = \frac{q(k,l,z)}{q_R(k,l,z)}$, where $q(k, l, z) = \frac{n(k,l,z)}{\sum_{k',l',z} n(k',l',z)}$. Here, $n(k, l, z) = |\{\sigma_{ij} | \sigma_{ij} \in \mathcal{K} \text{ and } \mathbb{I}(i) \neq \mathbb{I}(j)\}|$ is the number count of alpha edge contacts on protein interface involving residue type $k$ and residue type $l$ when coordination number is $z$. $\sum_{k',l',z} n(k', l', z)$ is the total number of all interfacial alpha contacts of any residue types with the same coordination number $z$. The random probability $q_R(k, l, z)$ is the probability that a pair of contacting residues is selected from surface residue of type $k$ and type $l$, when chosen randomly and independently. Here a surface residue is defined as that with more than 15% of its total solvent accessible surface area exposed in the model of a tri-peptide Gly-X-Gly [21]. We divide the range of coordination number $z$ into five intervals $[0, 3], [4, 6], [7, 9], [10, 12]$ and $[13, \infty)$ for all pair contact interactions.

The choice of random model or reference state for estimating $q_R(k, l, z)$ is critical for empirical potential. We use a random model or reference state, where there are no preferred contacts between any residue type $k$ and any residue type $l$, no preference for location of $k$ or $l$ to be on interface or on the rest of surface, and no preferential coordination number $z$ for any interfacial contact pair. For our random model, packing plays no direct roles for protein-protein interactions. We have:

$$q_R(k, l, z) = q_R(k, l) = 2 \cdot n_k n_l \cdot \left(\frac{1}{n(n-1)}\right), \quad \text{when} \quad k \neq l \quad (1)$$

and

$$q_R(k, l, z) = q_R(k, l) = n_k(n_k - 1) \cdot \frac{1}{n(n-1)}, \quad \text{when} \quad k = l, \quad (2)$$

where $n_k$ is the number of surface residues of type $k$, and $n$ is the total number of surface residues.

The alpha contact potential $U(i, j)$ of protein-protein interaction between residue $i$ and residue $j$ with coordination number $z$ is obtained as $U(\sigma_{ij}) = -\ln p(a(i), a(j), z_{ij})$ using $kT$ unit, where $a(i)$ and $a(j)$ are the residue types of residue $i$ and $j$, respectively. The overall energy of a protein-protein interface is calculated as:

$$E = \sum_{\substack{\sigma_{ij}, \\ \mathbb{I}(i) \neq \mathbb{I}(j)}} U(\sigma_{ij}), \quad (3)$$

To assess the importance of local packing environment as reflected by the coordination number $z$, we also develop a simpler scoring function which does not consider the local packing environment: $p(k, l) = \frac{q(k,l)}{q_R(k,l)}$, where $q(k, l) = \frac{n(k,l)}{\sum_{k',l'} n(k',l')}$, and $n(k, l) = |\{\sigma_{ij}|\sigma_{ij} \in \mathcal{K}, a(i) = k, a(j) = l \text{ and } \mathbb{I}(i) \neq \mathbb{I}(j)\}|$ is the number count of interfacial contact pairs between residue types $k$ and $l$. The random probability $q_R(k, l)$ is calculated using Equation (1) and (2).

**Dataset of nonredundant antibody-antigen complexes.** Contact potentials derived from one dataset may not be universal and fully transferable to other systems [22]. Therefore, the selection of a representative set of proteins is important for developing empirical potential. Because our goal is to design synthetic antibody for enhanced binding affinity or for creating novel binding, we collect a dataset of antibody-antigen complex structures. We select co-crystallized complex structures from the Protein Data Bank that satisfy the following criteria: the resolution of each chain should be less than 2.5 Å; each chain should have more than 30 amino acids; no pair of chains in protein complexes have a sequence identity larger than 25% to any other chain in the data set. Based on these three criteria, we collected a set of 34 antibody-antigen complexes.

## 3    Results

**Discriminating native antibody interfaces.** We first validate the empirical potential by testing whether it can identify interface residues on the native antibody. For each antibody-antigen complex, we first locate the interface residues on the antibody and on the antigen, respectively. These are residues connected by alpha edges across the two proteins. To model a random phage library where each residue has equal probability to be generated at each position of the peptide, we randomly substitute uniformly all the interfacial residues on the antibody with any of the 20 amino acid residues. The interface residues on the antigen are unchanged. For simplicity, we assume that the interface contacts and hence the coordination numbers for all contacting pairs remain unchanged. Given $m$ interface residues on the antibody, there are $20^m$ possible different sequences, where $m$ ranges typically from 4 to 26. We randomly generate a sample of 10,000 sequences to test the empirical potential.

We performed 34 leave-one-out tests. In each case, 33 of the antibody-antigen complexes are used to construct the empirical potential. The remaining antibody-antigen complex is used for testing. Table 1 shows that among 34 antibody-antigen complexes, 28 native interfaces rank among the top 10 interfaces among the corresponding 10,000 random interfaces. The median and average rankings of 34 native sequences among the 34 sets of 10,000 generated sequences are 1 and 20, respectively, and their $z$-scores are 4.79 and 4.38, respectively. The incorporation of the coordination number for local packing environment is important. Without such consideration, the median and mean rankings of the 34 native sequences are only 9 and 236, respectively, and their $z$ scores are 3.44 and 3.01, respectively. In terms of the mean value of ranking of native sequences, the performance improves more than 10 times when local-packing is considered. For comparison, results of discrimination using Miyazawa-Jernigan potential are also listed in Table 1.

**Recognition of binding surface patch of CAPRI targets.** The CAPRI (Critical Assessment of PRedicted Interactions) competition is designed to evaluate current protein docking algorithms. A blind docking prediction starts from two known crystallographic or NMR structures of unbound proteins and ends with a comparison to a solved structure of the protein complex, to which the participants did not have access. Since its inception in 2001, a total of 19 protein complexes have been used for blind docking. Among these, seven are antibody or antibody related proteins (*e.g.*, Fab fragment, T-cell receptor). We use these seven complex structures to evaluate the effectiveness of the empirical potential.

In docking, a *cargo* protein is docked to a *seat* protein. All surface patches as candidate at binding interface are sampled from the surface of an unbound structure. We therefore generate candidate sequences of binding peptide from the surfaces of *cargo* protein structure. Since our goal is not docking but to evaluate the performance of the potential function, we assume the knowledge of the binding interface on the *seat* protein. We further assume the knowledge

Table 1: Discrimination of Native Antibody Interfaces.

| Ab - Ag Complexes | Local packing dependent | Local packing independent | Miyazawa-Jernigan potential | [b]$N_{Ab}$ | [c]$N_{Ag}$ |
|---|---|---|---|---|---|
| 1i8k | [a]375/2.29 | 4770/0.04 | 3523/0.39 | 18 | 8 |
| 1nmb | 137/ 2.47 | 111/2.37 | 406/1.84 | 15 | 18 |
| 1e6j | 54/2.72 | 123/2.34 | 463/1.66 | 18 | 13 |
| 1jps | 35/2.93 | 364/1.86 | 142/2.31 | 21 | 22 |
| 1iqd | 15/3.19 | 1022/1.29 | 1764/0.94 | 26 | 17 |
| 1nsn | 7/3.30 | 502/1.69 | 410/1.70 | 20 | 21 |
| 1osp | 20/3.38 | 69/2.63 | 215/2.08 | 15 | 21 |
| 1nca | 10/3.47 | 23/2.81 | 1192/1.18 | 22 | 24 |
| 1qfu | 3/3.63 | 326/1.86 | 47/2.55 | 24 | 21 |
| 1kb5 | 5/3.78 | 50/2.54 | 1399/1.10 | 24 | 26 |
| 2jel | 1/3.86 | 181/2.19 | 1508/1.02 | 20 | 16 |
| 1eo8 | 5/3.90 | 201/2.11 | 861/1.39 | 19 | 19 |
| 1ai1 | 3/4.29 | 53/2.73 | 6582/-0.42 | 17 | 6 |
| 1dvf | 1/4.32 | 7/3.25 | 272/1.92 | 19 | 18 |
| 1wej | 1/4.35 | 2/3.97 | 2/4.16 | 13 | 11 |
| 1mpa | 1/4.60 | 4/3.57 | 4781/0.04 | 16 | 7 |
| 1ktr | 1/4.71 | 157/2.25 | 1486/1.04 | 17 | 4 |
| 1fe8 | 1/4.79 | 4/3.67 | 480/1.71 | 23 | 20 |
| 3hfm | 1/4.80 | 4/3.55 | 262/1.93 | 21 | 17 |
| 1f58 | 1/4.81 | 4/3.60 | 1022/1.28 | 18 | 10 |
| 3hfl | 1/4.84 | 7/3.64 | 45/2.80 | 19 | 16 |
| 1nby | 1/4.86 | 3/3.93 | 33/2.90 | 22 | 18 |
| 1jhl | 1/4.89 | 2/4.16 | 9/3.43 | 16 | 11 |
| 1fns | 1/4.94 | 6/3.80 | 191/2.16 | 16 | 12 |
| 1iai | 1/5.01 | 3/3.46 | 1147/1.21 | 21 | 23 |
| 1gc1 | 1/5.10 | 29/3.09 | 225/2.04 | 14 | 12 |
| 1g9n | 1/5.16 | 16/3.46 | 22/2.76 | 13 | 12 |
| 1a2y | 1/5.20 | 1/4.38 | 153/2.29 | 14 | 14 |
| 1jrh | 1/5.27 | 1/4.54 | 4/3.97 | 20 | 15 |
| 2iff | 1/5.28 | 10/3.42 | 28/3.03 | 20 | 16 |
| 2hrp | 1/5.63 | 7/3.46 | 303/1.88 | 17 | 8 |
| 1cu4 | 1/5.65 | 4/3.52 | 223/2.04 | 22 | 9 |
| 1a3r | 1/5.69 | 2/4.35 | 48/2.57 | 31 | 14 |
| 2ap2 | 1/5.83 | 1/4.22 | 1581/1.02 | 17 | 8 |
| Average | 20/4.38 | 236/3.01 | 907/1.88 | 19 | 15 |
| Median | 1/4.75 | 9/3.44 | 286/1.90 | 19 | 16 |

[a] The first number in each cell is the rank of native antibody interface and the second number is the $Z$ score. $Z$ score $= \overline{E} - E_{native}/\sigma$; $\overline{E}$ and $\sigma$ are the mean and standard deviation of the scores of 10,000 randomly generated peptides, respectively. [b]$N_{Ab}$: number of interfacial residues on the interface of antibody side. [c]$N_{Ag}$: number of interfacial residues on the interface of antigen side.

of the coordination number for interface residues.

We first partition the surface of the unbound *cargo* protein into candidate surface patches, each has the same size as the native binding surface of $m$

residues. A candidate surface patch is generated by starting from a surface residue on the *cargo* protein, and following alpha edges on the boundary of the alpha shape by breadth-first search, until $m$ residues are found. We construct $n$ candidate surface patches by starting in turn from each of the $n$ surface residue on the *cargo* protein. None of the candidate patches is identical to the native binding surface patch.

Second, we assume that a candidate surface patch on *cargo* protein has the same set of contacts as that of the native binding surface. The coordination number for each hypothetical contacting residue pair is also assumed to be the same. We replace the $m$ residues of the native surface with the $m$ residues from the candidate surface patch. There are $\frac{m!}{\prod_{i=1}^{20} m_i!}$ different ways to permute the $m$ residues of the candidate surface patch, where $m_i$ is the number of residue type $i$ on the candidate surface patch. A typical candidate surface patch has about 20 residues, therefore the number of possible permutation is very large. For each candidate surface patch, we take a sample of 1,000 random permutations. The expected binding energy $\bar{E}$ for a candidate surface patch is estimated as $\bar{E} = \sum_{k=1}^{1,000} E_k$, where $E_k$ is calculated using Equation (3) for the $k$-th permutation. The value of $\bar{E}$ is used to rank the candidate surface patches.

We assess the empirical potential by taking antibody/antigen protein in turn as the seat protein, and the antigen/antibody as cargo protein. The native interface on the seat protein is fixed and we test if our empirical potential can identify the correct surface patch on the *cargo* protein from the set of candidate surface patches plus the native surface patch. The results are listed in Table 2. Among the 14 native binding surfaces for 7 protein complexes, we can rank 11 native binding surfaces successfully as the top surface of the rank ordered list. The remaining 3 native binding surfaces all rank among the top 5, and the best ranking candidate surface patches for these three proteins all have over 50% native interfacial residues. By this criteria, our potential function can correctly recognize the native or near native binding surface patches of antibody and antigen complexes.

**Weighted Amino Acid Residue sequence Generator (WAARG)** We then develop an method to generate candidate antibody sequences for a given antigenic protein (called WAARG for Weighted Amino Acid Residue sequence Generator) based on the empirical potential. Again, we assume the knowledge of the binding surface on the seat protein, the size $m$ of the binding interface on the cargo protein, which is also taken as the length of peptide that needs to be generated. We further assume the same contact patterns as observed in the complex structure. The unknowns are the identities and sequence of the

Table 2: Recognition of Native Binding Surface of CAPRI Targets

| Target | Complex | [a]Antibody | | | Antigen | | |
|--------|---------|-------------|------|------|---------|------|------|
| | | [b]$R_{native}$ | [c]$O$ | [d]$N$ | $R_{native}$ | $o$ | $m$ |
| T02 | Rotavirus VP6-Fab | 1 | 0.65 | 283 | 1 | 0.72 | 639 |
| T03 | Flu hemagglutinin-Fab | 1 | 0.55 | 297 | 1 | 0.68 | 834 |
| T04 | α-amylase-camelid Ab VH 1 | 2 | 0.53 | 89 | 1 | 0.47 | 261 |
| T05 | α-amylase-camelid Ab VH 2 | 1 | 0.43 | 90 | 5 | 0.56 | 263 |
| T06 | α-amylase-camelid Ab VH 3 | 1 | 0.63 | 88 | 1 | 0.56 | 263 |
| T07 | SpeA superantigen TCRβ | 1 | 0.57 | 172 | 1 | 0.64 | 143 |
| T13 | SAG1-antibody complex | 3 | 0.64 | 286 | 1 | 0.68 | 249 |

[a] "Antibody": surface patches on the antibody molecule are scored, while the native binding surface on the antigen is kept unchanged. "Antigen": similarly defined as "Antibody". [b]$R_{native}$: Ranking of native binding surface among all $n$ candidate surface patches. [c]$O$: Percentage of overlap of residues from the best candidate patch with that of the native binding surface patch. [d]$m$: Number of surface residues. It is also the number of partitioned candidate surface patches.

residues that would best bind to the binding surface on the seat protein.

To generate biased sequences, the probability $\pi(i, a(i))$ of placing a residue of type $a(i)$ at position $i$ of the length $m$ sequence is set to be proportional to $\exp(\sum_{j,\sigma_{ij}} U(i, j, z_{ij}))$, where $\sigma_{ij}$ is an interfacial alpha edge across two proteins, and $U(i, j, z_{ij})$ is the empirical energy score. The value of $\pi(i, a(i))$ therefore depends on the residue types of the contacting residue pair $i$ and $j$, and the local packing environment reflected by the coordination number $z_{ij}$.

One way to specify the design of a biased peptide library is to provide a profile listing the favorable residues at each peptide position, along with the bias (or weight). Table 3 shows an example of such a profile for constructing a Protein A binding peptide library of length 7. The known native binding sequence is listed in the first row, followed by the profile consisting of the top 10 amino acid residues ranked by their weights at each position. At three residue positions, the wild type residues in the native binding surface are ranked first. For other positions, the wild type residues are ranked among the top 6, except for the position of GLY.

**Assessing WAARG performance.** To assess the overall quality of the peptide library generated computationally, we calculate similarity score of a designed sequence to the corresponding native sequence using the BLOSUM62 substitution scoring matrix. We found that in most cases, candidate peptides generated by the Weighted Amino Acid Residue sequence Generator (WAARG) have significantly higher sequence similarity than random sequences (Figure 1). To generate random sequences, we sample uniformly each of the

Table 3: Example of Weighted Sequence Library for Complex of Protein A and Antibody Fab (*1osp*).

| [a]Native Seq. | Y | S | D | Y | G | Y | R |
|---|---|---|---|---|---|---|---|
| 1 | [b]Y(0.47) | H(0.34) | Y(0.34) | Y(0.51) | R(0.32) | Y(0.58) | W(0.36) |
| 2 | W(0.27) | N(0.26) | E(0.25) | N(0.19) | K(0.23) | N(0.12) | S(0.22) |
| 3 | F(0.06) | M(0.09) | H(0.16) | W(0.12) | Y(0.12) | S(0.06) | T(0.12) |
| 4 | N(0.04) | Y(0.06) | D(0.09) | V(0.06) | V(0.08) | V(0.06) | Y(0.10) |
| 5 | S(0.03) | S(0.04) | R(0.02) | Q(0.05) | I(0.05) | W(0.05) | K(0.06) |
| 6 | V(0.02) | W(0.04) | N(0.02) | C(0.01) | W(0.04) | Q(0.02) | R(0.03) |
| 7 | C(0.02) | Q(0.03) | I(0.01) | R(0.01) | A(0.03) | R(0.01) | M(0.01) |
| 8 | R(0.02) | R(0.03) | L(0.01) | K(0.01) | C(0.03) | K(0.01) | G(0.01) |
| 9 | K(0.01) | L(0.02) | F(0.01) | G(0.01) | M(0.02) | G(0.01) | N(0.01) |
| 10 | G(0.01) | V(0.02) | S(0.01) | M(0.01) | H(0.02) | M(0.01) | A(0.01) |

[a]Native Seq.: The interfacial sequence from the heavy chain. [b]Y(0.47): residue type Y is to be chosen with a weight of 0.47. Underline: the chosen residue is the same as the residue at native interface.

20 amino acid residues for each of the $m$ positions. The average similarity between a native sequence of antibody interface and 1,000 random sequences $\overline{S}_{random}$ ranges from -30.14 to -11.95. The average similarity between a native sequence of antibody interface and 1,000 biased sequences generated by WAARG $\overline{S}_{weighted}$ ranges from -6.73 to 38.36. Figure 1(b) shows the distributions of similarity scores for designed and random sequences for N10-staphylococcal nuclease-antibody complex (*1nsn*). The $\overline{S}_{random}$ and $\overline{S}_{weighted}$ is -16.22 and 25.12, respectively. The overall distribution of similarity scores by WAARG has much higher similarity compared to the distribution of random sequences. These results shows the peptide library generated by WAARG will have significantly more enriched native-alike peptides than random.

Another method to assess the performance in generating biased library is to compare the number of sequences appeared before a sequence similar to that of the wild-type binding interface first occurs for both WAARG and random generators. This evaluation provides indication of the appropriate size of a peptide library to ensure inclusion of a number of good candidate sequences. We illustrate with the example of the binding interface between the heavy chain of NC10 antibody and influenza virus neuraminidase (*1nmbHN*) as a testing example. On the interface of this complex, there are seven residues on the binding surface of the antibody, and ten residues on the binding surface of the antigen. We design a library of peptides of length 7 that would bind to the antigen, and record the number of sequences appeared for the two methods before a candidate sequence with 4, 5, 6, and 7 identical residues to that of the wild-type binding surface peptide occur. If the sequence identity is 7, the candidate sequence is exactly the same as the native sequence. Table 4 shows

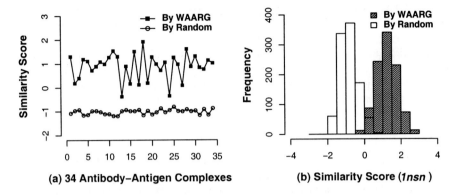

**(a) 34 Antibody–Antigen Complexes**    **(b) Similarity Score (*1nsn*)**

Figure 1: Evaluation of biased library. (a) Average similarity scores between one of the 34 native sequences of antibody interfaces and 1,000 sequences generated by random ($\overline{S}_{random}$) and 1,000 sequences by empirical potential ($\overline{S}_{weighted}$) ; (b) The similarity score distribution of $\overline{S}_{random}$ and $\overline{S}_{weighted}$ for antibody N10-staphylococcal nuclease complex (*1nsn*), one example among the 34 complexes. Every similarity score is normalized by the sequence length.

that the WAARGG can generate reasonably good candidate sequences with a much smaller library size than the random generator.

Table 4: Efficiency of WAARG

| Identity | Num by WAARG | Num by Random | Candidate Seq. by WAARG |
|---|---|---|---|
| 4 | 170 | 1721 | N N Y Y D W H |
| 5 | 2854 | 19417 | S N Y F Y Y G |
| 6 | 13645 | [a]$20^7/(7 \times 19)$ | S N Y Y Y Y G |
| 7 | 367288 | [a]$10^7$ | [b]S N Y Y D Y G |

[a]Expected number of sequences generated before an active peptide occurs. This number follows exponential distribution with the expectation $1/\lambda$, where $\lambda$ is the probability of a random sequence to be a required one. $\lambda = 7 \cdot (1/20)^6 \cdot (19/20) = (7 \times 19)/20^7$ when the identity is required to be six, and $\lambda = (1/20)^7$ when the identity is required to be seven. [b]: Wild type.

## 4   Discussion

We have developed a method for computational design of peptide library that can introduce useful bias to increase the efficiency in discovery of peptides binding to a target antigen protein. The key elements of our method is the alpha shape method to identify precise contact interactions, and an empirical potential for antibody-antigen interactions. We show that such a potential can be obtained by analyzing the alpha edges of known protein complexes. We find that it is important to consider the local packing environment, and the introduction of the coordination number in the empirical potential significantly improves the performance of the designed peptide library. Further development will need to consider the codon usage of the bacteria where the phage library is expressed.

# References

1. D.J. Rodi, L. Makowski, and B.K. Kay, *Curr Opin in Chem Biol* **6**, 92 (2002).
2. R.H. Hoess, *Chem. Rev.* **101(10)**, 3205 (2001).
3. M.D. Hughes, D.A. Nagel, A.F. Santos, A.J. Sutherland, and A.V. Hine, *J. Mol. Biol.* **331(5)**, 973 (2003).
4. I. Halperin, H. Wolfson, and R. Nussinov, *Protein Sci.* **12**, 1344 (2003).
5. A.T. Binkowski, L. Adamian, and J. Liang, *J. Mol. Biol.* **332**, 505 (2003).
6. T. Clackson and J.A. Wells, *Science* **267**, 383 (1995).
7. I. Halperin, H. Wolfson, and R. Nussinov, *Stucture* **12**, 1027 (2004).
8. X. Li, O. Keskin, B. Ma, R. Nussinov, and J. Liang, *J. Mol. Biol.* **In press.** (2004).
9. S. Miyazawa and R. Jernigan, *J. Mol. Biol.* **256**, 623 (1996).
10. X. Li, C. Hu, and J. Liang. *Proteins* **53** 792 (2003).
11. B. Krishnamoorthy and A. Tropsha. *Bioinformatics* **19(12)** 1540 (2003).
12. R. Samudrala and J. Moult. *J. Mol. Biol.* **275** 895 (1998).
13. H.Y. Zhou and Y.Q. Zhou, *Protein Sci.* **11** 2714 (2002).
14. H. Edelsbrunner. *Discrete Comput. Geom.* **13** 415 (1995).
15. M. Levitt. *J Mol. Biol.* **104** 59 (1976).
16. H. Edelsbrunner and P. Fu. *Rept. UIUC-BI-MB-94-01, Molecular Biophysics Group, Beckman Inst. Univ. Illinois, Urbana, IL*, (1994).
17. H. Edelsbrunner and E.P. Mücke. *ACM Trans. Graphics*, 13:43–72, (1994).
18. H. Edelsbrunner, M. Facello, P. Fu, and J. Liang. In *Proc. 28th Ann. Hawaii Int'l Conf. System Sciences*, volume 5, pages 256–264, Los Alamitos, California. IEEE Computer Scociety Press (1995).
19. M.A. Facello. *Computer Aided Geometric Design* **12** 349 (1995).
20. J. Liang, H. Edelsbrunner, P. Fu, P.V. Sudhakar, and S. Subramaniam. *Proteins* **33** 1 (1998).
21. G.D. Rose, A.R. Geselowitz, G.J Lesser, R.H. Lee, and M.H. Zehfus. *Science* **229** 834 (1985).
22. J. Khatun, S. Khare, and N. Dokholyan. *J. Mol. Biol.* **336** 1223 (2004).

# EXPLORING PROTEIN FOLDING TRAJECTORIES USING GEOMETRIC SPANNERS

D. RUSSEL and L. GUIBAS

*Computer Science Department*
*353 Serra Mall*
*Stanford, CA 94305, USA*
*E-mail: {drussel,guibas}@cs.stanford.edu*

We describe the 3-D structure of a protein using geometric spanners — geometric graphs with a sparse set of edges where paths approximate the $n^2$ inter-atom distances. The edges in the spanner pick out important proximities in the structure, labeling a small number of atom pairs or backbone region pairs as being of primary interest. Such compact multiresolution views of proximities in the protein can be quite valuable, allowing, for example, easy visualization of the conformation over the entire folding trajectory of a protein and segmentation of the trajectory. These visualizations allow one to easily detect formation of secondary and tertiary structures as the protein folds.

## 1 Introduction

There has been extensive work on visualizing the 3-D structure of proteins in ways that attempt to make the certain aspects of the structure more apparent. For example, commonly used software packages such as RasMol [10], ProteinExplorer [9], or SPV [8], among others, permit visualizations via hard-sphere models, stick models, and ribbon models that emphasize different aspects of the protein surface or secondary structure. Even more abstract visualizations have been used as a tool for understanding intra-molecular proximities, including contact maps and distance matrix images [13]. None of these approaches work very well, however, if the goal is to visualize proteins in motion and not just their static conformations.

Large corpora of molecular trajectories are becoming available through efforts such as *Folding@Home* [12] where molecular simulations are carried out on distributed networks of many thousands of computers. There is an increasing need to compare, classify, summarize, and organize the space of such protein trajectories with an eye toward advancing our understanding of protein folding by studying their ensemble behaviors. Most currently used methods for understanding such data revolve around computing a few summary statistics for each conformation, such as radius of gyration or number of native contacts and watching how these evolve during each trajectory. More similarly, the chemical distance, a statistic of an adjacency graph of the amino acids, was use to differentiate folded and unfolded states [4]. In this paper we explore the use of a more rich and abstract representation of the protein structure, based on spanners, which makes the task of understanding and exploring the space of protein motions easier.

Our basic idea is to take the continuous folding process and map it to a more discrete combinatorial representation. This representation focuses on higher-level geometric proximities that tend to form and be more stable over time rather than atom coordinates or specific aspects of secondary/tertiary structure. Specifically, we look at the formation of proximities between different parts of the protein across a range of scales, and track the changes of such proximities over time. Our more abstract description of the folding process is in terms of 'proximity events' — when certain proximities are formed or destroyed. Together, these characterize the folding process in a qualitative way and capture the important aspects of the *trajectory*, the sequence of conformations adopted by a protein in a particular folding path. Just as an algebraic topologist captures the essence of the connectivity of a continuous space in a few discrete invariants (the homology groups), we aim to capture the significant conformational changes during motion through a discrete representation of proximities that form and break.

We use *geometric spanners* to accomplish this goal. Starting from an abstract graph with weights on its edges, a spanner is a sparse subgraph (in the sense of having a number of edges roughly proportional to the number of vertices), such that all edges in the full graph can be well approximated by paths in the spanner (in the sense that the sum of the weights of edges of the path in the spanner is very close to the weight of the original graph edge). In the geometric setting the vertices in the original graph are points each pair of which is connected by an edge with weight equal to the Euclidean distance between the corresponding pair of points. The quality of the approximation can be controlled by varying the number of edges in the spanner.

Note that spanners are at once generalizations of contact maps as well as compressions of distance matrices. One can think of a spanner as a multiresolution contact map that allows an approximate reconstruction of the full distance matrix (and therefore the full 3-D structure as well).

We propose to use these combinatorial structures as a tool for capturing the important proximities of a protein conformation and, in this paper, for comparing and visualizing sequences of protein conformations from molecular trajectories. Key properties of the spanner that facilitate these goals include:

- Spanners are proximity based — this parallels proteins where local interactions determine the behavior.

- Spanners are discrete — they have a combinatorial structure whose description does not include any geometric coordinates.

- Spanners are controllable — we can produce descriptors that more loosely or more tightly capture the shape of the protein, converging to distance matrices as the approximation gets tighter.

- Spanners are uniform — there is only one type of combinatorial element, namely an edge. This makes comparison, processing and display simpler.

- Spanners can be made smooth — small changes in the protein conformation generally result in few large changes in the spanner, enabling tracking of the spanner structure over time.

- Spanners are local — the combinatorial features, edges, are affected by a small subset of the total point set. This means that changes in one part of the protein do not generally affect the spanner edges in other parts. As a result the edges can be assigned semantic meaning based on their endpoints, rather than on larger regions of the protein.

We use our spanners to investigate the folding of the protein BBA5 [11] using simulation data produce produced by the *Folding@Home* project. The spanners enable us to produce diagrams which show the formation (and sometimes dissolution) of secondary and tertiary structure during a whole folding trajectory and allow us to segment these trajectories into logical parts. We expect that the spanner approach will provide a valuable toolkit for the understanding and visualization of protein trajectories.

In the next sections we describe how we construct and smooth our spanner-based representation and how we use it to visualize trajectories. Then we discuss our how we have used spanners to try to understand the folding of BBA5. Finally we mention other promising applications of our spanner based representations.

## 2   Representing Proteins Using Spanners

We first provide a more rigorous definition of a geometric spanner. Let $P$ be a set of points in $\mathbb{R}^3$, Euclidean three-space, and $G$ be a Euclidean graph on $P$ (graph whose vertices are points from $P$ and whose edge weights are Euclidean distances between the endpoints of the edge). For a parameter $s > 1$, known as the *stretch factor*, $G$ is a spanner for $P$, if for all pairs of points $i$ and $j$ in $P$ with Euclidean coordinates $p_i$ and $p_j$, $\pi_G(i,j) \leq s\|p_i p_j\|$ where $\pi_G(i,j)$ denotes the shortest path distance between $i$ and $j$ in the graph $G$. Thus, the spanner represents the quadratic number of interpoint distances in $P$ by the much sparser set of edges in $G$. There is a vast literature on spanners that we will not attempt to review in here any detail; many different spanner constructions are possible. It has been shown that for $s$ arbitrarily close to 1, there exist spanners whose number of edges is proportional to the size of $P$. The reader is referred to a number of survey papers for background material and additional references [1, 6].

For simplicity, we only use the backbone atoms of the protein. This allows us to meaningfully identify each atom by its index along the backbone, so an edge $i, j$ connects the $i$th and $j$th atoms in the backbone. We can identify the edge $i, j$ with the point $(i, j)$ where $i < j$. We define the distance, between two edges $i_0, j_0$ and $i_1, j_1$ as the $\mathcal{L}_1$ distance between the points $(i_0, j_0)$ and $(i_1, j_1)$, namely $|i_0 - i_1| + |j_0 - j_1|$. We will write it $d(i_0, j_0, i_1, j_1)$. Two edges are close if they have a small $\mathcal{L}_1$ distance between them the corresponding points. The length as opposed to weight of an edge $l(i, j)$ is defined as $j - i$. Throughout the section $s$ will designate the stretch factor. A $s$-spanner is a spanner with stretch factor $s$.

## 2.1 Computation

We use what is known in the literature as the 'greedy' spanner. Its computation is conceptually very simple: starting with graph $G$ initially containing only the points $P$, test each of the $\binom{|P|}{2}$ interpoint *candidate edges* for inclusion, ordered from shortest to longest. For each candidate edge $i, j$, check if $s\|p_i p_j\| < \pi_G(i, j)$. If so, add the edge $i, j$ to the $G$. We call this test the *inclusion test*. The algorithm runs in $O(n^3)$ time due to the quadratic number of edges and the worst case linear time required to evaluate the inclusion test.

This greedy spanner construction has been shown to have asymptotically optimal complexity (number of edges) and weight (the sum of the lengths of the edges) as well as good practical complexity and weight [2]. Having low weight is important in our context since we want the spanner to consist of as many short edges as possible in order to capture local interactions. Euclidean spanners can be also be produced in $O(n \log^2 n)$ time with the same asymptotic edge count and weight bounds [3] although we have not implemented such methods.

If implemented naively, performing the inclusion test for long edges dominates the running time as it requires a nearly linear time graph search for each of these edges. However, such long edges are extremely unlikely to be in a spanner of a packed protein. If we maintain an upper bound on the graph distance, $d_G(i, j) \geq \pi_G(i, j)$ between each pair of points, $i, j$, then we can quickly eliminate any candidate edge for which $s\|ij\| > d_G(i, j)$. We can similarly prune many of the paths while searching.

These upper bounds can be maintained lazily as graph searches are performed. To tighten the upper bounds and further accelerate the process, it is advantageous to periodically compute the all atoms shortest path distances in the current spanner (an $O(n^2)$ process). In addition, we guide the search using the Euclidean distance as a lower bound on the graph distance to bias our search direction, as in the graph search algorithm $A*$. Using these heuristics, the 2-spanner of an 800 atom backbone of a protein can be computed in about a second.

The kinetic spanner proposed in [7] is a possible alternative. It can be cheaply maintained as the underlying points move around. However, it is non-canonical, making comparison between trajectories tricky and it has more long edges, which are hard to assign biological meaning.

## 2.2  Spanners of Proteins

Figure 1 shows spanners computed using different expansion factors for single protein and gives an estimate of the number of spanner edges per point for typical proteins in their native state.

| Expansion | 1.25 | 1.5 | 2.0 | 2.5 | 3.0 |
|-----------|------|-----|-----|-----|-----|
| $|G|/|P|$ | 4.5  | 1.5 | .52 | .31 | .21 |

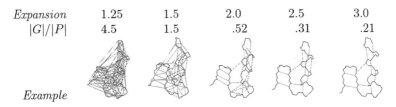

Example

Figure 1: Example spanners and average edge per point for various expansion factors. We mostly use 2-3 spanners for out computations and visualizations as spanners below 2 get very dense.

Secondary structure creates very well defined patterns in the spanner. If each edge is visualized as a point $(i, j)$, then $\alpha$ helices appear as a sequence of points $(i + kq, j + kq)$ where $k$ is a counter variable and $q$ is a stepsize which depends on the expansion factor. For a expansion factors between 2 and 3 the step size is 3, the edges are just longer when the expansion factor is larger. For a expansion factor of 1.5, the stepsize is still 3 but there are several edges leaving from each of the points. $\beta$ hairpins appear as series of points heading in an orthogonal direction to helices, namely, $(i + kq, j - kq)$. $k$ is 2 for 2 spanners and rises to 4 or 6 for 3 spanners (depending on how the hairpin twists). Both patterns are shown in Figure 2.

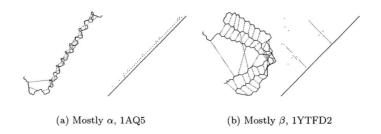

(a) Mostly $\alpha$, 1AQ5            (b) Mostly $\beta$, 1YTFD2

Figure 2: Helices and hairpins both give rise to a distinctive pattern of spanner edges: The 2-spanners and corresponding point patterns are shown for two proteins. The backbone edges $i, i + 1$ are displayed in addition to the other spanner edges.

## 3   Filtering Noise

Some addition processing must be done before we can use the spanner for detecting the creation and destruction of long-lasting structural patterns and proximities. Simply computing the spanner for each frame independently results in a sequence of different structures for each frame as the atoms vibrate. We need to be able to distinguish between edges which 'move' by a small amount between frames and edges which are entirely new at a given frame. Once we have done that we can filter the edges based on whether they represent long-lasting proximities.

### 3.1   Matching Spanners

We can set up the problem of associating edges from one frame with edges from the next as a bipartite matching problem. Such problems can be solved efficiently for a wide variety of similarity metrics. The similarity metric we used with the most success is to make the score for the edges $e_0$ and $e_1$ be $\min(l(e_0), l(e_1))/(1 + d(e_0, e_1))$. This allows longer edges to move more easily than shorter ones. In order to avoid spurious matches, we disallow matches of edges which are far from one another, specifically edges which are more than $\min(l(e_0), l(e_1))/3 + 2$ apart. This threshold allows helix and strand edges to be readily matched with one another over the expansion factors we use, but cuts off long distance matches.

Using this technique we can associate edges from successive pairs of frames. We call a sequence of paired edges a *proximity*. The *lifetime* of a proximity is how many different consecutive frames contain edges that contributed to that proximity. We can now talk about the creation and destruction of a proximity—exactly the type of proximity events we mentioned in the introduction.

## 3.2   Filtering Noise

Many of the proximities tracked using the previously mentioned technique will have very short lifetimes, perhaps only one frame. These short-lived proximities form a sort of topological noise, which can obscure the real signal in the data.

In order to remove this noise we turn to the idea of persistence [5]. A proximity is *persistent* if its lifetime exceeds a certain threshold. Persistent proximities represent stable aspects of the conformation of the protein backbone. We simply drop non-persistent proximities.

There are certain cases where persistence filtering removes too many edges. For example, if the protein conformation is such that two candidate edges, both of which independently pass the inclusion test, have nearly the same length, then small perturbations can cause either edge to be included in the spanner. The spanner can rapidly alternate between the two edges in succeeding frames, resulting in neither of them being persistent and both of the edges being dropped. Dropping them both looses too much information about the protein conformation. To solve this, in addition to matching edges from the current frame against the edges from the previous frame, we add in an edge from each proximity that was destroyed in a recent frame. As a result, in the previously mentioned example of two edges flipping back and forth, both will now be included.

In practice, we discard proximities which exist for fewer than 20 frames and allow edges to be matched against edges which disappeared up to 6 frames before.

## 3.3   Segmenting Trajectories

We can now define two spanners as being close if they have similar persistent proximities. This gives us a means of segmenting trajectories into logical components. To do this, assign each consecutive pair of trajectories a *spanner distance*—the sum of the lengths of the persistent proximities which are created or destroyed between this pair of trajectories. We can then smooth these distances in time and use local maxima to segment the trajectory. We will discuss the segmentations produced by this method more in the next sections once we present our technique for displaying spanners.

## 4   Encoding Spanners in One Dimension

Simply displaying the spanner along with the backbone helps make interactions easier to pick out, especially in less ordered states. However, by itself it does make it easy to visualize the whole trajectory since each frame must be displayed successively. To avoid that problem we want to encode the spanner as a one dimensional object so that we can lay out a whole series of them for inspection together.

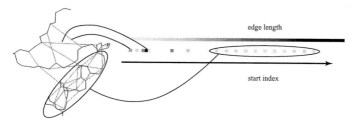

edge length

start index

Figure 3: Encoding a spanner into a strip spanner: a 2-spanner and corresponding strip spanner are drawn and some edge and structure correspondences are indicated.

To encode a spanner for a frame, take each spanner edge which is part of a persistent proximity and mark the first point of the edge with the length—i.e. for edge $i, j : i < j$ set $S[i] = l(i, j)$. We call the resulting vector, $S$ a *strip spanner*. An example is shown in Figure 3.

One drawback of the strip spanner encoding is that it cannot directly handle more than one spanner edge originating a single point. This can be remedied in one of several manners. One solution is to expand each backbone atom into a constant number of points, i.e. atom $i$ goes to $i - \epsilon, i, i + \epsilon$. This allows a constant number of edges per atom. Alternatively, spanner construction can enforce the one edge per point restriction. For expansion factors 2 and above, this does not significantly distort the spanner, however, it causes problems for factors much below 2 as there are too many edges. For the purposes of generating strip spanners for this paper we simply take the maximum length edge originating at each point. For a small protein such as BBA5 where there are few tertiary edges, this is quite adequate.

Strip spanners from successive frames can then be stacked creating a view of the trajectory as a whole. We call this stacking a *strip history*. An example is shown in Figure 4.

The strip history gives us a good way of displaying the trajectories which were segmented using the technique presented in Section 3.3. An example segmentation is shown in Figure 5. The results of segmentation corresponds

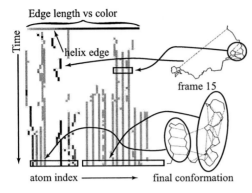

Figure 4: A strip history: In this trajectory, the protein does not fold completely, but forms significant secondary and tertiary structure. The patterns for $\alpha$ helical secondary structure are quite easily seen (lines 3 units apart) on the right half of the diagram. The $\beta$ hairpin pattern is slightly less easily picked out on the left (lines 2 units apart). Long tertiary edges show up as the darker pixels on the left half of the diagram.

quite well with manual segmentations we had previously performed using the strip history.

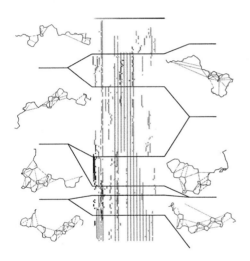

Figure 5: Subdivision of a trajectory using the strip history: A non-folding trajectory for BBA5 was divided into 7 phases using the strip history. The conformation for the middle frame of each phase is shown, always oriented with the $\alpha$-helical end to the right. During the first segment the protein has little to no secondary structure. In the second the first part of the alpha helix forms and stays fairly static through the third. In the fourth segment we see the hairpin forming, but part of the helix disintegrates. The hairpin disintegrates in the fifth. In the seventh the hairpin reforms, and the end of the helix forms for the first time.

## 5   Understanding BBA5 Trajectories

We applied our spanner descriptor to the task of trying to understand the folding process of the protein BBA5. It is a 23 residue protein which folds comparatively quickly. The native state of BBA5 consists of a $\beta$ hairpin, involving the first 8 residues and an $\alpha$ helix involving the last 10 residues.

The hairpin is packed against the helix, although not very tightly due to the presence of large sidechains between the two pieces of secondary structure.

Our data consists of 23 trajectories with frames every 200fs. Thirteen of the trajectories come within 3.1Å cRMSD$_{C_\alpha}$ of the native structure.

Figures 4 and 5 show strip histories for two example trajectories. The times of formation of the $\alpha$ helix are quite clearly visible. The $\beta$ hairpin formation is less clearly defined, but still can be seen on the left of the image, as can the occasional large backbone distance tertiary interactions, which are the points colored the darkest.

Perusal of the strip histories show that there are few constants among the folding trajectories, as might be expected given its small size.

- $\alpha$ helical secondary structure rarely went away once created, although there were a few cases where parts of the helix formed and then dissolved (such an example can be seen in Figure 5). This may be an artifact of how trajectories were selected for storage. Parts of the helix formed in all possible orders.

- $\beta$ hairpin structure was much less stable than the helix, however it did stabilize without the presence of tertiary interactions in several of the trajectories.

- tertiary interactions formed before secondary structure in some but after strong secondary structure in others.

In one of the strip histories a helix like pattern can be seen forming in the loop end of the hairpin. It persists for 30 frames. Direct inspection of the backbone structure confirmed that there is indeed a one and a half turn helix like structure there. This trajectory is not shown here. We did not observe similar structure formation in any of the other trajectories.

## 6   Future Work

Sidechains play an important role in the protein folding process. For example, in BBA5, the rings from a phenylalanine and a tyrosine occupy much of the space between the helix and the hairpin in the native state, making the helix and hairpin pack much less tightly than they would in the absence of such rings. The position of the sidechains is currently ignored in our calculations, although it may be important to the folding process and ignoring it makes the tertiary structure show up much less clearly. Simply adding the whole sidechains adds too many new edges to the spanner making the relevant data hard to extract and disrupts the linear order which we depend on for matching and visualization. A better approach may be to add a single point per sidechain which will capture its location without complicating the structure too much.

The strip spanner is a one dimensional descriptor which captures key aspects of the proteins conformation. Searching and matching of one dimensional structures is a much easier problem than matching three dimensional curves, suggesting that the strip spanner might have applications in protein structure motif searching and structure alignment. However, there are a number of issues with incorporating gaps which need to be resolved.

We are trying to apply spanners to the problem of understanding the parts of protein conformation space relevant to folding. The strip history based segmentation provides one way of dividing trajectories into chunks which could be matched against one another to find common paths through conformation space. There are a number of problems with measuring the distances between spanners which need to be resolved first. In addition, we suspect simple proteins such as BBA5 fold too quickly and have too small an energy barrier for its fold space to have significant structure. As a result we plan to apply the techniques to unfolding data.

### Acknowledgments

This work has been supported by NSF grants CARGO 0310661, CCR-0204486, ITR-0086013, ITR-0205671, ARO grant DAAD19-03-1-0331, as well as by the Bio-X consortium at Stanford.

The authors would like to thank Rachel Kolodny for many valuable suggestions regarding the project and Vijay Pande for providing the data.

### References

[1] S. Arya, G. Das, D. Mount, J. Salowe, and M. Smid. Euclidean spanners: short, thin, and lanky. In *Proceedings of the twenty-seventh annual ACM symposium on Theory of computing*, pages 489–498. ACM Press, 1995.

[2] G. Das, P. Heffernan, and G. Narasimhan. Optimally sparse spanners in 3-dimensional Euclidean space. In *Symposium on Computational Geometry*, pages 53–62. ACM Press, 1993.

[3] G. Das and G. Narasimhan. A fast algorithm for constructing sparse Euclidean spanners. In *Symposium on Computational geometry*, pages 132–139. ACM Press, 1994.

[4] Nikolay Dokholyan, Lewyn Li, Feng Ding, and Eugene Shakhnovich. Topolical determinants of protein folding. *Proceedings of the National Academy of Science*, pages 8637–8641, 2002.

[5] H. Edelsbrunner, D. Letscher, and A. Zomorodian. Topological persistence and simplification. *Discrete Compututational Geometry*, 28:511–533, 2002.

[6] D. Eppstein. Spanning trees and spanners. *Handbook of Computational Geometry*, pages 451–461, 2000.

[7] J. Gao, L. J. Guibas, and A. Nguyen. Deformable spanners and applications. In *Symposium on Computational Geometry*, pages 179–199, June 2004.

[8] N. Guex and M. C. Peitsch. Swiss-model and the swiss-pdbviewer: An environment for comparative protein modeling. *Electrophoresis*, 18:2714–2723, 1997. URL http://www.expasy.org/spdbv/.

[9] E. Martz. Protein explorer: Easy yet powerful macromolecular visualization. *Trends in Biochemical Sciences*, 27:107–109, 2002. URL http://proteinexplorer.org.

[10] Rasmol. URL http://www.umass.edu/microbio/rasmol/.

[11] Y.M. Rhee, E. J. Sorin, G. Jayachandran, E. Lindahl, and V. Pande. Simulations of the role of water in the protein-folding mechanism. In *Proceedings of the National Academy of Science*, volume 101, pages 6456–6461, 2004.

[12] M. Shirts and V. Pande. Screen savers of the world, unite! *Science*, 2000. URL http://folding.stanford.edu/.

[13] M. J. Sippl. On the problem of comparing protein structures. *Journal Molecular Biology*, 156:359–388, 1982.

# CHAINTWEAK: SAMPLING FROM THE NEIGHBOURHOOD OF A PROTEIN CONFORMATION

ROHIT SINGH and BONNIE BERGER[*†]

*Computer Science and Artificial Intelligence Laboratory*
*Massachusetts Institute of Technology*
*Cambridge MA 02139*
*E-mail: {rsingh, bab}@mit.edu*

When searching for an optimal protein structure, it is often necessary to generate a set of structures similar, e.g., within 4Å Root Mean Square Deviation (RMSD), to some *base* structure. Current methods to do this are designed to produce only small deviations ($< 0.1$Å RMSD) and are inefficient for larger deviations. The method proposed in this paper, ChainTweak, can generate conformations with larger deviations from the base much more efficiently. For example, in 18 seconds it can generate 100 backbone conformations, each within 1-4Å RMSD of a given 45-residue conformation. Moreover, each conformation has correct bond lengths, angles and omega torsional angles; its phi-psi angles have energetically favorable values; and there are rarely any backbone steric clashes. The method uses the insight that loop closure techniques can be used to perform compensatory changes of dihedral angles so that only a part of the conformation is changed. It is demonstrated, using decoys from the Decoys 'R Us data-set, that ChainTweak can be used to construct good decoys. It also provides a novel and intuitive way of analyzing the energy landscape of a protein. In addition, ChainTweak can improve the accuracy and performance of the loop modeling program RAPPER by an order of magnitude (1.1 min. vs. 36 min. for an 8-residue chain).
**Availability & Supp. Info.:** http://theory.csail.mit.edu/chaintweak

## 1. Introduction

A fundamental axiom of molecular biology is that the function of a protein is determined by its structure. In turn, most protein structure determination problems are, essentially, search problems. In some of these, e.g., homology modeling or protein re-design, the problem specification may restrict the search to the neighbourhood of some template structure. In other cases, restricting the search to the neighbourhoods of a set of candidate structures might just be a solution strategy (e.g., in the Rosetta[4] method for *ab-initio* folding). Here, the *neighbourhood* of a structure is the set of structures similar to it. For example, the set of all structures within, say, 4Å RMSD of a base structure could be defined as its neighbourhood[a].

---

[*]Corresponding author
[†]Also in the MIT Dept. of Mathematics
[a]Of course, the size of the neighbourhood and, consequently, the exact choice of a RMSD

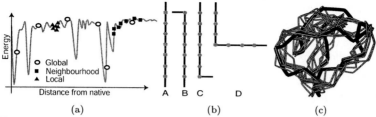

(a)  (b)  (c)

Figure 1: **(a)** A cartoon illustrating the space coverage differences between global, neighbourhood, and local search. Observe that local search techniques can only cover the basin on one local minima. **(b)** Cartoon illustrating that changes in dihedral angles near the terminal regions of a chain (A) result in small perturbations (B and C), while changing an angle in the middle of the chain results in a very large perturbation (D). **(c)** Example output from ChainTweak. Ten conformations from the neighbourhood of a 32-residue protein structure (PDB:1clv, chain I) were sampled and aligned with the original. The original structure is in black, the others are in gray.

Efficiently searching in the neighbourhood of a possible protein structure (conformation) is thus an important and frequently recurring problem. As the term "neighbourhood search" signifies, this search problem is different from global or local search problems (Fig 1a), even though it has usually been studied as an extreme case of these. This paper focuses on the sampling component of this search problem and presents a method, ChainTweak, for efficiently and representatively sampling from a given neighbourhood.

Many different approaches to neighbourhood sampling have been tried. High-temperature Molecular Dynamics (MD) methods have been used to generate structures with 2-4Å RMSD from the native[1]. Methods based on discrete off-lattice models[2,3] discretize the dihedral-angle space and try out different combinations. Similarly, in Monte Carlo (MC) search methods, various move-sets have been developed for making local moves. For example, fragment-swap MC in Rosetta[4] relies on using a database of polypeptide fragments to swap one fragment for another, as long as their ends match. Another set of approaches, such as in torsional dynamics[5], or the MC-based methods proposed by Ulmschneider & Jorgensen[6] and Cahill et al.[7] use geometric insights to perform such local modifications.

Our proposed neighbourhood sampling method, ChainTweak, has many advantages over existing methods. Rather than being closely tied to some search strategy (or an energy function), it is a stand-alone method that can be used by researchers as a black-box, allowing them to focus on other parts of the search problem (e.g., energy function design[8]). Moreover, ChainTweak

---

for a 50-residue protein, two conformations within 2Å of each other are considered almost identical. Thus, in this case, the threshold size should be $\geq$ 2Å.

enables fast generation of ensembles (sets of conformations) centered around any given base conformation. The flexibility of ChainTweak enables novel applications (e.g., energy function analysis) and enhances the performance of existing applications (see Section 5).

### 1.1. *Neighbourhood Sampling: The Right Representation?*

Almost all neighbourhood sampling methods work by perturbing the base conformation's structure to generate conformations in its neighbourhood. To model the structure, these methods use either an all-atoms Cartesian coordinates based model or a dihedral angles based model.

Most existing methods use the Cartesian coordinate based model. With this model, however, an energy minimization step is needed to restore correct bond lengths/angles in the perturbed structures. Efficiency and convergence issues with this step limit the size of a single perturbation step ($< 0.1\text{Å}^9$). Thus, only a small neighbourhood around the base can be explored. For larger deviations, successive perturb-and-minimize operations, using an MD-like approach[1], can be done. However, generating many MD trajectories, to ensure representative sampling, may become computationally expensive.

In contrast, representing the protein backbone by its dihedral angles offers distinct advantages. All conformations sampled from the neighbourhood will then have different dihedral angles but the same bond lengths/angles. Since the latter can always be set to their desired/ideal values, no minimization step is necessary. Hence, the restriction on small perturbation sizes is removed. However, modifying a dihedral angle at residue $i$ changes the positions of all residues $i + 1$ onwards. As a result, the perturbed structure may deviate so far from the base as to not be in the neighbourhood at all, especially if residue $i$ is in the middle of the chain (Fig 1b). This problem is the major stumbling block in using a dihedral angles based representation.

One way to solve this problem, e.g., in Torsional Dynamics (TD)[5], is compensatory modification of multiple torsional angles such that the overall structural deviation is acceptably small. However, the differential calculus-based methods used by TD algorithms work well only for small perturbations. Moreover, the sampling behavior is effected by the energy function chosen for the TD simulations. The reader might also notice the parallel here with the loop closure problem where one needs to find small chains joining two fixed ends. Indeed, our proposed algorithm, ChainTweak, exploits this parallel.

### 1.2. *Contributions*

ChainTweak is an algorithm for efficiently sampling from the neighbourhood of a given base conformation. It generates a set of backbone conformations such that each new conformation has the following properties: it lies in a neighbourhood of the base; it has the terminal (first and last) residues fixed in the same relative positions as the base; and it has bond lengths/angles set to their desired/ideal values. In Section 2 we describe a simple extension that allows the positions of terminal residues to vary as well.

ChainTweak iteratively perturbs the base conformation using the dihedral angle representation. A sliding window approach is used to successively move some atoms by 0–2Å while keeping all others fixed. Inside the window, loop closure methods are used to generate such perturbations. Moreover, residue-specific phi-psi angle preferences can be used to choose a perturbation.

We show that ChainTweak can explore large neighbourhoods efficiently. Given a conformation of a 45-residue protein, in 18 seconds it can generate 100 backbone conformations, each within 1–4Å RMSD of the base. Moreover, by running ChainTweak for more iterations larger neighbourhoods can be explored: for this protein, a conformation with RMSD of 12Å from the base can be found. In contrast, after 18 seconds, an MD simulation (run using TINKER[9]) produces a single conformation for the same protein (with 0.91Å RMSD). Even theoretically, ChainTweak's running time is asymptotically optimal— linear in the length of the chain and the number of samples desired.

We also describe some applications of ChainTweak (Section 4.2). It improves upon the performance of some existing applications (decoy generation and *ab-initio* loop-modeling using RAPPER) and also enables novel applications (energy function analysis in an intuitive way).

## 2. Algorithm

Here we present the algorithm ChainTweak that has the following input and output:

*Input:* A single backbone conformation $C_0$ described by its bond lengths, bond angles and dihedral angles.

*Output:* N conformations such that the RMSDs of these conformations w.r.t. $C_0$ roughly follow a desired distribution. For example, half of the output conformations are 0–2Å RMSD from the base while the rest are 2–4Å RMSD from the base. For each output conformation, the bond lengths, bond angles and the relative positions of the end-residues are the same as in $C_0$.

The initial restriction on preserving the relative positions of the end-residues can be adapted for flexible chain ends by pre-processing $C_0$ to produce a set of conformations with randomly sampled values for dihedral angles at the end-residues. Recall that modifying dihedral angles at the ends only results in local structure changes (Fig 1b). Each of these conformations then becomes the input to a separate ChainTweak instance.

Observe that by iteratively setting each output conformation as the input of a new ChainTweak problem, more solutions can be found for the original ChainTweak problem. Also, the problem can be *recursively* solved by splitting the input chain into two sub-chains and concatenating the respective solutions. We do this until we have a chain small enough to be solved using loop-closure techniques. The pre-processing step (moving the chain ends) mentioned previously is required only at the top level of recursion, i.e., for the full-length chain.

The loop closure problem was informally discussed by Robert Diamond[14] and was formally defined by Go and Scheraga[15]. The input in such a problem is the relative position of two fixed residues (anchors) at each end and the goal is to find different possible conformations for a polypeptide chain of length $m$ joining the fixed ends. For a problem instance with 6 unknown dihedral angles, i.e. 6 degrees of freedom (DOFs), the maximum number of possible solutions is 16. With more DOFs, the number of solutions is infinite. In the 6-DOF case, Manocha et al.[16] applied inverse kinematics techniques from robotics to numerically generate all possible 16 solutions. More recently, Wedemeyer and Scheraga[17] and Coutsias et al.[18] have also presented analytic solutions for the 6-DOF problem. ChainTweak can use any of these as a subroutine (Algorithm 3 in Supp. Info.).

ChainTweak iteratively calls the subroutine SLIDEWIN (Algorithm 2 in Supp. Info.). Given a starting backbone conformation, SLIDEWIN finds a new backbone conformation using a sliding window approach (Fig 2a). A window of 3 residues (9 points) is chosen. After fixing 3 points on both ends, this results in a 6-DOF loop closure problem. We use Manocha et al.'s algorithm when omega angles are unrestricted and Coutsias et al.'s algorithm when omega angles have to be restricted to particular values (say, 180°). A wrapper around these routines (LOOPCLSR6, Algorithm 3) suggests up to 15 alternative conformations for the conformation inside the window. Of these, we randomly select one conformation, biasing our choice towards a conformation that has phi-psi angles in favorable/acceptable regions of the Ramachandran Plot (Fig 3). Residue and secondary structure information can thus be encoded by designing appropriate phi-psi preference maps.

A single iteration of SLIDEWIN moves each residue by about 0.5–1.5Å. ChainTweak (Algorithm 1 in Supp. Info.) iteratively applies SLIDEWIN $K$ times to achieve a much larger deviation from the starting conformation; the output conformations of one iteration form the input for the next. Between each iteration, some conformations may be pruned out, depending on their RMSD from the original structure. The exact pruning policy is described by

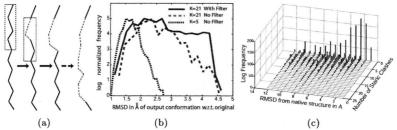

|     |     |     |
| --- | --- | --- |
| (a) | (b) | (c) |

Figure 2: **(a)** A cartoon describing SLIDEWIN. Inside each window, LOOPCLSR6 is used to perform the tweak. Observe that the first and last positions in the window are not changed, both in LOOPCLSR6 and SLIDEWIN (see Supp. Info.). **(b)** A plot showing the frequency distribution of ChainTweak-generated conformations vs. their RMSD w.r.t the base. The parameters $K$ and Filter can be used to control structural variation in the output set. A low value of $K$ (=5) results in conformations that are similar to the the original. $K = 21$ resulted in greater structural variation. Filter was used to ensure that the distribution was "more even". The frequencies of each distribution have been scaled so that the maximum is same across all three. **(c)** A plot showing the frequency distribution of ChainTweak-generated conformations, classified by the number of steric clashes per conformation and its RMSD from the native. 10000 backbone conformations from the neighbourhood of a 45-residue structure (PDB 1bh9:31-75) were generated. For each conformation, the number of backbone steric clashes, using a cutoff of 2Å, were counted. Most of the conformations, even those with large RMSDs from the base, have no steric clashes. Note that the frequencies are shown on a log scale.

the user-specified parameter Filter (described below) and helps in achieving a desired structural variation in the final solution set (Fig 2b).

## 3. Results

### 3.1. *Performance Analysis*

The size of the neighbourhood explored by ChainTweak, measured in RMSD from the base, is controlled by the number of iterations, $K$. In our simulations, we observed that this size increases from 2.5Å, for $K = 5$, to about 4.5Å, for $K = 21$ (Fig 2b). ChainTweak can explore rather large neighbourhoods: for a 45-residue protein it can generate a conformation with 12Å RMSD from the base.

Another user-specified parameter, Filter, can be used to control the structural variation in ChainTweak's output by describing a pruning policy. An example pruning strategy (Fig 2b) is to remove enough structures, after every $4^{th}$ iteration, such that the RMSDs (w.r.t. the base) of the remaining structures are uniformly distributed. Without any pruning, the output set's composition is skewed towards structures with low RMSD (approx 1-2Å) from the base. This is understandable— having performed a tweak operation on a conformation, a second tweak operation is as likely to take it further away from the original as it is to bring it back closer to the original. Analogously,

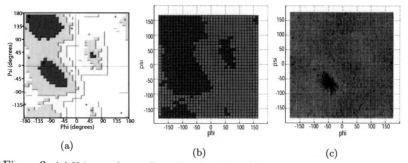

**Figure 3:** **(a)** Using a reference Ramachandran Plot, **(b)** we implemented a simple phi-psi priority scheme: (red: favorable, allowed) > (dark blue: generously allowed) > (light blue: others). **(c)** For 10000 conformations of a 45-residue protein (PDB: 1bh9,31-75) generated by ChainTweak, the phi-psi distributions match well with the specified priorities. This protein has 2 alpha-helices which explains the higher frequency of phi-psi angles in regions of the plot corresponding to alpha-helical structure.

recall that in a 1-D random walk, the probability of being at distance $d$ from the origin decreases exponentially with $d$.

Can ChainTweak representatively sample from the entire neighbourhood? Some recent theoretical work[19,20] on the folding of polygonal chains suggests that any two protein backbone conformations (with same bond lengths/angles) can be converted into each other by simply changing the dihedral angles. This suggests that ChainTweak can explore the entire neighbourhood. Also, observe that the "tweak" operation of SLIDEWIN is essentially a random walk in this neighbourhood. This, in turn, suggests that CHAINTWEAK's sampling is representative.

ChainTweak is efficient in both practice and theory: for a chain of length $n$ with $N$ output conformations, the running time of ChainTweak is $O(Nn)$. It is dominated by the approximately $KNn/3$ calls to LOOPCLSR6. The actual time spent per call of LOOPCLSR6 does not vary much (avg: 8.3 millisecs; std dev: 3.6 millisecs on a Pentium-4 2.4GHz PC). Also, observe that just writing the output ($N$ conformations, each of size $O(n)$) would take $O(Nn)$ time. Hence, ChainTweak is an *asymptotically* optimal algorithm.

ChainTweak has high numerical accuracy. Its implementation avoids error accumulation (see Supp. Info. for details). For example, the deviation of atom positions in the terminal residues is negligible: avg error = 0.001Å.

Conformations generated by ChainTweak have very few backbone steric clashes (Fig 2c). This is probably because, in all our experiments, the base conformation did not have any steric clashes and the output conformations are similar to the base. After the addition of sidechains to the generated backbone conformations, both new sidechain and old backbone steric clashes, if

| | ChainTweak + RAPPER | | | | RAPPER only | | | | |
| | Best generated (RMSD) | | | | Best generated (RMSD) | | | | |
| | Time | Backbone | | $C_\alpha$ | C Anchor | Time | Backbone | | $C_\alpha$ | C Anchor |
| Length | (min.)* | Global | Local | Local | (RMSD)† | (min.)‡ | Global | Local | Local | (RMSD) |
| 8 | 0.7(1.1) | 1.40 | 0.92 | 0.93 | 0.02 | 36.4 | 1.11 | 0.70 | 0.56 | 0.30 |
| 9 | 1.1(1.5) | 1.61 | 1.07 | 1.12 | 0.03 | 30.5 | 1.29 | 0.81 | 0.72 | 0.33 |
| 10 | 1.3(1.7) | 2.02 | 1.24 | 1.31 | 0.01 | 44.15 | 1.67 | 1.11 | 1.00 | 0.41 |
| 11 | 1.7(2.3) | 2.45 | 1.49 | 1.61 | 0.04 | 59.17 | 1.99 | 1.27 | 1.23 | 0.33 |
| 12 | 2.1(3.1) | 2.21 | 1.56 | 1.72 | 0.02 | 100.4 | 2.21 | 1.47 | 1.46 | 0.54 |

Table 1: ChainTweak can improve upon the performance of the loop modeling program RAPPER. The latter's performance in generating 1000 loop conformations for loops of various lengths has been measured using the FISER dataset[10]. From the same dataset, for chain lengths between 8-12 residues, we picked 20 chains each. For each of these, a representative set (in terms of their RMSDs from the native conformation) of 10 conformations was picked from the RAPPER-generated set. Using ChainTweak, 100 conformations in the neighbourhood of each such conformation were sampled. As in ref. (10), the quality of these 1000-conformation ensembles is measured in terms of the smallest Global (only loop ends aligned) and Local (whole chain aligned) RMSD of any conformation w.r.t. to the native (averaged across all 20 chains) and the deviation of C-terminal loop ends from the desired position. [*] The time in parentheses includes the estimated cost of generating 10 conformations using RAPPER. [†] Before running ChainTweak, the chosen RAPPER-generated conformations were fixed, if possible, so that their ends matched those of the native. [‡] To account for differences in processing power (2.4GHz for us vs. 900MHz in ref. (10)), these running times are one-fourth of the actual times reported in ref. (10).

any, can be relieved simultaneously. Hence, we decided against explicitly checking for steric clashes in ChainTweak. In case such checks become necessary, they can be done efficiently by taking advantage of ChainTweak's incremental modification approach and using kinetic data structures[22] or Lotan et al.'s hierarchical approach[23].

ChainTweak generates structures where most of the phi-psi angles have favorable values. A random 6-DOF chain with fixed ends has multiple alternative conformations. Residue and secondary-structure related phi-psi preferences can be used to pick the most appropriate alternative. We encoded a simple phi-psi preference map which accorded higher priority to any phi-psi combination lying in favorable and acceptable regions of the Ramachandran Plot. Even this simple map yielded impressive results (Fig 3).

### 3.2. Applications

**Loop Modeling:** ChainTweak can be used to supplement an *ab-initio* loop modeling program like RAPPER (DePristo et al.[10]). The latter generates loop conformations by sampling in a discretized phi-psi angle space and then using a dihedral angle-based minimizer to ensure that the position of loop ends is roughly unchanged. The method is computationally expensive

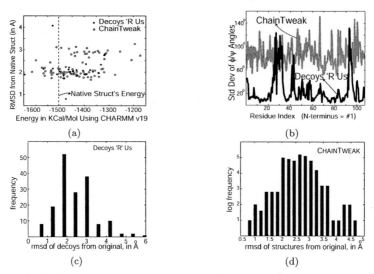

Figure 4: **(a,b)** Decoys of a protein domain (PDB 1mfa:1-111, ig_structal_hires in De-coys 'R Us) were extracted from the Decoys 'R Us Database (DB) and were also created using ChainTweak (CT). Both the sets, as well as the native, were minimized using the CHARMM v19 energy function and the TINKER package. **(a)** Post-minimization ener-gies and $C_\alpha$-RMSD from the native structure are plotted to illustrate that both DB and CT decoy-sets manage to "fool" the energy function and are structurally similar to the native structure. **(b)** To identify regions with local structural variation, we measured the standard deviation of phi-psi angles along the chain. This indicates that the CT set is more representative: its local structural variation is not limited to a few regions. **(c,d)** Comparison of decoys produced by DB (c) and CT (d) for the loop region 1vfa:158-166. The structural variation among the decoys, as measured by $C_\alpha$-RMSD to the native con-formation, is comparable across the two sets

because all conformations with incorrect positions of the loop ends have to be rejected. In contrast, ChainTweak only generates conformations that have the loop ends in the right positions.

ChainTweak can be used to efficiently expand a small ensemble generated by RAPPER, thus improving overall efficiency by an order of magnitude (Table 1). The ensembles generated by the two methods are of comparable quality (as measured by the RMSD to the native conformation). On one important criterion, that of fixing the positions of the loop ends, ChainTweak actually performs much better.

**Decoy Generation:** Decoys[2] are non-native structures that can be used to design energy functions[8] capable of distinguishing such structures from the native. We have found that ensembles generated by ChainTweak can be used to generate decoys, especially those that are globally similar to the

| | w.r.t. the native's minimum, # of minima with | | largest |
| Energy Function | < 0.1Å RMSD | < 0.5Å RMSD | RMSD |
|---|---|---|---|
| Amber '99 + GB/SA | 6 of 10 | 10 of 10 | 0.32Å |
| Charmm v19 + GB/SA | 5 of 10 | 10 of 10 | 0.16Å |
| OPLS + GB/SA | 3 of 10 | 10 of 10 | 0.25Å |
| Charmm v27 + GB/SA | 3 of 10 | 10 of 10 | 0.39Å |
| Charmm v19 | 2 of 10 | 5 of 10 | 4.38Å |
| Charmm v27 | 0 of 10 | 4 of 10 | 2.33Å |
| Amber '99 | 0 of 10 | 4 of 10 | 1.72Å |
| OPLS | 0 of 10 | 0 of 10 | 3.17Å |

Figure 5: **Figure:** This figure illustrates how ChainTweak can be used for analyzing energy landscapes. (A) Given some decoy and two candidate energy functions, (B) Chain-Tweak can sample from the neighbourhood of the decoy. (C) The generated conformations are minimized and the distribution of local minima provides information about the energy landscape. Note that the same set of conformations is used for each energy function.
**Table:** ChainTweak was used to generate 10 conformations similar (within 0.003-0.13Å RMSD) to an alpha-helix (PDB 2gib:22-37). Eight different energy functions were used to minimize the ensemble conformations, resulting in 10 local minima per function (see Supp. Info. for details). For each function, the RMSDs of these local minima from the local minimum corresponding to the native structure were measured. An energy function ranked higher if it had more local minima with a very low (< 0.1Å) RMSD from the minimum corresponding to native structure. As can be seen, the addition of a solvation term (GB/SA[24]) improves the performance of these energy functions.

native structure but have significant local differences from it. As the use of homology modeling to predict structure increases, the need for such decoys will increase. We used some loop decoy-sets and some homology modeling based decoy-sets (HM) from the the Decoys 'R Us database (Samudrala and Levitt[12]) to evaluate ChainTweak-generated decoys.

ChainTweak's HM decoys are comparable to the database decoys in terms of their energy-vs-RMSD profile. CT decoys are more representative, i.e, their local structural variation is not limited to a few small regions (Fig 4a, 4b). With HM decoys, the use of homology forces biased sampling: the local structural variation across them is limited to a few regions. With Chain-Tweak, the user has the option to either emulate such behavior (by applying ChainTweak only on specific parts of the chain) or have equal local variation through-out the entire chain (Fig 4b).

We also compared ChainTweak-generated loop decoys against some loop decoy-sets from the database. The former performed comparably with database decoys in terms of their structural characteristics (Fig 4c, 4d). They performed significantly better on the criteria of preserving the positions of loop ends (∼ 0.01Å deviation vs. ∼ 0.5Å deviation).

**Energy Landscape Analysis:** As discussed by Keasar and Levitt[21], well-designed functions should have wide basins and few local minima so that structurally similar conformations are minimized to the same local minima (Fig 5). Ensembles generated by ChainTweak can be used to analyze the energy landscape of any energy function $f$ around a protein structure $b$: after each conformation in the ensemble is minimized, the distribution of these local minima and their proximity to the base provide direct information about the energy landscape. Observe that such analysis does not require that the native structure be known. This is an important advantage of ChainTweak: it can be used in homology modeling to pick the right energy function.

Using a ChainTweak-generated ensemble around an alpha-helix, we compare different energy functions and demonstrate, in a direct way, the value of incorporating solvation effects (Table in Fig 5).

## 4. Discussion

In this paper, we have presented a formulation of the neighbourhood sampling problem that is independent of any search problem or energy function. Our proposed method for this problem, ChainTweak, can be used as a tool in many different applications and also enables novel applications like analysis of the energy-landscape around a particular conformation.

ChainTweak provides significant performance improvements over existing methods. Unlike discrete off-lattice phi-psi angle models[2,3], it does not generate (and reject) infeasible solutions. Its perturbation size (and, thus, efficiency) is much larger than what is possible with MD-like methods[1,5] and, unlike these methods, it can also modify partial structures. With database-based methods, e.g., fragment-swap MC[4], a small database size restricts the number of solutions that can be found while a larger database reduces efficiency. ChainTweak, in contrast, is fast and explores all possible local perturbations at each step.

Like ChainTweak, some MC-based methods[6,7] also make local moves by compensatory modification of dihedral angles. While ChainTweak's modular design allows easy emulation of these methods' local-modification approaches, its currently chosen methods for loop-closure[16,18] allow larger local perturbations (i.e., more efficient space coverage) and the ability to get multiple possible alternative local moves at each step, at no extra cost. Thus, unlike existing methods, per-residue phi-psi preferences can be easily supported.

A goal of this paper has been to demonstrate the usefulness of a stand-alone neighbourhood sampling program. In future work, we hope to further explore the use of ChainTweak in problems where it might enable new analyses and methods. For example, ChainTweak-generated ensembles could be used to further analyze energy functions and add entropic terms to them. Using ChainTweak, conformational propensities of disordered regions in proteins[25] and conformational variation across sets of re-engineered[28,29] or homologous[30] proteins could be studied. It could be used in conjunction with existing methods[26,27] to analyze the ligand-protein docking process. We are also considering extending the algorithm to handle sidechain rotamer preferences and covalently-modified resiudues (e.g., phosphorylation).

**Acknowledgments:** The authors thank Phil Bradley, Amy Keating and Michael Levitt for their suggestions; Jean-Claude Latombe for pointing out reference 18; and Nathan Palmer and Allen Bryan for their comments.

## References

1. Huang ES *et al. Using a hydrophobic contact...* J Mol Biol, 257(3):716-25, 1996
2. Park B *et al. Energy functions that discriminate...* J Mol Biol, 258(2):367-92, 1996
3. Kolodny R *et al. Small libraries of protein ...* J Mol Biol, 323(2):297-307, 2002
4. Chivian D *et al. Automated prediction....* Proteins, 53:524-533, 2003
5. Guntert P *et al. Torsion angle dynamics...* J Mol Bio, 273:283-298, 1997
6. Ulmschneider JP, Jorgensen WL. *Polypeptide folding...* J Am Chem Soc 18;126(6):1849-57, 2004
7. Cahill M, Cahill S, Cahill K. *Proteins wriggle* Biophys J, 82(5):2665-70, 2002
8. Krishnamoorthy B, Tropsha A. *Development of a...* Bioinformatics 19:1540-8, 2003
9. Dudek MJ, Ponder JW. *Accurate Modeling of...* J Comp Chem, 16:791-816, 1995
10. DePristo MA *et. al. Ab initio construction...* Proteins, 1:51(1):41-55, 2003
11. Berman HM *et al. The Protein Data Bank.* Nucl Acids Res, 28:235-242, 2000
12. Samudrala R, Levitt M. *Decoys 'R Us ...* Protein Sci, 9(7):1399-401, 2000
13. Branden C, Tooze J. *Introduction to Protein Structure.* Garland Pub, NY, 1991
14. Diamond R. *Personal communication with Michael Levitt*
15. Go N, Scheraga H. *Ring closure....* Macromolecules, 3(2):178-187, 1970
16. Manocha D *et al. Conformational analysis....* Comp App of Bio Sci, 11(1):71-86, 1995
17. Wedemeyer WJ, Scheraga HA. *Exact Analytical...* J Comp Chem, 20:819-844, 1999
18. Coutsias EA *et. al. A kinematic view....* J Comp Chem 25(4):510-28, 2004
19. Aloupsis G *et al. Flat-state connectivity of linkages under dihedral motion.* In Proc 13th Intl Sym on Alg and Comp, 369-380, 2002
20. Biedl T *et al. Locked and unlocked polygonal chains in 3D.* In Proc 10th ACM-SIAM Symposium on Discrete Algorithms, 866-867, 1999
21. Keasar C, Levitt M. *A novel approach...* J Mol Biol, 23;329(1):159-74, 2003
22. Basch J *et al. Data structures for mobile data.* In Proc of 8th ACM-SIAM Symp Discrete Algo, 747-756, 1997
23. Lotan I *et al. Efficient Maintenace....* In Proc Symp Comp Geom, 2002
24. Qiu D *et al. The GB/SA Continuum...* J Phys Chem A, 101:3005-3014, 1997
25. Dunker K *et al. Intrinsically disordered protein.* J Mol Grpa Mod, 19:26-59, 2001
26. Edelsbrunner H *et al. Anatomy of protein...* Prot Sci, 7:1884-1897, 1998
27. Apaydin MS *et al. Studying Protein-Ligand...* Bioinformatics, 18(2):18-26, 2002
28. Babbitt PC, Gerlt JA. *New Functions from...* Adv in Prot Chem, 55:1-28, 2000
29. Mooney SD *et al. Conformational Preferences of...* Biopolymers, 64(2):63-71, 2002
30. Gerstein M, Altman RB. *Using a measure...* Comp App Bio, 11(6):633-44, 1995

# COARSE AND RELIABLE GEOMETRIC ALIGNMENT
# FOR PROTEIN DOCKING*

Y. WANG[†] P. K. AGARWAL[‡] P. BROWN[§] H. EDELSBRUNNER[¶] and J. RUDOLPH[||]

Duke University, Durham, North Carolina

**Abstract.** We present an efficient algorithm for generating a small set of coarse alignments between interacting proteins using meaningful features on their surfaces. The proteins are treated as rigid bodies, but the results are more generally useful as the produced configurations can serve as input to local improvement algorithms that allow for protein flexibility. We apply our algorithm to a diverse set of protein complexes from the Protein Data Bank, demonstrating the effectivity of our algorithm, both for bound and for unbound protein docking problems.

## 1. Introduction

Protein-protein docking is the computational approach to predicting interactions between proteins. In this paper, we contribute to this field by describing an algorithm for generating a small set of coarse alignments between protein structures.

**Motivation.** Highly organized transient or static assemblies of proteins control most cellular events. A better understanding of the protein-protein interactions involved in these assemblies would help elucidate how individual proteins form complexes and dynamically function in concert to generate the cell circuitry and its time-dependent responses to external stimuli. Protein structures determined at atomic resolution by X-ray crystallography, nuclear magnetic resonance, and increasingly by computer modeling provide one basis for the study of protein interactions. However, given the relative wealth of structural details for monomeric proteins compared to multimeric protein complexes, there exists a need for computational tools and thus for the field of protein docking.

**Prior work.** Current research on protein-protein docking focuses on either bound

---

*All authors are supported by NSF under grant CCR-00-86013. JR and HE are also supported by NIH under grant R01 GM61822-01. PA is also supported by NSF under grants EIA-01-31905 and CCR-02-04118 and by the U.S.-Israel Binational Science Foundation.
[†] Department of Computer Science.
[‡] Departments of Computer Science and Mathematics.
[§] Department of Computer Science.
[¶] Departments of Computer Science and Mathematics, and Raindrop Geomagic.
[||] Departments of Biochemistry and Chemistry.

docking (the reassembly of known complexes from their constituents), or unbound docking (the assembly of as yet unknown complexes under the assumption of only small protein conformational changes). Most approaches to unbound docking consist of two stages[20]: the *rigid docking stage* produces a set of potential docking configurations by considering only rigid motions, and the *refinement stage* locally improves the docking configuration, possibly allowing for a limited amount of flexibility. The two essential components in both stages are: a scoring function that discriminates near-native from incorrect docking configurations and a search algorithm to find (approximately) the best configuration for the scoring function.

Approaches to the rigid docking stage rely mainly on geometric complementarity. Some are based on uniform discretizations of the space of rigid motions, which they search exhaustively[3]. This approach has been accelerated using the fast Fourier transform (FFT)[16], which forms the basis of the docking software FTDock[13], 3D-Dock[18], GRAMM[21], and ZDock[6]. Others sample a small number of rigid motions non-uniformly from the space by aligning feature points found on the molecular surfaces[14,17]. This idea goes back to Connolly[11], who proposed to use the minima and maxima of a function related to mean curvature, now known as the *Connolly function*. An example of this method has been described by Fischer *et al.*[12], who use geometric hashing to align critical points of a variant of the Connolly function. The refinement stage is usually modeled as an energy minimization problem, with the scoring function focusing on the thermodynamic aspects of the interaction. The difficulty of the problem increases with the dimension of the search space or, equivalently, the degree of freedom, which is large even if we keep the back-bone rigid and consider only side-chain flexibility. Recently, Vajda *et al.* have proposed a hierarchical, progressive refinement protocol[4,5], which seems to reliably converge to a near-native docking configuration starting with initial configurations up to $10\,\text{Å}$ root-mean-square-distance away from the native configuration. Little success has been reported on including backbone conformational changes[19]. Since each step in the refinement stage is costly, it is essential that the set of potential configurations generated in the rigid docking stage is small and reliably contains configurations not too far from the native configuration. Current solutions to the rigid docking stage fall short on at least one of the two requirements.

**New work.** In this paper, we present an efficient algorithm for the rigid docking stage. We use geometric complementarity to guide the search for a small set of rigid motions so that the two proteins fit loosely into each other. Such a set of potential configurations can be further refined to obtain more accurate docking predictions[5,9]. We remark that for the case of unbound docking, it is especially important to start with coarse (not tight) fits between proteins to take advantage of flexibility in the later refinement stage.

We describe our algorithm in Section 2. It relies on a novel approach to describe protrusions and cavities on molecular surfaces using a succinct set of point pairs computed from the elevation function[1]. We then align such pairs and evaluate the resulting configurations using a simple and rapid scoring function. Compared to similar approaches that align feature points[12,17], our algorithm inspects orders of magnitude fewer configurations. This is made possible by using slightly more complicated features that contain information useful in assessing their significance. We exploit this extra information twice, first to ignore insignificant features and second to be more discriminant in matching up features from different proteins. In Section 3, we demonstrate the efficacy of our approach by testing a set of 25 bound protein complexes from the Protein Data Bank[2]. We demonstrate that a combination of our algorithm with the local improvement procedure described in[9] efficiently finds near-native docking positions for all but two cases without creating false positives. In addition, we test our algorithm on the unbound protein docking benchmark[7]. In particular, we demonstrate that the algorithm generates poses sufficiently close to the native configuration such that refinement methods that take into account protein flexibility will succeed in bringing them within an acceptable neighborhood of the correct solution. We conclude and discuss future work in Section 4.

## 2. Methods

We represent each protein by a set or union of finitely many balls in three-dimensional Euclidean space, which we denote by $\mathbb{R}^3$. Specifically, we are given two proteins, $A = \{A_1, A_2, \ldots, A_n\}$ and $B = \{B_1, B_2, \ldots, B_m\}$, where $A_i$ is the ball with center $a_i \in \mathbb{R}^3$ and van der Waals radius $r_i \in \mathbb{R}$ and $B_j$ is the ball with center $b_j$ and van der Waals radius $s_j$. We fix $A$ in $\mathbb{R}^3$ and describe an algorithm that finds a small set of candidate transformations for $B$. Each transformation is a rigid motion $\mu$ that produces a candidate configuration $(A, \mu(B))$. We begin by describing the scoring function that assesses the fit between two proteins.

**Scoring function.** A good scoring function favors near-native configurations over configurations that are far from the native. Letting $d_{ij} = \|a_i - b_j\| - r_i - s_j$ be the distance between the balls $A_i$ and $B_j$, we define

$$\text{contact}(i, j); \text{collision}(i, j) = \begin{cases} -3; 1 & \text{if} \quad d_{ij} < 0, \\ 1; 0 & \text{if } 0 \leq d_{ij} \leq \lambda, \\ 0; 0 & \text{if } \lambda < d_{ij}, \end{cases}$$

where $\lambda$ is a constant we refer to as the *contact-threshold*. The score of $(A, B)$

is based on the total *number of contacts*, $\#\mathrm{cont}(A, B) = \sum_{i,j} \mathrm{contact}(i,j)$, and the total *number of collisions*, $\#\mathrm{coll}(A, B) = \sum_{i,j} \mathrm{collision}(i,j)$. We call the configuration $(A, B)$ *valid* if $\#\mathrm{coll}(A, B) \leq \chi$, where the constant *collision-threshold* $\chi$ defines the maximum number of collisions we tolerate. The score ignores invalid configurations and equals the number of contacts for valid configurations. This notion of score is similar to the ones in[3,6] but different because our score penalizes collisions twice, first by counting them toward a possibly invalid configuration and second by reducing the contact number. The reason for this difference is that we aim at coarse alignments and thus are more tolerant to collisions, using $\chi = 60$ rather than $\chi = 5$, as in[3]. The second penalty counteracts the usual increase in contact number that goes along with an increase in collision number. In other words, it seeks to avoid a bias toward configurations with higher collision number without unfairly discriminating against them.

**Features.** Our algorithm generates rigid motions from feature sets $\Phi_A$ and $\Phi_B$ obtained by analyzing the shapes of the two proteins. We compute these features from (approximately) smooth surfaces representing the two shapes[8,10]. Letting $\mathbb{M}$ be the surface representing $A$, we briefly review the function $\mathrm{Elevation} : \mathbb{M} \to \mathbb{R}$ that underlies our definition of feature. To first approximation, it resembles the elevation on Earth, which is the height difference of a point and the mean sea level at that point. This definition makes sense on Earth, where we have a natural choice of origin (the center of mass) and mean sea level (a level set of the gravitational potential), neither of which exists for general surfaces. In the absence of both concepts, we associate each point $x \in \mathbb{M}$ with a canonically defined *partner* $y \in \mathbb{M}$, with same normal direction $\mathbf{n}_y = \mathbf{n}_x$, and define $\mathrm{Elevation}(x)$ as the absolute height difference between $x$ and $y$ in that common direction. For more details, in particular on how to define the canonical pairing, we refer to[1]. Loosely speaking, $x$ is the top of a protrusion or the bottom of a cavity in the direction $\mathbf{n}_x$ and the pairing partner $y$ is the saddle point that marks where the protrusion or cavity starts. It is also possible that the roles of $x$ and $y$ are reversed. For the purpose of protein docking, we are interested in points with locally maximal elevation, as they represent locally most significant features. Almost all points $x$ on $\mathbb{M}$ have exactly one partner $y$, but most maxima arise at positions where the partner is ambiguous. More specifically, for a generic surface there are four types of maxima describing different types of features, as illustrated in Figure 1.

By definition, a *feature* consists of two points, $u$ and its partner $v$, with common surface normal, $\mathbf{n}_u = \mathbf{n}_v$, and common elevation, $\mathrm{Elevation}(u) = \mathrm{Elevation}(v)$. Its *length* is the Euclidean distance between the two points, $\|u - v\|$. Each maximum of the elevation function is defined by $k \in \{2, 3, 4\}$

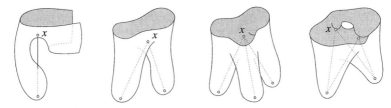

Figure 1. Left: a one-legged maximum characterized by $x$ having a unique partner. Middle left: a two-legged maximum in which $x$ has two partners, both with the same normal direction and the same height difference to $x$. Middle right: a three-legged maximum in which $x$ has three partners, again sharing the same normal direction and the same height difference to $x$. Right: a four-legged maximum in which $x$ has two partners and both partners have the same two partners each.

points and gives rise to $\binom{k}{2}$ features in $\Phi_A$. For example, the 3-legged maximum (third from the left in Figure 1) consists of $k = 4$ points defining $\binom{k}{2} = 6$ point pairs each forming a feature in $\Phi_A$. The length and elevation of a feature are used to estimate its importance, and both together with the normal direction are used to pair up features from the sets $\Phi_A$ and $\Phi_B$.

**Coarse alignment.** Given two proteins $A$ and $B$ together with their feature sets $\Phi_A$ and $\Phi_B$, our algorithm computes a set of potential coarse alignments $\Gamma$:

```
for every α ∈ Φ_A and every β ∈ Φ_B do
    if α, β form a plausible alignment then
        μ = Align(α, β);
        compute the contact and collision numbers for (A, μ(B));
        if (A, μ(B)) is valid then add μ to Γ endif
    endif
endfor;  sort Γ by contact number.
```

The rationale behind the algorithm is that good fits between the input proteins have aligned features, such as a protrusion of $A$ fitting inside a cavity of $B$, or vice versa. If we pair up all features of $A$ with all features of $B$, we surely cover all good fits. On the other hand, the information that comes with each feature can be used to discriminate between pairs and gain efficiency by filtering out alignments we deem not important or implausible. Specifically, we introduce an *importance filter* that eliminates features from $\Phi_A$ and $\Phi_B$ whose lengths or elevations are below threshold. The remaining features form pairs $(\alpha, \beta)$ which pass the *plausibility filter* provided $\alpha$ and $\beta$ are not too different in length and they represent complementary types (a protrusion and a cavity). The constants used in the importance filter are given in the caption of Table 1.

Assuming $(\alpha, \beta)$ passes the importance and the plausibility filters, we compute an aligning rigid motion $\mu$ as follows. Writing $\alpha = (u, v, \mathbf{n}_\alpha)$ and $\beta = (p, q, \mathbf{n}_\beta)$ for the points and normals, we define the bi-normals $b_\alpha = \mathbf{n}_\alpha \times \frac{v-u}{\|v-u\|}$ and $b_\beta = \mathbf{n}_\beta \times \frac{p-q}{\|q-p\|}$. We obtain the rigid motion in three steps:

1. translate $\beta$ so that the two midpoints coincide: $\frac{u+v}{2} = \frac{p+q}{2}$;
2. rotate $\beta$ about the common midpoint so that $u, v, p, q$ are collinear;
3. rotate $\beta$ about the common line so that $b_\alpha = b_\beta$.

We note that there is an ambiguity in Step 2, allowing for two different alignments distinguished by having $v - u$ and $q - p$ point in the same or in opposite directions. We are interested in both but simplify the description by pretending that Function Align returns only one rigid motion, instead of two as it really does. Observe that Step 3 positions the two features to maximize the angle between the two normal vectors. Given $\mu$, we compute the score and the number of collisions using a hierarchical data structure storing $A$ and $B$. Letting $n$ and $m$ be the number of balls in the two sets, this takes time $O((n + m) \log(n + m))$.

## 3. Results

In this section, we present the results of testing our algorithm on twenty-five bound docking problems obtained from the Protein Data Bank[2] and on forty-nine unbound docking problems from the benchmark in[7]. We begin with a detailed study of a well known protein complex.

**A case study.** We use the barnase/barstar complex (pdb-id 1BRS, chains A and D, with 864 and 693 atoms) as a sample system to introduce the capabilities of our algorithm. We generate molecular surfaces of the two chains with the MSMS software (available as part of the VMD software distribution[15]) and obtain triangulations with 8,959 and 7,248 vertices. In Table 1, we show the total number

| | chain A, # legs | | | chain D, # legs | | |
|---|---|---|---|---|---|---|
| | 2 | 3 | 4 | 2 | 3 | 4 |
| total # of features | 1,044 | 696 | 156 | 828 | 510 | 154 |
| #s after importance filter | 112 | 205 | 50 | 68 | 160 | 49 |

Table 1. Compare the total number of features obtained from two-, three- and four-legged maxima for chains A and D of 1BRS with the number of features that pass the importance filter, having length at least 3.0Å and elevation at least 0.2Å. (There are no one-legged maxima for this data set.)

of features generated from the maxima of the elevation function and the number of features that survive the importance filter. The latter form the input to our coarse alignment algorithm. We note that a substantially larger number of features obtained from 3-legged maxima are retained than features obtained from 2- and 4-legged maxima.

Given the two sets of input features, our algorithm takes about three minutes on a single processor PIII 1GHz computer to generate a family $\Gamma$ of 5,021 valid configurations with contact number larger than or equal to 150. Each configuration in $\Gamma$ corresponds to a transformation $\mu$ for chain D. We use the root-mean-

| before local improvement | | | | after local improvement | | |
|---|---|---|---|---|---|---|
| rank | #cont | #coll | RMSD | rank | #cont' | RMSD |
| 12 | 327 | 24 | 3.23 | 1 | 359 | 0.54 |
| 5 | 342 | 48 | 2.42 | 2 | 338 | 0.80 |
| 1 | 427 | 23 | 1.59 | 3 | 328 | 0.72 |
| 4 | 353 | 49 | 3.57 | 4 | 314 | 0.80 |
| 2 | 391 | 39 | 1.70 | 5 | 311 | 0.91 |
| 59 | 269 | 12 | 2.84 | 6 | 310 | 0.78 |
| 3 | 373 | 29 | 2.32 | 7 | 307 | 1.50 |
| 11 | 339 | 18 | 3.07 | 8 | 281 | 1.47 |
| 15 | 318 | 16 | 3.00 | 9 | 251 | 2.09 |
| 76 | 263 | 29 | 39.39 | 10 | 213 | 39.96 |

Table 2. Top ten configurations after local improvement and their ranks before local improvement. The first nine have small RMSD and may be considered near-native configurations. We use different definitions for the number of contacts before and after the local improvement: #cont is defined as in Section 2, and #cont' is as computed by the local improvement algorithm, which is the number of non-overlapping spheres at distance at most 1.5Å.

square-distance (RMSD) between the centers of the matching atoms in D and $\mu(D)$ to measure how close the configuration is to the native one. Ranking by score, the top configuration in $\Gamma$ has an RMSD of 1.59 Å, and six of the top ten configurations have RMSD smaller than or equal to 4.0 Å. Letting $\Gamma^*$ be the subset of top $N = 100$ configurations, we refine each one using the local improvement heuristic of Choi et al.[9] We then re-rank the configurations in $\Gamma^*$ based on the new scores, limiting ourselves to configurations with collision number at most five. The results in Table 2 show that our algorithm generates multiple coarse alignments that are useful, in the sense that the local improvement heuristic succeeds in refining them to near-native configurations.

**More bound protein complexes.** We extend our experiments to a collection of twenty-five protein complexes obtained from the Protein Data Bank. Each complex consists of two chains, and we generate a set of features for each. For a typical chain, the number of features that survive the importance filter is on the same order of magnitude as the number of atoms. In Table 3, we show a low-RMSD configuration for each protein complex, as well as its rank in the list of configurations output by our algorithm (using contact-threshold $\lambda = 2.0\,\text{Å}$ and collision-threshold $\chi = 60$). With only one exception (1JAT), we have at least one low-RMSD configuration ranked among the top one hundred. The last column shows the running time for the coarse alignment algorithm, which does not include the time to compute the triangulated surface and the maxima of the elevation function.

| pdb-id | chains | rank | #coll | RMSD | time |
|--------|--------|------|-------|------|------|
| 1A22 | A, B | 2 | 23 | 2.75 | 20 |
| 1BI8 | A, B | 12 | 43 | 2.48 | 26 |
| 1BRS | A, D | 1 | 11 | 1.52 | 3 |
| 1BUH | A, B | 5 | 14 | 1.85 | 2 |
| 1BXI | B, A | 3 | 34 | 2.54 | 8 |
| 1CHO | E, I | 1 | 14 | 2.71 | 3 |
| 1CSE | E, I | 2 | 22 | 2.21 | 9 |
| 1DFJ | I, E | 78 | 11 | 3.09 | 27 |
| 1F47 | B, A | 15 | 1 | 1.49 | 1 |
| 1FC2 | D, C | 5 | 49 | 4.13 | 6 |
| 1FIN | A, B | 11 | 44 | 3.70 | 41 |
| 1FS1 | B, A | 1 | 29 | 1.62 | 5 |
| 1JAT | A, B | 522 | 20 | 1.20 | 9 |
| 1JLT | A, B | 8 | 23 | 3.64 | 10 |
| 1MCT | A, I | 1 | 27 | 3.49 | 3 |
| 1MEE | A, I | 1 | 23 | 1.33 | 9 |
| 1STF | E, I | 1 | 43 | 1.18 | 8 |
| 1TEC | E, I | 9 | 54 | 3.07 | 7 |
| 1TGS | Z, I | 1 | 46 | 2.61 | 6 |
| 1TX4 | A, B | 2 | 4 | 3.35 | 14 |
| 2PTC | E, I | 1 | 18 | 4.55 | 6 |
| 3HLA | A, B | 1 | 19 | 1.87 | 16 |
| 3SGB | E, I | 1 | 38 | 3.21 | 5 |
| 3YGS | C, P | 6 | 7 | 1.07 | 6 |
| 4SGB | E, I | 10 | 33 | 2.33 | 4 |

Table 3. For each protein complex, we show data for the highest ranking configuration with RMSD at most 5.0Å. The running time of the coarse alignment algorithm is given in minutes.

Next, we apply the local improvement heuristic[9] to the top $N = 100$ configurations of each complex (except 1JAT, for which we need $N = 522$ to get a near-native configuration) and re-rank them based on the new scores. Eliminating all configurations with more than 5 collisions, Table 4 shows before and after data for the configuration that is ranked at the top after local improvement. In all but two cases, the top ranked configuration is near-native, and in one of the two exceptional cases, the second ranked configuration is near-native. In the remaining exceptional case (1BI8), we can obtain a near-native configuration by relaxing the threshold of allowed collisions to eight. In summary, for 23 of the 25 test complexes, our coarse alignment algorithm combined with the local improvement heuristic[9] predicts a near-native configuration without false positives.

| pdb-id | before local improvement | | | | after local improvement | | |
|---|---|---|---|---|---|---|---|
| | rank | #cont | #coll | RMSD | rank | #cont' | RMSD |
| 1A22 | 2 | 363 | 23 | 2.75 | 1 | 475 | 1.08 |
| 1BI8 | 62 | 324 | 10 | 30.00 | 1 | 234 | 29.88 |
| 1BRS | 12 | 327 | 37 | 3.23 | 1 | 349 | 0.54 |
| 1BUH | 5 | 311 | 14 | 1.85 | 1 | 256 | 0.61 |
| 1BXI | 16 | 261 | 21 | 5.59 | 1 | 289 | 0.63 |
| 1CHO | 1 | 375 | 14 | 2.71 | 1 | 305 | 0.99 |
| 1CSE | 23 | 276 | 36 | 2.57 | 1 | 317 | 0.82 |
| 1DFJ | 78 | 273 | 11 | 3.09 | 1 | 220 | 1.28 |
| 1F47 | 15 | 238 | 1 | 1.49 | 1 | 221 | 0.56 |
| 1FC2 | 5 | 323 | 49 | 4.13 | 2 | 200 | 1.33 |
| 1FIN | 34 | 361 | 54 | 9.94 | 1 | 413 | 0.61 |
| 1FS1 | 2 | 402 | 27 | 1.59 | 1 | 326 | 0.89 |
| 1JAT | 522 | 203 | 21 | 1.20 | 1 | 288 | 0.87 |
| 1JLT | 3 | 362 | 14 | 6.17 | 1 | 310 | 1.77 |
| 1MCT | 84 | 280 | 34 | 3.57 | 1 | 322 | 0.32 |
| 1MEE | 1 | 542 | 23 | 1.33 | 1 | 372 | 0.57 |
| 1STF | 1 | 444 | 43 | 1.18 | 1 | 314 | 0.79 |
| 1TEC | 10 | 334 | 51 | 4.51 | 1 | 304 | 1.28 |
| 1TGS | 2 | 373 | 13 | 2.71 | 1 | 348 | 0.44 |
| 1TX4 | 80 | 296 | 25 | 4.34 | 1 | 355 | 0.36 |
| 2PTC | 1 | 346 | 18 | 4.55 | 1 | 314 | 0.66 |
| 3HLA | 1 | 402 | 19 | 1.97 | 1 | 416 | 0.70 |
| 3SGB | 1 | 364 | 38 | 3.21 | 1 | 257 | 2.24 |
| 3YGS | 6 | 315 | 7 | 1.03 | 1 | 209 | 0.85 |
| 4SGB | 10 | 298 | 33 | 2.33 | 1 | 266 | 2.50 |

Table 4. For each protein complex, we locally improve the $N$ top ranked configurations and show the data for the highest re-ranked configuration with small RMSD. After local improvement we admit only configurations with at most five collisions, as usual. The number of contacts before and after the improvement, #cont and #cont', are computed as described in the caption of Table2.

It is interesting to compare the data in Tables 3 and 4 and notice that the highest ranked configuration after local improvement is the highest ranked configuration with small RMSD before the local improvement in only slightly more than half the cases. Consider for example 1FIN, which has a configuration at 3.70 Å RMSD with 44 collisions but the one that leads to the best final configuration has RMSD = 9.94Å and #coll = 54.

**Unbound docking benchmark.** We further test our algorithm on the protein-protein docking benchmark provided in[7]. We omit the seven complexes classified as difficult in[7] because they have significantly different conformations in the unbound vs. bound structures. We also omit complexes 1IAI, 1WQ1 and 2PCC for which we had difficulties to generate surface triangulations of required quality. Of the remaining forty-nine complexes, twenty-five are so-called *bound-unbound*

| | | bound-unbound | | | | | | unbound-unbound | | | |
|---|---|---|---|---|---|---|---|---|---|---|---|
| C-id | #hits | min* | rank | size | min | C-id | #hits | min* | rank | size | min |
| 1ACB | 20 | 3.70 | 3,951 | 14,426 | 1.75 | 1MLC | 7 | 3.71 | 6,949 | 29,747 | 3.32 |
| 1AVW | 8 | 5.51 | 4,698 | 23,565 | 5.42 | 1WEJ | 3 | 6.27 | 4,659 | 18,194 | 5.86 |
| 1BRC | 35 | 4.66 | 1,629 | 12,770 | 4.66 | 1BQL | 11 | 6.98 | 10,388 | 23,308 | 4.39 |
| 1BRS | 7 | 1.60 | 426 | 11,607 | 1.60 | 1EO8 | 1 | 2.31 | 11 | 45,512 | 2.31 |
| 1CGI | 5 | 3.04 | 695 | 10,135 | 3.04 | 1FBI | 8 | 6.49 | 11,783 | 26,036 | 2.30 |
| 1CHO | 27 | 2.35 | 92 | 11,815 | 2.35 | 1JHL | 18 | 3.47 | 14,185 | 32,091 | 2.61 |
| 1CSE | 7 | 3.15 | 15,271 | 21,068 | 2.74 | 1KXQ | 2 | 5.99 | 1,495 | 37,218 | 5.99 |
| 1DFJ | 2 | 6.44 | 1,433 | 35,231 | 6.44 | 1KXT | 12 | 4.52 | 153 | 39,240 | 4.52 |
| 1FSS | 2 | 7.65 | 10,721 | 25,609 | 5.15 | 1KXV | 7 | 2.48 | 321 | 46,368 | 2.48 |
| 1MAH | 4 | 2.78 | 1,561 | 25,402 | 2.78 | 1MEL | 8 | 2.21 | 73 | 17,741 | 2.21 |
| 1TGS | 18 | 5.27 | 543 | 11,383 | 5.27 | 1NCA | 7 | 1.75 | 621 | 49,600 | 1.75 |
| 1UGH | 3 | 7.95 | 8,268 | 14,656 | 7.16 | 1NMB | 7 | 7.18 | 14,202 | 42,066 | 2.72 |
| 2KAI | 26 | 6.55 | 2,560 | 13,478 | 3.41 | 1QFU | 4 | 1.97 | 12 | 47,693 | 1.97 |
| 2PTC | 32 | 4.55 | 4,983 | 13,929 | 4.16 | 2JEL | 19 | 3.46 | 115 | 34,072 | 3.46 |
| 2SIC | 27 | 4.04 | 76 | 20,065 | 4.04 | 2VIR | 11 | 1.08 | 1 | 40,813 | 1.08 |
| 2SNI | 10 | 6.34 | 4,894 | 15,830 | 4.58 | 1AVZ | 8 | 4.06 | 4,243 | 7,895 | 3.52 |
| 1PPE | 10 | 4.13 | 37 | 7,660 | 4.13 | 1L0Y | 2 | 2.75 | 1,136 | 34,044 | 2.75 |
| 1STF | 8 | 1.41 | 1 | 15,082 | 1.41 | 2MTA | 40 | 2.91 | 19,167 | 36,903 | 2.07 |
| 1TAB | 3 | 3.78 | 48 | 8,296 | 3.78 | 1AOO | 3 | 5.95 | 3,950 | 9,113 | 4.35 |
| 1UDI | 3 | 4.50 | 1,124 | 21,133 | 4.50 | 1ATN | 8 | 1.52 | 1 | 50,729 | 1.52 |
| 2TEC | 5 | 1.42 | 6 | 21,134 | 1.42 | 1GLA | - | - | 25,307 | 33,879 | 2.82 |
| 4HTC | 2 | 5.94 | 396 | 14,032 | 5.94 | 1IGC | 3 | 2.48 | 3,260 | 25,303 | 2.06 |
| 1AHW | 1 | 9.38 | 2,781 | 32,919 | 4.37 | 1SPB | 3 | 2.83 | 617 | 13,728 | 2.83 |
| 1BVK | 5 | 1.95 | 1,189 | 24,611 | 1.95 | 2BTF | 2 | 5.02 | 10,132 | 33,480 | 3.28 |
| 1DQJ | 7 | 4.59 | 710 | 28,694 | 4.59 | | | | | | |

Table 5. Twenty-five bound-unbound cases on the left plus twenty-four unbound-unbound cases on the right. From left to right: the complex identification, the number of configurations in $\Gamma^*$ with RMSD* less than or equal to 10.0Å, the smallest RMSD* value of any configuration in $\Gamma^*$ (min*), the rank of this configuration within $\Gamma$, the number of configurations in $\Gamma$, and the smallest RMSD* value of any configuration in $\Gamma$ (min).

cases, in which one of the components is rigid. For each complex, we fix one chain as A, which is the rigid chain for each bound-unbound case and the receptor for each unbound-unbound case. We generate $\Gamma$, a set of the potential configurations, each corresponding to a rigid motion $\mu$ applied to the other chain, B. For each $\mu$, we measure the root-mean-square-distance between the matching interface $C_\alpha$ atoms of B and $\mu(B)$, and refer to it as RMSD*. Similar to the bound docking case, this value is a good estimate for the distance to the native configuration since the benchmark provides the unbound structures superimposed onto their corresponding crystallized bound structures. For each complex, we let $\Gamma^*$ be the subset of top $N = 2,000$ configurations in $\Gamma$. We show the results of our experiments in Table 5, demonstrating a number of favorable characteristics of our coarse alignment algorithm:

1. Within the relatively small set of 2,000 top-scoring configurations, $\Gamma^*$, about 78% of the complexes yield a configuration below 6.0 Å RMSD and about 98% yield a configuration below the 10.0 Å cut-off needed as input for the hierarchical, progressive refinement protocol in [4,5].

2. For most complexes, our algorithm generates multiple hits, implying that a local refinement is not likely to get trapped in a local minimum and instead find a near-native configuration.

3. Within the set of all generated configurations, $\Gamma$, about 96% of the complexes yield a configuration below 6.0 Å, typically within the top 10,000 scores. All 49 complexes generate at least one configuration below 7.5 Å within the top 25,000 scores.

We remark that there are at least two ways to further improve the results: use a different ranking mechanism that moves more low-RMSD configurations into the top ranks, and reduce the size of $\Gamma$ by clustering similar configurations [12].

## 4. Discussion

We conclude this paper with a brief comparison of our results with prior work on bound and unbound docking. We classify the bound docking methods by how they sample the search space of rigid motions. Methods that sample densely and more or less uniformly predict more accurate rigid docking configurations, but at a high computational cost. To adapt these methods to unbound docking, we may run the algorithms at low resolution or select a small set of promising candidate configurations for further refinement. As of today, neither approach has produced a workable solution to the problem of unbound docking. Methods that sample the space of rigid motions in a biased manner rely on some sort of shape analysis, aimed at detecting locally complementary configurations. All prior work is based on point features marking protrusions and cavities. Alignments are created by matching the points, e.g. all pairs from one set with all pairs from another. The running time is often improved using geometric hashing, as in [12].

Our algorithm belongs to the second class of methods but differs from prior work in the nature of the features, which are point pairs with extra information useful in estimating the scale level and in finding promising matches. Using this information, we generate significantly sparser samples of the search space. Our experiments provide evidence that despite the lower density, we always get candidates that can be refined to near-native configurations. The algorithm is reasonably fast and improvements are still possible.

**Acknowledgement.** The authors would like to thank Vicky Choi for the local improvement software used in our experiments.

## References

1. P. K. Agarwal, H. Edelsbrunner, J. Harer and Y. Wang. Extreme elevation on a 2-manifold. In *Proc. 20th Ann. Sympos. Comput. Geom.*, 357–365, 2004.

2.  H. Berman, J. Westbrook, Z. Feng, G. Gilliland, T. Bhat, H. Weissig, I. Shinkdyalov and P. E. Bourne. The protein data bank. *Nucleic Acid Res.*, 28:235–242, 2000.

3.  S. Bespamyatnikh, V. Choi, H. Edelsbrunner and J. Rudolph. Accurate protein docking by shape complementarity alone. Manuscript, Duke Univ., Durham, NC, 2004.

4.  C. J. Camacho, D. W. Gatchell, S. R. Kimura and S. Vajda. Scoring docked conformations generated by rigid-body protein-protein docking. *Proteins: Struct. Funct. Genet.*, 40:525–537, 2000.

5.  C. J. Camacho and S. Vajda. Protein docking along smooth association pathways. *Proc. Natl. Acad. Sci.*, 98:10636–10641, 2001.

6.  R. Chen, L. Li, and Z. Weng. ZDOCK: an initial-stage protein docking algorithm. *Proteins: Struct. Funct. Genet.*, 52:80–87, 2003.

7.  R. Chen, J. Mintseris, J. Janin and Z. Weng. A protein-protein docking benchmark. *Proteins: Struct. Funct. Genet.*, 52:88–91, 2003.

8.  H.-L. Cheng, T. K. Dey, H. Edelsbrunner and J. Sullivan. Dynamic skin triangulation. *Discrete Comput. Geom.*, 25:525–568, 2001.

9.  V. Choi, P. K. Agarwal, H. Edelsbrunner and J. Rudolph. Local search heuristic for rigid protein docking. In *Proc. 4th Intl. Workshop Alg. Bioinform.*, 2004, to appear.

10. M. L. Connolly. Analytic molecular surface calculation. *J. Appl. Crystallogr.*, 6:548–558, 1983.

11. M. L. Connolly. Shape complementarity at the hemo-globin albl subunit interface. *Biopolymers*, 25:1229–1247, 1986.

12. D. Fischer, S. L. Lin, H. Wolfson and R. Nussinov. A geometry-based suite of molecular docking processes. *J. Mol. Biol.*, 248:459–477, 1995.

13. H. A. Gabb, R. M. Jackson and M. J. Sternberg. Modeling protein docking using shape complementarity, electrostatics and biochemical information. *J. Mol. Biol.*, 272:106–120, 1997.

14. B. B. Goldman and W. T. Wipke. QSD: quadratic shape descriptors. 2. Molecular docking using quadratic shape descriptors (QSDock). *Proteins*, 38:79–94, 2000.

15. W. Humphrey, A. Dalke and K. Schulten. VMD—Visual Molecular Dynamics. *J. Mol. Graphics*, 15:33–38, 1996.

16. E. Katchalski-Katzir, I. Shariv, M. Eisenstein, A. Friesen, C. Aflalo and I. Vakser. Molecular surface recognition: determination of geometric fit between protein and their ligands by correlation techniques. *Proc. Natl. Acad. Sci. (USA)*, 89:2195–2199, 1992.

17. H. Lenhof. An algorithm for the protein docking problem. In *Bioinformatics: From Nucleic Acids and Proteins to Cell Metabolism*, eds. D. Schomburg and U. Lessel, Wiley, 1995, 125–139.

18. G. Moont and M. J. E. Sternberg. Modelling protein-protein and protein-dna docking. In *Bioinformatics: From Genomes to Drugs*, ed. T. Lengauer, Wiley, 2002, 361–404.

19. B. Sandak, R. Nussinov and H. J. Wolfson. A method for biomolecular structural recognition and docking allowing conformational flexibility. *J. Comput. Biol.*, 5:631–654, 1998.

20. G. R. Smith and M. J. E. Sternberg. Prediction of protein-protein interactions by docking methods. *Curr. Opin. Struct. Bio.*, 12:29–35, 2002.

21. I. A. Vakser. Protein docking for low-resolution structures. *Protein Engin.*, 8:371–377, 1995.

# BIOMEDICAL ONTOLOGIES

O. BODENREIDER

*U.S. National Library of Medicine*
*8600 Rockville Pike, MS 43, Bethesda, Maryland, 20894, USA*
*E-mail: olivier@nlm.nih.gov*

J. A. MITCHELL

*University of Missouri, Department of Health Management & Informatics,*
*Columbia, Missouri, 65211, USA*
*E-mail: MitchellJo@health.missouri.edu*

A. T. MCCRAY

*U.S. National Library of Medicine*
*8600 Rockville Pike, MS 52, Bethesda, Maryland, 20894, USA*
*E-mail: mccray@nlm.nih.gov*

In the "post-genomic era", biomedical ontologies are becoming increasingly popular in the computational biology community as the focus of biology has started to shift from mapping genomes to analyzing the vast amount of information resulting from functional genomics research. In fact, biomedical ontologies play a central role in integrating the information about various model organisms, acquired under different conditions and stored in heterogeneous databases. The need for a controlled vocabulary to annotate gene products certainly explains the success of the Gene Ontology™ (GO), which has become a *de facto* standard in this domain.

The presence of research focused on or enabled by biomedical ontologies in molecular biology conferences illustrates the increasing role of ontologies in biological research. At the Pacific Symposium of Biocomputing (PSB), for example, the place of biomedical ontologies has grown from one paper in 1998 to 41 papers submitted to our session this year. Similarly, a large number of papers presented at the 12th conference on Intelligent Systems for Molecular Biology (ISMB/ECCB 2004) focused on some aspect of biomedical ontology. Finally, events such as the workshop on Bio-Ontologies collocated with ISMB each year since 1998 and the success of the Standards and Ontologies for Functional Genomics conferences are another testimony to the importance of ontologies to biologists.

While the purpose of biomedical *terminology* is to collect the names of entities (i.e., substances, qualities and processes) employed in the biomedical domain, the purpose of biomedical *ontology* is to study classes of entities in reality which are of biomedical significance. Beyond names, ontology is concerned with the principled definition of biological classes and the relations among them. In practice, as they are more than lists of terms but do not necessarily meet the requirements of formal organization, the many products developed by biomedical terminologists and ontologists often constitute an "ontology gradient". Gene Ontology is one such structure lying between terminology and ontology.

Biomedical ontology research encompasses a variety of entities (from dictionaries of names for biological products, to controlled vocabularies, to principled knowledge structures) and processes (i.e., acquisition of ontological relations, integration of heterogeneous databases, use of ontologies for reasoning about biological knowledge). This session reflects many aspects of this research. Not surprisingly, a large number of submissions focus on the Gene Ontology.

A first group of papers investigates foundational issues in biomedical ontology as well as the creation of ontological resources. Hoffman et al. discuss the extension of existing clinical vocabularies to include molecular diagnostics and cytogenetics concepts, using information from the RefSeq database. Following-up on the study on the compositional structure of GO terms they presented last year at PSB, Ogren et al. reflect on the implications of such properties on the curation and usage of GO. In addition to lexical properties, Bodenreider et al. show that statistical methods applied to annotation databases can also help reveal associative relations among GO terms. Finally, Spasić et al. present a method for measuring similarity among biomedical terms, which not only utilizes ontological relations but can also contribute to identifying additional relations.

The second group of papers focuses on the role played by ontologies in integrating disparate biomedical resources. Marshall et al. explore five levels of constraint for matching biological entities and the links among them. Foreseeing what a biomedical Semantic Web would require, Bechhofer et al. developed a system which automatically adds semantic annotations to existing web resources, enabling the dynamic integration of such resources. In the tradition of the Microarray Gene Expression Data (MGED) ontology, Orchard et al. investigate the resources required for describing the complex experimental procedures used in proteomics and sharing the corresponding data. A practical application of integration is presented by Gennari et al., visualizing in the same information space anatomical data and various genomic resources.

The remaining papers have a somewhat different perspective on biomedical ontologies. While some of these papers make a limited use of the rich structure of ontologies and draw essentially on their terminology component, none of these papers could have existed without the standardization fostered by ontologies. Most papers, however, take advantage – to some degree – of the relations recorded in biomedical ontologies.

Two papers exploit the information contained in various annotation databases to investigate the relations among biological entities. Tari et al. study the properties of functionally-related gene networks. Xiong et al. use hyperclique patterns to identify functional modules in protein complexes. Conversely, Yamakawa et al. analyze the common features in sets of genes using their annotations, through gene-GO term bipartite graphs.

Two papers focus on predicting the functional annotations of biological entities. Hayete and al. use decision trees to learn the associations between GO terms and protein domains. Lu and al. show that subcellular localization information can be predicted from molecular function information. Finally, with the GenesTrace system, Cantor et al. analyze the relations between diseases and genes as represented in existing databases integrated through terminology and ontology resources.

This session reflects the diversity of the biomedical ontology community. Topics of interest range from the foundational issues in defining the entities existing in biological reality to the formalisms required to represent these entities and their interrelations. Other topics involve the use of ontologies to enable sharing complex biological information and the integration of heterogeneous databases, as well as the various applications made possible by such integrated data repositories.

As better ontological resources are developed, such applications will increasingly enable complex reasoning about biomedical knowledge. As standard formalisms and communication protocols emerge, the use of heterogeneous resources will become more dynamic and automatic to biologists. Ultimately, the applications supported by biomedical ontologies will not only make it possible for biologists to keep up with an increasing amount of information, but hopefully also free them from the least interesting tasks. Beyond the *personal digital assistants* of today, which store our agendas and email messages, the *digital research assistants* of tomorrow will scan online information sources for us, summarize their content and organize the related knowledge into research hypotheses.

# ONTOLOGY DRIVEN DYNAMIC LINKING OF BIOLOGY RESOURCES

S. K. BECHHOFER, R. D. STEVENS AND P. W. LORD

*Department of Computer Science*
*University of Manchester*
*Oxford Road*
*Manchester*
*UK, M13 9PL*
*E-mail: seanb@cs.man.ac.uk*

Biologists were early adopters of the Web and continue to use it as the primary means of delivering data, tools and knowledge to their community. The Web is made by the links between pages, yet these links have many limitations: they are static and maintained by hand; they can only link one lexical item to another single resource; ownership is necessary for the placement of link anchors and the link mechanism is essentially inflexible. Dynamic linking services, supported by ontologies, offer a mechanism to overcome such restrictions. The Conceptual Open Hypermedia Service (COHSE) system enhances web resources through the dynamic addition of hypertext links. These links are derived through the use of an ontology and associated lexicon along with a mapping from concepts to possible link targets. We describe an application of COHSE to Bioinformatics, using the Gene Ontology (GO) as an ontology and associated keyword mappings and GO associations as link targets. The resulting demonstrator (referred to here as GOHSE) provides both glossary functionality and the possibility of building knowledge based hypertext structures linking bioinformatics resources.

## 1. Introduction

This paper investigates the use of ontology driven open hypermedia within bioinformatics. Using this technology it is possible to separate links from document resources and consequently provide a rich, dynamically linked collection of biology oriented documents. By driving the dynamic formation of links through an ontology we are taking advantage of the common understanding provided by an ontology. The relationships between the concepts in the ontology add a further dimension to the flexibility of linking, which can help to enhance the Web, the primary mechanism used within biology for delivery of data, tools and knowledge. As a discipline, bioinformatics relies on the knowledge held within its documents (Web pages, database entries, books or articles). Query by navigation, via links between these documents and others is still fundamental to practical bioinformatics. It is the links between biology documents that provide the utility to both humans and machines. Common usage of the Web involves embedding links within documents. There are, however, a number of limitations to this approach, that can be extended to other representations of linked documents, such as PDF.

**Hard Coding:** Links are hand-crafted and hard coded in the HTML encoding of a page. An anchor is placed around the source object in the originating page and the location of the end-point is included in the link. This end-point, or target, of a link can be a page, an anchor placed within a page, or perhaps some dynamically evoked service such as a query. The link is a static, inflexible entity that is intimately bound with the source node.

**Format Restrictions:** Documents need to be written in a particular format (e.g. HTML or PDF) in order to support the addition of links.

**Ownership:** Ownership of the page is required in order to place an anchor in a page. It is, of course, possible to point to targets on other pages without ownership, but in order to insert a link source anchor, ownership is required.

**Legacy resources:** It can be difficult to deal with legacy material – when the view of a world changes, old pages might need to be updated with new links.

**Maintenance:** There is a weight of maintenance in creating and updating links in pages. This is due in part to the hard coding and ownership issues described above.

**Link targets:** Current Web links are restricted to point to point linking; there is only one target. Web links are essentially unary with no explicit inverse link (although browsers offer a "back" button that will take the user back to the originating point). Binary or n-ary links would allow greater flexibility in linking by offering more choice of targets for each link.

Dynamic linking services, supported by ontologies, offer a mechanism to overcome such restrictions. The Conceptual Open Hypermedia Service (COHSE)[3] system enhances document resources through the dynamic addition of hypertext links. These links are derived through the use of an ontology and associated lexicon along with a mapping from concepts to possible link targets. We describe an application of the COHSE architecture to Bioinformatics, using the Gene Ontology (GO)[12] as an ontology and GO associations as link targets. The resulting demonstrator (referred to here as GOHSE) provides both glossary functionality and the possibility of building dynamic hypertext structures linking bioinformatics documents. Ontology driven dynamic linking offers a vision of biology documents dynamically linked to multiple resources based on a common understanding of a domain based upon ontologies.

The following scenario describes the added value that the COHSE system can provide. A biologist is reading a Web page about cellular structure, e.g. http://users.rcn.com/jkimball.ma.ultranet/BiologyPages/C/CellularRespiration.html. When viewed using a traditional browser, she will see static links (as inserted by the author) contained within the page. She then employs the COHSE agent (making use of the cellular component of GO) to assist browsing. Now, as well as the static links contained within the page, the lexical items within the page corresponding to GO cellular component terms or their synonyms are also highlighted as link sources – for example, the term "cytochrome c oxidase", which is a synonym of the GO term respiratory chain complex IV (sensu Eukarya) [GO:0005751]). Next to this highlighted term is an icon indicating that a definition is available. She clicks on this icon and sees a pop up definition of the term, taken from the GO. Also shown are a number of further link targets, taken from a link base, offering her a range of resources related to the term, such as COX3_YEAST in the SwissProt[11] database. In addition, if the number of resources found for the term is below a threshold, the agent will use the taxonomic structure of the GO cellular component ontology to find more general or more specific terms that may provide appropriate resources. Clicking upon COX3_YEAST, she is taken to the appropriate UniProt/SWISS-PROT entry. This resource can itself then be dynamically linked to terms etc. within the GO cellular component ontology.

In the following sections, we discuss how this scenario is realised. In Section 2 we describe the Gene Ontology and how it makes a suitable resource for the open, dynamic linking of biology document resources. We then introduce the COHSE system in Section 3 and describe the combination with GO in Section 4. We conclude with discussion and pointers to future work.

## 2. GO

The Gene Ontology (GO)[a] is a collaborative effort to address the need for consistent descriptions of the major attributes of gene products in different databases[12]. Figure 1 shows a portion of the cellular component ontology from GO. Each term has an associated textual definition describing the term along with subsumption and partitive relationships with other terms in the ontology. Each term within GO also has associated synonyms that represent alternative, equally valid terms for the concept within the ontology. In addition, a number of mappings between other vocabularies or classification systems and GO are available[b].

```
respiratory chain complex IV (sensu Eukarya)
  Accession:  GO:0005751
  Aspect:  cellular_component
  Synonyms:
            o cytochrome c oxidase
            o GO:0005752
  Definition:
            o A part of the respiratory chain, containing the 13
            polypeptide subunits of cytochrome c oxidase, including
            cytochrome a and cytochrome a3. Catalyzes the oxidation of
            reduced cytochrome c by dioxygen (O2). Found in eukaryotes.
  hierarchy
  * GO:0003673 : Gene_Ontology ( 146200 )
    o GO:0005575 : cellular_component ( 79199 )
      + GO:0005623 : cell ( 56534 )
        # GO:0005622 : intracellular ( 46101 )
        * GO:0005737 : cytoplasm ( 35977 )
          o GO:0005739 : mitochondrion ( 12311 )
            + GO:0005740 : mitochondrial membrane ( 979 )
              # GO:0005743 : mitochondrial inner membrane ( 775 )
                * GO:0005746 : mitochondrial electron transport chain ( 211 )
                  o GO:0005751 : respiratory chain complex IV (sensu Eukarya) ( 54 )
          * GO:0045277 : respiratory chain complex IV ( 55 )
            o GO:0005751 : respiratory chain complex IV (sensu Eukarya) ( 54 )
        # GO:0016020 : membrane ( 13431 )
        * GO:0019866 : inner membrane ( 803 )
          o GO:0005743 : mitochondrial inner membrane ( 775 )
            + GO:0005746 : mitochondrial electron transport chain ( 211 )
              # GO:0005751 : respiratory chain complex IV (sensu Eukarya) ( 54 )
        * GO:0005740 : mitochondrial membrane ( 979 )
          o GO:0005743 : mitochondrial inner membrane ( 775 )
            + GO:0005746 : mitochondrial electron transport chain ( 211 )
              # GO:0005751 : respiratory chain complex IV (sensu Eukarya) ( 54 )
```

Figure 1.   GO Ontology fragment

Collaborating databases annotate their gene products with appropriate GO terms, providing the consistency of annotation needed for reliable querying of databases. All the entries from the 16 collaborating databases either contain GO identifiers (that map to GO terms) or their equivalent mappings to internal database keyword lists or GO synonyms. In addition, other Web pages about biology, articles in on-line databases such as

---

[a]http://www.geneontology.org
[b]See http://www.geneontology.org/GO.indices.html

Pubmed, etc. also use these GO terms, synonyms and keyword mappings. Finally, there are tools, such as the GO Amigo browser, that are also oriented towards the Gene Ontology and offer further mechanism for linking via the GO terminologies.

GO offers a huge resource of community knowledge, but no one organisation owns all the "documents" that use the GO vocabularies and definitions. The use of COHSE together with GO offers a mechanism by which any Web page or other document containing lexical items matching a GO term or any of its equivalents can be automatically linked to not only a definition of the term from GO, but also a legion of other resources, based upon GO. As a consequence, we can dynamically generate a rich, flexible web of biology resources.

## 3. COHSE

Detailed descriptions of COHSE[c] can be found elsewhere[3], but we give here a brief overview of the basic approach and architecture.

Open Hypermedia Systems[6,9] seek to solve some of the problems outlined in Section 1. Rather than embedding links in the documents, which is inflexible, links are considered *first class citizens*. They are stored and managed separately from the documents and can thus be stored, transported, shared and searched separately from the document itself. The Distributed Link Service (DLS)[4], developed by the University of Southampton is a service that adopts this approach, and provides dynamic linking of documents. Links are taken from a link base, and can be either *specific*, where the source of the link is given by addressing a particular fragment of a resource, or *generic*, where the source is given by some selection, e.g. a word or phrase. Documents and linkbases are dynamically brought together by the DLS, which then adds appropriate links to documents.

COHSE extends the DLS with *ontological services*, providing information relating to an ontology. These services include mappings between concepts and lexical labels (synonyms). For example, GO tells us that cytochrome c oxidase is a synonym of respiratory chain complex IV (sensu Eukarya (See Figure 1). In this way, the terms and their synonyms in the ontology form the means by which the DLS *generically* finds lexical items within a document from which links can be made. The services also provide information about relationships, such as sub- and super-classes – here respiratory chain complex IV (sensu Eukarya) is a sub-class of respiratory chain

---

[c]http://cohse.semanticweb.org

complex IV. The use of an ontology helps to bridge gaps[1] between the terms used in example web pages (e.g. in this case cytochrome c oxidase), and those used to index other bioinformatics resources, such as the more specialised or generalised GO terms. We can loosen the restriction of linking only to Eukaryote complexes, and consider also linking to all complexes.

COHSE thus extends the notion of generic linking – the key point being that the ontology provides the link service with more opportunities for identifying link sources. As the ontology contains the terms that inform the DLS about the lexical items that may become links, there is no longer a need to own the page in order to make the link from the source to the target – this is taken care of by the DLS. Furthermore, the effort in providing the source of links moves from the document author to the creator(s) of the ontologies that are used by COHSE. In this case, these are the creators of GO – an ontology supported by a wide community of biologists and database curators that form a consensus in the ontological representation of domain understanding.

The system is implemented as a *COHSE agent*, along with two supporting services: the *Ontology Service* (OS) and *Annotation Service* (AS). The agent augments documents with links based on the semantic content of those documents. The Ontology Service delivers ontological information (as introduced above) in a dynamic fashion[2] to the DLS. The Annotation Service associates concepts with resources and provides mechanisms for querying those associations.

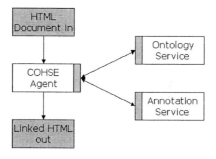

Figure 2.   COHSE Architecture

In the implementation, the agent is attached to a proxy through which all HTTP requests are routed. The agent first contacts the OS to obtain a collection of relevant lexical items. As documents come through the proxy, the agent then looks for these items. Any that are found in the documents provide potential link sources. For each source, a link is then added that includes:

(1) The concept to which the lexical item resolves (in the case of GOHSE this will be the GO term).
(2) A description of the term.
(3) A collection of link targets associated with that term.

Items 1 and 2 are supplied by the OS. The targets in 3 are supplied via calls to the AS. The concepts in the ontology are used to determine appropriate targets for links out of the given document. Within COHSE, the AS plays two roles. It maps resources or documents to concepts (via explicit annotations that have been made), and maps concepts to documents. It is this latter functionality that allows us to provide potential link targets once a link source has been found. For the concept to document mapping, we can either rely on a reversal of the explicit document to concept mapping, or provide targets through the use of external resources. For example, the AS provides potential link targets through queries to the GO database. Once a link source and associated GO term has been identified, we can query the GO database for proteins that have been annotated with that term, in the UniProt/SWISS-PROT database. Given the identifiers for those proteins, we can then produce URLs allowing browsing of those proteins via UniProt's web front end. In addition, we can provide links to the AmiGO or MGI GO browsers. Alternative mechanisms that we have explored for target retrieval include using external search engines (such as Google or the Amazon catalogue) with the query being based on keywords associated with a concept.

Central to the COHSE agent is the provision of an *editorial component* within the agent. This component uses information within the ontology (such as hierarchical classification) in order to either determine whether the generated links are suitable or to expand or cull the set of possible targets. Figure 2 shows a simplified view of the basic architecture of the system.

Both OS and AS are presented to the COHSE agent using simple CGI interfaces. This allows the system to make use of existing protocols and results in a relatively lightweight, loosely-coupled, and open architecture. In this demonstrator, we use a proxy to give access to COHSE. There are potential disadvantages in terms of scalability, but the use of a proxy allows users to access the demonstration without requiring local installation of software or browser plugins and avoids problems of developing bespoke implementations for different browsing platforms.[d].

## 4. COHSE+GO = GOHSE

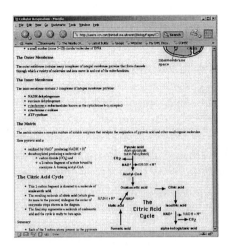

Figure 3.   Before Proxy Linking

For the purposes of the GOHSE demonstration, the cellular component hierarchy of GO provides the ontology, while link targets are derived from the GO annotations in a variety of resources.

The concept taxonomy, along with term synonyms, is loaded into the COHSE OS. Annotation retrieval in the AS is implemented by returning UniProt/SWISS-PROT GO associations. For any given GO term, the AS returns URIs providing access to a number of potential targets:

- The AmiGO browser focused on the term.
- The Gene Ontology Browser focused on the term
- Any UniProt/SWISS-PROT entries known to be annotated with the term.

Both the ontology and the annotations are obtained dynamically from a (local) copy of the GO database. The ontology is produced via an on-the-fly translation to the recently developed W3C Web Ontology Language

---

[d]A Mozilla plug-in providing the COHSE agent functionality is, also available. See http://cohse.man.ac.uk/ for details of available software.

OWL[8] (the format expected by the Ontology Service) and annotations are obtained via appropriate queries.

To use the system, the user first configures their web browser to use the GOHSE proxy. The user can then set up appropriate options concerning the behaviour of the proxy, including the number of times an individual link source should be identified. When the user requests a particular Web page, the proxy will attempt to find words or terms in the page that correspond to GO terms (via the lexical information held in the OS). For any matches to terms or term variants found, a link is added to the page providing access to the term and targets as described in Section 3. This presentation of the information is as a "linkbox" which pops up when the link is selected.

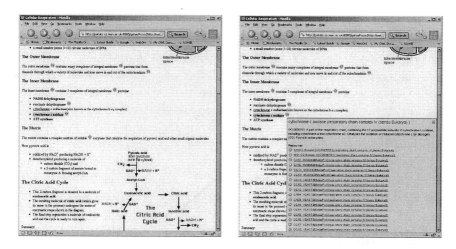

Figure 4.   After Proxy Linking

The resulting additional links can help in providing both explanations of relevant terms and links to relevant materials. A key aspect of the system is its open-ness – we do **not** need to have resources under our control in order to add links to them. Thus third-party resources can be enhanced with additional links, allowing us to construct hypertexts using existing web resources. Figures 3 and 4 show the effects of the proxy – the scenario here is exactly as described in Section 1. In Figure 3 we see the original web page. In Figure 4 (left hand image), terms have been identified and marked (with small icons). On clicking on an icon (the right hand image) we see that the term maps to the GO term respiratory chain complex IV (sensu Eukarya) [GO:0005751]), and that a number of resources relating to this are available.

## 5. Discussion

We have described an application of the COHSE infrastructure to support ontology driven browsing of biology document resources on the Web. In particular, the *dynamic* nature of the linking process helps to alleviate some of the problems with traditional Web linking, which can be static, restricted and inflexible.

Linking is based upon a conceptual model provided by an ontology, where the definitions and structure of the ontology, together with the lexical labels drive the consistency of link provision and dynamic aspects of the linking. The Gene Ontology has already been developed and encapsulates a great deal of shared knowledge about important concepts in the domain. Although this was not necessarily the purpose for which GO was designed, by using GO within this application, we are able to access a wealth of domain knowledge and gain added value "for free". The ontological resource that drives the linking of documents has already been created and its existence independent of the documents it links means that linking is consistent between documents (for a given version of the ontology).

The use of GO in this context illustrates one of the benefits of the Semantic Web approach: a computationally amenable representation of the content and facilities of documents and services[7]. GO provides an encoding of some domain knowledge (concept synonyms and taxonomy) in a *machine processable* fashion. By making this information available to applications, we are able to use this to support the presentation and browsing of resources. Note also that the approach used in COHSE is generic – we are not bound to the use of the Gene Ontology, but could use any other ontology appropriate for the domain, for example MGED[10]. Indeed, given an ontology relating to *any* domain, we can use COHSE to dynamically link suitable resources.

We note that the separation of the maintenance of link data from the underlying content enables us to manage the task of updating databases (or at least their web representations) as new knowledge becomes available. For example, while UniProt/SWISS-PROT links directly to the Gene Ontology, its predecessor, SWISS-PROT, did not. Using GOHSE, we can synthesize these links before the underlying data source provides them. Similarly, by extending GOHSE to recognise UniProt/SWISS-PROT identifiers, we can link between free text resources, such as PubMed, GO concepts, and the underlying protein data sources. This feature of open hypermedia systems in general, and GOHSE in particular, is of particular relevance to biological data where cross-linking is known to be fragile[13]. As a discipline, bioinformatics relies on access to knowledge held in its databases. A system such as COHSE, augmented by ontologies such as GO, provide a knowledge model to drive the linking of diverse, distributed resources according to that knowledge.

It is clear that one of the reasons that this approach works here is because we have a well-defined domain of interest, a community and (to a certain extent) agreement on the important terms and concepts within that community. This is where we believe Semantic Web technology will have its initial successes – within well-defined communities.

In the current implementation, the identification of potential link sources is done in a rather naive fashion – effectively through a straight lexical match. In the GOHSE setting, this produces reasonable results (from the technical point of view), largely because (as discussed above), GO tries to use terms commonly used in the domain and these provide a clear set of lexical items to act as potential link sources. We can, however, encounter problems when, for example, formatting information is included in the source, making the identification of lexical items harder. We are investigating the use of the GATE[5] framework in order to gain access to effective text processing components. These will provide the DLS greater flexibility in its use of the terms provided by the OS.

Once concepts have been identified within the page, navigation of the ontology is driven by the COHSE Agent rather than the user. The agent decides if sufficient link targets are available, and whether or not to traverse the hierarchy to obtain more candidates. It may be more profitable to allow the user to explicitly navigate or explore the ontology at this point. However, there are then questions as to how one exposes the ontological structure to the user. In a similar vein, COHSE is clearly a system in which *personalization* can play a part – different users will want to use different ontologies or annotation collections. The current architecture is rather inflexible in this respect, and we are investigating support for more effective personalization.

GOHSE does not provide us with any new knowledge – it simply allows us to organise and present what is already there. Nor is it, at present, a particularly sophisticated implementation and improvements can certainly be made. It does, however, allow us to link together diverse biology resources, including those not in our control, in a consistent fashion based upon a community understanding of the domain.

## Acknowledgments

Phil Lord is supported by the <sup>my</sup>Grid EPSRC E-science pilot (EPSRC GR/R67743). The original COHSE system was developed in collaboration with the University of Southampton. The authors would like to thank John Kimball for permission to use his pages in our examples.

## References

1. Marcia J. Bates. Indexing and Access for Digital Libraries and the Internet: Human, Database and Domain Factors. *JASIS*, 49(13):1185–1205, 1998.
2. Sean Bechhofer and Carole Goble. Delivering Terminological Services. *AI\*IA Notizie, Periodico dell'Associazione Italiana per l'intelligenza Artificiale.*, 12(1), March 1999.
3. L. Carr, S. Bechhofer, C. A. Goble, and W. Hall. Conceptual Linking: Ontology-based Open Hypermedia. In *Proceedings of WWW10, Tenth World Wide Web Conference*, Hong Kong, May 2001.
4. L. Carr, D. De Roure, W. Hall, , and G. Hill. The Distributed Link Service: A Tool for Publishers, Authors and Readers. *World Wide Web Journal*, 1(1):647–656, 1995.
5. H. Cunningham, D. Maynard, K. Bontchev, and V. Tablan. GATE: A Framework and Graphical Development Environment for Robust NLP Tools and Applications. In *Proceedings of the 40th Anniversary Meeting of the Association for Computational Linguistics (ACL'02)*, Philadelphia, July 2002.
6. K. Grønbæk, L. Sloth, and Orbæk P. Webvise: Browser and Proxy Support for Open Hypermedia Structuring Mechanisms on the WWW. In *Proceedings of the Eighth International World Wide Web Conference*, pages 253–268, 1999.
7. J. Hendler. Science and The Semantic Web. *Science*, page 24, Jan 2003.
8. D. L. McGuinness and F. van Harmelen. OWL Web Ontology Language Overview. W3C Recommendation, World Wide Web Consortium, 2004. http://www.w3.org/TR/owl-features/.
9. K. Osterbye and U Wiil. The Flag Taxonomy of Open Hypermedia Systems. In *Proceedings of the 1996 ACM Hypertext Conference*, pages 129–139, 1996.
10. Chris Stoeckert and Helen Parkinson. The MGED Ontology: A framework for describing functional genomics experiments. *Comparative and Functional Genomics*, 4(1):127–132, 2002.
11. SWISS-PROT Annotated Protein Sequence Database. http://www.expasy.org.
12. The Gene Ontology Consortium. Gene Ontology: a tool for the unification of biology. *Nature Genetics*, 25:25–29, 2000.
13. Jonathan D. Wren. 404 not found: the stability and persistence of URLs published in MEDLINE. *Bioinformatics*, 20(5):668–672, 2004.

# NON-LEXICAL APPROACHES TO IDENTIFYING ASSOCIATIVE RELATIONS IN THE GENE ONTOLOGY

OLIVIER BODENREIDER

*U.S. National Library of Medicine*
*8600 Rockville Pike, MS 43, Bethesda, Maryland 20894, USA*
*E-mail: olivier@nlm.nih.gov*

MARC AUBRY

*Unité de Génétique Humaine, UMR 6061 CNRS*
*Avenue du Pr Léon Bernard 35043 Rennes Cedex, France*

ANITA BURGUN

*Laboratoire d'Informatique Médicale, Université de Rennes I*
*Avenue du Pr Léon Bernard 35043 Rennes Cedex, France*

The Gene Ontology (GO) is a controlled vocabulary widely used for the annotation of gene products. GO is organized in three hierarchies for molecular functions, cellular components, and biological processes but no relations are provided among terms across hierarchies. The objective of this study is to investigate three non-lexical approaches to identifying such associative relations in GO and compare them among themselves and to lexical approaches. The three approaches are: computing similarity in a vector space model, statistical analysis of co-occurrence of GO terms in annotation databases, and association rule mining. Five annotation databases (FlyBase, the Human subset of GOA, MGI, SGD, and WormBase) are used in this study. A total of 7,665 associations were identified by at least one of the three non-lexical approaches. Of these, 12% were identified by more than one approach. While there are almost 6,000 lexical relations among GO terms, only 203 associations were identified by both non-lexical and lexical approaches. The associations identified in this study could serve as the starting point for adding associative relations across hierarchies to GO, but would require manual curation. The application to quality assurance of annotation databases is also discussed.

## 1. Introduction

The Gene Ontology™ (GO) is an important resource that has transformed the functional annotation of gene products by providing the curators of model organism databases with a controlled vocabulary which has rapidly become a *de facto* standard. GO has over 17,000 terms and is organized in three hierarchies for molecular functions, cellular components, and biological processes. However, if hierarchical relations (*is a*, *part of*) constitute the backbone of ontologies, GO is essentially a skeleton because it completely lacks associative relations across its three hierarchies. Such associative relations would indicate, for example, that a cellular component is the location of a biological process and that a molecular function is involved in a biological process.

The lack of representation in GO of the relations existing among functions, processes, and components severely limits the power of reasoning based on GO. This issue has been recognized by Bada et al. as they developed the *Gene Ontology Annotation Tool* (GOAT) [1]. One major task in GOAT and its companion project *Gene Ontology Next Generation* (GONG) is the acquisition of such relations and their formal representation in the Ontology Web Language (OWL) [2].

The approach taken in GOAT for acquiring associations between GO terms has been to mine the annotation database *Gene Ontology Annotation* (GOA) for co-occurrence of GO terms. 600,000 associations were obtained by this method, excluding unreliable associations and the hierarchical relations explicitly represented in GO [1]. Another approach to identifying relations among GO terms draws on the compositional structure of these terms. Ogren et al. found that 65% of all GO terms contain another GO term as a proper substring [3]. Finally, in a previous study, we suggested that association rule mining could be applied to identifying dependence relations among GO terms [4]. Kumar et al. successfully applied association rule mining techniques to the annotation databases of six bacterial genomes from The Institute for Genome Research (TIGR) and evaluated their findings in light of formal ontological principles [5].

In this study, rather than identifying all dependence relations, we concentrate specifically on associations among GO terms across ontologies. The primary objective of this study is to investigate three non-lexical approaches to identifying such associative relations in the Gene Ontology (GO) and compare them to lexical approaches. Our three approaches are: computing similarity in a vector space model, statistical analysis of co-occurrence of GO terms in annotation databases, and association rule mining. A secondary objective is to analyze the consistency of the associations discovered across five model organism databases. In other words, the major contribution of this study is not to define novel non-lexical methods for studying term-term associations, but rather to compare multiple existing approaches among themselves and to traditional lexical methods, systematically and across several model organism databases.

## 2. Datasets

The three approaches under investigation in this study take advantage of the existing annotation databases created for various model organisms. These databases, made publicly available in a common format by the GO Consortium[*], describe gene products that have been annotated with GO terms by each collabo-

---

[*] http://www.geneontology.org/GO.current.annotations.shtml

rating group. The annotation databases used in this study correspond to the major model organisms and were downloaded from the GO website[†]:

1. **FlyBase** (*Drosophila melanogaster*)
2. Human subset of **GOA** (*Homo sapiens*)
3. **MGI** (*Mus musculus*)
4. **SGD**™ (*Saccharomyces cerevisiae*)
5. **WormBase** (*Caenorhabditis elegans*)

Details about these datasets are provided in Table 1.

Table 1 – Detail of the datasets used in this study

| Dataset | Developed by | Web site | Dated |
|---------|-------------|----------|-------|
| FlyBase | FlyBase Consortium | http://flybase.bio.indiana.edu/ | 5/22/2004 |
| GOA-Human | European Bioinformatics (EBI) | http://www.ebi.ac.uk/GOA/ | 6/4/2004 |
| MGI | Jackson Laboratory | http://www.informatics.jax.org/ | 6/4/2004 |
| SGD | Stanford University | http://www.yeastgenome.org/ | 6/11/ 2004 |
| WormBase | WormBase Consortium | http://www.wormbase.org/ | 5/11/2004 |

Table 2 – Number of unique gene products, GO terms, and gene product-term pairs in the five annotation databases under investigation

| Annotation DB | # gene products | # GO terms | # GP-term pairs |
|---------------|-----------------|------------|-----------------|
| FlyBase | 9,090 | 3,597 | 38,089 |
| GOA-Human | 22,720 | 4,247 | 92,658 |
| MGI (Mouse) | 14,471 | 3,616 | 65,571 |
| SGD (Yeast) | 6,457 | 2,412 | 25,278 |
| WormBase | 10,534 | 1,540 | 36,695 |

The version of GO used throughout this study is the June 2004 monthly release, available from the GO website. The GO terms present in the annotation databases but not in the ontology were replaced by current terms whenever possible. For example, the term *amine oxidase (flavin-containing) activity* (GO:0004041) is no longer present and was replaced by *amine oxidase activity* (GO:0008131), with which it is currently asserted to be synonymous. The annotations for which no current GO term existed were ignored. Also ignored were the annotations for which the evidence supporting the association between a gene product and a GO term is insufficient. In practice, we filtered out all annotations inferred from electronic annotation (with 'IEA' as evidence code), because they are not reviewed by curators. We did not include either the negative associations, marked with 'NOT' in the *Qualifier* field of the annotation files. The number of unique gene products, GO terms, and gene product-term pairs in each annotation

---

[†] http://geneontology.org/

database is given in Table 2. These counts reflect the substitutions and filtering mentioned above.

In addition to GO and the annotation databases, the evaluation relies in part on the Unified Medical Language System® (UMLS®) Metathesaurus®. The UMLS‡ is a terminology integration project developed at the U.S. National Library of Medicine. The UMLS Metathesaurus integrates many biomedical terminologies, including the Gene Ontology [6]. Although no relations across ontologies are defined in GO, such relations – contributed by other sources – may be present in the Metathesaurus. More specifically, associative relations asserted in other source vocabularies are found in the MRREL table. For example, the GO terms *chloroplast* and *photosynthesis* are also defined in the Medical Subject Headings (MeSH), where they are cross-referenced. This "see also" relationship is recorded in the Metathesaurus between the two concepts. Similarly, the co-occurrence of MeSH descriptors in the MEDLINE database is recorded in the MRCOC table of the Metathesaurus. The edition of UMLS used in this study is 2004AA (April 2004).

## 3. Methods

The three approaches to identifying associative relations in GO, presented in detail below, can be summarized as follows:
1. A vector space model in which each GO term is described by a vector of gene products corresponding to the annotations of this product in the annotation database for a given organism.
2. Statistical analysis of co-occurrence of GO terms in the annotations of gene product, where the observed frequency of co-occurrence of two GO terms is compared to the frequency expected under the hypothesis of independence of GO terms.
3. Association rules mined from the sets of GO terms extracted from annotation databases, where each transaction corresponds to the annotations of a given gene product in a given annotation database.

In all three cases, the associations identified are restricted to associations across GO ontologies (e.g., molecular function to biological process) by filtering out the association within hierarchies.

Common to the three approaches is the assumption that the dependence relations identified (e.g., among frequently associated GO terms) should reflect *ontological* relations (i.e., among entities whose existence depend on one another), themselves possibly corresponding to *biological* relations. Based on different mathematical principles, the three approaches are expected to identify different sets of dependence relations. While the three methods essentially rely

---

‡ http://umlsks.nlm.nih.gov/ (free license required)

on the frequency of association between two GO terms, they use different criteria for determining which associations are significant.

### 3.1. *Similarity in the vector space model*

Vector space models (VSMs) are frequently used in information retrieval for computing the similarity between documents described as vectors of keywords [7]. A collection of gene products annotated with the controlled vocabulary provided by GO is in fact analogous to a collection of scientific articles indexed with the MeSH controlled vocabulary. Although the primary use of a collection of indexing terms for documents (or annotation terms for gene products) is to compute the similarity among documents (or gene products), our interest here is to compute the similarity among terms. Therefore, we have to transpose the matrix of gene products by GO terms in order to obtain a matrix of GO terms by gene products. As usual in the VSM paradigm, the similarity between two vectors is represented by the angle between these vectors, measured by the dot product of the two (normalized) vectors.

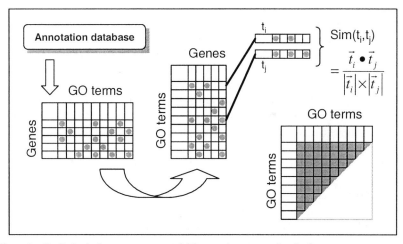

Figure 1 – Similarity in the vector space model from a given annotation database

As illustrated in Figure 1, the original matrix (gene products by GO terms) consists of binary values indicating the presence (1) or absence (0) of an association between a gene product and a GO term in a given annotation database. One such matrix is created for each model organism. The matrix is then transposed. Not represented in the figure is the weighting step. A weight is applied to each binary association in order to lower the importance of an association between a GO term and a gene product when a given gene product is associated with many

GO terms[§]. This weighting scheme, shown in Eq. (1), is known as inverse document frequency (idf) in information retrieval. Here, the weight of each association between a GO term and gene product $j$ is inversely proportional to the ratio of the number of annotations for this gene product ($n_j$) to the total number of distinct gene products in the corresponding annotation database (N). Then, each vector is normalized in order to compensate for differences in the number of genes associated with GO terms. Once the vectors are normalized, their dot product varies between 0 and 1 and measures the similarity between them (see Figure 1). A value of 0 corresponds to no similarity, while 1 indicates complete similarity. Term-term similarity is computed pairwise for all GO terms present, resulting in a half-matrix for each model organism database. We use an arbitrary threshold of .5 for the dot product in order to select the pairs of terms exhibiting a high drgree of similarity.

$$idf_j = \log \frac{N}{n_j} \tag{1}$$

### 3.2. Co-occurrence in annotation databases

In probability theory, two events $E_1$ and $E_2$ are independent when the probability of occurrence of the two events simultaneously, $P(E_1 \cap E_2)$, is not greater than the product of the probabilities of occurrence for each event, $P(E_1) \cdot P(E_2)$. Conversely, when $P(E_1 \cap E_2) > P(E_1) \cdot P(E_2)$, $E_1$ and $E_2$ are not independent. What we are interested in identifying here are pairs of "non-independent" GO terms, whose frequency of co-occurrence (i.e., simultaneous presence in the annotation of a gene product) is higher than would be expected if the two terms had been used independently by the curators. For a given pair of GO terms ($A,B$), information about their association in gene product annotations can be summarized in a two-way contingency table and analyzed statistically [8]:

☐ $n_{AB}$, the number of gene products annotated with both term $A$ and term $B$
☐ $n_{Ab}$, the number of gene products annotated with term $A$ but not term $B$
☐ $n_{aB}$, the number of gene products annotated with term $B$ but not term $A$
☐ $n_{ab}$, the number of gene products annotated with neither term $A$ or term $B$

The chi-square test of independence (or Pearson's chi-square) is often used to test independence between two categorical variables (here, the presence or absence of a given term in the annotations of genes). The chi-square ($x^2$) statistic

---

[§] The weighting step could probably be omitted in the case of a matrix of GO terms by gene products, because each gene has a limited number of annotations. It would, however, be crucial for computing gene-gene similarity in a matrix of gene-products by GO terms.

relies on the difference between observed frequencies ($n_{ij}$) for the four events listed above and the frequencies expected under the hypothesis of independence. The $x^2$ statistic has a chi-square distribution, specified by its degrees of freedom. There is one degree of freedom in the case of two-way contingency tables for binary variables. A large value of the $x^2$ statistic indicates a deviation from the expected frequencies. In this case, i.e., when the corresponding P-value is lower than the usual .05 threshold, the hypothesis of statistical independence is rejected and the association is considered statistically significant. One limitation of the chi-square test is that all expected frequencies are required to be 5 or more. In practice, this condition cannot be met if the frequency of the terms is small.

An alternative to the Pearson's chi-square test is the likelihood ratio test (also called G-test or G-square test). The $G^2$ statistic compares the maximum of the likelihood function under two circumstances: 1) under the hypothesis of independence and 2) under the general, observed conditions. Like the $x^2$ statistic, the $G^2$ statistic has a chi-square distribution (also with one degree of freedom in our setting). Interestingly, the $G^2$ statistic does not have the minimum expected frequency requirements imposed by the $x^2$. However, for the $G^2$ statistic to be computed, all observed frequencies must be greater than 0.

In practice, for each pair of terms, we first attempt to compute a $G^2$ statistic. A $x^2$ statistic is used instead when the requirements are not met for $G^2$. Finally, the association is ignored if it fails to meet both $G^2$ and $x^2$ requirements. Because of the low frequency of co-occurrence of the terms in this case, identifying their association is of little interest anyway.

While both $x^2$ and $G^2$ indicate the existence of an association between two variables, neither one describes the strength of the association. Several similarity coefficients have been developed for this purpose [9], which could be used to select the pairs of terms exhibiting a strong association. In this study, however, we simply included all pairs of terms for which the test indicated a statistically significant association, regardless of the strength of the association.

### 3.3. Association rule mining

Association rules capture the association between two sets of events of arbitrary size and are expressed in the form: $A \Rightarrow B$, where $B$ is the set of events that can be predicted from $A$ [10]. Historically, the identification of association rules was applied to analyzing grocery buying patterns, with rules such as {*bread, milk*} $\Rightarrow$ {*sugar*} expressing that customers buying bread and milk also often buy sugar. By applying association rule mining techniques to annotation databases, we expect to discover that genes annotated with the GO term $T_1$ are also frequently annotated with $T_2$. The set of GO terms annotating a gene product is called a transaction in association rule mining parlance.

We used Christian Borgelt's implementation of the *apriori* algorithm[**] to mine association rules. Since our objective is to identify pairs of related GO terms, we restricted the size of the sets under investigation to two. The two major parameters in the algorithm are *support* and *confidence*. Support for the rule $T_1 \Rightarrow T_2$ represents the proportion of genes annotated with both $T_1$ and $T_2$. Confidence for the same rule represents the proportion of genes annotated with both $T_1$ and $T_2$ among those annotated with $T_1$. In order to restrict rules to almost systematic associations, we required confidence to be at least 90%. The minimum support was set to a low value (.05%) simply to eliminate "accidental" associations. We use the product of support by confidence to describe the strength of the association.

### 3.4. *Evaluation*

The first step of the evaluation consists in comparing the results of the three approaches. Associations identified independently by several approaches simultaneously are expected to be stronger and therefore more important. Finally, the presence of the association in the annotation databases of several organisms suggests that this association is stronger than isolated associations. What is evaluated here is essentially the statistical significance of the associations. Evaluating the ontological and biological significance of these associations is beyond the scope of this study.

Additionally, we compared the results of our three approaches to lexical associations and to associations present in the UMLS Metathesaurus.

**Lexical relations**. Using the method proposed by Ogren et al., we identified all pairs of GO terms where one term is nested as a substring in the other term [3]. In order to reveal additional lexical relations, we did a second run after systematically removing the word 'activity' from terms in the molecular function hierarchy.

**UMLS relations**. We searched the MRREL table for the presence of associative relations[††] among concepts present in GO. Similarly, we searched the MRCOC table for the presence of co-occurrence relations among GO concepts (co-occurrence of MeSH descriptors in MEDLINE records).

---

[**] http://fuzzy.cs.uni-magdeburg.de/~borgelt/apriori.html

[††] Their relationship type (REL) is 'RO' for "other related concepts".

# 4.   Results

## 4.1. *Associations identified*

Examples of association identified specifically by each method are presented in Table 3. The first three are the methods under investigation: VSM (vector space model), COC (co-occurrence in annotation databases), and ARM (association rule mining). The others are the methods used in the evaluation: LEX (lexical relations), REL (associative relations in UMLS), and MDL (co-occurrence in MEDLINE). Quantitative results are presented in Table 4 where the number of associations identified by each method is broken down by category of association.

Table 3 – Examples of association identified specifically by each method

| Method | Association | |
|--------|-------------|---|
| **VSM** | MF: *ice binding* | [GO:0050825] |
| | BP: *response to freezing* | [GO:0050826] |
| **COC** | MF: *chromatin binding* | [GO:0003682] |
| | CC: *nuclear chromatin* | [GO:0000790] |
| **ARM** | MF: *carboxypeptidase A activity* | [GO:0004182] |
| | BP: *proteolysis and peptidolysis* | [GO:0006508] |
| **LEX** | MF: *mannosyltransferase activity* | [GO:0000030] |
| | CC: *mannosyltransferase complex* | [GO:0000136] |
| **REL** | CC: *cell-matrix junction* | [GO:0030055] |
| | BP: *cell adhesion* | [GO:0007155] |
| **MDL** | CC: *synaptic vesicle* | [GO:0008021] |
| | BP: *exocytosis* | [GO:0006887] |

Table 4 – Number of associations identified by each method for each category of association (MF: molecular function; CC: cellular component; BP: biological process)

| | VSM | COC | ARM | LEX | REL | MDL |
|--------|-----|-----|-----|-----|-----|-----|
| **MF-CC** | 499 | 893 | 362 | 917 | 0 | 0 |
| **MF-BP** | 3057 | 1628 | 577 | 2523 | 0 | 1 |
| **CC-BP** | 760 | 1047 | 329 | 2053 | 22 | 469 |
| **Total** | 4316 | 3568 | 1268 | 5493 | 22 | 470 |

## 4.2. *Overlap*

A total of 13,398 associations were identified by at least one method, 7,665 by at least one of the three major methods (VSM, COC, and ARM) and 5,963 by at least one of the evaluation methods (LEX, REL, and MDL). Of these, only 230 associations were identified by both major and evaluation methods. Examples of associations identified independently but simultaneously by several methods are presented in Table 5. As illustrated in Figure 2, 12% of the associations identified by the three major methods were identified by more than one method. In contrast, only a few lexical associations are also present in the UMLS.

Out of the 7,665 associations identified by at least one method, 5,950 (78%) came from only one annotation database. In 1,116 cases (16%), the association was simultaneously identified in two annotation databases, 6% in three, and 2% in two. Only 41 associations (less than 1%) were present in all five databases. Pairwise, after normalizing by the number of annotations in each database, the highest rates of overlap are between MGD and GOA-Human and MGD and WormBase, the lowest between SGD and GOA-Human and SGD and MGD.

Table 5 – Examples of association identified simultaneously by several methods

| Association | | VSM | COC | ARM | LEX | REL | MDL |
|---|---|---|---|---|---|---|---|
| MF: *potassium channel activity* <br> BP: *potassium ion transport* | [GO:0005267] <br> [GO:0006813] | X | X | X | | | |
| MF: *chemokine activity* <br> BP: *immune response* | [GO:0008009] <br> [GO:0006955] | | X | X | | | |
| CC: *hemoglobin complex* <br> BP: *oxygen transport* | [GO:0005833] <br> [GO:0015671] | X | X | | | | |
| MF: *taste receptor activity* <br> BP: *perception of taste* | [GO:0008527] <br> [GO:0050909] | X | | X | | | |
| MF: *metal ion transporter activity* <br> BP: *metal ion transport* | [GO:0046873] <br> [GO:0030001] | X | | X | X | | |
| CC: *transport vesicle* <br> BP: *transport* | [GO:0030133] <br> [GO:0006810] | | | | | X | X |
| CC: *gap junction* <br> BP: *cell communication* | [GO:0005921] <br> [GO:0007154] | X | X | | | | X |

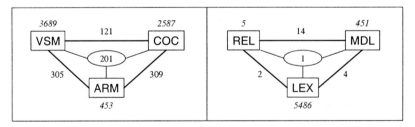

Figure 2 – Number of associations specific to each method (italic) or common to several methods

## 5. Discussion

**Applications.** The major application of our methods is of course to help enrich GO with associative relations across ontologies. Ontology creation and extension is a daunting task. However, by automatically extracting candidate relations from annotation databases, the approaches investigated in this study can significantly reduce the human effort required. We recommend that the associations identified in this study serve as the starting point for adding associative relations across hierarchies to GO. The associations we identified could also be used for quality assurance purposes, i.e., to assess the consistency and completeness of annotation databases. Analogously, knowledge of frequently associated terms could be presented to curators in annotation environments.

**Advantages and limitations.** Many associations identified by our approaches cannot be found by lexical methods. The performance of lexical methods could be improved by factoring in term variation (inflection, derivation) and using more rigorous parsing of the terms; it would, however, remain poor due to the limited number of synonyms available in GO for each term. By imposing constraints such as minimum frequency and statistical significance of co-occurrence and minimum confidence for association rules, our approaches are more selective than the unrestricted methods used in GOAT. Limitations – not specific to our approaches – include the fact that what is identified is the presence of associations between GO terms, not their nature. Moreover, the manual curation of the associations identified remains necessary in order to assess their biological significance.

**Evaluation.** The limited overlap between associations identified by our major methods and the evaluation methods was somewhat unexpected. The lexical relation between, for example, *transport* and *transport vesicle* is ontologically valid but never present in annotations. Although the biomedical literature plays a role in both approaches, the limited overlap between annotation databases and MEDLINE co-occurrences may have the following explanations. Many annotations are derived from sources other than the literature (e.g., inferred from sequence or structural similarity) and MEDLINE co-occurrences are not guaranteed to relate to the same gene when several genes are discussed in an article.

**Generalization.** As shown in earlier studies [4,5], dependence relations can be found both within and across the three GO ontologies. Although this study is purposely restricted to the identification of associative relations across GO ontologies, our methods actually identified almost as many dependence relations *within* ontologies (not reported on here). The lexical method captures five times as many associations within ontologies than across, including a majority of direct parent-child associations. Because curators are unlikely to use both a parent term and its child in the annotation of a gene, the associations within ontologies captured by our methods are essentially between distinct subtrees of GO hierarchies (e.g., between *metallopeptidase activity* [*catalytic activity* subtree] and *zinc ion binding* [*binding* subtree]). Finally, our approaches could be applied to other domains (e.g., for identifying relations among terms of a clinical terminology using clinical databases indexed with this terminology).

**Future directions.** Many interesting aspects of the association between GO terms are beyond the scope of this paper. Those issues, which we expect to address in the near future, include the redundancy of associations across species and applications to the functional interpretation of experimental results.

## Acknowledgments

Marc Aubry's contribution is funded in part by the Conseil Régional de Bretagne. The authors thank Kelly Zeng who provided technical support for computing similarity in the vector space model.

## References

1. Bada, M., Turi, D., McEntire, R. & Stevens, R. Using Reasoning to Guide Annotation with Gene Ontology Terms in GOAT. *SIGMOD Record* **33**(2004). http://www.acm.org/sigmod/record/issues/0406/04.Bada_Turi_McEntire_Stevens.pdf

2. Wroe, C.J., Stevens, R., Goble, C.A. & Ashburner, M. A methodology to migrate the gene ontology to a description logic environment using DAML+OIL. *Pac Symp Biocomput*, 624-35 (2003)

3. Ogren, P.V., Cohen, K.B., Acquaah-Mensah, G.K., Eberlein, J. & Hunter, L. The compositional structure of Gene Ontology terms. *Pac Symp Biocomput*, 214-25 (2004)

4. Burgun, A., Bodenreider, O., Aubry, M. & Mosser, J. Dependence relations in Gene Ontology: A preliminary study. *Workshop on The Formal Architecture of the Gene Ontology - Leipzig, Germany, May 28-29, 2004* (2004). http://mor.nlm.nih.gov/pubs/pdf/2004-go_workshop-ab.pdf

5. Kumar, A., Smith, B. & Borgelt, C. Dependence relationships between Gene Ontology terms based on TIGR gene product annotations. *Proceedings of the 3rd International Workshop on Computational Terminology (CompuTerm 2004)*, 31-38 (2004)

6. Bodenreider, O. The Unified Medical Language System (UMLS): integrating biomedical terminology. *Nucleic Acids Res* **32 Database issue**, D267-70 (2004)

7. Baeza-Yates, R. & Ribeiro-Neto, B. *Modern information retrieval*, 513 p. (ACM Press ; Addison-Wesley, New York; Harlow, England, 1999).

8. Agresti, A. *An introduction to categorical data analysis*, xi, 290 p. (Wiley, New York, 1996).

9. Duarte, J.M., dos Santos, J.B. & Melo, L.C. Comparison of similarity coefficients based on RAPD markers in the common bean. *Genet. Mol. Biol.* **22**, 427-432 (1999)

10. Agrawal, R., Imielinski, T. & Swami, A. Mining association rules between sets of items in large databases. *Proceedings of the 1993 ACM SIGMOD*, 207-216 (1993)

# GENESTRACE: PHENOMIC KNOWLEDGE DISCOVERY VIA STRUCTURED TERMINOLOGY

MICHAEL N. CANTOR*

*Department of Medicine, Beth Israel Medical Center, New York, NY 10003*

INDRA NEIL SARKAR*

*Division of Invertebrate Zoology, American Museum of Natural History, New York, NY 10024
& Department of Biomedical Informatics, Columbia University, New York, NY 10032*

OLIVIER BODENREIDER

*Lister Hill National Center for Biomedical Communications,
National Library of Medicine, National Institutes of Health, Bethesda, MD 20894*

YVES A. LUSSIER [§]

*Departments of Biomedical Informatics and Medicine,
Columbia University, New York, NY 10032
Email: yves.lussier@dbmi.columbia.edu*

The era of applied genomic medicine is quickly approaching accompanied by the increasing availability of detailed genetic information. Understanding the genetic etiology behind complex, multi-gene diseases remains an important challenge. In order to uncover the putative genetic etiology of complex diseases, we designed a method that explores the relationships between two major terminological and ontological resources: the Unified Medical Language System (UMLS) and the Gene Ontology (GO). The UMLS has a mainly clinical emphasis; Gene Ontology has become the standard for biological annotations of genes and gene products. Using statistical and semantic relationships within and between the two resources, we are able to infer relationships between disease concepts in the UMLS and gene products annotated using GO and its associated databases. We validated our inferences by comparing them to the known gene-disease relationships, as defined in the Online Mendelian Inheritance in Man's morbidmap (OMIM). The proof-of-concept methods presented here are unique in that they bypass the ambiguity of the direct extraction of gene or disease term from MEDLINE. Additionally, our methods provide direct links to clinically significant diseases through established terminologies or ontologies. The preliminary results presented here indicate the potential utility of exploiting the existing, manually curated relationships in biomedical resources as a tool for the discovery of potentially valuable new gene-disease relationships.

The GenesTrace system may be accessed at the following URL:

http://phene.cpmc.columbia.edu:8080/genesTrace/index.jsp

---

* These authors contributed equally to the work
§ Corresponding author

## 1. Introduction

Much of biological hypothesis generation follows the proverbial model of trying to discover the "golden needle in the haystack." Discovering significant genes is becoming a daunting task as various genome projects are creating sequence data information at accelerating rates. Thus, it is quickly becoming an intractable problem for the biomedical scientist to stay abreast all putative genes that may hold hidden keys toward the understanding of disease. Barring exhaustive wet-lab and in vivo experimentation, the use of knowledge bases may offer some insight to this task. Novel bioinformatic methods are required to elucidate putative genes that may be related to the etiology of disease. We describe one such method that may offer such insight using relationships that exist within and among terminologies and ontologies of biomedical knowledge.

The emergence of the Gene Ontology™ (GO) [1] and its related databases is an important advance in discovering elusive gene-disease relationships, as it provides a standardized, easily searchable repository of biological information. A great deal of research is focused on developing or improving methods for gene and sequence annotation (see below); however, relatively little research has looked at the equally, if not more, complex idea of relating GO to the level of clinical diseases. This project attempts to bridge that gap by exploring links between GO and its annotation databases to clinical concepts that are in the Unified Medical Language System® (UMLS®) [2]. Ultimately, this research aims to highlight possible gene-disease relationships via the mappings of structured terminologies (both clinical and biological) contained in the UMLS.

A large amount of biomedical knowledge is represented in free-text form, such as MEDLINE abstracts. Extracting important information from these resources is an extremely active area of research. PubGene, for example, is a database for gene-expression analysis, extracted from a weighted network of gene co-occurrence data in MEDLINE [3]. PubGene's gene-gene relationships were validated by comparison with the Online Mendelian Inheritance in Man (OMIM) database [4]. The utility of literature associations were also validated by a comparison to microarray data [3]. The authors noted problems with the ambiguity of gene names and symbols while creating the PubGene network. Additionally, the network did not attempt to characterize the relationships represented in the co-occurrence network.

Fuzzy set theory has also been used as an attempt to characterize candidate disease genes. This approach exploits relationships between two types of annotations, MeSH headings for MEDLINE articles and GO terms for protein sequences. The goal of two methods based on fuzzy set theory is to "derive relationships between pathological conditions and terms describing protein function." [5,6] The evaluation of one system showed an association between the authors' scoring system and the likelihood of a gene-disease association [5].

Similarly, the creators of the MedGene database looked at co-occurrences between genes and MeSH disease terms, and then using those relationships, analyzed microarray data [6].

Several groups have also looked at various automated methods for annotating genes and protein sequences with gene ontology functions. The Gene Ontology Annotation (GOA) project uses manual mappings between GO terms and either protein domains or SWISS-PROT keywords to automatically assign GO terms to a sequence containing previously annotated domains [7]. Raychaudhuri et al. [8] used statistical document classification methods to analyze abstracts from the medical literature. From this analysis, they were able to associate a set of GO terms to the genes mentioned in the abstracts. Building on their previous work [5], Perez et al. [9] developed a system to associate keywords to genes or protein sequences using mappings between SWISS-PROT keywords, MESH terms associated with MEDLINE abstracts, and GO terms. Their system demonstrated better performance with mapping SWISS-PROT terms that GO terms. The difference was attributed to the ambiguity introduced into GO mappings by its ontological structure.

Here, we describe a novel method that builds on previous attempts at bridging biomedical terminologies to infer putative genes implicated in disease etiology. Our method, GenesTrace, uses biological and clinical terminologies contained in the UMLS to induce modal relationships. We hypothesize that this putative phenomic network can further be filtered and mined to reveal buried knowledge. The method we propose is fundamentally different from previously cited ones. For example, instead of using fuzzy logic or statistical methods, our proposed method infers genes-to-diseases relationships by constructing an original network of relationships between curated ontologies and databases and then selecting paths in the network, which fulfill valid semantic constraints. From this proof-of-concept study, we will further describe how GenesTrace's inferences may provide some guidance towards subsequent investigations of the genetic etiology of complex diseases.

## 2. Methods

### 2.1. *Materials*

*UMLS Database.* We used the 2003AB version of the UMLS Metathesaurus®, which contains approximately 900,000 concepts from 102 biomedical source

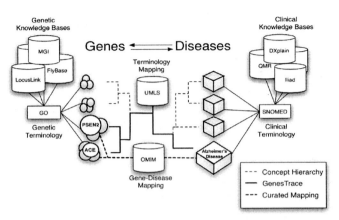

**Figure 1:** *Gene-Disease Mapping*. Gene Terms from Genetic Knowledge Bases, such as MGI, LocusLink, or Flybase, are codified using genetic terminologies, such as GO (left). Disease Terms from Clinical Knowledge Bases, such as DXplain, QMR, or Iliad, are codified using clinical terminologies, such as SNOMED (right). Collectively, these terminologies are mapped as concepts in a hierarchical manner, as in the UMLS (center). Manually curated knowledge databases, like OMIM, map some gene product concepts with disease concepts (bold dashed line). The proposed method, GenesTrace, exploits terminology mappings, which relate disease concepts to gene concepts (bold solid line).

terminologies[2]. Many of these source terminologies have a strictly clinical focus. No single terminology in UMLS 2003AB is as specific to molecular biology as GO. With the inclusion of GO, the UMLS spans all the various knowledge representation levels of structural and functional concepts in biomedicine – as previously conceptualized by Blois – from molecular information to clinical and disease terms [10].

In the process of integrating GO into the UMLS, many GO terms became new concepts. In many instances, however, related concepts already existed in the Metathesaurus, to which these new concepts were not necessarily linked by explicit relationships. When appropriate, we mapped the new GO concepts to their existing relatives. We also used a table, as supplied by one of the authors, of gene products in the GO databases that were directly represented in the UMLS as an alternate target for disease relationships. Our mappings and inferences were based on the April 2003 distribution of GO [1] and its associated databases. The GO databases include genes and gene products and their associated GO terms, which are represented in a number of important biological databases. We searched four model organism databases (fly, yeast, worm, mouse) and used SwissProt-TREMBL for human genes and products.

---

[2] http://umlsks.nlm.nih.gov

*OMIM Database.* The Online Mendelian Inheritance in Man database (OMIM) was used as the gold standard for our validation steps. OMIM contains over 14,000 detailed free-text entries about human genes and genetic disorders [4]. Additionally, OMIM provides *morbidmap,* which gives "the chromosomal location, gene symbol, method(s) of mapping, and disorder(s) related to the specific gene [4]," as well as specific mutations of identified genes. In this study, we used the April 2003 versions of OMIM and its *morbidmap.*

### 2.2. *General Steps of the GenesTrace Method*

The proposed method, **GenesTrace,** reveals relationships (*traces*) between a disease and a gene according to the following three-step process: (see *Figure 1*).

1. **Identify a Disease** that exists in the UMLS as a concept;

2. **Determine Relationships** between a UMLS disease and other UMLS concepts, such as those in biological terminologies such as GO, using both the symbolic relationships (hierarchical and associative) and the statistical relationships (co-occurrence information);

3. **Identify Putative Genes** that use these related concepts and then use the terminology to determine the list of putative genes through links to valuable knowledge contained in biological databases (e.g., FlyBase, WormBase, MGI, etc.).

### 2.3. *Inferring Gene-to-Disease Relationships*

To take advantage of the knowledge resources presented by the UMLS, GO, and GO's associated databases (hereafter referred to as "GODB"), we developed a series of methods that related clinical concepts of disease, represented in the UMLS, to gene products represented in GODB.

The first step in our analysis was to select the set of source concepts in the UMLS. We were able to take advantage of a previously described method that transforms UMLS's full graph organization in a directed acyclic graph to obtain all entries that were descendants of the concept "disease" (C0012634), which led to a data set of approximately 200,000 concepts [11]. For each "disease" concept in this list, we then obtained a set of related concepts, using two knowledge source tables from the UMLS: MRREL and MRCOC. MRREL consists of semantically related concepts; MRCOC contains co-occurences between concepts in MEDLINE. We set a minimum threshold of five co-occurrences in MEDLINE as represented in MRCOC to be considered as a significant entry.

Next, we obtained the subset of the related concepts that were represented in GO. We used two sources: the GO terms already represented officially as concepts in the UMLS and the set of experimental mappings of gene products to

UMLS concepts. In order to provide the most inclusive set of results, we used all relevant concepts for GO terms, since there was some overlap among the new, GO-specific concepts and previous concepts (i.e. "acetylcholinesterase", C0001044 and "acetylcholinesterase activity" [from GO], C1149827). Once a merged set of relevant GO terms was established, we systematically obtained the gene products that were associated with each GO term from the GO databases. This was done using SQL queries based on the Perl GO Application Programming Interface. Valid gene-to-disease relationships were then inferred by extracting associations supported by the highest levels of evidence in GO, *IDA* (Inferred from Direct Assay), and *TAS* (Traceable Author Statement).

From the results of each disease-specific query, we created a database wherein each row contained a disease concept, the concept related to the disease and to a GO term, the GO term and accession number represented by the related concept, and additional descriptive information for each gene product (i.e., gene name, gene unique identifier, source database). We also noted the source of the relationship, which was either statistical (co-occurrence), semantic, or both. For semantic relationships, we noted the type(s) of relationship represented (i.e., parent, sibling, etc.)

A second method consisted in finding concepts that were gene products related to the "disease" concepts. For the gene products that were directly represented in the UMLS, according to the experimental mapping mentioned above, we incorporated similar relevant information (e.g., the disease concept, the related concept, the GO product represented by the related concept, gene name, unique identifier, and source database).

## 2.4. *Evaluation*

We created a sub-table in our database consisting of the concepts corresponding to OMIM diseases represented in the UMLS. These concepts were obtained using string matching (both exact and normalized) with semantic checking, between entries in *"omim.txt"* and the set of UMLS concepts.

*Gold Standard.* Next, we compared – using lexico-semantic mapping techniques – the set of genes listed in each disease's *OMIM morbidmap* entry to the set of corresponding genes and gene products that had been associated with the disease in our database. We then tabulated the number of genes associated with each disease. True Positives (TP) were defined as instances where the gene was associated with a concept it is related to in *OMIM's morbidmap*. False Positives (FP) were defined as instances where the gene was associated with any other concept. False Negatives (FN) were instances where genes that were known to be associated with a disease were not retrieved using GenesTrace. The precision and recall of our system was measured as TP/(TP+FP), and TP/(TP+FN), respectively.

## 3. Results

### 3.1. *Sample Trace*

The genes trace for UMLS concept C0002395, "Alzheimer's Disease", is presented here as an illustrative example of a GenesTrace and its efficacy using OMIM's *morbidmap*. From the UMLS, we retrieved 128 distinct concepts related to Alzheimer's Disease using MRREL. An additional 993 concepts were retrieved from MRCOC. The relationships from MRREL were divided among seven semantic classes. Mapping these concepts to the GO database led to 102 distinct GO terms: 62 were found in the molecular function axis, 25 in the biological process axis, and 16 in the cellular component access. From GO, GenesTrace found 102 associated GO terms annotating 10,774 distinct molecular products. For this specific trace, we noted that all of the associations were the results of relationships in MRCOC. Of the 12 genes associated with Alzheimer's disease found in *OMIM's morbidmap*, GenesTrace returned 3; however, only 6 of the 12 genes from *OMIM's morbidmap* were represented in the GODB. 242 other genes were also associated with this source concept. Thus we assessed the number of TP to be 3; FN to be 3, and FP to be 242, giving a precision of 1.2%, and a recall (sensitivity) of 50% Of note, this is close to the lowest value of precision in our range of results.

### 3.2. *Validation*

Out of the 200,000 disease concepts in the UMLS Metathesaurus, 1,407 were associated with at least one associated GO term. We found at least one gene in the database related to 142 of the 1,407 disease concepts. Globally, we retrieved 124 distinct genes in the context of being related to their specific disease concept, and 290 distinct genes erroneously associated with concepts, for a precision of 30% and recall of 8.8%. Overall, there were an average of 3.1 gene products per disease concept in this dataset (range 1-30; 89% had 1-3 genes). Of note, only 978 of the genes in OMIM's morbidmap existed in the GO databases. For specific diseases, our system had a wide range of precision and recall, from 100% each (1 TP, no FP's or FN's) for several diseases (Multiple Endocrine Neoplasia type 1, Neurofibromatosis type 2), to a precision of 1% with a recall of 100% (Ovarian Tumors), to a precision of 100% with a recall of 50% (Hurler Syndrome, Familial Hypobetalipoproteinemia). *Figure 2* shows the distribution

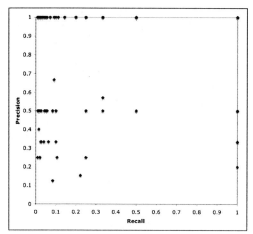

**Figure 2:** *Distribution of Precision and Recall Values.* All values, including overlapping points, shown for retrieving pertinent genes for diseases contained in OMIM.

of the precision and recall for all the diseases examined. Interestingly, almost 30% of diseases have a precision and recall of 100%. Overall, average precision for diseases was 51%, with an average recall of 79%. Two cases are illustrative of the complexity of the results. The results for Alzheimer's Disease, mentioned above, are typical of results for complex diseases. Of note, among the gene products were *ACE, ACE1,* and *PSEN2,* all of which were related to Alzheimer's Disease in OMIM's *morbidmap*. On the other end of the spectrum, for concept C0027832, "Neurofibromatosis 2", we retrieved only 2 gene products, *GOT1* and *NF2*; however, as *NF2* is the only related gene in *morbidmap* this led to a precision and recall of 100%.

### 3.3. GO-UMLS-OMIM Specific Results

The results of our gene-disease mappings also revealed some interesting properties of the relationship between GO and the UMLS. As shown in *Tables 1 to 4*, GenesTrace found different types of relationships between either GO terms and UMLS diseases. While few GO terms are associated with 100 diseases or more, 75% of them are linked to 8 diseases or less (*Table 1*). Similarly, 60% of the diseases are associated with only one GO term, but 116 diseases are linked to 10 GO terms or more (*Table 2*). Similar findings were observed between gene products and diseases. While 23 gene products are associated with 200 diseases or more, 75% of them are linked to 25 diseases or less (*Table 3*). Similarly, 80 diseases are associated with more than 1000 gene products, but most diseases are linked to at most 100 gene products (*Table 4*).

**Table 1:** Percentage of GO Terms associated with individual diseases.

| GO Terms (%) | Disease(s) |
|---|---|
| 1-25 | 1 |
| 26-50 | 1-3 |
| 51-75 | 3-8 |
| 76-100 | 8-132 |

**Table 2:** Percentage of individual diseases associated with individual GO terms

| Diseases (%) | GO Term(s) |
|---|---|
| 1-25 | 1 |
| 26-50 | 1 |
| 51-75 | 1-2 |
| 76-100 | 2-102 |

**Table 3:** Percentage of Gene Products associated with individual diseases.

| Gene Products (%) | Disease(s) |
|---|---|
| 1-25 | 1-3 |
| 26-50 | 3-13 |
| 51-75 | 13-25 |
| 76-100 | 25-331 |

**Table 4:** Percentage of individual diseases associated with gene products.

| Diseases (%) | Gene Product(s) |
|---|---|
| 1-25 | 1-4 |
| 26-50 | 4-15 |
| 51-75 | 15-70 |
| 76-100 | 70-10,774 |

## 4. Discussion

Our methods expand on those of other authors in several ways. Perhaps most importantly, we bypass the ambiguity involved in extracting gene names or symbols directly from MEDLINE articles by using pre-defined concepts in the UMLS and GO. Additionally, we use more robust statistical (co-occurrence) data than direct extraction from MEDLINE. Our methods use co-occurrence data that it is calculated on the conceptual, rather than textual, level. In addition to imposing a minimum threshold on the frequency of co-occurrence, the quality of the associations selected is ensured in part by the fact that co-occurrences recorded in the Metathesaurus are limited to the "starred" (major) MeSH descriptors in MEDLINE. Since we also use manually-curated semantic relationships extracted from the UMLS, the characteristics of the relationships we use are also richer than pure co-occurrence data. GenesTrace leverages the above annotation methods and their results, in order to exploit the knowledge contained in GO, the GO annotation databases, and the UMLS. In doing so, the purpose of GenesTrace can be seen in parallel to the annotation projects mentioned above; instead of annotating sequences, however, GenesTrace provides automated methods of annotating diseases.

Inherently, using the UMLS and GO as knowledge sources produces a very large search space. Of the approximately 200,000 initial concepts from the UMLS, for example, 22,040 (11%) had at least one corresponding entry in the GO database. The entire GenesTrace database was comprised of approximately 3 million rows, with 1,326 distinct GO terms and 16,984 distinct gene products, with an average of 129 products for each disease entry. Theoretically, there would be a maximum of approximately 21.5 million gene-disease combinations in the OMIM sub-table of the database, considering the 15,294 distinct gene products and 1,407 distinct diseases represented there. Our methods significantly reduce the combinatorial space for these gene-disease relationships, (i.e., 688,126 rows in the OMIM sub-table) however, by more than an order of magnitude, and therefore provide a more efficient starting point for high-throughput analysis of the complicated genetic interactions underlying complex diseases.

Our results not only reveal the potential utility for the GenesTrace system, but also point to the limitations of the process. The success of the traces is heavily based on the quality and accuracy of annotation for the corresponding gene products [7]. Indeed, the precision and recall values reported in this study may be affected by varying annotations for homologous genes across multiple organisms. Perhaps the most significant current limitation, however, is the need for filtering of results for complex diseases, such as the almost 11,000 gene products retrieved for Alzheimer's Disease. Investigating the most commonly occurring gene products within the set is the most basic approach, but due to the likely low signal-to-noise ratio (e.g., 3 of the top 4 products for the Alzheimer disease query (Aryl Hydrocarbon Receptor; Uracil-DNA Glycosylase; and E2F-1 transcription factor) are involved in neuronal development or apoptosis, but are not specifically implicated in Alzheimer's disease), other methods would need to be applied. Further complicating methods such as ours is the lack of uniformity in gene names or gene product representation, as evidenced by the relatively small number of genes in *OMIM's morbidmap* also present in the GO databases.

Several factors also explain our relatively wide-ranging values for precision and recall. As mentioned above, one of the major reasons may be, the lack of uniform naming of genes. This was particularly evident in mappings between *morbidmap* and the GODB. For example, *morbidmap* has entries for both *VRNF* and *NF1*, which are the same gene, while only *NF1* is in the GODB. Additionally, even with the sub-table of OMIM diseases, the large search space of genes and gene products makes false positives much more likely. Due to the limitations of gene annotations, another possibility is that a set of our FP's are not easily verified FP's, but instead have "buried or undiscovered" relationships to the diseases. For example, of the 242 genes returned as FP's for Alzheimer's disease, 97 return at least one entry in a PubMed search for "[*gene*] AND dementia", where '[*gene*]' is the query gene symbol, and thus may represent examples of previously discovered, yet buried, knowledge. In this case, buried would indicate that the relationship is difficult to retrieve in high throughput, as it is not explicitly annotated in either OMIM or PubMed. However, besides these examples, other inferred relationships may be undiscovered altogether. Similarly, of the 129 FP's for hepatocellular carcinoma, 69 return at least one article (buried knowledge) for a search on "[*gene*] AND hepatoma", where '[*gene*]' is the query gene symbol.

This study has several limitations. We did not exploit the graph structure of the knowledge sources we used. Nor did we contrast putative results as "buried" (if found in PubMed) and "undiscovered" (if not found in PubMed). Appreciation of the graph structure would have enabled us to modify our results based on recursively navigating up or down levels of MRREL, MRCOC, MeSH or GO. We intend to investigate the recursive use of the relationships to expand the database. An additional relevant study that we have initiated is to use statistical comparisons to compare groups of genes mapped to diseases. For instance, housekeeping genes are likely to be associated non-specifically with a large number of diseases. The top five products in terms of frequency, for example, consisted of two stress response genes (*SKN7, SKI1*), two transcription regulators (*HNT1, E2F1*) and one metabolic regulator (*IATP*). Future work will include the usage of established information retrieval weighting techniques to stratify results such as term frequency * inverse document frequency (TF*IDF).

The strengths of the GenesTrace system stems from its ability to combine several sources of manually curated information into a high-throughput system for gene discovery. Additionally, very few systems provide genome-wide approaches to phenotypic discovery, and tools, such as GenesTrace, are foundational steps to comprehend the phenome as they provide an incremental discovery path using established knowledge bases. Furthermore, GenesTrace performs searches based both on semantic and statistical information, and has the ability to exploit the ontological properties of the source materials. The system is also easy to update, as the source materials are updated either quarterly or monthly. Finally, though the system explores direct, clinically diverse gene-disease relationships, providing a high-level view, it is also expandable to virtually any source annotated by GO, down to the sequence level, as well as any source incorporated into the UMLS, such as SNOMED CT.

## 5. Conclusion

The GenesTrace methodology presented here is a proof-of-concept study that mines gene-disease relationships across biomedical databases. By design, the prototypal method was unfiltered. Consequently, while the results of this study provide sufficient accuracy that attests to the validity of the GenesTrace principle, it yet remains inadequate for active use by researchers. However, as the work progresses in planned complementary studies, we expect the methods presented here to have potential for significant accuracy improvements, which may yield a powerful tool for research and discovery. Specific future studies will include machine learning or filtering metrics such as the frequency of co-occurrences of the intermediating knowledge, and convergence of distinct gene-disease traces ("knowledge pathways").

Understanding the genetics behind complex diseases is one of the principal goals of genomics research. Many complex genetic phenomena are encoded in biological knowledge bases. Similarly, many clinical manifestations of disease are represented in clinical knowledge bases. Multiple levels of both explicit and implicit pathophysiologic and phenomic knowledge may be buried in mappings across mappings between clinical and biological knowledge bases. By examining these mappings, one can envisage automated systems, like GenesTrace, that may help elucidate testable genetic hypotheses by tracing the links between these clinical and biological knowledge bases.

## Acknowledgements
These studies were supported in part by the Department of Medicine, BIMC and the following grants: NIH/NLM 1K22 LM008308-01, NIH/NIAID 5U54 AI057158-02, and NIH/NLM LM-07079-09. The authors thank Inderpal Kohli and Jianrong Li for the development of the GenesTrace interface and database system, respectively.

## References
1. Ashburner, M., et al., *Gene Ontology: tool for the unification of biology.* Nature Genetics, 2000. **25**: p. 25-9.
2. Bodenreider, O. *The Unified Medical Language System (UMLS): integrating biomedical terminology.* Nucleic Acids Res 2004;**32**, D267-70.
3. Jenssen, T.K., et al., *A literature network of human genes for high throughput analysis of gene expression.* Nature Genetics, 2001. **28**:p. 21-28.
4. Hamosh, A., et al., *Online mendelian inheritance in man.* Human Mutation, 2000. **15**: p. 57-61.
5. Perez-Iratxeta, et al., *Association of genes to genetically inherited diseases using data mining.* Nature Genetics, 2002. **31**: p. 316-9.
6. Hu, Y., et al., *Analysis of genetic and proteomic data using advanced literature mining.* Journal of Proteome Research, 2003. **2**: p. 405-12.
7. Camon, E., et al., *The Gene Ontology Annotation [GOA] Project: Implementation of GO in SWISS-PROT, TrEMBL, and InterPro.* Genome Research, 2003. **13**: p. 662-72.
8. Raychaudhuri, S., et al., *Associating genes with Gene Ontology codes using a maximum entropy analysis of biomedical literature.* Genome Research, 2002. **12**: p. 203-14.
9. Perez, A.J., et al., *Gene annotation from scientific literature using mappings between keyword systems.* Bioinformatics, 2004. (April 1) [epub].
10. Blois, M.S., *Information holds medicine together.* MD Computing, 1987. **4**: p. 42-6.
11. Bodenreider, O., *Circular hierarchical relationships in the UMLS: etiology, diagnosis, treatment, complications, and prevention.* Proc AMIA Symp, 2001: p. 57-61.

# INTEGRATING GENOMIC KNOWLEDGE SOURCES THROUGH AN ANATOMY ONTOLOGY

JOHN H. GENNARI, ADAM SILBERFEIN, & JESSE C. WILEY

*Biomedical & Health Informatics, The Information School, & Comparative Medicine, University of Washington, Seattle, WA, 98195, USA*

Modern genomic research has access to a plethora of knowledge sources. Often, it is imperative that researchers combine and integrate knowledge from multiple perspectives. Although some technology exists for connecting data and knowledge bases, these methods are only just beginning to be successfully applied to research in modern cell biology. In this paper, we argue that one way to integrate multiple knowledge sources is through anatomy—both generic cellular anatomy, as well as anatomic knowledge about the tissues and organs that may be studied via microarray gene expression experiments. We present two examples where we have combined a large ontology of human anatomy (the FMA) with other genomic knowledge sources: the gene ontology (GO) and the mouse genomic databases (MGD) of the Jackson Labs. These two initial examples of knowledge integration provide a proof of concept that anatomy can act as a hub through which we can usefully combine a variety of genomic knowledge and data.

## 1    The Problem: Overwhelming, Distributed Genomic Knowledge

Modern biology researchers are hampered by the need to integrate information from rapidly developing and diverse knowledge sources. As a general problem, researchers in computer science and informatics have developed methods for combining and integrating data and knowledge bases. However, these methods are only beginning to be applied to practical problems in biology research. In this paper, we present an example of knowledge integration where we use an anatomy ontology as a hub through which we have connected data sources for two disparate views of cell biology.

One can view modern cell biology research as having two branches. In one core branch of research, molecular biologists and biochemists use model systems such as cell culture and yeast based assays to examine general principles of cell biology. These researchers describe the structure and function of the abstract cell, irrespective of how that cell participates in larger systems. Alternatively, researchers also study anatomy-specific aspects of biology, such as developmental biology and disease pathology. Necessarily, the latter approach is tissue-specific; however, their work must also understand and be consistent with the more generic approach to cell biology. Specifically, if the generic model of the cell specifies all possible genetic interactions and functions, then a tissue specific model must account only for those genes that are expressed in the tissue.

An ideal informatics solution would allow researchers who focus on either approach to see and understand results from both these views. Since anatomy underlies all of biological research, we propose an anatomy-based platform for integrating data sources. The long-term goal of this platform is to allow researchers to associate genes with cellular function from both viewpoints—both at the level of an 'abstract cell' and in tissue-specific fashion.

Our motivation for this work is linked to the goals and informatics needs of a current genomic research effort. One of the co-authors (JCW) is also a member of the Comparative Mouse Genomics Centers Consortium (CMGCC). The primary goal of the CMGCC is to identify, and produce genetic mouse models for, human genetic variants of genes believed to be 'environmentally sensitive' and linked to human pathological conditions (www.niehs.nih.gov/cmgcc/). The goal is to use the mouse models developed to explore genetically conferred diseases prevalence. The mouse consortium is building genetic mouse models of low frequency variants of genes involved in cell cycle control and DNA repair mechanisms—two target biological processes widely believed to be of significance in determining disease prevalence. In this research context, linking together functional genetic information with anatomy is of particular significance. Although a given mouse model may be developed with a specific pathological condition in mind, the genetic manipulation will affect biological processes distributed across the entire organism and may have different manifestations across different tissue types.

In this paper, we provide two proof-of-concept demonstrations where anatomy is the central hub for both tissue-specific and for abstract cellular genetic data. For anatomic concepts, we use the Foundational Model of Anatomy (FMA), a comprehensive, well-structured ontology of anatomy [1,2]. In our first demonstration, we show how the FMA can support the investigation of the abstract cell, by connecting the FMA ontology of cellular anatomy (the structure and sub-components of the cell) with the Gene Ontology (GO) hierarchy of cell component terms [3]. Given this connection, our tool can browse GO-annotated databases from within the FMA viewer.

As a second demonstration, we have also connected the FMA to a database of tissue-specific gene expression results. Specifically, we used the Mouse Genomic Database from Jackson Labs that includes anatomic information about gene expression results in the mouse [4]. To scope our effort, we looked at tissues of the brain, using the FMA organization and description of brain regions and components. In Section 3, we provide details and screenshots from both of these demonstration projects. However, before giving these details, we first describe the relevant bioinformatics resources and efforts in knowledge sharing.

## 2 Standards and Knowledge Bases for Bioinformatics Sharing

Our work builds on a number of important bioinformatics resources. In general, there are many groups working to standardize bioinformatics knowledge and data. For example, the Microarray Gene Expression Data Society aims to facilitate the sharing of microarray data through the development of standards for experimental design descriptions and data descriptions (www.mged.org/). Our work is designed to include other bioinformatics resources as they become available.

### 2.1 Gene Annotation and The Gene Ontology

The Gene Ontology (GO) is a structured, controlled vocabulary to allow molecular biologist to better share data and knowledge about the roles of gene products [3]. The usual use of the GO is to provide researchers with a standard language for annotating a gene—providing information about a gene product, such as its molecular function. It is organized into three hierarchies: (1) a set of terms for the molecular functions of a gene product, (2) terms that describe the larger-scale biological processes that may involve the gene product, and (3) terms that include the cellular components that may be important or relevant for the gene product. The GO views all cell types equally, and does not include any tissue-specific information. The GO is developed collaboratively and is an evolving vocabulary, with new versions released monthly.

An important aspect of the GO is that a set of databases exist that contain information about particular gene products, in particular research species, that have been studied and annotated with the GO controlled vocabulary. For example, the GO was developed by the groups developing *Flybase*, the *Saccharomyces* Genome Database, and the Mouse Genome Database (MGD), so the genes in all of these sources are annotated with appropriate GO terms. Currently the GO web pages list more than 30 such annotated databases contributed by about 15 groups worldwide. GO provides a uniform search capability: researchers can use GO terms to retrieve related gene products across these multiple databases.

Some researchers have pointed out that the GO has some problems and weaknesses, if viewed as a formal ontology rather than a controlled vocabulary [5,6]. However, in our current work, we focus on the GO as a simple entry point into the annotation databases such as the MGD. Our assumption is that the GO will improve over time, allowing for better, higher quality inferences and search capabilities.

### 2.2 Standards and Ontologies for Anatomy

For our work, we are interested in efforts to standardize gene expression annotations about anatomy—information about the source of a tissue sample. One group that is focused on anatomy for gene expression is the Standards and Ontologies for Functional Genomics (SOFG) group (www.sofg.org/). This group recognizes a problem faced by a genetic researcher working with animal models. One would like to annotate results with anatomic information that is consistent across both the animal model of disease (e.g., mouse) and human, the eventual target for therapies. Even within a single species, there is variability and inconsistency in anatomic labeling. As with other forms of meta-data, the tendency is for scientists to use informal, natural language terms, rather than terms from a controlled vocabulary or ontology.

As steps toward ameliorating this situation, the SOFG group has identified a number of on-line resources for anatomy ontologies, and developed a short "SOFG anatomy entry list" (SAEL). This list of about 100 anatomic terms represents the most commonly used anatomic terms for annotating gene expression results. The resource list includes anatomy ontologies developed by groups devoted to functional genomics in particular species (e.g., an anatomy ontology for the mouse), as well as long-standing efforts to describe human anatomy—either for medicine and pathology (OpenGalen, www.opengalen.org/), or in purely structural terms, by the Foundational Model of Anatomy (see below). The expectation is that each of these anatomy resources would match or map their anatomic terms to the entry list, and then this entry list could be used as a way to align or link together different ontologies of anatomy [7]. As we show in Section 3, we share this aim of aligning ontologies via anatomy.

## 2.3 *The Foundational Model of Anatomy(FMA)*

The Foundational Model of Anatomy is a long-standing project to describe all of human anatomy as a symbolic ontology of concepts and relationships [1,2]. The FMA was not designed especially for the genomics domain, nor indeed for any particular biomedical viewpoint. Rather, it was designed with the idea that if one could represent in a principled, formal manner the truth about biological structure, then the resulting knowledge base could become a reference ontology for all of biomedical informatics [1]. The FMA is already fairly comprehensive (with more than 70,000 concepts), and it continues to grow and evolve.[1]

In contrast to anatomy ontologies that are designed specifically for annotating tissue-specific genetic data, the FMA is designed for and by anatomists, and this perspective has both advantages and problems. On the one hand, the FMA represents a

---

[1] The FMA is related to, but different from the UW Digital Anatomist (UWDA) terminology of the UMLS. The UWDA was the predecessor of the FMA, and within the UMLS this terminology was kept in sync with the FMA only through the Summer of 2002. Since then, the FMA team has added thousands of new classes, including many of the sub-cellular anatomy terms that are our focus.

much more comprehensive view of anatomy—both broader in scope and with greater detail than is necessary for current research in cell biology. However, this wealth of detail and information can overwhelm and obscure a simpler view of anatomy—as exemplified by the SAEL list of only 100 terms that covers ~80% of current anatomic annotations [7]. Additionally, the FMA contains no explicit information about the function or physiology of anatomy. Therefore, it also contains no information about dysfunction, or pathology, which is often a focus for the genomic researcher.

The FMA uses the Protégé environment for knowledge-based systems to store, retrieve, and manage its anatomic concepts and relationships [8]. Protégé is an open-source, extensible environment with a large user base. These characteristics make it easy to adapt the system for special needs. As we present in the next section, we were able to modify the Protégé user interface to connect directly to specific bioinformatics databases.

## 3 Results: an Anatomy Ontology as an Information Portal

Our claim is that a comprehensive anatomy ontology such as the FMA can be fruit-fully used as a hub to integrate a variety of genomic and cell-biology knowledge sources. In this section, we present two examples: (1) We connected the FMA concepts of cellular anatomy to the GO annotation databases by linking to the GO terms in its cellular components tree, and (2) We connected gene expression results from MGD to the concepts in the FMA that describe brain anatomy. Figure 1 shows our anatomy ontology as a hub with tissue-specific knowledge on the left, and generic GO knowledge on the right. As we show, the user interacts with a single Protégé user interface that integrates the multiple knowledge sources into a single view.

For both systems, we leveraged the extensibility of the Protégé environment. For each linking system, we built a Java plug-in component that modified the behavior of the default user interface. In response to user browsing actions, our plug-ins access the relevant mapping tables, and then use this information to construct JDBC calls to locally stored copies of the relevant DBs. Our plug-ins then display the data returned from these calls directly within the FMA viewer.

### 3.1 Mapping Cellular Anatomy: the FMA and GO

As we mentioned earlier, the GO describes function, process, and cell structure independent of anatomy, tissue, or cell type. In contrast, the FMA looks at anatomy structurally, independent of function or physiology. Thus, the only way to directly link the FMA and GO knowledge sources is through cell structure.

Of course, the anatomic names and concept organization for cellular structure in the FMA do not always match the terms and organization used in GO's cell compo-

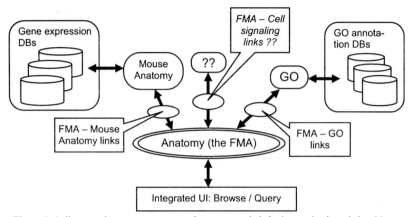

**Figure 1.** A diagram where our anatomy ontology acts as a hub for integrating knowledge. Mappings to GO and to mouse anatomy are described in Section 3.1 and 3.2; mappings to cell-signaling knowledge sources are future work.

nent hierarchy. If our goal is to view the GO knowledge sources from within the broad FMA anatomy ontology, we must connect terms in the FMA to terms in GO. As an initial proof-of-concept, we hand-built a database table that connects about 150 terms in the FMA's ontology of cell parts to the corresponding terms in the GO. With this table, we then implemented a viewer within Protégé that allows direct access to the GO annotation databases from within the FMA Protégé viewer. Figure 2 shows an example of this connection. Within the FMA, we have browsed to the concept "wall of lysosome", and we can then see both the corresponding GO term ("lysosomal membrane") and information from the GO databases (Flybase & MGI, in this case) that show information about the gene products marked as associated with the lysosomal membrane. Currently, our interface also provides a simple table with the details about each of these GO annotations. (Users can view this additional information by selecting a particular GO annotation and hitting the "V" button.)

By itself, this system provides a new view onto the GO databases—it organizes the information according to the FMA's formal definitions of cellular anatomy. However, the real strength of our approach is that this viewer can be combined with other anatomy-centric viewers of genetic information, as we describe below.

### 3.2 Mapping Gene Expression Data: MGD, FMA, & GO

In contrast to studying cellular behavior across all cell types, one may want to study a specialized set of cells—ones associated with a particular organ or tissue type. Gene expression data is one way of understanding which genes are active in which anat-

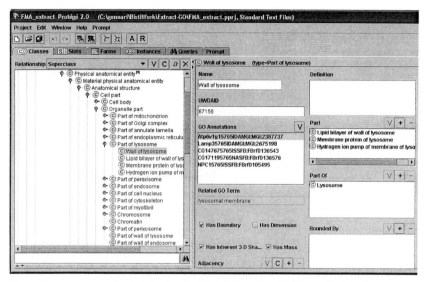

**Figure 2.** A screen from Protégé showing the FMA information about the wall of the lysosome and highlighting the gene products that are annotated with the GO term "lysosomal membrane". In this case, information from MGI and Flybase are displayed. Additional details about each annotated object can be retrieved via the "V" (view) button.

omic parts. Therefore, scientists need an understanding of anatomy to effectively use and organize this data. To date, there is not a standard source for anatomic knowledge for annotation of gene expression results—as we described earlier, this is the concern and work of the SOFG group. As a concrete example, we have connected the FMA with one source of gene expression data for the mouse: the Mouse Genome Database (MGD) [4]. As with the GO—FMA mappings described in section 4.1, we created a table connecting anatomy terms as used by the MGD to concepts defined in the FMA.[2] To scope our work, we choose to focus on the Brain regions. (Fortunately, there are few anatomic differences between mouse brain regions and human ones.) With our mapping tables, we can connect FMA brain regions with data in the MGD that lists relevant gene expression results for that region. Figure 3 shows an example interface, with the hippocampus selected, and indicating that one can retrieve 615 gene expression results for that region.

The connection between the FMA, the GO, and the MGD data provides interesting capabilities for the cell biology researcher. For example, the ability to view gene

---

[2] We understand (personal communication from M. Ringwald) that the MGD is in the process of updating/changing the anatomic terms used to be more consistent with the SOFG efforts

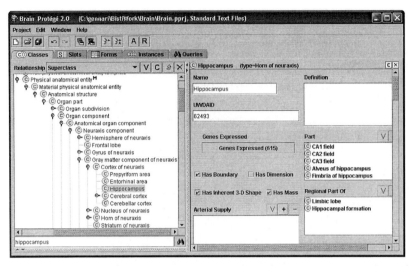

**Figure 3**. A screen showing FMA brain regions and the link to MGI data about gene expression results for that region. Selecting the "Genes Expressed" button produces a table of 615 results (in this case), and for each of these, its associated GO terms. As a next step, for one of these genes, one can select a specific GO term to further explore, as shown in Fig 4.

expression data by both GO term and tissue type will be of utility for the CMGCC. As we described in the introduction, this consortium studies specific pathologies via animal models, whose genetic alteration may extend beyond the intended target pathology. As an example of how genetic modifications can have different effects in different tissues, consider one target of interest for the CMGCC: the cell cycle regulatory gene cyclin D1. It is well known that cyclin D1 expression is related to colorectal cancer [9], centrocytic lymphomas [10], and mammary adenocarcinoma development [11]. However, cyclin D1 expression has also been linked with the induction of apoptosis within the post-mitotic neurons in the brain [12]. Consequently, research groups focused on cancer biology would not necessarily notice the impact of genetic manipulations of cyclin D1 in the brain as the anticipated phenotype would be the exact opposite observed in other tissues—namely, cyclin D1 is associated with cell proliferation in the context of cancer research and neuronal degeneration and death within the brain. Thus, to fully explore the impact of human genetic variants within mouse models, it is essential to possess information about both the tissue specificity of expression and the biological process within which that gene is involved.

From the interface in Figure 3, one can retrieve a table (not shown) of all the gene expression results for the selected region. If the user chooses to look at cyclin

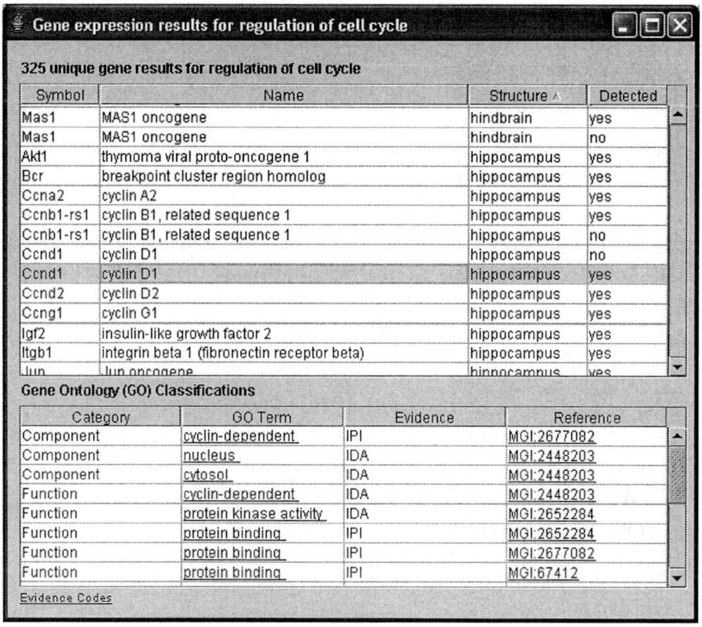

**Figure 4** A screen shot showing the set of genes that (1) have results associated with the hippocampus, and (2) are annotated with the GO process term "regulation of cell cycle" . The lower half of the Figure shows the full set of GO terms associated with cyclin D1.

D1, then one also sees all the GO annotations for that gene. Since the consortium is interested in the GO process labeled "regulation of cell-cycle", we choose that GO term to further browse. Figure 4 then shows all genes expressed within the hippocampus that are also involved in that particular GO process. One can then continue to iteratively explore the GO annotations of Cyclin D1 by selecting other terms (see bottom half of Figure 4), and then seeing the other expressed genes organized by brain structure.

## 4   Related Work in Knowledge Base & Data Base Integration

The systems we have described are small prototypes—they are limited by the size of the mapping tables we built between the FMA concepts and the other knowledge sources. However, the general task of linking or integrating knowledge bases and databases has been well-studied in computer science. The challenge we face in linking knowledge sources into a central anatomy hub is a specific example of the more general task of knowledge integration in bioinformatics. For example, the Bio-

Mediator project has developed methods for answering queries about genetic tests across multiple sources [13,14]. Their approach does not rely on static tables of maps between terminologies, and instead builds more dynamic rules that describe how terms match across sources.

More generally still, in knowledge-base research, there are methods for semi-automatically determining the set of matching concepts across two ontologies. For example, within the Protégé environment, there is a plug-in tool known as Prompt, that allows users to link any two related ontologies [15]. Outside of Protégé, researchers have used similar methods to align a small portion of the FMA with the Open-Galen medical and anatomic terminology [16]. Eventually, we may want to augment our mapping work by using these sorts of methods and tools. However, compared to anatomy as a whole, cellular anatomy may not need that large of an ontology—e.g., the entire GO cell structure tree consists of less than 1300 terms. These more automated approaches will be important in the longer term, as the number and size of new genomic knowledge sources increases.

## 5    Future Work and Discussion

We have argued that an anatomy ontology (the FMA) can function as a unifying hub for integrating knowledge spanning from subcellular components to macro-anatomical features. We recognize that our demonstration of two examples does not suffice to show the broad applicability of our ideas. However, the ability to browse clustered data by GO terms and tissue expression distribution (shown in Figure 4), rather than hunting and searching gene by gene, is of immediate utility to at least the CMGCC community. We also acknowledge that we have only demonstrated our approach with the subset of data for brain regions. With other anatomic concepts there may be more significant differences across species. Future work will look at a more complete mapping across species, and as we described in the previous section, we may leverage tools like Prompt for helping to create these mappings.

We plan to connect other sources into our anatomy hub. For example, we know of other work that provides tissue-specific gene expression data. We may be able to link knowledge from the Gene Expression Atlas (symatlas.gnf.org/SymAtlas/), or to information reported about tissue-specific mouse transcriptions on RIKEN cDNA microarrays [17]. Of course, crossing species boundaries to the mouse raises a separate set of issues, especially when the anatomy is significantly different [18].

In the longer term, we believe that an important resource to link to our anatomy hub is knowledge about cell-signaling pathways (see middle spoke of Figure 1). The distinction we made between general-principle driven cellular biology research and tissue-specific biology research, has the same significance in the context of modeling cellular signaling. That is, signaling pathways should be understood both from the view of the abstract cell, and in a tissue-specific manner, accounting for genetic conditions present in specific cell types. Thus, our next research step is to combine anatomy (as in the FMA) with a comprehensive ontology for cell signaling. We have already begun work designing a cell-signaling ontology and are examining related work in this area, such as that by the BioPax group (www.biopax.org). The combination of anatomy with signaling knowledge should provide significant benefits to the biology researcher who is looking for a unified view of genomic and proteomic knowledge.

The ultimate success of our work depends on several assumptions. The primary assumption is that it is possible to align diverse ontologies about cell biology, and that it will be useful to do so. The second assumption is anatomy is a useful organizing hub to connect the various sorts of functional genetic information that currently resides in distributed databases. Our results so far are an encouraging proof of principle exploration of these key assumptions. Specifically, our success in mapping parts of the GO cellular component information and the Jackson lab's mouse anatomy ontology onto the FMA suggests that commonly utilized biological structure ontologies can be linked together and viewed within an anatomy ontology.

Integrating and connecting bioinformatics knowledge resources is a long-term research task. However, given the flood of information and the need to combine data and knowledge, we believe this goal is an important one. Unfortunately, knowledge about anatomy, cellular physiology, and pathology is unlikely to be codified into a single ontology that all stake-holders can agree on. For the foreseeable future there will be different ontologies with different perspectives, and researchers must be able to appropriately combine knowledge from these different views and ontologies.

## Acknowledgments

We thank Cornelius Rosse and Jose Mejino for explanations and insights into the FMA. Partial funding for this work was provided by the BISTI planning grant (#P20 LM007714) from the National Library of Medicine.

## References

[1]     Rosse C, Mejino JLV. A Reference Ontology for Bioinformatics: The Foundational Model of Anatomy. *Journal of Biomedical Informatics* 2003; **36**:478-500.

[2]     Rosse C, Mejino JLV, Modayur BR, Jakobovits RM, Hinshaw KP, Brinkley JF. Motivation and Organizational Principles for Anatomical Knowledge Representation: The Digital Anatomist Symbolic Knowledge Base. *Journal of the American Medical Informatics Association* 1998; **5** (1):17-40.

[3]     GeneOntologyConsortium. Creating the gene ontology resource: design and implementation. *Genome Res* 2001; **11** (8):1425-1433.

[4]    Blake J, Richardson J, Bult C, Kadin J, Eppig J. MGD: The Mouse Genome Database. *Nucleic Acids Research* 2003; **31**:193-195.

[5]    Smith B, Williams J, Schulze-Kremer S. The ontology of the gene ontology. *Proceedings, AMIA Annual Symposium*, Washington, D.C., 2003. 609-613.

[6]    Wroe CJ, Stevens R, Goble CA, Ashburner M. A Methodology to Migrate the Gene Ontology to a Description Logic Environment Using DAML+OIL. *Pacific Symposium on Biocomputing*, 2003. 624-235.

[7]    Aitken S, Baldock R, Bard J, et al. SAEL -- The SOFG anatomy entry list. *ISMB 2004*, Glasgow, Scotland, 2004.

[8]    Gennari JH, Musen MA, Fergerson RW, Grosso WE, Crubezy M, Eriksson H, Noy NF, Tu SW. The evolution of Protégé: An environment for knowledge-based systems development. *International Journal of Human-Computer Studies* 2003; **58** (1):89-123.

[9]    Bartkova J, Lukas J, Strauss M, Bartek J. The PRAD-1/cyclin D1 oncogene product accumulates aberrantly in a subset of colorectal carcinomas. *Int J Cancer* 1994; **58** (4):568-573.

[10]    Lovec H, Grzeschiczek A, Kowalski MB, Moroy T. Cyclin D1/bcl-1 cooperates with myc genes in the generation of B-cell lymphoma in transgenic mice. *Embo J* 1994; **13** (15):3487-3495.

[11]    Wang TC, Cardiff RD, Zukerberg L, Lees E, Arnold A, Schmidt EV. Mammary hyperplasia and carcinoma in MMTV-cyclin D1 transgenic mice. *Nature* 1994; **369** (6482):669-671.

[12]    Kranenburg O, van der Eb AJ, Zantema A. Cyclin D1 is an essential mediator of apoptotic neuronal cell death. *Embo J* 1996; **15** (1):46-54.

[13]    Mork P, Halevy A, Tarczy-Hornoch P. A Model for Data Integration Systems of Biomedical Data Applied to Online Genetic Databases. *Proceedings of the Annual AMIA Fall Symposium*, Washington, D.C., 2001. 473-477.

[14]    Shaker R, Mork P, Barclay M, Tarczy-Hornoch P. A Rule Driven Bi-Directional Translation System for Remapping Queries and Result Sets Between a Mediated Schema and Heterogeneous Data Sources. *Proceedings, AMIA Fall Symposium*, San Antonio, TX, 2002. 692-696.

[15]    Noy NF, Musen MA. The PROMPT suite: Interactive tools for ontology merging and mapping. *Int J of Human-Computer Studies* 2003; **59** (6):983-1024.

[16]    Zhang S, Bodenreider O. Aligning representations of anatomy using lexical and structural methods. *Proceedings, AMIA Fall Symposium*, Washington, D.C., 2003. 753-757.

[17]    Bono H, Yagi K, Kasukawa T, et al. Systematic expression profiling of the mouse transcriptome using RIKEN cDNA microarrays. *Genome Res* 2003; **13** (6B):1318-1323.

[18]    Travillian RS, Rosse C, Shapiro LG. An Approach to the Anatomical Correlation of Species through the Foundational Model of Anatomy. *Proceedings, AMIA Fall Symposium*, Washington, D.C., 2003. 669-673.

# GOTREES: PREDICTING GO ASSOCIATIONS FROM PROTEIN DOMAIN COMPOSITION USING DECISION TREES

BORIS HAYETE[1,2] AND JADWIGA R. BIENKOWSKA[1,3]

[1]*Serono Reproductive Biology Institute,*
*One Technology Pl, Rockland MA 02370*
[2]*Bioinformatics Program and [3]BioMedical Engineering Department,*
*Boston University*
*36 Cummington St. Boston MA 02215*
*E-mail: jadwiga.bienkowska@serono.com*

The Gene Ontology (GO) offers a comprehensive and standardized way to describe a protein's biological role. Proteins are annotated with GO terms based on direct or indirect experimental evidence. Term assignments are also inferred from homology and literature mining. Regardless of the type of evidence used, GO assignments are manually curated or electronic. Unfortunately, manual curation cannot keep pace with the data, available from publications and various large experimental datasets. Automated literature-based annotation methods have been developed in order to speed up the annotation. However, they only apply to proteins that have been experimentally investigated or have close homologs with sufficient and consistent annotation. One of the homology-based electronic methods for GO annotation is provided by the InterPro database. The InterPro2GO/PFAM2GO associates individual protein domains with GO terms and thus can be used to annotate the less studied proteins. However, protein classification via a single functional domain demands stringency to avoid large number of false positives. This work broadens the basic approach. We model proteins via their entire functional domain content and train individual decision tree classifiers for each GO term using known protein assignments. We demonstrate that our approach is sensitive, specific and precise, as well as fairly robust to sparse data. We have found that our method is more sensitive when compared to the InterPro2GO performance and suffers only some precision decrease. In comparison to the InterPro2GO we have improved the sensitivity by 22%, 27% and 50% for Molecular Function, Biological Process and Cellular GO terms respectively.

## 1    Introduction

With genomic data available in large volume for many organisms, assigning a function to a sequence has become the new challenge for genomics. Various computational methods provide insights into properties of a novel or poorly studied protein. Relevant to this study are the homology-motivated methods that describe the protein function in terms of *functional domains.* Databases such as InterPro (Mulder, Apweiler et al. 2003), PFAM (Bateman, Birney et al. 2002) and others identify functional domains and describe the domain function. Knowledge of a protein's domain content is crucial to understanding the protein's role. However, proteins often contain multiple domains, some of which may be shared with proteins playing a different role in the cell (Fig. 1). Thus, domain identification cannot become the final point of any annotation. This issue can be addressed by mapping functional domains to Gene Ontology (Ashburner, Ball et al. 2000) terms and by using the Gene Ontology for protein annotation.

127

The goal of the Gene Ontology Consortium (Ashburner, Ball et al. 2000) is to produce a controlled biological vocabulary that can be applied to all organisms even as the knowledge of gene and protein roles in cells is accumulating and changing. GO provides three structured networks of defined terms to describe gene product attributes: biological process, cellular component and molecular function. GO is one of the controlled vocabularies of the Open Biological Ontologies (OBO.) It is one that is most advanced in its development and will likely serve as a reference for other proposed biological ontologies in the OBO family. The curators of GO annotate a gene's associations with GO terms and the evidence of the association is recorded in the GO database. Contributions to GO come from various public databases and model organism consortia such as the FlyBase (The FlyBase Consortium 2003), Saccharomyces Genome Database (SGD) (Issel-Tarver, Christie et al. 2002), WormBase (Harris, Lee et al. 2003), SwissProt (Boeckmann, Bairoch et al. 2003). Information stored in the GO database is extremely valuable as it is created in a controlled manner and provides a compilation of knowledge about genes from various organisms. The hierarchical nature of the Gene Ontology allows for an elegant representation of both knowledge and uncertainty in understanding of the biological role of a protein. For example, let us imagine a novel protein known to be a kinase whose target remains to be determined. Gene Ontology can store 'kinase activity' as the current known molecular function without requiring further specification. Thus each GO term has parents that represent a less certain annotation than the child. The Gene Ontology offers two properties essential for protein annotation: completeness and breakdown by generalization.

A number of methods have been developed to annotate gene products with GO terms electronically[1]. The electronic methods fall into the categories of:

- Text mining, such as literature mining (Chiang and Yu 2003) and (Raychaudhuri, Chang et al. 2002), and pattern of annotation mining (King, Foulger et al. 2003).

- Analysis of experimental and sequence data:

  - Sequence similarity-motivated methods such as The Institute Of Genomic Research (TIGR) annotations for *T. brucei* (El-Sayed, Ghedin et al. 2003) and *Arabidopsis* (Buell, Joardar et al. 2003; Haas, Delcher et al. 2003; Wortman, Haas et al. 2003), annotations by (Hennig, Groth et al. 2003), the BLAST-based approach of (Khan, Situ et al. 2003), and others.

  - Methods based on protein domains, such as (Schug, Diskin et al. 2002).

  - Methods using gene-expression datasets as in (Lagreid, Hvidsten et al. 2003), (Hvidsten, Komorowski et al. 2001), (Hvidsten, Laegreid et al. 2003).

  - Methods using protein-protein interaction data, such as (Letovsky and Kasif 2003).

  - Multi-source and multi-approach (integrative methods):

  - Database-driven EBI GOA (Camon, Magrane et al. 2003) and Mouse Genome Informatics annotations (Hill, Davis et al. 2001).

---

- Other multi-source methods, such as (Xie, Wasserman et al. 2002), which use text mining, domain information, sequence homology and other approaches.

Electronic annotation methods can take advantage of mappings between various existing databases and GO[2]. The following types of mappings to GO from several datasets currently exist:

- Keyword/concept: spkw2GO, which maps UniProt (Apweiler 2004) keywords; genprotec2GO, which maps GenProtEC (Riley 1998; Serres and Riley 2000) function names to GO; tigr2GO, mapping TIGR roles (Haft, Selengut et al. 2003) and others.
- Protein family: tigrfam2GO maps TIGR protein families (Haft, Selengut et al. 2003).
- Pathway: metacyc2go maps MetaCyc (Karp, Riley et al. 2002) metabolic processes and functions .
- Domain: InterPro2GO, which maps InterPro (Mulder, Apweiler et al. 2003) entries to GO; PFAM2GO, derived from InterPro2GO, which maps PFAMs, and others.

In some cases, such as in the case of InterPro2GO/PFAM2GO, the map between the PFAM domains and GO terms can be used to associate proteins with GO terms directly.

A number of methods combine various sources of information to predict the GO assignments. For example, a Bayesian method developed by (Troyanskaya, Dolinski et al. 2003) integrates multiple data sources such as protein-protein interactions and gene expression data and creates functional groupings of *S. Cerevisiae* genes. Probabilistic decision trees have been used by (Syed and Yona 2003) to predict protein function based on biochemical properties of proteins coupled with sequence database analysis. (Syed and Yona 2003) noted the possibility of using domain content for function prediction. (Hvidsten, Komorowski et al. 2001) have combined gene expression and ontology data to predict protein function using rough set theory. (Raychaudhuri, Chang et al. 2002) have implemented a text-mining algorithm for GO annotation.

There are about 100,000 genes currently annotated by the GO consortium and only a small fraction of those, about 10,000, are human and mouse genes. Most of the proteins for which there exists an experimentally validated annotation come from simple uni-cellular organisms like bacteria or yeast. Thus for many human and mammalian genes the association with the GO terms has to be predicted by comparison with simpler proteins. Given the complexity of the mammalian protein domain architectures, a comparison with the simpler proteins is not straightforward. In order to address this challenge we propose a method for inferring the GO annotation of a protein from the protein domain composition.

We chose proteins' domain content as a model, and use PFAM domain annotations as our principal data source. Representation of proteins in terms of

---

2 The complete, up-to-date listing of mappings to GO can be found at the Gene Ontology Consortium's web site at http://www.geneontology.org/GO.indices.html

their functional domain content creates a multi-dimensional attribute space, where each domain is a dimension. From here on, we will refer to a protein's domain composition as to its domain content or domain vector (in the space of all known domains). We will use these terms interchangeably. The task of GO term annotation becomes that of mapping a multi-dimensional attribute vector to a set of labels (GO terms). Broadly speaking, we restate protein annotation as a classification problem. As documented in the literature, for example, in (Krishnan and Westhead 2003), the decision tree paradigm can work well for this type of problems.

We compare two different representations of proteins by their domain composition, integer and binary one. We evaluate their performance in associating proteins and GO terms. We also compare the performance of our method to GO term association using the InterPro2GO mapping. We demonstrate that using a protein's entire domain vector over a single domain significantly enhances the sensitivity of annotation for all three GO networks at the expense of a relatively small decrease in precision.

## 2 Systems and Methods

We need to tackle several obstacles that confound the classification problem as formulated in the decision tree context:

- A protein can be associated with many GO terms. Owing to the complex structure of Gene Ontology, the number of "true" labels describing a protein is much larger than the number of GO terms. This is because, in principle, almost any subset of GO terms should be treated as a unique label. The number of such labels is combinatorially large and precludes any classification attempt. Thus GO terms must have individual classifiers.

- The space of attributes for our classification problem is quite large. Currently there are about 7,000 PFAMs identified. Generally, classifiers don't perform well in high-dimensional attribute spaces. Fortunately, most GO terms describe relatively small numbers of proteins. Thus it is feasible to construct for each GO term a training set that is represented in a subspace with much lower dimensionality.

- Our classifier needs to take into account not only the known associations between proteins and GO terms but also, cautiously, the absence of such associations. We can make a generally true assumption that proteins not already annotated with a term ought to stay dissociated from that term for a good reason. In other words, some proteins not associated in GO with a particular GO term may be treated as negative examples in that term's training set. However, due to the incompleteness of data in GO we cannot always take this to be true and therefore we should not accept blindly all such proteins as negative examples.

## 2.1 Details of Protein Representation

We have chosen PFAM domains as units of a domain content. PFAM domains become the single dimensions of the attribute space, with each protein represented as a sparse vector in that space. We distinguish between binary and integer attribute vectors, as illustrated in Figure 1 and Table 1. In addition to PFAM domain composition, we analyze all proteins from the training and testing sets for the presence of the signal peptide, transmembrane regions (Krogh, Larsson et al. 2001) and coiled coil (Berger, Wilson et al. 1995) regions. These three indicators are treated as extra dimensions in the attribute space.

Naturally, all proteins associated with a GO term constitute a **true positives** set. The definition of the negatives is more elaborate. Let us define all proteins associated with a particular GO term $t$ as $\{P_t\}$. Let us denote as $\{D_t\}$ the set of all domains belonging to proteins described by that GO term. Let us further denote as $\{D_P\}$ the domains of some protein $P$. If $\{D_t\}$ and $\{D_P\}$ have non-empty intersection and $P$ is not assigned to a parent of $t$, then we call $P$ a **true negative**. This definition reflects the fact that $P$, although similar in domain composition to proteins in $\{P_t\}$, was nonetheless annotated by a human expert to fall outside of this GO term, presumably for a biological reason. This definition of true negatives substantially reduces the dimensionality of the space of attributes for each GO term.

For each GO term we have also created a set of 'synthetic' negative protein examples. These 'synthetic' proteins are single domain proteins made of all domains from the true negative examples but only those domains that are not present in the positive set's parent terms. These **synthetic negatives** are always assigned binary values for their domain counts.

In addition to synthetic proteins we have also designed one artificial '**supernegative**' protein per GO term which is composed of all domains that are not present in the protein set representing parents of this term. The supernegative served as the reduced representation for the entire set of proteins eliminated from the training set by the selection of true positives and negatives. Dimensionality of supernegatives is much higher than that of regular proteins since $\{D_{supernegative}\} \approx \{D_{data\_set} - D_t\}$.

## 2.2 Algorithm

Several well-known decision-tree-learning algorithms are available (e.g. CART (Breiman, Friedman et al. 1984)). For this work we have chosen the OC1 decision tree package (Murthy, Kasif et al. 1994). OC1 allows for oblique as well as axis-parallel splits[3] in the space of attributes, which gave us more flexibility in our data representation.

The OC1 classifier was configured to test both axis-parallel and oblique splits in the integer attribute space, and axis-parallel splits only in the binary space. Following some initial experimentation with different goodness criteria supported by OC1, we have selected the twoing rule (Breiman, Friedman et al.

---

3 The OC1 is comparable to C4.5 when run in axis-parallel mode (see the OC1 paper and user manual)

1984), although similar (but slightly inferior) results were obtained using information gain during this initial testing. In the oblique mode, the maximal number of iterations was set to 50 and minimal to 20. We have always used the axis-parallel mode for the GO terms that had more than 1000 positive examples in the training set due to the runtime constraints.

## 2.3 Training and Testing Sets

We have used two different sets of proteins from SwissProt database for training and testing the validity of the decision tree approach.

- Training set: SwissProt proteins from human, mouse and yeast annotated by GO – 6367 proteins, 1756 domains and 3375 GO terms.
- Testing set: SwissProt proteins from fly (*D. Melanogaster)* and worm (*C.Elegans)* annotated by GO – 1024 proteins, 640 domains and 1138 GO terms.

The training set was selected to encompass the eukaryotic evolutionary tree. The testing set was selected to represent organisms that are evolutionarily distant from the training set organisms.

## 2.4 Benchmark

As a benchmark we used a reference list generated by the InterPro database. InterPro2GO associates the protein domains with GO terms (Camon, Magrane et al. 2003; Mulder, Apweiler et al. 2003). InterPro2GO is the only available approach that uses domain information to predict GO terms. InterPro2GO uses a simple association rule: a protein domain is associated with a GO term if all proteins associated with the term have that domain. We have used the InterPro2GO list to assess the performance of our method.

## 2.5 Measures of Performance

Each GO term has a unique decision tree trained using the set of examples from the training set. For each term we have a number of true positive associations (TP) predicted by the method, a number of all true associations in the test set (T), a number of positive associations (P) predicted by the method, a number true negative associations predicted (TN), and a total number of negative (non-existing) associations possible (N). We measure the overall performance of a method by Sensitivity=TP/T, Specificity=TN/N and Precision=TP/P where the numbers TP, T, P, TN, N are summed over all predicted GO term associations.

We also can measure the average per term performance of the classifier. Given the limited size of training and testing sets, not all GO terms were trained or tested and in consequence not all tested GO terms have positive examples. Thus to estimate the average term performance over the complete set of GO proteins we also calculate the performance averaged over the GO terms that do have positive examples. We split the set of GO terms into parents, those terms that do have as children more specific terms linked with them and leaves, the terms that are the last nodes in the GO diagram and represent the most specific protein descriptors in GO. We calculate the averages for 'parent' and 'leaf'[4] terms separately to check whether the method performs differently on generic and specific term types and whether it is affected by a small number of positive examples.

## 3   Results

In the initial testing of the decision tree approach we limited our predictions to the molecular function of the GO ontology. We have tested 12 possible settings of the method that split into two categories:

1. using synthetic negatives and synthetic supernegative examples
2. using binary versus integer domain representation and axis-parallel or oblique attribute space partition.

Results from those tests are presented in Table 2. We completed only one of four tests with the oblique mode in the time limit that we have set for the algorithm runtime (4 days). Performance of this test convinced us that the oblique search mode is not improving the performance. The standard deviations for Precision and Sensitivity are about 1-1.5%. The standard deviations for specificity are less then 0.5%. Results presented in Table 2 show that the use of synthetic negatives improves the precision of the method by about 15%. It is also clear that the binary representation of the protein is sufficient for the best resolution of the GO terms and is also computationally less demanding. The differences in performance of axis-parallel, oblique and binary domain methods are not statistically significant. The use of supernegative alone is detrimental to the performance of the classifier. However, the supernegative in combination with synthetic negatives marginally improves the precision and sensitivity. The time runs for the algorithm are only approximate since the processors were not uniquely reserved for our application.

Given relatively small training set, some GO terms have quite a small number of positive examples. Table 3 shows the average numbers of the positives, negatives, and synthetic negatives (taken from the training set of proteins) for the GO terms that are in the testing set. We need to note that these distributions are not normal. As expected, the average number of positives is much smaller for the leaf terms than the parent terms. However, distributions of sensitivity, specificity and precision do not show strong dependency on the number of positives or the dimensionality of the attribute space (data not shown). To check whether the performance is different for leaf and parent terms we have calculated precision, sensitivity and specificity averaged over terms in each of those two categories. We discuss below results for the Molecular Function branch of the GO graph. For the binary domain representation and axis-parallel method the average performance for leaves (168 leaves with defined testing set positives) was: Precision = 93.8 ± 21.0,

---

4 Here we mean the 'leaves' in the sense of 'leaves of the Directed Acyclic Graph', or those terms that are the most specific in their hierarchy and don't have any children. 'Parents' are 'all nodes which are not leaves'.

Sensitivity=88.6±25.3, Specificity=99.99 ± 0.07. The average performance for parents (321 parents with defined testing set positives) is: Precision = 88.0 ± 24.0, Sensitivity = 81.7 ± 25.3, Specificity = 99.6 ± 4.1. This results show that the performance of the classifier is not adversely affected by a small number of positive examples. In fact the leaf terms have on average better performance as we can expect from the more detailed level of description given by the leaf terms. Similar results are obtained for the other two branches of the GO graph.

The comparison of our results to InterPro2GO mappings is shown in Table 4. We evaluated both methods using the same test set of proteins. The decision trees were trained using the binary domain representation with synthetic negatives and supernegative examples. The InterPro2GO mapping used for comparison was derived from the complete set of GO proteins but GO terms absent from our training set of proteins were not considered. The results demonstrated that the Decision Tree approach was more sensitive than the InterPro2GO mapping. The greatest improvement in sensitivity, a 50% increase, was achieved for the cellular component network of the GO ontology. This good result is not surprising since, in addition to PFAM domain information, we have included as attributes the information about signal peptides, transmembrane helices and coiled coil regions which are good predictors of cellular localization. Sensitivity was also substantially improved for two other networks: for the Biological Process the sensitivity was improved by 26.5% and for Molecular Function the sensitivity was improved by 22.6%. The precision of our method is lower than that of InterPro2GO. This, too, is not unexpected since the InterPro2GO mapping uses a very conservative assignment of domains to GO terms. The values of precision and sensitivity averaged over leaves and parents (see above) suggest that the performance over a complete set of proteins characterized in GO should have a better precision than the ones listed in Table 4.

## 4    Discussion

The results of our initial investigation show that the Decision Tree classification approach is a valid and effective method for assigning GO ontology terms to proteins based on the domain composition. The method tested here favorably compares with the InterPro2GO approach, currently the only available method that analyses proteins by their domain composition. A prototype web server for assignment of GO ontology to proteins has been designed. Figure 2 shows the results of assignment to GO terms based on the small training set described above. We note that our definition of true negatives could introduce some false negative assignments in the training set. This is unavoidable due to incompleteness of curation and the sparseness of experimental evidence. The excessive assignment of training examples to the set of true negatives should increase the count of false negatives, decreasing both sensitivity and precision of the algorithm. This may account for some of classification mistakes (see Fig. 3), however the very good performance of the classifier justifies our definition of true negatives.

We plan to run the domain assignment for all GO terms with reviewed annotations. This set of proteins will be used for a final training of the GO decision trees. Ten–fold cross-validation procedure will be used to estimate the performance of each tree trained for individual GO terms. The performance data from those experiments will be used to estimate the performance of each decision tree as exemplified in Figure 2. The final trees will be trained on the whole set of GO proteins.

## 5    Tables and Figures

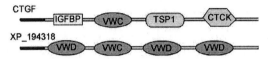

Figure 1.  An example of two proteins sharing common domains: Connective Tissue Growth Factor (CTGF) and XP_194318 (similar to kielin).  Domains are: Insulin-like Growth Factor Binding Protein (IGFBP), Von Wlillebrand Factor C domain (VWC), Von Willebrand factor D domain (VWD), Trombospondin-domain 1, c-terminal cys-knot domain (CTCK).

```
⊟GO:0003673 Gene_Ontology sens: 95.59 spec: 25 prec: 99.69
   ⊟GO:0003674 molecular_function sens: 95.59 spec: 25 prec: 99.69
      GO:0005215 transporter activity sens: 81.3 spec: 84.46 prec: 41.67
      GO:0016209 antioxidant activity sens: 100 spec: 87.07 prec: 2.22
      ⊟GO:0005488 binding activity sens: 82.86 spec: 75.95 prec: 78.38
         ⊟GO:0046872 metal ion binding activity sens: 71.43 spec: 98.35 prec: 71.43
            GO:0005509 calcium ion binding activity sens: 77.78 spec: 99.4 prec: 77.78
      GO:0003824 enzyme activity sens: 88.7 spec: 81.91 prec: 80
      GO:0005198 structural molecule activity sens: 68.24 spec: 86.58 prec: 31.52
      GO:0030528 transcription regulator activity sens: 93.33 spec: 81.48 prec: 32.68
```

Figure 2. The prototype GO ontology assignment server.

|           | Binary model        | Integer model       |
|-----------|---------------------|---------------------|
| CTGF      | {1, 1, 1, 1, 0}     | {1, 1, 1, 1, 0}     |
| XP_919318 | {0, 1, 0, 0, 1}     | {0, 2, 0, 0, 2}     |

Table 1.  Example of two different models of domain composition. The proteins show are the same as in Figure 1. The attribute space for these two proteins is D={IGFBP,VWC,TSP1, CTCK, VWD}

|                                     | Integer domains Axis-parallel | Integer domains Oblique | Binary domains Axis-parallel |
|-------------------------------------|-------------------------------|-------------------------|------------------------------|
| No synthetics no supernegative      | prec = 67.3 ± 1.0<br>sens = 79.3 ± 1.3<br>spec = 99.8 ± 0.1 | Longer than the runtime limit | prec = 65.9 ± 1.0<br>sens = 82.7 ± 1.3<br>spec = 99.8 ± 0.1 |
| Using synthetics no supernegative   | prec = 80.9 ± 1.3<br>sens = 78.5 ± 1.3<br>spec = 99.9 ± 0.1 | Longer than the runtime limit | prec = 82.2 ± 1.3<br>sens = 80.8 ± 1.3<br>spec = 99.9 ± 0.1 |

| No synthetics using supernegative | prec = 26.8 ± 0.4<br>sens = 80.5 ± 1.3<br>spec = 99.1 ± 0.1 | Longer than the<br>runtime limit | prec = 28.0 ± 0.4<br>sens = 82.3 ± 1.3<br>spec = 99.1 ± 0.1 |
|---|---|---|---|
| Using synthetics and supernegative<br>(timed[5]) | prec = 81.1 ± 1.3<br>sens = 79.9 ± 1.3<br>spec = 99.9 ± 0.1<br><br>time : 27 hours | prec = 81.4 ± 1.3<br>sens = 78.8 ± 1.3<br>spec = 99.9 ± 0.1<br><br>time : 66 hours | prec = 82.6 ± 1.3<br>sens = 81.0 ± 1.3<br>spec = 99.9 ± 0.1<br><br>time : 20 hours |

*Table 2. Data models' effect on GO term prediction*

|  | all | Leaves | Parents |
|---|---|---|---|
| Positives | 28 | 4 | 67 |
| Negatives | 792 | 601 | 1113 |
| synthetics | 205 | 164 | 275 |

Table 3. The average numbers of positive, negative and synthetic negative examples for different types of GO terms.

|  | InterPro2GO | Decision Tree |
|---|---|---|
| Biological Process | prec = 91.5 ± 1.8<br>sens = 42.6 ± 0.7<br>spec = 99.9 ± 0.2 | prec = 82.9 ± 1.3<br>sens = 69.1 ± 1.1<br>spec = 99.9 ± 0.1 |
| Cellular Component | prec = 99.8 ± 3.4<br>sens = 34.8 ± 1.0<br>spec = 99.9 ± 0.4 | prec = 85.0 ± 1.8<br>sens = 84.9 ± 1.8<br>spec = 99.8 ± 0.2 |
| Molecular Function | prec = 98.9 ± 2.0<br>sens = 58.4 ± 1.0<br>sepc = 99.9 ± 0.2 | prec = 82.6 ± 1.3<br>sens = 81.0 ± 1.3<br>spec = 99.9 ± 0.1 |

*Table 4. Comparison of performance of the Decision Tree classifier (binary domain representation and synthetic negative examples) to that of InterPro2GO PFAM-based assignments*

## 6 Acknowledgements

We thank S. Kasif for making available to us the OC1 software and S.K Murthy and S. Salzberg for developing the method.

5 Time required to finish testing and training of Decision Trees for the sets using synthetic negative and supernegative examples using 20 600MHz SGI R3800 CPUs during training stage. The computational times for tests not using synthetic examples were shorter.

# 7 References

3. Apweiler (2004). "UniProt: the Universal Protein Knowledgebase." Nucleic Acids Res. 32(in press.).
4. Ashburner, M., C. A. Ball, et al. (2000). "Gene ontology: tool for the unification of biology. The Gene Ontology Consortium." Nat Genet 25(1): 25-9.
5. Bateman, A., E. Birney, et al. (2002). "The Pfam protein families database." Nucleic Acids Res 30(1): 276-80.
6. Berger, B., D. B. Wilson, et al. (1995). "Predicting coiled coils by use of pairwise residue correlations." Proc Natl Acad Sci U S A 92(18): 8259-63.
7. Boeckmann, B., A. Bairoch, et al. (2003). "The SWISS-PROT protein knowledgebase and its supplement TrEMBL in 2003." Nucleic Acids Res 31(1): 365-70.
8. Breiman, L., J. Friedman, et al. (1984). Classification and Regression Trees, Wadsworth International Group.
9. Buell, C. R., V. Joardar, et al. (2003). "The complete genome sequence of the Arabidopsis and tomato pathogen Pseudomonas syringae pv. tomato DC3000." Proc Natl Acad Sci U S A 100(18): 10181-6.
10. Camon, E., M. Magrane, et al. (2003). "The Gene Ontology Annotation (GOA) project: implementation of GO in SWISS-PROT, TrEMBL, and InterPro." Genome Res 13(4): 662-72.
11. Chiang, J. H. and H. C. Yu (2003). "MeKE: discovering the functions of gene products from biomedical literature via sentence alignment." Bioinformatics 19(11): 1417-22.
12. El-Sayed, N. M., E. Ghedin, et al. (2003). "The sequence and analysis of Trypanosoma brucei chromosome II." Nucleic Acids Res 31(16): 4856-63.
13. Haas, B. J., A. L. Delcher, et al. (2003). "Improving the Arabidopsis genome annotation using maximal transcript alignment assemblies." Nucleic Acids Res 31(19): 5654-66.
14. Haft, D. H., J. D. Selengut, et al. (2003). "The TIGRFAMs database of protein families." Nucleic Acids Res 31(1): 371-3.
15. Harris, T. W., R. Lee, et al. (2003). "WormBase: a cross-species database for comparative genomics." Nucleic Acids Res 31(1): 133-7.
16. Hennig, S., D. Groth, et al. (2003). "Automated Gene Ontology annotation for anonymous sequence data." Nucleic Acids Res 31(13): 3712-5.
17. Hill, D. P., A. P. Davis, et al. (2001). "Program description: Strategies for biological annotation of mammalian systems: " Genomics 74(1): 121-8.
18. Hvidsten, T. R., J. Komorowski, et al. (2001). "Predicting gene function from gene expressions and ontologies." Pac Symp Biocomput: 299-310.
19. Hvidsten, T. R., A. Laegreid, et al. (2003). "Learning rule-based models of biological process from gene expression time profiles using gene ontology." Bioinformatics 19(9): 1116-23.

20. Issel-Tarver, L., K. R. Christie, et al. (2002). "Saccharomyces Genome Database." Methods Enzymol 350: 329-46.
21. Karp, P. D., M. Riley, et al. (2002). "The MetaCyc Database." Nucleic Acids Res 30(1): 59-61.
22. Khan, S., G. Situ, et al. (2003). "GoFigure: Automated Gene Ontology(TM) annotation." Bioinformatics 19(18): 2484-5.
23. King, O. D., R. E. Foulger, et al. (2003). "Predicting gene function from patterns of annotation." Genome Res 13(5): 896-904.
24. Krishnan, V. G. and D. R. Westhead (2003). "A comparative study of machine-learning methods to predict the effects of single nucleotide polymorphisms on protein function." Bioinformatics 19(17): 2199-209.
25. Krogh, A., B. Larsson, et al. (2001). "Predicting transmembrane protein topology with a hidden Markov model: application to complete genomes." J Mol Biol 305(3): 567-80.
26. Lagreid, A., T. R. Hvidsten, et al. (2003). "Predicting gene ontology biological process from temporal gene expression patterns." Genome Res 13(5): 965-79.
27. Letovsky, S. and S. Kasif (2003). "Predicting protein function from protein/protein interaction data: a probabilistic approach." Bioinformatics 19 Suppl 1: I197-I204.
28. Mulder, N. J., R. Apweiler, et al. (2003). "The InterPro Database" Nucleic Acids Res 31(1): 315-8.
29. Murthy, S. K., S. Kasif, et al. (1994). "A System for Induction of Oblique Decision Trees." Journal of Artificial Intelligence Research 2: 1-32.
30. Raychaudhuri, S., J. T. Chang, et al. (2002). "Associating genes with gene ontology codes using a maximum entropy analysis of biomedical literature." Genome Res 12(1): 203-14.
31. Riley, M. (1998). "Genes and proteins of Escherichia coli K-12." Nucleic Acids Res 26(1): 54.
32. Schug, J., S. Diskin, et al. (2002). "Predicting gene ontology functions from ProDom and CDD protein domains." Genome Res 12(4): 648-55.
33. Serres, M. H. and M. Riley (2000). "MultiFun, a multifunctional classification scheme for Escherichia coli K-12 gene products." Microb Comp Genomics 5(4): 205-22.
34. Syed, U. and G. Yona (2003). Using a mixture of probabilistic decision trees for direct prediction of protein function. Proceedings of the 7th annual international conference on CMB. Berlin, Germany, ACM Press: 289--300.
35. The FlyBase Consortium (2003). "The FlyBase database of the Drosophila genome projects" Nucleic Acids Res 31(1): 172-5.
36. Troyanskaya, O. G., K. Dolinski, et al. (2003). "A Bayesian framework for combining heterogeneous data sources for gene function prediction (in Saccharomyces cerevisiae)." Proc Natl Acad Sci U S A 100(14): 8348-53.
37. Wortman, J. R., B. J. Haas, et al. (2003). "Annotation of the Arabidopsis genome." Plant Physiol 132(2): 461-8.
38. Xie, H., A. Wasserman, et al. (2002). "Large-scale protein annotation through gene ontology." Genome Res 12(5): 785-94.

# THE CLINICAL BIOINFORMATICS ONTOLOGY: A CURATED SEMANTIC NETWORK UTILIZING REFSEQ INFORMATION

M. HOFFMAN , C. ARNOLDI, I. CHUANG

*Cerner Corporation, 2800 Rockcreek Parkway*
*Kansas City, MO 64117, USA*
*{mhoffman1,carnoldi,ichuang}@cerner.com*

Existing medical vocabularies lack rich terms to describe findings that are generated by modern molecular diagnostic procedures. Most bioinformatics resources were designed primarily to support the needs of the research community. We describe the development of a curated resource, the Clinical Bioinformatics Ontology (CBO), a semantic network appropriate for describing clinically significant genomics concepts. The CBO includes concepts appropriate for both molecular diagnostics and cytogenetics. A standardized methodology based on consistent application of RefSeq information is applied to the curation of the CBO in order to provide a reproducible and reliable tool. Challenges related to this curation process are discussed in this paper. At the time of submission the CBO included 4,069 concepts, associated by 8,463 relationships.

## 1. Introduction

The practice of medicine increasingly utilizes methods based on recent advances in genomics. Diagnostic tests based on the detection of single nucleotide polymorphisms, cytogenetic observations or gene expression patterns are utilized to confirm, classify or monitor inherited or pathological conditions. Clinical information systems will be expected to manage these results and to support the exchange of data based on genomics-related findings. Extension of clinical information systems to manage genomics finding should include the ability to standardize information using a controlled vocabulary and the ability to support genomics-based inference, ideally using ontology-based reasoning.

Current medical vocabularies lack expressions capable of describing the granular findings generated by molecular tests in sufficient detail. SNOMED Clinical Terms® (SNOMED-CT®)[a] is an ontology but lacks concepts appropriate for the detailed description of chromosomal structures below the level of arm and lacks terms for the accurate description of molecular findings. Additionally, there are errors in the hierarchical organization of those genomics-related concepts that are included in SNOMED-CT; for example, "Genome" is positioned as a child of "Gene". The Logical Observation and Identification Codes (LOINC®) (1) resource has recently incorporated a limited number of

---

[a]www.snomed.org

molecular diagnostics-related concepts but is not structured as an ontology and thus was not designed to support conceptual-inferencing.

Of the many bioinformatics resources provided by the National Center for Biotechnology Information (NCBI)(2), the Online Mendelian Inheritance in Man™ (OMIM)(3), is the most clearly clinically oriented. However, OMIM is not provided in a machine readable format suitable for integration with relational databases. Furthermore, entries in OMIM apply inconsistent methods of describing positional information. For example, the description of the CFTR gene uses the first position of the DNA sequence in the reference sequence file, 132 bp before the beginning of the coding sequence, for referencing the position of mutations[a]. The HADHA gene content references mutations relative to the beginning of the coding sequence[b], the approach that is consistent with the standard naming conventions of the Human Genome Organization (HUGO)(4;5).

Other bioinformatics resources were not designed to support clinical applications or were not implemented in a machine readable format that is publicly available. The Gene Ontology (GO) (6) has become a valuable tool for the research community and includes concepts describing molecular processes, molecular functions and cellular components. However, GO lacks expressions related to the practice of medicine, for example terms appropriate for the description of chromosomal observations or expressions appropriate for describing human genetic variations of clinical significance. Another bioinformatics resource, the Human Gene Mutation Database (HGMD)(7), provides a curated set of uniquely identified expressions associated with human gene variations but is not available in a machine readable format to the general public. Likewise, the most widely utilized human gene mutation database, dbSNP(8), was designed primarily to support the needs of the research community and lacks the clarity needed to support the clinician.

There have been a variety of efforts to integrate existing bioinformatics resources and medical vocabularies (9-11). While useful for promoting the mining of research data for research purposes, these projects have not filled the gaps and inaccuracies in existing resources. To satisfy the need for a clinically oriented genomics vocabulary, we have developed a curated semantic network, the Clinical Bioinformatics Ontology™ (CBO). This resource combines the attributes of a medical vocabulary with the positional information offered by bioinformatics resources, especially RefSeq (12).

---

[a] http://www.ncbi.nlm.nih.gov/entrez/dispomim.cgi?id=602421
[b] http://www.ncbi.nlm.nih.gov/entrez/dispomim.cgi?id=600890

## 2. Materials

### 2.1. *RefSeq*

RefSeq is a part of the NCBI collection of bioinformatics resources (2). Characterized genes are assigned RefSeq identifiers which link to GenBank sequence files containing a DNA sequence generated by reverse transcription of the mRNA product of the gene, protein sequence and often a genomic contig file.

### 2.2. *SNOMED-CT*

SNOMED-CT (version 20030108) was evaluated using the CLUE Browser, version 5.5 (Clinical Information Consultancy, UK)[a].

### 2.3. *LExScape - Health Language Incorporated*

The CBO was managed using LExScape, a nomenclature management program produced by Health Language Incorporated (HLI) (Aurora, Colorado). LExScape controls the versioning, re-parenting and retirement of concepts. LExScape also assigns a global unique identifier (GUID) to each concept. In this paper we use the naming conventions associated with this system – concepts, facets and terms. Concepts are uniquely identified entities or ideas. Facets are associated with concepts and provide further descriptive information. Terms are alternative names (synonyms, abbreviations) for concepts.

### 2.4. *Protégé-2000*

Protégé-2000 (Stanford University)[b] is a publicly available frames-based ontology management tool.

### 2.5. *Controlled Medical Terminology (CMT)™*

The Cerner Controlled Medical Terminology (CMT)™ (Cerner Corporation, Kansas City, MO) is a meta-vocabulary associating most of the common medical vocabularies, including SNOMED and LOINC. CMT enables integration of concept-based capabilities into the Cerner Millennium® suite of clinical software. Cerner Pathnet Helix™ was used to demonstrate the capture of genetic test results in a laboratory information system (LIS) using the CBO.

---

[a] http://www.clininfo.co.uk/clue5/index.htm
[b] http://protege.stanford.edu

## 3. Methods

### 3.1. *Design*

The CBO was designed to reflect a structure similar to that of SNOMED-CT in order to favor compatibility with the phenotypic information that is codified in clinical information systems. Hierarchical "IS_A" relationships ("SUBSUMES" in HLI) were combined with lateral definitional relationships (Table 1) to create the semantic network.

**Table 1 – Definitional relationships**

| Full name | Alias |
|---|---|
| Has Chromosomal Location | HAS_CHRM_LOC |
| Arm of | ARM_OF |
| Band of | BAND_OF |
| Has Arm Location | HAS_ARM_LOC |
| Has Band Location | HAS_BAND_LOC |
| Exon of | EXON_OF |
| Intron of | INTRON_OF |
| Allele of | ALLELE_OF |
| Has Constituent Variant | HAS_CONS_VAR |
| Nucleotide Variant of | NUC_VAR_OF |
| Has Effect | HAS_EFFECT |
| Transcript of | TRANSCRPT_OF |
| RT Product of | RT_PROD_OF |
| Amino Acid Variant of | AA_VAR_OF |
| Encodes | ENCODES |
| Has Phenotype | HAS_PHENOTYP |
| Has Location | HAS_LOC |
| Includes Band | INCLUDS_BAND |
| Mode of Inheritance | MODE_OF_INH |
| Had CH3 Status | HAS_CH3_STAT |
| Source Band | SOURCE_BAND |
| Target Band | TARGET_BAND |

Some significant design decisions included:

- **Method of describing mutations and polymorphisms**. Associating a concept representing a mutation to the gene of origin would violate "IS_A" logic. Therefore, we applied definitional relationships to associate a mutation to its gene. Most of the mutations described in the CBO are documented relative to the cDNA sequence associated with the RefSeq. The model defines a series of relationships that must be traversed to associate a genomic gene concept to a mutation concept.

- **Consistent use of Reference Sequence data.** In order to create a reproducible methodology, we associate gene concepts with a facet that provides the RefSeq identifier for the mRNA derived DNA sequence (12). The sequence content of a Reference Sequence can change as corrections or revisions are submitted, therefore we include a facet with the "gi" identifier of the GenBank sequence in order to capture the version of the gene utilized at the time that content was created (13).
- **Allele naming.** We adopted a generalized allele naming approach that was based on the OMIM style of naming gene alleles {Gene.allele identifier}[a]. CBO allele concepts are associated with OMIM allele identifiers through facets storing the OMIM allele identifier.
- **Description of chromosomal structures.** We applied the standards of the International System for Human Cytogenetic Nomenclature (ISCN)(14) to the naming of chromosomal structures (arms, centromeres, telomeres and bands). To address the multiple resolutions applied to clinical observation of cytogenetic structures (400, 550 and 850), we prefixed chromosomal band concepts with the resolution at which the bands are observed. High resolution bands were associated to lower resolution bands using the BAND_OF definitional relationship. For example, the "550.7q31.2" concept is related to the "400.7q31" concept using the "BAND_OF" relationship.
- **Directionality of definitional relationships.** The nomenclature modeling tool that we utilized does not require explicit statement of inverse relationships because one attribute of a relationship (as modeled in LExScape) is whether or not that relationship automatically includes an implicit inverse.

### 3.2. Curation

Controlling the scope of the CBO was an important design parameter. The primary approach to the selection of concepts that are clinically significant was a detailed analysis of the test catalogues offered by molecular diagnostic laboratories and interviews with these laboratories. Additional clinically significant concepts were identified by reviewing GeneTests (15) and the primary literature. Concepts, as well as their associated terms and facets, were created by a content analyst. The curation process involved utilizing a documented methodology in order to ensure consistency. The white paper describing this methodology will be posted with the content on a publicly available web site (URL to be provided during final edit). After new content was created, a reviewer examined the content for accuracy, consistency with the methodology and completeness.

---

[a] http://www.ncbi.nlm.nih.gov/Omim/omimfaq.html#numbering_system

### 3.3. *HLI import and export*

All concepts, relations, facets and terms were imported into LExScape using a java program. After LExScape assigned GUIDs and formally defined the structure of the ontology, the content was exported from LExScape into CSV files. These files were imported into CMT and the Cerner Millennium system using a program written in the Cerner Command Language (CCL)® , a variant of the Structured Query Language (SQL) that is optimized for the clinical information systems developed by Cerner Corporation. The CSV files were also imported into Protégé (Stanford University), with an intermediate conversion to RDF, to further demonstrate the portability of the content.

### 4. Discussion

Clinical information systems will need an appropriate controlled vocabulary to manage information gathered during clinical genomics-based diagnostic procedures. In order to address the gaps identified in current medical vocabularies and bioinformatics resources, we developed the Clinical Bioinformatics Ontology (CBO). The CBO is structured to provide the benefits of a controlled vocabulary as well as the advantages of a semantic network. The scope of the CBO was limited to concepts currently applied in clinical practice, thus concepts related to microarray-based diagnostics, proteomics and other emerging technologies have not yet been incorporated into the CBO.

**Table 2 – Summary of CBO elements (July 2004)**

| Element | Number |
|---------|--------|
| Concept | 4069 |
| Nucleotide variant | 286 |
| Facet | 2110 |
| Term | 460 |
| Relationships | 8463 |

Central to the CBO is the use of a content curation process. Resources created through automated annotation can manage high volumes of information, however the resulting output is often heterogeneous in quality; in contrast a curated resource offers the opportunity to enforce a minimal level of annotation and apply a deliberate strategy to the growth of the resource (16). In order for the CBO to be useful in clinical settings, where there are high quality control expectations, we chose to apply a curation approach, despite the limited rate of growth associated with human curation.

**Challenges**

A variety of challenges were identified during the design of the model, examples of which are discussed below.

1. *Describing mutations using "IS_A" compliant methods.*

During the initial design of the CBO, models in which mutations would be children of the gene to which they are associated were evaluated. The inability to represent mutations in this way and maintain "IS_A" relationships led us to create a relatively flat model in which definitional relationships are used to describe clinically significant mutations. In our model all single nucleotide polymorphisms are children of the "Human Nucleotide Variant" concept, which is a child of "Human Nucleic Acid Variant" (of which the other children are "Human Allele" and "Human Haplotype").

2. *The lack of a consistent resource providing the number of exons in a gene.*

Many molecular diagnostic laboratories have begun to perform diagnostic sequencing. Often only a few of the exons for a particular gene are sequenced. The majority of the genes represented in the CBO are associated to concepts representing their constituent exons and introns. The efforts to generate these concepts were often hindered by the lack of a consistent resource providing the number of exons for a gene. Our approach was to utilize the RefSeq tables provided by the UCSC (17) combined with a review of literature relating to the gene being curated. The UCSC values are based on a computed number of exons generated by automated alignment of the genomic DNA sequence with the RefSeq mRNA sequence (18). Often the values determined by this method do not agree with published information (Table 3). When the UCSC value agreed with the published literature we added exon and intron concepts to the CBO. When they did not agree we deferred the creation of the exon and intron concepts.

**Table 3 – Disparities in exon count**

| Gene | UCSC exons | Published value |
|------|-----------|-----------------|
| CFTR | 27 | 27 (19) |
| MTHFR | 12 | 11 (20) |
| RHD | 11 | 10 (21) |

A further issue in naming exons is the use of suffixes to describe short exons (5a, 5b). We applied an absolute approach in which each exon, regardless of size, is assigned a unique integer value relative to the beginning of the gene. The convention used for naming exons in the CBO is "Gene name".e."exon number", with a similar approach for naming introns. Thus the 3rd exon for the CFTR gene is named "CFTR.e.3".

*3. Reconciling inaccuracies in existing resources and the literature.*

Another challenge that we identified were significant discrepancies between descriptions of mutations (or polymorphisms) in the literature and the reference information provided by RefSeq. One of our goals was to adopt a methodology which could be applied by any user yet result in the same concept naming conventions. On occasion the application of this approach generated concepts whose names disagree with values in the published literature.

For example, the human platelet antigen ITGA2 has polymorphisms that result in antigenic diversity that are factors in neonatal alloimmune thrombocytopenia. The paper describing the nomenclature of these antigens refers to the RefSeq NM_002203 and cites the polymorphism 1600G>A with the amino acid substitution result of E505K (22). Analysis of the DNA sequence for the coding region (CDS) associated with this RefSeq file (gi:6006008) clearly shows an "a" at this position:

```
                                             *
1561 aagaaagagg aaggaagagt ctacctgttt actatcaaaa agggcatttt
```

In order to comply with our methodology we created the a concept to describe the mutation relative to the RefSeq, "ITGA2.c.1600A>G", and the associated amino acid substitution concept, "ITGA2.p.K534E". These concept names are accurate relative to the RefSeq but will require close communication with clinicians accustomed to the naming used in the literature. A log of such exceptions was created and is included with the CBO.

*4. The lack of a widely accepted system for describing alleles.*

Patient results are often reported to clinicians in terms that describe an aggregate of mutation or polymorphism findings for a particular gene. This required us to adopt an approach to allele naming that could be generalized. While many genetic systems, including the HLA locus (23)and the CYP2D6 gene[a] have locus or gene specific allele naming conventions, the only resource that offers a fairly generic approach to allele description was OMIM. However, many clinically significant alleles do not have OMIM allele identifiers; for example many of the CYP2D6 gene alleles are under-represented in OMIM. Therefore, we adopted an OMIM-like approach to allele naming in which the allele is named using the convention: [Gene name][.][unique allele number]. The allele concepts are related to their constituent nucleic acid variations through the definitional relationship type of "HAS_CONS_VAR". This approach can be extended to describe haplotypes, in which a haplotype is described by definitional relationships to the constituent alleles.

**Applications of the CBO**

Importing the CBO into a commercial healthcare information system allowed us to demonstrate that the general benefits of a controlled vocabulary

---

[a] www.imm.ki.se/CYPalleles/cyp2d6.htm

and a semantic network are applicable in the context of a clinical genomics software system. Discrete results captured in the PathNet Helix LIS system are associated to a CBO GUID. The benefits of this approach include:

- The comparison of results between patients is facilitated.
- The retrieval of historic results is enabled.
- The exchange of data is greatly simplified.

By associating discrete clinical genomics results to a CBO GUID, the user is free to utilize any display name for a genetic test result. For example, molecular diagnostics labs often perform allele-specific PCR to detect mutations in genes, such as the CFTR gene. Laboratories often report these DNA-based findings in terms of the amino acids substitutions that they cause. For example, the mutation most commonly associated with cystic fibrosis, the deletion of nucleotides 1522-1524 of the CFTR gene, results in the deletion of the phenylalanine at position 508 in the CFTR protein. Users can display their results using expressions such as: _508, del 508, 508 del, delF508 or any other name, but the underlying association between the discrete result and the GUID for the concept "CFTR.c.1522-1524del" remains.

Our implementation of the CBO within a commercial HIS also allowed us to demonstrate the benefits of an ontology applied within this context. Areas in which the semantic network aspects of the CBO have been demonstrated include:

- **Simplification of the database implementation process**. Using the series of definitional relationships connecting a gene concept to associated mutation concepts, users representing an orderable test procedure in the system can quickly access likely discrete results for that procedure and reduce the amount of effort needed to build the representations of these discrete results in the system.
- **Utilize the CBO as reference data**. Software applications designed to utilize the CBO can be implemented to access the CBO as a source of reference data. For example, the "MODE_OF_INHERITANCE" relationship can enable or disable display behaviors that are appropriate for X-linked conditions.
- **Flexible queries**. Most importantly, the semantic network created by the CBO allows research-oriented users to efficiently design queries that apply the CBO model for questions such as "Retrieve all results related to chromosome 7" or "retrieve all results related to arm p of chromosome 10" or "retrieve all results related to exon 28 of the von Willebrand factor gene".

A subset of the CBO is provided in Table 4 in order to demonstrate how the combination of "IS_A" (SUBSUMES) and definitional relations were used to associate concepts related to the CFTR gene, including the mutation discussed above.

**Table 4 – Example relationships associating CFTR related concepts**

| Source Concept | Relationship Type | Target concept |
|---|---|---|
| Human Gene | SUBSUMES | (gDNA).CFTR |
| (gDNA).CFTR | HAS_CHRM_LOC | Chromosome 7 |
| (gDNA).CFTR | HAS_ARM_LOC | 7q |
| (gDNA).CFTR | HAS_BAND_LOC | 850.7q31.2 |
| 850.7q31.2 | BAND_OF | 550.7q31.2 |
| 550.7q31.2 | BAND_OF | 400.7q31 |
| Human mRNA | SUBSUMES | (mRNA).CFTR.0 |
| (mRNA).CFTR.0 | TRANSCRPT_OF | (gDNA).CFTR |
| Human Protein | SUBSUMES | (AA).CFTR.0 |
| (mRNA).CFTR.0 | ENCODES | (AA).CFTR.0 |
| Human cDNA | SUBSUMES | (cDNA).CFTR.0 |
| (cDNA).CFTR.0 | RT_PROD_OF | (mRNA).CFTR.0 |
| (gDNA).CFTR | MODE_OF_INH | Autosomal |
| Human Exon | SUBSUMES | CFTR.e.11 |
| Human Amino Acid Variant | SUBSUMES | CFTR.p.508delF |
| CFTR.p.508delF | AA_VAR_OF | (AA).CFTR.0 |
| Human Nucleotide Variant | SUBSUMES | CFTR.c.1522_1524del |
| CFTR.c.1522_1524del | NUC_VAR_OF | (cDNA).CFTR.0 |
| CFTR.c.1522_1524del | HAS_EFFECT | CFTR.p.508delF |
| CFTR.c.1522_1524del | HAS_LOC | CFTR.e.11 |

Queries designed to use these relationships can thus support a wide variety of research applications using clinical results codified using the CBO.

While the primary focus of this work was the implementation of the CBO in a commercial HIS, we have also converted the ontology into formats that will support wider utilization, including RDF and Protégé.

We have developed a resource that extends the scope of existing medical vocabularies and benefits from the ability to standardize positional genomic information utilizing RefSeq information. The combined attributes of a controlled vocabulary and a semantic network allow the CBO to be useful for both the delivery of care and research applications. We continue to expand the scope and depth of the content provided in the CBO. Areas of current effort include representation of pathogen-related concepts, including those needed to describe HIV genotype findings, representation of cytogenetic abnormalities and adding further support for the representation of splice variants. The CBO content is available in CSV and RDF formats at www.cerner.com/cbo at no cost to non-commercial users.

149

**Acknowledgments**

We acknowledge the contributions of Siew-Tien Lai, Kevin Power, Ginger Kuhns, Chris Finn, Roll-Jean Cheng, Daniel Bellissimo, Barbara Zehnbauer, Arthur Hauck, David McCallie, Doug McNair and Daniel Peterson. This work was funded in its entirety by Cerner Corporation.

**References**

1. Huff SM, Rocha RA, McDonald CJ, et al. Development of the Logical Observation Identifier Names and Codes (LOINC) vocabulary. J Am Med Inform Assoc. 1998;5:276-92.
2. Wheeler DL, Church DM, Edgar R, et al. Database resources of the National Center for Biotechnology Information: update. Nucleic Acids Res. 2004;32 Database issue:D35-40.
3. Hamosh A, Scott AF, Amberger J, Bocchini C, Valle D, McKusick VA. Online Mendelian Inheritance in Man (OMIM), a knowledgebase of human genes and genetic disorders. Nucleic Acids Res. 2002;30:52-5.
4. Antonarakis SE. Recommendations for a nomenclature system for human gene mutations. Nomenclature Working Group. Hum Mutat. 1998;11:1-3.
5. den Dunnen JT, Antonarakis SE. Mutation nomenclature extensions and suggestions to describe complex mutations: a discussion. Hum Mutat. 2000;15:7-12.
6. Harris MA, Clark J, Ireland A, et al. The Gene Ontology (GO) database and informatics resource. Nucleic Acids Res. 2004;32 Database issue:D258-61.
7. Stenson PD, Ball EV, Mort M, et al. Human Gene Mutation Database (HGMD): 2003 update. Hum Mutat. 2003;21:577-81.
8. Sherry ST, Ward MH, Kholodov M, et al. dbSNP: the NCBI database of genetic variation. Nucleic Acids Res. 2001;29:308-11.
9. Lussier YA, Sarkar IN, Cantor M. An integrative model for in-silico clinical-genomics discovery science. Proc AMIA Symp. 2002;469-73.
10. Cantor MN, Lussier YA. A knowledge framework for computational molecular-disease relationships in cancer. Proc AMIA Symp. 2002;101-5.
11. Cantor MN, Lussier YA. Putting data integration into practice: using biomedical terminologies to add structure to existing data sources. Proc AMIA Symp. 2003;125-9.
12. Pruitt KD, Maglott DR. RefSeq and LocusLink: NCBI gene-centered resources. Nucleic Acids Res. 2001;29:137-40.
13. Andreas Baxevanis BFFO. Bioinformatics: A practical guide to the analysis of genes and proteins. ed. John Wiley & Sons, Inc, 2001:
14. AnonymousAn International System for Human Cytogenetic Nomenclature (1985) ISCN 1985. Report of the Standing Committee on Human Cytogenetic Nomenclature. Birth Defects Orig Artic Ser.

1985;21:1-117.
15. Pagon RA, Tarczy-Hornoch P, Baskin PK, et al. GeneTests-GeneClinics: genetic testing information for a growing audience. Hum Mutat. 2002;19:501-9.
16. Pruitt KD, Katz KS, Sicotte H, Maglott DR. Introducing RefSeq and LocusLink: curated human genome resources at the NCBI. Trends Genet. 2000;16:44-7.
17. University of California Santa Cruz. http://genome.ucsc.edu/cgi-bin/hgText.
18. Terry Furey, personal communication. (GENERIC)
19. Claustres M, Laussel M, Desgeorges M, et al. Analysis of the 27 exons and flanking regions of the cystic fibrosis gene: 40 different mutations account for 91.2% of the mutant alleles in southern France. Hum Mol Genet. 1993;2:1209-13.
20. Goyette P, Pai A, Milos R, et al. Gene structure of human and mouse methylenetetrahydrofolate reductase (MTHFR). Mamm Genome. 1998;9:652-6.
21. Wagner FF, Gassner C, Muller TH, Schonitzer D, Schunter F, Flegel WA. Molecular basis of weak D phenotypes. Blood. 1999;93:385-93.
22. Metcalfe P, Watkins NA, Ouwehand WH, et al. Nomenclature of human platelet antigens. Vox Sang. 2003;85:240-5.
23. Marsh SG. Nomenclature for factors of the HLA system, update March 2004. Tissue Antigens. 2004;64:108-9.

# GO MOLECULAR FUNCTION TERMS ARE PREDICTIVE OF SUBCELLULAR LOCALIZATION

Z. LU AND L. HUNTER

*Center for Computational Pharmacology*
*University of Colorado Health Sciences Centre*
*School of Medicine, Denver, CO*

A protein's function is closely linked to its subcellular localization. Use of Gene Ontology (GO) molecular function terms to extend sequence-based subcellular localization prediction has been previously shown to improve predictive performance. Here, we explore directly the relationship between GO function annotations and localization information, identifying both highly predictive single terms, and terms with large information gain with respect to location. The results identify a number of predictive and informative GO terms with respect to subcellular location, particularly nucleus, extracellular space, membrane, mitochondrion, endoplasmic reticulum and Golgi. There are several clear examples illustrating why the addition of function information provides additional predictive power over sequence alone. Other interesting phenomena can also be seen in the results. Most predictive or informative terms are imperfect, and incorrect prediction may often call out significant biological phenomena. Finally, these results may be useful in the GO annotation process.

## 1. Introduction

High-throughput sequencing technology has heightened the need for automatic annotation of uncharacterized genes and gene products. As part of this annotation process, a number of systems have been developed that support automated prediction of subcellular localization of proteins. Most such methods are *sequence-based*, that is, they predict location based on features calculated from the amino acid sequence of a protein, such as degree of match to motifs constructed from short N-terminal signal peptides, or global amino acid composition. Recently, Chou & Cai[3] and Gardy et al.[7] have demonstrated that the addition of protein function information improves the performance of pure sequence-based predictions of protein subcellular localization. Chou & Cai reported the best performing predictive system, as measured on a carefully constructed gold standard, used a hybrid input containing both sequence patterns and protein annotations based on the molecular function terms from the Gene Ontology (GO)[1].

Computational methods to predict subcellular localization are a crucial bioinformatic task[5]. Several broad classes of predictive information have been brought to bear on this task. One class of informative information is the global amino acid composition of the protein. For example, NNPSL[13] used neural networks with amino acid composition inputs, and SubLoc[8] took a similar approach using support vector machines as the induction method. A second class of predictive information is the presence of signal peptides, which are short sub-sequences of approximately 15 to 60 amino acids shown to play functional role in protein transport. Various computational methods have been used to characterize and recognize instances of these signals, and then use them to predict specific cell locations. An example of this approach is TargetP[6]. A third class of predictive information arises from protein homology, which as the breadth of annotation grows, becomes increasingly useful. The recent LOCkey[11] system offered an unusual twist on homology-based approaches, identifying putatively homologous proteins by sequence similarity search, using natural language processing technology to extract textual features from the annotations of the homologs, and using those features as inputs to an inductive classifier for prediction of location. Approaches that combine multiple classes of predictive information, such as Grady et al.'s and Chou & Cai's actually have a long history in this area, going back more than a decade to Nakai & Kanehisa's classic PSORT system[12].

Although the Chou & Cai results demonstrate the utility of inclusion of GO annotations in the set of information used for location prediction, the direct relationship between macromolecular function terms in the GO and subcellular location has, to our knowledge, not been previously explored. The main motivation for exploring this relationship is to understand the nature of the contribution that molecular function (and its existing annotation) makes to the prediction of subcellular localization; however, there are other aspects of this work that are also significant. As described below, there is a linkage in the GO annotation process between molecular function annotation and subcellular localization annotation; the assessment made in this study may be useful in improving that process. Also, most predictive or informative function terms are imperfectly discriminating; the minority (or mispredicted) localizations often call out significant biological phenomena.

## 2. Methods

### 2.1. *Source of annotations*

The GO provides controlled vocabularies in three broad categories: molecular function, biological process, and subcellular localization. Only protein annotations from the GO molecular function category are included in this analysis, and they are compared with two distinct subcellular localization gold standards. One localization dataset is derived from annotations from the SwissProt database[2], the other from a recent hand-curated set described in Nair & Rost[11].

We did *not* use the GO subcellular localization annotations for this work. The reason is that the GO location and function are annotated in an intentionally dependent manner. GOA curation guidelines specify that subcellular location annotations can be inferred by the curator directly from molecular function information when the curator cannot find any relevant evidence in the biological literature. The assessment of the true degree of correlation between molecular function and subcellular localization would be confounded by this practice if we used GOA's localization data. The use of the independently curated annotations for localization disentangles the annotation process from any actual biological relationship.

The GO biological process category was not assessed for relationship to location because it is intentionally designed to integrate sets of molecular functions. For example, the process of signal transduction inherently involves signals from the extracellular space interacting with receptors at the membrane, which in turn modify secondary messengers in the cytoplasm, which often end up causing changes in transcription occurring in the nucleus. While each of these activities is carried out by molecules which can reasonably be expected to have their own function annotation and location, the process itself cannot be said to have a subcellular location.

### 2.2. *Localization-specific GO Terms*

The GO molecular function terms and relationships used in this analysis were taken from the molecular function ontology flat file on the GO website[a] from May 05 2004, version 1.28. Protein annotations for both SwissProt generated localization (the SUBCELLULAR LOCALIZATION field in CC lines) and GO terms (GO field in DR lines) were taken from SwissProt

---

[a]http://www.geneontology.org/

release 42. Only GO terms that appeared in the molecular function flat file were used in the analysis.

Note that both of the above-mentioned fields are optional, and many proteins are missing one or the other field, or both. SwissProt release 42 includes 135,850 proteins; only 6,686 ($\sim$ 5%) have relevant annotations for both location and molecular function. Generally, the issue is lack of coverage of GO annotations; for example, 3,655 proteins are annotated as localized in the nucleus, but only 937 ($\sim$ 25%) of those have GO molecular function terms associated with them.

Certain subcellular localization is not well represented in SwissProt. For example, almost no GO molecular function terms are associated with proteins in chloroplast, which is an organelle typically found in green plants. Although GO annotation for plants is underway by several model organism database groups, particularly for *Arabidopsis thaliana* and the agriculturally important cereal grasses, these annotations are not yet very prevalent in SwissProt, where the priority for the ongoing GO annotation effort is on human proteins.

We took two different approaches to identifying molecular function terms that are associated with localization. The first approach calculated the proportion of occurrences of each molecular function term to the most frequent location for that term and, separately, for the term and all of its subtypes ("is-a" children in the GO DAG). Any term (or term and its subtypes) that is associated with one location over a threshold proportion (set here at 70%) is reported in Table 1.

In order to make this calculation efficient, this specific approach was taken. Define the set of locations of interest to be L $\equiv$ {nucleus, extracellular space, membrane, mitochondrion, endoplasmic reticulum and Golgi} and the set F to be all of the GO terms that are children of Molecular Function.

- Initialize four accumulators for each combination of location ($l \in L$) and function ($f \in F$): $acc_{f,l}, acc_{f,\sim l}, acc_{f,l}^{incl}, acc_{f,\sim l}^{incl}$ to zero.
- For each location $l \in L$, do
    - Let $P_l$ be the set of proteins annotated to have location $l$. For each $p_l \in P_l$, do
        * Let $F_{p_l}$ be the set of molecular function annotations associated with the protein $p_l$. For each $f_{p_l} \in F_{p_l}$, increment $acc_{l,f}$.
    - For each $l$ and $f$ such that $acc_{l,f} > 0$, let $P_{f,\sim l}$ be the set of

proteins annotated with function f but with a location other than $l$. For each $p_{f,\sim l} \in P_{f,\sim l}$, increment $acc_{f,\sim l}$

- Beginning from the leaves of the GO DAG and working up is-a links to the root node for the molecular function hierarchy, traverse each $f \in F$ and set $acc_{f,l}^{incl}$ to be the sum of $acc_{f,l}$ and the $acc_{f_{child},l}$ for each child $f_{child}$ of function node f. Similarly for $acc_{f,\sim l}^{incl}$

- Use depth-first search starting from the root of the molecular function DAG to identify the highest-level terms whose ratios $acc_{f,l}/(acc_{f,l} + acc_{f,\sim l})$ or $acc_{f,l}^{incl}/(acc_{f,l}^{incl} + acc_{f,\sim l}^{incl})$ are greater than threshold

Caching of partial results is used to avoid redundant calculations. The approach avoids doing any work on the large portion of the proteins that are not annotated in a way that is relevant to the particular calculation.

### 2.3. *Identification of discriminative terms by information gain*

The informativeness of GO molecular function terms with respect to location can also be quantified by *information gain*, a measure of the amount of information (in bits) of that the knowledge of a feature (here, molecular function) contributes to knowledge of the class of the entity (here, location)[10]. An information gain of 1 means that knowledge of the molecular function is a perfect predictor of location, and an information gain of 0 means that knowledge of the molecular function provides no information regarding location.

In order to address concerns about the representativeness of the SwissProt dataset, we did this test on a completely independent set of annotations, that used by Nair & Rost[11]. This dataset classifies 1161 proteins into 10 different locations. These proteins are annotated with 207 different GO molecular functions. The information gain of the presence or absence of each molecular function term with respect to the distribution of locations was calculated. Information gain calculated in this way does not indicate *which* location is associated with a particular molecular function, or even that a function is associated with a single location. The implication of positive information gain is that "purity" of the locations associated with a set of proteins separated by the presence or absence of a particular function is higher than would be expected from a random division into the same size sets. This entropic measure has cleaner formal qualities, but is less directly useful, than the more ad hoc first method.

## 3. Results

The results of the first method, the ad hoc generation of localization-specific molecular function terms, found terms that either on their own or including annotations from their descendent terms provided ≥ 70% specificity for at least one of the 6 locations. Nineteen terms were over threshold for nuclear localization: 52 for membrane, 16 for extracellular, 2 for endoplasmic reticulum, 4 for mitochondrion, and 4 for Golgi. A sample of the highest-scoring terms is shown in Table 1. The complete set of results is available on a supplementary web site http://compbio.uchsc.edu/Hunter_lab/Zhiyong/psb2005.

Table 1. Selected highly discriminating terms (including children) for the six subcellular localization derived from the search for location-specific terms in SwissProt.

| Location | Predictive GO Molecular Function terms |
|---|---|
| nucleus | GO:0003676 Nucleic acid binding |
| | GO:0008134 Transcription factor binding |
| | GO:0030528 Transcription regulator activity |
| membrane | GO:0004872 Receptor activity |
| | GO:0015267 Channel/pore class transporter activity |
| | GO:0008528 Peptide receptor activity, G-protein coupled |
| extracellular | GO:0005125 Cytokine activity |
| | GO:0030414 Protease inhibitor activity |
| | GO:0005201 Extracellular matrix structural constituent |
| mitochondria | GO:0015078 Hydrogen ion transporter activity |
| | GO:0004738 Pyruvate dehydrogenase activity |
| | GO:0003995 Acyl-CoA dehydrogenase activity |
| | GO:0015290 Electrochemical potential-driven transporter activity |
| E.R. | GO:0004497 Monooxygenase activity |
| | GO:0016747 Transferase activity, transferring groups other than amino-acyl groups |
| Golgi | GO:0016757 Transferase activity, transferring glycosyl groups |
| | GO:0015923 Mannosidase activity |
| | GO:0005384 Manganese ion transporter activity |

The results from the information gain method were largely comparable with the location-specificity measure, even though the datasets were completely independent and the methods are quite different. 194 of the 207 terms had positive information gain, although only ten had information gains of .01 or greater, with the greatest gain at 0.047. These numbers are relatively low because while the presence of a function (such as DNA binding) may be quite specifically associated with a particular location, many other proteins without that function are likely to be in the same location, driving down the information gain substantially. The top ten results are shown in Table 2, and the complete results are available from the supplementary web site (URL above).

Table 2. The ten highest location information gain GO molecular function terms.

| Information Gain | GO Molecular Function terms |
|---|---|
| 0.047 | GO:0003677 DNA binding |
| 0.024 | GO:0005179 hormone activity |
| 0.024 | GO:0003676 nucleic acid binding |
| 0.022 | GO:0003700 transcription factor activity |
| 0.016 | GO:0008270 zinc ion binding |
| 0.015 | GO:0004129 cytochrome-c oxidase activity |
| 0.015 | GO:0003735 structural constituent of ribosome |
| 0.013 | GO:0008009 chemokine activity |
| 0.011 | GO:0008083 growth factor activity |
| 0.010 | GO:0016491 oxidoreductase activity |

## 4. Discussion

### 4.1. *Comparability of the methods*

Most of the high information gain terms appear in the list of the location-specific terms, and some of the differences are methodological. For example, the high information gain set includes two terms which have a parent/child relationship ("GO:0003677 DNA binding" and "GO:0003676 Nucleic acid binding") while only the parent term would make it into the location-specific list generated by the ad hoc method. Other differences arise from use of different datasets: e.g., "GO:0008528 Peptide receptor activity" is a highly predictive function in the ad hoc method, but is not associated with any of the 1161 proteins in the Nair & Rost dataset used to calculate information gain.

However, some of the differences are more interesting. For example, two of the top information gain terms, "GO:0008200 Zinc ion binding" and "GO:0005179 Hormone activity" have relatively low location specificity scores. These terms strongly favor a biased subgroup of locations, rather than a single location (e.g. the molecular function "Hormone activity" is associated with proteins annotated to both extracellular and membrane locations, not exclusively one or the other).

## 4.2. *Why function is complementary to sequence in location prediction*

As described above, all published location prediction methods are at least in part sequence-based, but two recent methods[3],[7] that use functional information in addition to sequence outperform other methods. Using the set of proteins from our experiments that were labelled with highly predictive functions, we found several examples where sequence-only methods made inaccurate predictions. For example, MBL_DROME in Swiss-Prot is a protein associated with terminal differentiation of photoreceptor cells in Drosophila. It is annotated as a nuclear protein with GO function "GO:0003676 nucleic acid binding". However, it was not predicted as a nuclear sequence by predictNLS[4], a widely used server for identifying nuclear localization signals in sequence. Furthermore, TMHMM[9], a popular method for recognizing transmembrane helices in proteins, incorrectly predicted it as a membrane protein, since a putative transmembrane helix region was identified in the sequence. However, "nucleic acid binding" is highly predictive of nuclear localization, and was never observed (in our data) as a function of membrane proteins.

## 4.3. *Biological interpretations*

Several of the predictions (and prediction failures) have interesting biological interpretations. Focusing on the molecular function terms that are both predictive and occur in many annotations identifies useful biological knowledge implicit in the annotations. It may be possible to exploit these regularities in other applications, e.g. automated methods for constructing knowledge-bases. Here are several illustrative examples:

- Nucleus: The molecular functions that are predictive of nuclear localization are mostly related to nucleic acids. The most predictive term is "GO:0003676 Nucleic acid binding", which is also among the high information gain function terms. Other predictive terms include "GO:0030528 Transcription regulator activity", "GO:0008134 Transcription factor binding", and "GO:0004386 Helicase activity". However, none of these terms are associated solely with the nucleus; each annotates proteins that are localized elsewhere as well. For example, aside from the nucleus, "GO:0003700 Transcription factor activity", a child term of "GO:0030528 Transcription regulator activity" is commonly associated with proteins annotated as cytoplasmic. This duality reflects the fact that transcription factors are often found in inactive form in the cytoplasm, and transported to the nucleus when activated. For example, consider the major signal transduction family Rel/NF-kappaB (NF-kB, KBF1_HUMAN in Swiss-Prot), which is involved in the control of a large number of normal cellular and organismal processes, such as apoptosis and inflammation. The interaction of NF-kB with (among other proteins) IkBa both inhibits its ability to bind to DNA and plays a role in maintaining its cytoplasmic localization. When a cell receives any of a multitude of extracellular signals, NF-kB dissociates from IkBa, is activated functionally and is transported to the nucleus. Many other proteins annotated with transcription factor activity are processed similarly, leading to the observed "imperfect predictiveness." However, an alternative view is that the annotations fail to capture the dynamism in the localization of these proteins.

- Membrane: More function terms are associated with membrane localization that with any other location. This is in part because, unlike in the case of the nucleus, the most abstract functions of membrane proteins (e.g. "GO:0005215 Transporter activity") are not specific to membrane, and more specific children of these terms must be used to achieve over-threshold predictiveness. For example, note that "GO:0005489 Electron transporter activity" is associated mainly with cytoplasmic proteins (although not enough so to be predictive at the 70% level), while "GO:0015267 Channel/pore class transporter activity" is predictive of membrane proteins.

- Mitochondrion: Mitochondrion carries out oxidative phosphorylation and produces most of the ATP in eukaryotic cells. We found 4 highly predictive terms for localization to this organelle, including two that are commonly characterized as specific to energy metabolism ("GO:0004738 Pyruvate dehydrogenase activity" and "GO:0003995 Acyl-CoA dehydrogenase activity"). The other two terms ("GO:0015078 Hydrogen ion transporter activity" and "GO:0015290 Electrochemical potential-driven transporter activity") describe functions not generally characterized in textbooks as specific to mitochondria. However, these are among the strongest associations found in this study ("GO:0015078 Hydrogen ion transporter activity" is associated with 81 mitochondrial proteins and no non-mitochondrial proteins).

- Endoplasmic Reticulum (ER): The difference between the biological process carried out in a particular location and specificity of the functions involved in that process is illustrated in the results for the ER. The processes associated with the ER are primarily the synthesis of lipids and secretory proteins. However, the two molecular function terms that are predictive of ER localization, "GO:0004497 Monooxygenase activity" and "GO:0016747 Transferase activity, transferring groups other than amino-acyl groups", are not obviously representative of these processes.

### 4.4. *Implications for GO annotators*

As described in the introduction, GO provides an ontology of subcellular localization terms itself, and GO annotators sometimes infer such localizations on the basis of molecular function. Such inferences are assigned the evidence code IC (Inferred by curator), although it is not clear if all the localization annotations with IC evidence codes are done on the basis of molecular function alone.

We believe it might be possible to extend this kind of annotation on the basis of these results. For example, we did not find any localization annotations for the mitochondria with IC evidence codes. However, as noted above, the molecular function "GO:0015078 Hydrogen ion transporter activity" (among others) seems a very strong predictor for that localization. Several other instances also seem to provide evidence that could be used by curators to infer localizations.

### 4.5. *Conclusion*

The biological relationships among molecular function and subcellular localization are at least partially reflected in protein annotations. These parallel relationships can be demonstrated both in measures of information gain and in the development of effective ad hoc predictors of location. These results provide an explanation of why hybrid prediction methods perform better than sequence-based methods alone, but also suggest potential improvements that might be made in the annotation process, and illustrate important biological phenomena.

### Acknowledgments

This work was supported by NIAAA grant 5U01 AA13524-03 (LH) and NIGMS graduate training grant T32-GM07635-25 (ZL).

# References

1. M. Ashburner, C. A. Ball, J. A. Blake, D. Botstein, H. Butler, J. M. Cherry, A. P. Davis, K. Dolinski, S. S. Dwight, J. T. Eppig, M. A. Harris, D. P. Hill, L. Issel-Tarver, A. Kasarskis, S. Lewis, J. C. Matese, J. E. Richardson, M. Ringwald, G. M. Rubin, and G. Sherlock. Gene Ontology: tool for the unification of biology. *Nature Genet.*, 25(1):25–29, 2000.

2. B. Boeckmann, A. Bairoch, R. Apweiler, M. C. Blatter, A. Estreicher, E. Gasteiger, M. J. Martin, K. Michoud, C. O'Donovan, I. Phan, S. Pilbout, and M. Schneider. The SWISS-PROT protein knowledgebase and its supplement TrEMBL in 2003. *Nucleic Acids Research*, 31(1):365–370, 2003.

3. K. Chou and Y. Cai. A new hybrid approach to predict subcellular localization of proteins by incorporating gene ontology. *Biochem Biophys Res Commun.*, 311(3):743–747, 2003.

4. M. Cokol, R. Nair, and B. Rost. Finding nuclear localization signals. *EMBO Rep*, 1:411–415, 2000.

5. F. Eisenhaber and P. Bork. Wanted: subcellular localization of proteins based on sequence. *Trends in Cell Biology*, 8(4):169–170, 1998.

6. O. Emanuelsson, H. Nielsen, S. Brunak, and G. von Heijne. Predicting subcellular localization of proteins based on their n-terminal amino acid sequence. *Journal of Molecular Biology*, 300(5):1005–1016, 2000.

7. J. L. Gardy, C. Spencer, K. Wang, M. Ester, G. E. Tusnády, I. Simon, S. Hua, K. deFays, C. Lambert, K. Nakai, and F. S. Brinkman. PSORT-B: Improving protein subcellular localization prediction for gram-negative bacteria. *Nucleic Acids Research*, 31(13):3613–3617, 2003.

8. S. Hua and Z. Sun. Support vector machine approach for protein subcellular localization. *Bioinformatics*, 17(9):721–728, 2001.

9. A. Krogh, B. Larsson, G. von Heijne, and E. L. Sonnhammer. Predicting transmembrane protein topology with a hidden markov model: application to complete genomes. *Journal of Molecular Biology*, 305:567–580, 2001.

10. T. Mitchell. *Machine Learning*. McGraw-Hill, 1997.

11. R. Nair and B. Rost. Inferring sub-cellular localization through automated lexical analysis. *Bioinformatics*, 18:S78–S86, 2002.

12. K. Nakai and M. Kanehisa. A knowledge base for predicting protein localization sites in eukaryotic cells. *Genomics*, 14(4):897–911, 1992.

13. A. Reinhardt and T. Hubbard. Using neural networks for prediction of the subcellular location of proteins. *Nucleic Acids Research*, 26(9):2230–2236, 1998.

# LINKING ONTOLOGICAL RESOURCES USING AGGREGATABLE SUBSTANCE IDENTIFIERS TO ORGANIZE EXTRACTED RELATIONS

BYRON MARSHALL, HUA SU, DANIEL MCDONALD, HSINCHUN CHEN

Email: {byronm,hsu,dmm,hchen}@eller.arizona.edu

*MIS Department, University of Arizona, McClelland Hall 430  1130 East Helen Street*
*Tucson, Arizona 85721, USA*

Systems that extract biological regulatory pathway relations from free-text sources are intended to help researchers leverage vast and growing collections of research literature. Several systems to extract such relations have been developed but little work has focused on how those relations can be usefully organized (aggregated) to support visualization systems or analysis algorithms. Ontological resources that enumerate name strings for different types of biomedical objects should play a key role in the organization process. In this paper we delineate five potentially useful levels of relational granularity and propose the use of aggregatable substance identifiers to help reduce lexical ambiguity. An aggregatable substance identifier applies to a gene and its products. We merged 4 extensive lexicons and compared the extracted strings to the text of five million MEDLINE abstracts. We report on the ambiguity within and between name strings and common English words. Our results show an 89% reduction in ambiguity for the extracted human substance name strings when using an aggregatable substance approach.

## 1  Introduction

In the past few years several systems that extract biological relations from biomedical texts have been created. These systems are intended to help researchers leverage vast and growing collections of research literature. Extraction results are usually evaluated for accuracy but this is only one of several important considerations. Consumers of the extracted information (both humans and analysis algorithms) benefit when information is meaningfully organized. That is, (a) multiple references to the same biological substance or process are indexed, (b) substance references are marked so that they can be correctly associated with existing databases and resources, and (c) the context and granularity of the information is specified. These issues have not been ignored by other bioinformatics researchers. They are reflected in the development of semantic classes and ontologies [1], the use of existing lexicons for entity identification [2], the inclusion of context-related information in semantic frames [3], and the representation of extracted information in both binary and nested relations [4]. Still, techniques for aggregating and associating extracted information deserve additional investigation.

Creating a system to process biomedical texts and produce a useful network of relational information is a multi-faceted task [5]. As part of the GeneScene project (http://ai.bpa.arizona.edu) researchers have developed text mining tools to automatically extract regulatory pathway relations from MEDLINE abstracts [6, 7]. The extracted information is used to create network visualizations and to support various data mining efforts. The work reported in this paper is part of an ongoing effort to improve the usefulness of extracted relations but the resulting aggregation methodologies are potentially interesting for the output of other information extraction systems as well. In particular, we evaluate an extensive aggregatable substance lexicon. Considerable attention is paid to lexical ambiguity which would hinder the matching of relation extraction system output to specific elements in existing lists of biological substances.

## 2 Relation Output Formats

The design of a relation aggregation system depends on the format of extracted relations. While relation extraction systems vary significantly in their output formats, labeled relational triples are frequently produced by extraction systems and are useful in analysis algorithms [8]. Output formats include predicate relations like those captured by GENIES and EDGAR [4, 9], inhibition relation pairs [10], labeled relational triples with negation extracted by the GeneScene Parser and the Arizona Relation Parser [6, 7], and binary object pairs with categorized relations in [11]. Binary biomedical pathway relations are frequently displayed as conceptual graphs and can be used with many analysis algorithms. More complex nested predicate relations can be stored as a knowledge base and queried but may be more difficult to visualize unless they are first reduced to a binary format. The balance of Section 2 describes a few of these systems.

Initial implementations of the GENIES system extracted simple binary relations such as *X activates Y*. Later the system was expanded to handle more complex relations with the output expressed as nested predicates. This format is precise and can express much of the relational information contained in a text. Relations extracted by GENIES are integrated into the larger GeneWays system after they are "unwound" into simple binary statements (e.g., "Interlukin-2 binds Interlukin-2 receptor" [4]).

The GeneScene parser [6], the Arizona Relation Parser (ARP) [7], and a system developed by Palakal et al. [11] extract relational triples with labeled links connecting labeled entities. The GeneScene parser focuses on closed class words to identify important relations while ARP uses a hybrid syntax and semantic parser. Both extract negation indicators. In the ARP results, the link labels consist primarily of verbs or verb phrases and a negation indicator. Entities are phrases (generally noun phrases) extracted from the text. Table 1 shows 4 examples of relations extracted by ARP from MEDLINE abstracts. ARP output does directly support relation nesting. However, the second entity in relation (4), *E1A-induced apoptosis,* includes a substance (*E1-A*), a function (*apoptosis*), and an associator (*-induced*). ARP frequently extracts these complex entities, capturing important information beyond mere substance identification. Multiple relations can also be extracted from a single sentence as shown in the two relations extracted for the example sentence (4). This behavior represents a kind of relation nesting.

**Table 1.** Arizona Relation Parser Output

| Original Sentence | Resulting Relation | | | |
| --- | --- | --- | --- | --- |
| | Entity 1 | Negation | Connector | Entity 2 |
| (1) wild-type p53 tumor suppressor protein, which induces […] apoptosis… | wild-type p53 tumor suppressor protein | False | induces | Apoptosis |
| (2) Wt p53 also induced significant apoptosis | Wt p53 | False | also induced | significant apoptosis |
| (3) oncogene mutant p53 suppresses apoptosis | oncogene mutant p53 | False | suppresses | apoptosis |
| (4) mutant p53 blocked E1A-induced apoptosis | mutant p53 | False | blocked | E1A-induced apoptosis |
| | E1-A | False | Induced | apoptosis |
| (5) mutant p53 […] does not induce […] apoptosis | mutant p53 | True | does not induce | apoptosis |

The system described in [11] adds tags to text marking the boundaries of identified biological objects. Relations are specified by a subject, an object, and a relational classification. Classifications include directional relationships such as "binds" and "inhibits" as well as a hierarchical relation "same" to designate an "is a" relationship. Relationships are expressed between simple objects such as *KiAA1009 protein* but can also involve more complex objects such as *nuclear mitotic apparatus protein* or *cellular transfer RNA for tryptophan*. In addition to reporting on their object and relation extraction, [11] proposes an algorithm for grouping object synonyms. The grouping process does limited co-reference resolution and expands contractions to identify multiple references to the same object within a document. This process reduced the number of unique objects by 24.7% with 92% recall and 82% grouping specificity. That is, 18% of the groupings made were incorrect. The paper included a number of suggestions for improving the grouping process.

## 3    Five Levels of Relational Granularity

Based on feedback from researchers using GeneScene's visualization tools and experience employing extracted information for data mining applications we have developed a list of five potentially useful levels of relational granularity. These levels define how extracted relations can appropriately be combined in consolidating a network of extracted relations. They are intended to correspond to levels of aggregation that can be applied to extracted relations with reasonable accuracy. Thus, our selections were influenced by the needs of various users, the nature of available ontological resources, and the characteristics of the extracted relations. Once relations have been processed for aggregation it should be possible to provide a flexible query interface over various levels of granularity, extract simple pathway relation results for use in data mining and other analysis systems, and provide higher levels of analysis (e.g., identifying conflicting relations).

Our approach targets relations that describe how two biological objects interact. Each relation consists of two labeled entities (biological substances or functions) and a labeled connector (verbs). Entities may be assigned multiple features such as species, mutation, cellular component, and substance type. The intuition behind this approach is highlighted in a biologist's observation that we should combine references to wt MDM2, wild-type MDM2, and non-mutant MDM2. Each reference can be correctly understood as a non-mutated form of MDM2. Although some extraction systems may extract this kind of feature information as a relation, for our purposes we would want to assign the feature non-mutated to the entity and recognize the substance as a gene or protein related to the gene MDM2. Connectors are combined by grouping verbs. Appropriate methodologies need to be studied but we plan to begin our analysis using two fairly simple approaches: different morphological forms of a verb are combined, as in *induces* vs. *induce*, and a higher level of semantics can be applied to group verbs such as *inhibits* and *abrogates* into a common connector *inhibit*. Thus our aggregation approach allows for aggregation of complete relations as well as individual entities. We plan to support five levels of granularity in our aggregation system.

1) **Baseline** – the full text of the entity or connector labels must match
2) **Feature Match**
   a) all identified entity features must match
   b) morphological forms of the same connector verb are combined
3) **Typed Substance**
   a) entities with different identifiable substance types are not matched
   b) morphological forms of the same connector verb are combined
4) **Aggregatable Substance**
   a) references to a gene and its gene products are matched
   b) morphological forms of the same connector verb are combined
5) **Simple Pathway**
   a) references to a gene and its gene products are matched
   b) connector verbs are classified into one of 4 categories

**Baseline aggregation** is used to provide the maximum amount of differentiation in a visualization. We expect that it is most useful as a baseline for comparison of aggregation system results. Baseline aggregation makes no attempt to combine equivalent objects unless they are labeled with exactly the same words. Thus neither the relation nor any of the elements of *Mdm2 – inhibits – apoptosis* would be matched with *Mdm2 genes – are involved in – regulation of apoptosis.* Baseline aggregation minimizes information loss but accomplishes very little consolidation.

**Feature Match** aggregation increases network consolidation by comparing feature values assigned to an entity. If a substance has been identified as mutated or recognized as present in a particular tissue type or cellular domain, it is matched only with similarly identified items. For instance, mdm2 *antisense oligodeoxynucleotide – induces – Apoptosis* and *anti-sense MDM2 — induces – apoptosis* would be aggregated because the connectors and second entity (*induces – apoptosis*) match and both MDM2 entities can be identified as mutated (*antisensed*) forms of the substance MDM2. This level of aggregation may be useful for detailed pathway analysis.

**Typed Substance** aggregation is somewhat comparable to the granularity found in some manually created databases. In a network aggregated at this level entities with different substance types are not matched thus relations involving protein P53 would not be combined with references to the TP53 gene. In this case, *tumour suppressor gene p53 – induces – apoptosis* would be aggregated with *p53 tumour suppressor gene – is known to induce – apoptosis* but not with *p53 protein – induces – apoptosis* because the gene would not be considered equivalent to the protein.

**Aggregatable Substance** aggregation assigns equivalence to references to a gene and its related gene products. At this level of aggregation no attempt is made to distinguish between interactions related to a particular gene and interactions for the protein that gene encodes. This is partly a practical matter. Across a set of abstracts the exact same phrase is frequently used to refer to a gene and to the related protein making it difficult to distinguish these references. Analysis of nearby words and other cues in the document can help address but not eliminate this ambiguity. Also, as a matter of application, a researcher studying effects of the gene TP53 might well be interested in references to the protein p53 because the presence of the protein is related to expression of the gene. Entities here may also be biological functions and connectors at this level and morphological forms of a verb are matched (e.g., *MDM2 – inhibits – apoptosis* and *MDM2 oncoprotein – has been shown to inhibit – apoptosis* are considered to be equivalent relations).

**Simple Pathway** aggregation creates an overview of the information extracted from a text. This kind of relation can be used for example, as input into a data mining algorithm. Connecting verbs would be classified as belonging to one of 4 categories: induce, inhibit, directional association, non-directional association. Relations at this level might be comparable to relations extracted by parsers that identify only single semantic types of relations as in the work reported in [8] where only inhibition relations are extracted. Relations could be filtered so that only items with two recognizable substances and particular types of interaction are included. For a researcher, this representation of a pathway relation can be viewed as an outline or backbone of the regulatory network. For example, each of these relations,

*MDM2 inhibits apoptosis*

*MDM2 oncoprotein abrogates apoptosis*

*Human MDM2 interferes with p53-mediated cell death*

can be aggregated into a simple relation *MDM2 inhibits apoptosis*. Gene *MDM2* and its product *MDM2 oncoprotein* are matched to each other as the aggregatable substance *MDM2*, the verbs *inhibits, abrogates* and *interferes with* would all belong to the category inhibit, and *apoptosis* and *cell-death* would be identified as equivalent functions.

To implement our five level aggregation strategy, we need to identify multiple references to the same aggregatable substances and establish a method of linking those aggregatable substance references to other biomedical resources. Fortunately, several existing resources list and cross-reference genes, proteins, and other gene products. One primary task in building our aggregation system will be to construct an aggregatable substance lexicon from existing resources.

## 4    Merging Ontological Resources to Support Aggregation

Ontologies, lexicons, and controlled vocabularies are crucial to the aggregation process. Ontologies such as GO list concepts and concept classes and sometimes enumerate class instances (as GO does for biological functions). Other resources such as RefSeq, HUGO, LocusLink, and SGD also contribute to an overall ontology for the domain. For example, the concept class "Gene" is enumerated in the many human genes listed in LocusLink. Gene products such as proteins and mRNA are associated with Genes in the RefSeq repository. We will refer to these resources together as "ontological resources". Many different resources containing lists of biological object name strings have been created to support various tasks. Combining several lists will improve the coverage of an extraction system, but creating a lexicon merging process is a non-trivial task. One key merging problem is ambiguity. Ambiguity occurs (1) within a resource when a single name string is associated with more than one biological object, (2) between resources when two different resources associate the same name string with two or more different biological objects, and (3) when listed name strings are also commonly used English words such as *an, by, killer,* or *for.*

Two recent efforts to combine lists from multiple sources are documented in [12] and [2]. In [12] entries from selected mouse, fly, worm, and yeast lexicons were combined. Three kinds of ambiguity were measured: multiple name strings for the same gene, multiple genes identified by a single name string, and overlap with common English words. Ambiguity within each database was minimal, between 0% and 2.5%, except for the fly dataset with ambiguity found for 10% of the references. Across datasets there was significantly more ambiguity, ranging between 4% and 20%. [2] combined name strings from four existing resources (HUGO, SWISSPROT, OMIM, and TREMBL) into an unambiguous list of name references. They describe their generally automatic but manually adjusted process for lexicon curation. To test their lexicon, they selected a set of documents from which relations had been manually extracted for the TRANSPATH database. They report that they matched 94% of the substance names and that dictionary curation improved precision from 78% to 90%.

To support our five level aggregation approach, we created an aggregatable substance lexicon consisting of substance name strings from several ontological resources including RefSeq, LocusLink, HUGO, and SGD. RefSeq is a comprehensive repository of curated reference sequences for transcripts, proteins and genomic regions [13]. LocusLink provides an interface to curated sequences and descriptive information about genes with links to gene-related resources [13]. Entrez Gene was recently deployed as a replacement for LocusLink; in future work we will adjust to the new input data format. HUGO [14] and SGD [15] are standard databases for human and yeast genes respectively. We used the following information: from LocusLink official gene names, official symbols and aliases, and products and aliases; from RefSeq gene/protein names and synonyms; from Hugo previous symbols, aliases, and previous gene names; and from the SGD yeast gene collection symbols, names, and synonyms. We used LocusLink IDs for Human and RefSeq accession numbers for other species as the primary identifier of an aggregatable substance. A few erroneous entries (noise) inevitable in large data sources were removed from the lexicon.

Many name strings occur in several of the ontological resources. In some cases this reflects ambiguity, in other cases simple redundancy. Table 2 documents the degree of name string overlap in the resources we used. In some cases overlap was very high. For example, 68.5% of the HUGO name strings were also found in the RefSeq database. Still, each database seems to have added some new name strings to the list. Because human and yeast abstracts are our primary targets at this time, we report on overlap for those species independently.

For our purposes, a name string is only useful if it actually appears in the analyzed text. We began by measuring how frequently the name strings in our resources occur in MEDLINE abstracts. To implement this overlap test, we preprocessed the texts and the name strings to normalize word boundaries and used left-to-right, longest-first phrase matching. The name string list (NSL) included 214,862 unique, unfiltered words or phrases extracted from RefSeq, LocusLink, HUGO, and SGD. We evaluated the NSL list as a collection of lists, one for each resource and as a combined list with all duplicates removed. These name strings were compared to more than five million MEDLINE abstracts. This set was prepared by excluding non-English MEDLINE records and records with no abstract. As shown in Table 3, we found 35,289 of the unique (combined) NSL items in the abstracts. Only a small portion (35,289 / 214,862 or 16.4%) of the available name strings is directly useful in recognizing substance references in our target texts.

**Table 2.** Unique name string overlap between resources

| | RefSeq | LocusLink | HUGO | SGD | Other Databases |
|---|---|---|---|---|---|
| RefSeq | 169,312 a | 24.2% | 8.0% | 1.1% | 24.6% |
| LocusLink | 58.8% | 69,550 a | 26.6% | 2.3% | 66.4% |
| HUGO | 68.5% | 94.0% | 19,961 a | 3.7% | 94.8% |
| SGD | 14.1% | 11.1% | 5.1% | 14,089 a | 15.6% |

a - Diagonal entries list the number of unique name strings found e.g., 169,312 for RefSeq

The non-diagonal cells represent that portion of the entries from the row's lexicon also occurring the column's lexicon

The last column (Other Databases) shows the percentage of the entries from the row's lexicon that are found in any of the other resources

**Table 3.** Unique name strings found in 5 million MEDLINE abstracts by source and species

| | Human | Yeast | Other | All Species |
|---|---|---|---|---|
| RefSeq | 20,036 | 631 | 16,296 | 20,634 |
| LocusLink | 23,755 | | | 23,755 |
| HUGO | 9,070 | | | 9,070 |
| SGD | | 6,192 | | 6,192 |
| Combined | 25,570 | 6,588 | 16,296 | 35,289 |

While most of this overlap represents legitimate substance references, some of the matches are incorrect in that they result from overlap between name strings and common English words. We compared the extracted name strings to the MOBY list of common English words. Previous results show an overlap between 0 and 2.4% for substance names and common English words using the MOBY list and selected fly, worm, yeast, and mouse substance names [12]. Our results are comparable as shown in Table 4. We found overlap ranging from 0.16% to 3.19% between various resources and the MOBY list. Table 5 reports on only those NSL entries that actually occurred in MEDLINE abstracts and were common English words. Overlap between NSL and MOBY for human substance name strings has risen to 4.8% from the 1.29% reported in Table 4 for this important subset. Several systems have been created to identify biomedical substance references in free text (e.g. [16],[17]). These systems can use contextual cues to assign entity name boundaries; this might help them avoid identifying common English words as substances. One such system is the PROPER system [16]. This kind of processing may filter out some incorrect references (such as occurrences of common English words). Because PROPER is readily available on the internet, we also compared the NSL to a set of entities extracted by the PROPER system. Because some additional abstract preparation was required to run PROPER, we used a smaller set of 87,903 abstracts in this comparison. These abstracts were selected by searching MEDLINE for documents related to the p53 tumor suppressor gene. The PROPER system extracted 419,302 unique name strings from those abstracts. Of those, only 2.5%

(10,580 / 419,302) were found in the NSL. Table 6 breaks down the overlap by resource and species.

**Table 4.** Percentage of name strings that are common English words

|  | Human | Yeast | Other | All Species |
|---|---|---|---|---|
| RefSeq | 1.59% | 0% | 0.77% | 0.83% |
| LocusLink | 1.68% |  |  | 1.68% |
| HUGO | 3.19% |  |  | 3.19% |
| SGD |  | 0.16% |  | 0.16% |
| Combined | 1.29% | 0.16% | 0.77% | 0.74% |

**Table 5.** Percentage of unique name strings from MEDLINE that are also common English words

|  | Human | Yeast | Other | All Species |
|---|---|---|---|---|
| RefSeq | 5.3% | 0% | 5.3% | 5.2% |
| LocusLink | 4.7% |  |  | 4.7% |
| HUGO | 6.8% |  |  | 6.8% |
| SGD |  | 0.4% |  | 0.4% |
| Combined | 4.8% | 0.4% | 5.3% | 4.2% |

**Table 6.** Number of unique PROPER entity names found in the NSL

|  | Human | Yeast | Other | All Species |
|---|---|---|---|---|
| RefSeq | 6,933 | 62 | 5,532 | 8,543 |
| LocusLink | 8,073 |  |  | 8,073 |
| HUGO | 3,730 |  |  | 3,730 |
| SGD |  | 1,661 |  | 1,661 |
| Combined | 8,492 | 1,688 | 5,532 | 10,580 |

## 5    Aggregatable Substances and Lexical Ambiguity

In addition to common English word ambiguity, we tabulated situations where a single name string is used to reference multiple substances in a single species either in the same resource or in different resources, and where a single name string refers to substances in different species. Ambiguity values of 2-20% have been reported in previous research [2, 12]. Our results are shown in Table 7. The 11.7% reported for human name strings in RefSeq represent cases where a name string is associated with more than one substance. An example of this kind of ambiguity involves *RAB38* which is associated with RefSeq id *NP_071732* (a protein) and *NM_022337* (an mRNA). The "combined" row of Table 7 counts cases where a name string is associated with different substances across multiple resources. For example, the 39.4% combined human ambiguity includes the name string *p53* which is associated with an mRNA in RefSeq and with a gene in LocusLink. The "All Species" column includes examples such as *ACP1* which represents a particular human gene in LocusLink and a completely different yeast gene in SGD. The name strings that actually occur in MEDLINE abstracts are much more ambiguous as compared to the list of all available name strings. This is likely due to the common practice of referring to a gene and the protein it encodes with the same one word name string. These entries make up a substantial percentage of the items actually observed in MEDLINE abstracts.

**Table 7.** Percentage of unique name strings associated with multiple substances

|  | All NSL Entries | | | | Only entries found in MEDLINE | | | |
|---|---|---|---|---|---|---|---|---|
|  | Human | Yeast | Other | All Species | Human | Yeast | Other | All Species |
| RefSeq | 11.7% | 10.3% | 5.0% | 15.6% | 21.7% | 14.4% | 11.4% | 51.1% |
| LocusLink | 3.0% |  |  | 3.0% | 6.8% |  |  | 6.8% |
| HUGO | 3.1% |  |  | 3.1% | 5.6% |  |  | 5.6% |
| SGD |  | 2.5% |  | 2.5% |  | 3.8% |  | 3.8% |
| Combined | 39.4% | 3.8% | 5.0% | 23.1% | 73.2% | 4.7% | 11.4% | 62.0% |

Our system implements a single aggregatable substance ID for a gene and its gene products. This was accomplished primarily through the cross-reference information in RefSeq. RefSeq lists gene products and frequently includes the LocusLink identifier of the related gene. When this occurs we record both identifiers but use the LocusLink ID as the aggregatable substance ID. For example RefSeq provides the name string PIRB as a synonym for 3 mRNA transcription variants associated with LocusLink ID 29990. Thus, our list includes both PILRB (the official symbol) and PIRB (the synonym) as name strings associated with LocusLink ID 29990. To see how much this consolidation reduces name string ambiguity, we recalculated the percentage of ambiguous name strings where a single name string refers to a substance outside of its aggregatable substance. Table 8 shows that ambiguity is substantially reduced using the aggregatable substance approach. In Table 7 we showed that 39.4% of the human name strings found in the combined NSL list were associated with more than one substance. When considering only those items found in MEDLINE, ambiguity is even higher at 73.2%. Using our aggregatable substance approach we found only 6.1% of the name strings to be ambiguous, an 85% improvement. An even higher (89%) improvement was recorded for items that occur in MEDLINE with only 7.8% of the name strings ambiguous as to aggregatable substance. Similar improvement (84%) was found for NSL items that were also extracted by PROPER from our p53–related abstract collection. An aggregatable substance lexicon can be used to combine entities at the aggregatable substance and simple pathway levels of a five level framework of relational granularities. Our results show that considering existing lexicons from this perspective substantially reduces lexical ambiguity.

Table 8. Reduction in ambiguity using the aggregatable substance approach.

| | Name Strings Associated with Multiple Substances | Name Strings Associated with Multiple Aggregatable Substances | Improvement |
|---|---|---|---|
| NSL Entries | 39.4% | 6.1% | 85% |
| NSL Entries found in MEDLINE abstracts | 73.2% | 7.8% | 89% |
| NSL Entries Found by PROPER | 78.6% | 12.4% | 84% |
| * This table reflects human substance name strings only | | | |

## 6    Discussion and Future Directions

Our merged NSL associates name strings with substances that would appear at different levels of a biomedical ontology. Some resources list genes, some proteins, and others mRNA. The close relationship that exists between certain substances is reflected in the words authors use in referring to those substances and in the name strings associated with those substances in biomedical lexicons. To address this resulting ambiguity we have proposed the use of a single aggregatable substance ID for a gene and its products. Our results show that this approach substantially reduces lexical ambiguity. Unfortunately, using this approach, a substance name matcher would be unable to differentiate between genes and proteins. This problem can be addressed a number of other ways. In many cases, nearby words are likely to provide useful clues that can be used to differentiate references so a substance reference could be matched to an aggregatable substance list and the substance type could then be clarified using other techniques. Furthermore, there are likely to be many tasks for which this differentiation is not essential. For example, a researcher studying the relationship between p53 and apoptosis might well be interested in literature connecting apoptosis to either the p53 protein or the TP53 gene. The aggregatable substance notion presented here will be an integral part of a larger system that organizes extracted relations to support human visualization and algorithmic analysis of information from biomedical texts. In future work we plan to consider extending the aggregatable substance notion to account for homologous genes and we have already begun development of an aggregation system to support our five level aggregation approach.

## 7    Acknowledgments

This project was supported by the following grant: NIH/NLM, 1 R33 LM07299-01, 2002-2005, "Genescene: a Toolkit for Gene Pathway Analysis"

## References

1. A. Rzhetsky et al., "A knowledge model for analysis and simulation of regulatory networks", *Bioinformatics* **16**, 1120-1128 (2000)
2. D. Hanisch et al., "Playing biology's name game: identifying protein names in scientific text", PSB 2003
3. R. Gaizauskas et al., "Protein structures and information extraction from biological texts: the PASTA system" , *Bioinformatics* **19**, 135-143 (2003)
4. C. Friedman et al., "GENIES:a natural-language processing system for the extraction of molecular pathways from journal articles", *Bioinformatics* **17**, 74-82 (2001)
5. A. Rzhetsky et al., "GeneWays: a system for extracting, analyzing, visualizing, and integrating molecular pathway data", *J. of Biomedical Informatics* **37**, 43-53 (2004)
6. G. Leroy, H. Chen, and J. D. Martinez, "A shallow parser based on closed-class words to capture relations in biomedical text", *J. of Biomedical Informatics* **36**, 145-158 ( 2003)
7. D. M. McDonald et al., "Extracting gene pathway relations using a hybrid grammar: the Arizona Relation Parser", *Bioinformatics* Forthcoming (2004)
8. G. R. G. Lanckriet et al., "Kernel-based data fusion and its application to protein function prediction in yeast", PSB 2004
9. T. C. Rindflesch et al., "EDGAR: extraction of drugs, genes and relations from the biomedical literature", PSB 2000.
10. J. Pustejovsky et al., "Robust relational parsing over biomedical literature: extracting inhibit relations", PSB 2002
11. M. Palakal et al., "Identification of biological relationships from text documents using efficient computational methods", *J. of Bioinformatics and Comp. Biology* **1**, 307-342 (2003)
12. O. Tuason et al., "Biological nomenclatures: a source of lexical knowledge and ambiguity", PSB 2004.
13. D. L. Wheeler et al., "Database resources of the National Center for Biotechnology Information: update", *Nucleic Acids Res.* **32**, D35-40 (2004)
14. H. M. Wain, "Genew: the Human Gene Nomenclature Database, 2004 updates", *Nucleic Acids Res.* **32**, D255-7 (2004)
15. K. R. Christie et al., "Saccharomyces Genome Database (SGD) provides tools to identify and analyze sequences from Saccharomyces cerevisiae and related sequences from other organisms", *Nucleic Acids Res.* **32**, D311-4 (2004)
16. K. Fukuda et al., "Toward information extraction: identifying protein names from biological papers" , PSB 1998
17. L. Tanabe and W. J. Wilbur, "Tagging Gene and Protein Names in Full Text Articles, Natural Language Processing in the Biomedical Domain", July 2002, Assoc. for Comp. Ling.

# IMPLICATIONS OF COMPOSITIONALITY IN THE GENE ONTOLOGY FOR ITS CURATION AND USAGE

PHILIP V. OGREN

*University of Colorado at Boulder, Dept. of Computer Science, Boulder, CO USA*

K. BRETONNEL COHEN, LAWRENCE HUNTER

*Center for Computational Pharmacology, University of Colorado Health Sciences Center, School of Medicine, Aurora, CO USA*

In this paper we argue that a richer underlying representational model for the Gene Ontology that captures the implicit compositional structure of GO terms could have a positive impact on two activities crucial to the success of GO: ontology curation and database annotation. We show that many of the new terms added to GO in a one-year span appear to be compositional variations of other terms. We found that 90.2% of the 3,652 new terms added between July 2003 and July 2004 exhibited characteristics of compositionality. We also examine annotations available from the GO Consortium website that are either manually curated or automatically generated. We found that 74.5% and 63.2% of GO terms are seldom, if ever, used in manual and automatic annotations, respectively. We show that there are features that tend to distinguish terms that are used from those that are not. In order to characterize the effect of compositionality on the combinatorial properties of GO, we employ finite state automata that represent sets of GO terms. This representational tool demonstrates how ontologies can grow very fast, and also shows that small conceptual changes can directly result in a large number of changes to the terminology. We argue that the curation and annotation findings we report are influenced by the combinatorial properties that present themselves in an ontology that does not have a model that properly captures the compositional structure of its terms.

## 1. Introduction

There have been several papers in recent years that address the need to redesign the underlying model for representing concepts in the Gene Ontology [1,2]. For example, Verspoor et al. propose mapping GO to a lexical semantic network [3]. The Gene Ontology Next Generation project [4] seeks to represent GO using description logics, while Yeh et al. [5] show how GO could be maintained in the frame-based Protégé environment. Joslyn et al. [6] suggest representing GO as a poset. The Open Bio-Ontology Language (OBOL) [7] describes GO terms by means of a Prolog grammar that helps generate and maintain logical definitions. These efforts at suggesting varying representational schemes for GO have been productive. For example, a number of these efforts have detected missing terms and relations. The Gene Ontology Annotation Tool [8] makes use of description logic constructs to help facilitate annotation of gene and protein databases by use of constraints that help human annotators choose multiple concepts from different GO sub-ontologies that are logically consistent. Verspoor et al. [9] and Joslyn et al. [6] demonstrate some of the benefits that can accrue in natural language processing and high-throughput gene expression data analysis from other representations. In this paper we do not promote or propose any specific model but instead present experimental data that suggests that any new representation of GO should explicitly model the compositional nature of GO terms, which in the current incarnation of the ontology is only implicit.

### 1.1. *Compositional Structure of GO Terms*

GO terms exhibit underlying compositional structures that are not represented explicitly in the ontology. In previous work, we demonstrated that GO terms often contain other GO terms as proper substrings [10]. For example, *activated T-cell proliferation* (GO:0050798) contains the term *T-cell proliferation* (GO:0042098), and these embeddings can continue through several levels of ontological structure. In our earlier work, strings like *activated* that are added to other terms to form a new term were labelled *complements;* the term that is found inside another term is labelled a *contained term.* Where our previous work focused on the complements that are added to previously existing terms to create new terms, work by Verspoor et al. [3] looked closely at the terms to which complements are added, and showed that insights into the relations between these terms can be gained by considering the various derivational processes by which new terms are created. Taken together, these two papers address compositionality from two complementary perspectives and establish that compositionality is a pervasive and prevalent phenomenon in GO.

### 2. Methods

### 2.1. *Assessing implications of compositionality for curation*

To test the hypothesis that the compositional nature of GO terms significantly contributes to the growth of GO, we examined the 3,652 new terms added between July 2003 and July 2004. Our methods for detecting compositional terms are based on the observation that composed terms will be very similar to other terms, or will have contained terms, or both. Since no single measure will detect every compositional term, we performed two separate experiments to estimate the number of new terms added to GO in the past twelve months that appear to be compositional. For both experiments we compared the new terms with previously existing terms as well as other new terms.

The first experiment is based on the observation that compositionally derived terms are very similar to other terms (see Section 2.3). We measured similarity by calculating the minimum edit distance (MED) between each new term and all other terms in the ontology (including other new terms). The MED is the minimum number of edits needed to convert one string into another. An edit can be either an insertion, a deletion, or replacement of one string by another [11]. While typically this algorithm is applied on the character level to compare two words, we applied the algorithm on the word level to compare terms. We used the Levenshtein weighting [12], in which insertion, deletion, and substitution all have equal weights of one. With this weighting, the terms *trophectoderm cell proliferation* (GO:0001834) and *natural killer cell proliferation* (GO:0001787) have a MED of two because a transformation from the former to the latter can be accomplished with one string replacement (*trophectoderm* → *natural*) and one insertion ($\varnothing$ → *killer*). We then counted the number of new terms that are at a MED of exactly one from some other term[a]. Of the 3,652 new terms added to GO between July 2003 and July 2004, we found that 2,696 or 73.2% of the new terms had a MED of exactly one from at least one other term.

The second experiment looks for terms that contain another term as a proper substring (a string $x$ is a proper substring of the string $y$ if all of $x$ is in $y$ and $x$ is not identical to $y$) [13]. For example, *ligase activity, forming nitrogen-metal bonds, forming coordination complexes* (GO:0051002) contains the term *ligase activity, forming nitrogen-metal bonds* (GO:0051003). The longer term, which is a new term, is clearly compositionally related to the shorter term, but would not have been detected by the MED test, since the MED between them is three. We found that 2,507 or 68.6% of the new terms contained another term.

---

[a] With this weighting, all pairs of two-word terms, e.g. *symporter activity* (GO:0015293) and *hydrolase activity* (GO:0016787) have a minimum edit distance of one. However, in these cases the low minimum edit distance probably does not reflect a derivational relationship, so we calculated the minimum edit distance only when one of the terms is longer than two words.

We then determined the intersection and the union of the sets of terms discovered by the two experiments. Table 1 summarizes the results of the two experiments and the union and intersection of the two sets of terms. The union of the sets of terms discovered by both experiments contains 3,294 terms or 90.2% of the new terms. Thus, a large majority of the new terms have at least one characteristic of compositional terms. The number of terms identified by both measures is 1,909 or 52.5% of the new terms. Thus, a small majority of the new terms have two characteristics.

|  | Count | Percentage |
|---|---|---|
| Total New Concepts | 3652 | 100% |
| MED1 | 2696 | 73.2% |
| CT | 2507 | 68.6% |
| MED1 _ CT | 1909 | 52.5% |
| MED1 U CT | 3294 | 90.2% |
| Neither | 358 | 9.8% |

Table 1. MED1 – new concepts that have a minimum edit distance of one from another term. CT – new concepts that have a contained term.

## 2.2. Assessing implications of compositionality for annotation

### Annotations per term

Analysis of how GO is used to annotate gene and protein databases reveals that much of GO is either barely used or not used at all. We downloaded all annotations available at the GO website and counted for each GO term how many annotations (or usages) are associated with it[b]. We split the annotations into three broad categories: human curated, computer generated, and all annotations combined[c]. For each GO term we counted the number of annotations that contain that term. We then ranked the GO terms based on their frequency of usage. Figure 1 shows the data for the

---

[b] The annotations were downloaded on July 5, 2004 from
http://www.geneontology.org/GO.current.annotations.shtml.
[c] For the human curated annotations we used evidence codes IC, IDA, IEP, IGI, IMP, IPI, ISS, and TAS. For automatically generated annotations we used the evidence code IEA. For the *all* category we included every evidence code including NAS, NR, and ND.

manually curated annotations on both absolute and logarithmic scales[d]. Table 2 highlights some of the data across all three categories.

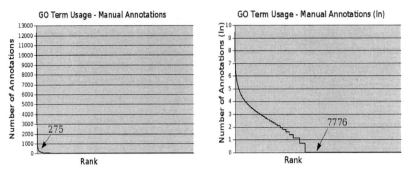

Figure 1. Distribution of manually curated annotations. These graphs are histograms that show for each rank how many annotations there were. The graph on the left gives the data on an absolute scale. The 275th most frequently used term is used in 200 annotations and is pointed out in the graph. The graph on the right gives the data on a logarithmic scale. Terms ranked 7,776 or higher are not used at all.

| Usage Count (ln) | Usage Count | Manual n | % | Automated n | % | All n | % |
|---|---|---|---|---|---|---|---|
| 0 | 0 or 1 | 9,568 | 57.3% | 9,272 | 55.5% | 8,044 | 48.2% |
| (0-2) | 2-7 | 2,876 | 17.2% | 1,286 | 7.7% | 1,602 | 9.59% |
| [2-4) | 8-54 | 3,378 | 20.2% | 2,590 | 15.5% | 3,069 | 18.4% |
| [4-6) | 55-401 | 768 | 4.60% | 2,259 | 13.5% | 2,603 | 15.6% |
| [6-∞) | 402-208K | 113 | 0.68% | 1,296 | 7.76% | 1,385 | 8.29% |

Table 2. Read the fourth row as "terms in the '[2-4)' group were used 8 to 54 times. For the manual annotations there were 3,378 terms in this group, which is 20.2% of all GO terms." The percent columns sum to 100.

### Relationship between term characteristics and annotation usage

While we are not surprised to observe a Zipfian distribution of term usage, the extreme skewness of the graphs in Figure 1 raises a broad question: is it possible to characterize terms that are used versus ones that are not? In an attempt to answer it, we test the following hypotheses: (1) there is a relationship between hierarchical depth and frequency of usage by annotators; (2) terms that are frequently used are

---

[d] We make the simplifying assumption that ln(0)=0.

less likely to be compositional than terms that are used infrequently. We first grouped the terms by their frequency of usage by creating five groups that correspond to even intervals on the log scale. These groups are given in Table 2. Figure 2 shows the relationship between frequency of usage and depth in the hierarchy. The average hierarchical depth of the terms in each group is shown. Statistically significant differences between these averages were determined using Kruskal-Wallis with a Bonferroni multiple comparison correction[e]. More frequently used terms tend to be higher in the hierarchy than infrequently used ones. Figure 3 shows the relationship between frequency of usage and compositionality. Terms that are more frequently used are less likely to have a contained term than terms that are less frequently used. We also examined the relationship between usage frequency and term length; the graph is similar to Figure 3 and is omitted for reasons of space. More frequently used terms tend to be shorter than infrequently used ones.

### 2.3. *Quantifying combinatorial effects*

A well-known difficulty of creating and maintaining a controlled terminology is the combinatorial explosion effects that present themselves when attempting to thoroughly represent all of the terms necessary to cover a domain [14]. Consider twenty of the terms that contain the words *T-cell* and *proliferation* shown in Figure 4. The blocked data highlights repeated data. It is apparent that the prefixes *regulation of, positive regulation of* and *negative regulation of* were added to five *T-cell proliferation* terms creating fifteen additional terms.

To better characterize this kind of combinatorial behavior we use a slightly modified finite state automaton (FSA) representation to represent a concise view of a set of GO terms[f]. Figure 5 represents all and only the 21 terms that contain the words *T-cell* and *proliferation*. No additional terms are represented. Any path that begins at a start state and ends in an end state corresponds to a subset of the graph's terms. A start state is represented by a single solid border, an end state is represented by double solid borders, and nodes with a dashed border are neither start states nor end states. The set of strings in a node represents a choice. A single GO

---

[e] All averages were found to differ with statistical significance (p=.05) except for the hierarchical depth averages for the groups labeled '0' and '(0-2).'

[f] FSAs are commonly used for representing regular expressions and grammars. We use rectangles instead of circles to reduce graph size.

180

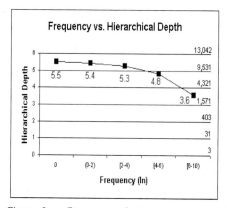

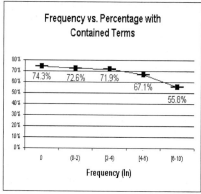

Figure 2. Frequency of term usage vs. hierarchical depth. Terms in the '[2-4)' group have an average hierarchical depth of 5.3. The cumulative number of terms at a given depth is shown on the right hand side of the graph, e.g. there are 1,571 terms at depths zero through three.

Figure 3. The y-axis shows the percentage of terms in a bin that have a contained term.

term is represented by a path for which a choice at each node has been made. The

| | | T-cell proliferation |
| | | activated T-cell proliferation |
| | | alpha-beta T-cell proliferation |
| | | gamma-delta T-cell proliferation |
| | | T-cell homeostatic proliferation |

| | regulation of | T-cell proliferation |
| | regulation of | activated T-cell proliferation |
| | regulation of | alpha-beta T-cell proliferation |
| | regulation of | gamma-delta T-cell proliferation |
| | regulation of | T-cell homeostatic proliferation |

| positive | regulation of | T-cell proliferation |
| positive | regulation of | activated T-cell proliferation |
| positive | regulation of | alpha-beta T-cell proliferation |
| positive | regulation of | gamma-delta T-cell proliferation |
| positive | regulation of | T-cell homeostatic proliferation |

| negative | regulation of | T-cell proliferation |
| negative | regulation of | activated T-cell proliferation |
| negative | regulation of | alpha-beta T-cell proliferation |
| negative | regulation of | gamma-delta T-cell proliferation |
| negative | regulation of | T-cell homeostatic proliferation |

Figure 4. Twenty GO terms that contain *T-cell* and *proliferation*. The redundancy of data introduced by appending five terms with three modifiers is highlighted by the blocked text.

number of terms represented by a graph is given by:

This is a summation over the number of terms represented by each path (p) in the graph (G). The number of terms represented by a path is the product of the number of choices in each node (n) of the path. For example, the number of terms represented in Figure 5 is (2*1*3*1*1) + (2*1*1*1) + ... = 21. We note that each term represented by this graph has a minimum edit distance of one from at least one other term in the graph.

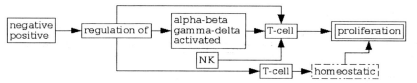

Figure 5. This FSA represents a concise view of all and only the 21 terms that contain the words *T-cell* and *proliferation*. The are two *T-cell* nodes because placing the *homeostatic* node between the other *T-cell* node and the *proliferation* node would license terms such as *alpha-beta T-cell homeostatic proliferation* that do not exist in GO.

This graphical representation is introduced only as a descriptive tool for visualizing terms in GO, not as a proposed solution for modeling GO. This representation, though inferior to other approaches for modeling GO, provides a very clear way to visualize and count terms and is sufficient for illustrating the combinatorial properties present in GO. We use this graphical representation to make two points: the number of terms licensed by simple compositional building blocks can be very large, and small conceptual changes to GO can result in large numbers of term changes.

A more complex example given in Figure 6 is a graphical term representation of all and only the 51 GO terms that contain the word *proliferation*. There are a number of observations about this set of terms that are easy to make when the 51 terms are displayed in this format. For instance, there are places where on could add an edge. For example, *regulation of neuroblast proliferation* is a reasonable GO term that could be represented by adding an edge from the *regulation of* node to the *neuroblast* node. Such observations would likely prove difficult to make by viewing a flat list of these 51 terms. Other edits that might be made to this graph include consolidating nodes and adding choices to individual nodes.

We next demonstrate that making small conceptual changes to the ontology results in a large number of terminological changes by observing the effect of making edits to the graph on the total number of terms represented by the graph. Consider the effect of making a conceptual change to the ontology such as "require differentiation and activation to appear in the same contexts as proliferation." To do this the strings *differentiation* and *activation* could be added to the two *proliferation* nodes in the graph. The effect of adding two choices to the rightmost node in the graph would nearly triple the number of terms represented by the graph to 151. To make an equivalent change to GO one must individually add or verify the existence of 100 terms. We also point out that each of these 100 terms differs by a single word from a proliferation term.

Coupling differentiation and activation to proliferation in this way is probably an undesirable oversimplification of the relationship between these three processes. We assessed the extent to which the terms that result from the kind of conceptual change described are valid. To do this we added the strings *differentiation* and *activation* to the rightmost node, creating 100 new terms. Of these 100 terms, 55 exist in GO. Of the 45 that do not, 14 or 31% of these were judged to be biologically meaningful and a reasonable addition to GO by two domain experts from the GO Consortium. Thus 69 of the 100 total "new" terms were biologically meaningful. This is evidence that there are relationships between proliferation, differentiation, and activation terms that require them to appear in similar contexts. It appears that the curation process for ensuring that these relationships are consistently applied when creating terms is error prone. In this case, reasonable terms were omitted. This may be due to the fact that the relationships themselves are not explicitly encoded or editable.

The graph in Figure 6 also suggests the possibility of simplifying the representation of *proliferation* terms by consolidating the various nodes that contain cell type descriptions by using a general *cell type* node as in Figure 7. The number of possible cell types that might fill in the *cell type* node is an open question. However, the OBO cell type ontology [15] contains 678 terms. Using this ontology in place of the *cell type* node along with the two additional choices in the rightmost node gives a graph that represents 8,136 terms! While not all of these 8,136 terms would be biologically meaningful, it would likely be difficult for a human curator to maintain a large subset of the 8,136 terms without some structure to manage the compositional building blocks that make up these terms.

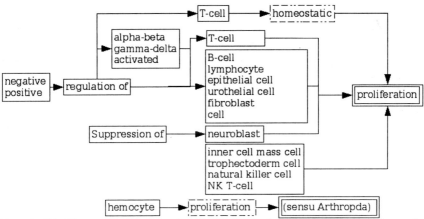

Figure 6. This FSA represents a concise view of all and only the 51 terms that contain the word *proliferation*.

Figure 7. Simplification of the finite state automaton shown in Figure 6. The many cell types reflected in Figure 6 have been collapsed to a single *Cell Type* node.

The preceding examples from GO show that the possible number of terms can be incredibly large, and also demonstrate how such a collection of terms can be unwieldy to maintain. We note that we have found many other examples similar to the *proliferation* example that exhibit similar combinatorial behavior[g].

## 3. Discussion

Decades of research on relational databases have resulted in well-understood and widely accepted best practices for modeling relational data. Central to relational database design is the idea of table normalization. The key idea is to reduce the redundancy of data by thoroughly analyzing relationships between entities with respect to cardinality and directionality. Such analyses result in tables that have

---

[g] The terms containing *amino acid* provide a nice example because they are found in a different ontology (molecular function) and show that term variation may be at both the front and end of the terms.

reduced storage requirements and are much easier to populate and maintain consistently. We argue that the current representational model of the Gene Ontology is somewhat analogous to a database consisting of a single non-normalized table. Although GO's model is simple and easily understood, the terminological data is highly redundant and unnecessarily large due to the combinatorial effects demonstrated above. We argue that these characteristics have significantly impacted two activities vital to the success of the GO community: ontology curation and the use of GO for annotating databases. While we would not argue that reducing the terminological size is desirable or possible, it seems favorable to have a highly normalized model from which the terms are generated or derived. This gives curators and users two perspectives of GO. The first is a compositional model: well-normalized, compact, and complex. The second is the ontological model as it is: non-normalized and large, but easy to understand. The compositional model would presumably allow annotators to spend more time thinking about the relationships between terms rather than having to fill in the combinatorial possibilities that new or modified relationships license. For annotators, the compositional model would provide an alternate and smaller "search space" to navigate through when looking for terms.

The OBOL project attempts to address these issues in GO curation by creating a Prolog grammar to represent GO term compositionality. The grammar is proposed to be used to find missing relationships and terms in the ontology. In contrast, our approach focuses on the conceptual structure of GO, remaining neutral with respect to the representational format. We suggest that priority should be given to defining: (1) the proper compositional building blocks, (2) constraints that license their combinations, and (3) the domain relationships that parallel these combinations and constraints. Only after these aspects of the ontology are well understood do we think it makes sense to select a representational scheme (e.g. a rule-based grammar versus a frame-based system) on the basis of which best fits the conceptual structure of GO.

## Acknowledgments

The authors thank Midori Harris, Jane Lomax, Chris Mungall, Sonia Leach, and Andrew Dolbey.

**References**

1.  Gene Ontology Consortium, "Gene Ontology: tool for the unification of biology." *Nature* 25:25-29 (2000).
2.  Gene Ontology Consortium, "Creating the Gene Ontology resource: design and implementation." *Genome Research* 11:1425-1433 (2001).
3.  C.M. Verspoor, C. Joslyn and G.J. Papcun, "The Gene Ontology as a source of lexical semantic knowledge for a biological natural language processing application," *Participant notebook of the ACM SIGIR'03 workshop on text analysis and search for bioinformatics,* pp. 51-56 (2003).
4.  C.J. Wroe, R. Stevens, C.A. Goble and M. Ashburner, "A methodology to migrate the Gene Ontology to a description logic environment using DAML+OIL." *Pacific Symposium on Biocomputing 2003.*
5.  I. Yeh, P.D. Karp, N.F. Noy and R.B. Altman, "Knowledge acquisition, consistency checking and concurrency control for Gene Ontology (GO)." *Bioinformatics* 19(2):241-248, 2003.
6.  C.A. Joslyn, S.M. Mniszewski, A. Fulmer and G. Heaton, "The Gene Ontology categorizer." *Bioinformatics,* in press.
7.  C. Mungall, "Open Bio-Ontology Language." 7th Annual Bio-Ontologies Meeting (2004), http://bio-ontologies.man.ac.uk/obol-glasgow.ppt.
8.  M. Bada, R. McEntire, C. Wroe and R. Stevens, "GOAT: The Gene Ontology Annotation Tool." *Proceedings of the 2003 UK e-Science All Hands Meeting,* 514-519 (2003).
9.  K.M.Verspoor, J. Cohn, C. Joslyn, S. Mniszewski, A. Rechtsteiner, L.M. Rocha and T.Simas, "Protein annotation as term categorization in the Gene Ontology using word proximity networks." *BMC Bioinformatics* in submission.
10. P.V. Ogren, K.B. Cohen, G.K. Acquaah-Mensah, J. Eberlein, and L. Hunter, "The compositional structure of Gene Ontology terms." *Proceedings of the Pacific Symposium on Biocomputing 2004,* pp. 214-225.
11. D. Jurafsky and J.H. Martin, *Speech and language processing.* Prentice Hall (2000).
12. V.I. Levenshtein, "Binary codes capable of correcting deletions, insertions, and reversals." *Cybernetics and control theory* 10(8):707-710 (1966).
13. B.H. Partee, A. ter Meulen, and R.E. Wall, *Mathematical methods in linguistics,* corrected 1st edition. Kluwer Academic Publishers (1987).
14. A.L. Rector, "Clinical terminology: why is it so hard?" *Methods of Information in Medicine.* 38:239-52 (1999).
15. J. Bard and M. Ashburner, Cell Type Ontology, http://obo.sourceforge.net/ (2004).

# THE USE OF COMMON ONTOLOGIES AND CONTROLLED VOCABULARIES TO ENABLE DATA EXCHANGE AND DEPOSITION FOR COMPLEX PROTEOMIC EXPERIMENTS

SANDRA ORCHARD

*Sequence Database Group, EMBL-European Bioinformatics Institute*
*Wellcome Trust Genome Campus*
*Hinxton*
*Cambridge*
*CB10 1SD UK*

LUISA MONTECCHI-PALAZZI

*Universita Tor Vergata,*

*Rome, Italy*

HENNING HERMJAKOB

*Sequence Database Group, EMBL-European Bioinformatics Institute*
*Wellcome Trust Genome Campus*
*Hinxton*
*Cambridge*
*CB10 1SD UK*

ROLF APWEILER

*Sequence Database Group, EMBL-European Bioinformatics Institute*
*Wellcome Trust Genome Campus*
*Hinxton*
*Cambridge*
*CB10 1SD UK*

*Controlled vocabularies provide a roadmap through complex biological data. Proteomic data is increasing in volume and is currently poorly served by public repositories due to the large number of different formats in which the data is generated and stored. The Human Proteome Organization Proteome Standards Initiative is establishing standards for data transfer and deposition. These standards utilize ontologies and controlled vocabularies to describe experimental procedures and common processes such as sample preparation This paper will discuss the development of such ontologies by the user community and their current utilization in the fields of protein:proein interactions and mass spectrometry.*

# 1. Introduction

Ontologies and controlled vocabularies are being established by many groups to provide roadmaps through the confused mass of data currently being generated from increasingly large-scale experimental biological experiments. The world of protein chemistry is no exception to this rule, with GO having lead the field by providing a framework in which individual molecules and complexes can be defined by their process, function and subcellular location [1]. The world's leading protein sequence database, UniProt [2], whilst incorporating and adding to the GO annotation of molecules described within UniProt-Swiss-Prot and UniProt-TrEMBL, also has its own defined keyword section that allows users to perform searches across the database using a standard nomenclature consistent to all entries. However, whilst the description of the function of these molecules is well served by established controlled vocabularies, the experimental techniques and procedures by which much of the functional information has been generated, has largely been ignored.

Proteomics is often described as the study of the protein translation products of the genome of a given organism but, in reality, this definition should be expanded to an understanding of the expression pattern and state of all proteins transcribed under a given set of conditions and the alteration of these parameters in response to a specific change to these conditions. The proteome of a cell encompasses the identity, subcellular location, post-translational modifications and protein:protein interactions made by the spectra of proteins expressed at any one moment in time and also how all these effect the function of both an individual protein and the cell as a whole. In order to map this, a multitude of experimental techniques have been developed. Proteins have first to be isolated and separated from a given biological sample, the latter usually either by 2-dimensional gel electrophoresis or by HPLC. The analytes are then ionized in the gas phase and the mass of the resulting peptide fragments measured by mass spectrometry. The resulting spectra are processed and specialized software used to match these fragments to known proteins. Such analyses will provide an expression map of the protein content of the cell under the defined experimental conditions – further techniques have been developed to provide further detail of the state of these proteins and their actual location within the cell. To fully understand the biological processes and pathways in which any one protein molecule may be involved, it is necessary to be aware of the interactions that molecule makes with other proteins, nucleic acids and small molecules within the cell. Experimental procedures by which these can be observed have been established for many years but these too are becoming more high-throughput and the rate of data generation is rapidly increasing.

The context-sensitive nature of proteomic data necessitates the capture of a larger set of metadata than is normally required for sequencing, where knowledge of the organism of origin will suffice. Not only is information of sample source, handling, stimulation and eventual preparation for analysis required but also the detail of the analysis itself will need to be recorded. For example, to compare images of 2D-gels knowledge of their mass and charge ranges are required, and this information will need to be retrieved by users wishing to perform meaningful analysis of this experiment. Whilst the results and conclusion drawn from this data is frequently published in great detail, the underlying data is often only available as supplementary material, or is stored in author maintained databases or on websites. These databases and websites tend only to exist for the lifespan of the underlying project or grant, are often poorly maintained and the data within is difficult to access for downloading [3]. Published web addresses may lead nowhere [4]. Even when a stable database has been established, comparison between different datasets has proven difficult, in part due to the wide variety of terms and spellings used to describe a common experimental process. As an example, yeast two hybrid technology is a well known and widely used methodology for identifying protein interaction partners [5], a technique which has proved ideally suited to scaling up to increase throughput and data output. However, there are more than 10 different spellings for this term, e.g. Y2H, 2H, two-hybrid. While all of these are easily human understandable, non-standardized use of key terms makes systematic searches in large databases very difficult. Data interchange between databases sharing a common philosophy and even common data formats is being hampered by the lack of common terminology to describe identical processes. Even when a high level term can be used to describe a technique, for example mass spectrometry, data exchange and integration may require detail of instrumentation and data handling in order to give a complete picture of the conditions under which the data was generated – essential information for a full understanding of the importance of a particular dataset

## 2. The Human Proteome Organization Protein Standards Initiative

The Human Proteome Organisation (HUPO) was formed in 2001 with the aim of consolidating national and regional proteome organizations into a single worldwide body. The Proteome Standards Initiative was established by HUPO with the remit of standardizing data formats within the field of proteomics to the end that public domain databases can be established where all such data can be deposited, exchanged between such databases or downloaded and utilized by laboratory workers [6]. The HUPO-PSI organized a series of meetings at which it was decided to develop a single data model that would describe and encompass central aspects of a proteomics experiment. This model would contain different sub-domains which will allow it to handle specific data types, for example 2-D electrophoresis gels or HPLC. Common processes would be described by a number of controlled vocabularies or ontologies. Where these processes are also relevant to micro-array data, for example in the area of sample preparation, this could be done in collaboration with the MGED consortium MGED (the micro-array gene expression data group), thus facilitating the comparison of proteomic with transcript data. Each sub-domain would then support a PSI-approved interchange format, which would permit the handling of data from many different sources. In the interests of making the task more manageable, the PSI agreed to concentrate their resources on two potential sub-domains, mass spectrometry and protein:protein interactions, whilst concurrently developing the encompassing proteomics data model, MIAPE [7].

## 3. Molecular Interactions and the Establishment of Common Vocabularies

A number of both commercial and academic molecular interaction databases already exist (IntAct [8], BIND [9], DIP [10], MINT [11], Hybrigencs [12], HPRD [13], MIPS [14]) which are wholly or partially in the public domain but, as discussed above, it was previously impossible to download or exchange data from any two of these databases in a common format. All these database providers, however, are committed to making their data more easily accessible and useful to the user community and actively supported the establishment of a common interchange format.

The HUPO-PSI MI format has been developed using a multi-level approach similar to that used by the Systems Biology Markup Language (SBML) [15]. Level 1, published early in 2004 [16], provided a basic format suitable for representing the majority of all currently available protein-protein interaction data. Level 2, released later in the year, allowed the description of protein interactions with nucleic acids or with small molecules. It allows the representation of both binary and n-ary interactions but does not contain detailed data on interaction mechanisms or full experimental descriptions. While a common data exchange format is a key requirement for an efficient exchange of protein interaction data, it does not by itself guarantee data compatibility. It is essential to ensure standardized use of the data attributes through documentation and controlled vocabularies. The PSI-MI format contains detailed documentation within the XML schema itself, which is automatically extracted as an easily accessible web page and accompanied by detailed documentation on the HUPO-PSI MI context (http://psidev.sourceforge.net/mi/xml/doc/user/). To standardize the contents of data attributes, the PSI-MI format makes extensive use of controlled vocabularies or ontologies. External systems such as the Gene Ontology and the NCBI taxonomy, are referenced where possible. Detailed controlled vocabularies have been developed for the PSI-MI format for several key molecular interaction data attributes, such as the experimental methodology by which molecules are demonstrated to interact. Since this format already has a large user-base, it is intended to maintain the ontologies in GO format for the farseeable future, however it is recognized that, to become incorporated into the wider aspects of the General Proteomics Standards activities, it may become necessary to migrate to a more formal ontology language, such as OWL.

The PSI-MI format is designed for data exchange by many data providers. It is therefore important to ensure that both the data format (syntax) and the meaning of the data items (semantics) are consistent and well-defined. Without the standardization of data items as part of a community standard, data sets which are generated by the combination of data from different sources will quickly become difficult to search and to use. To address this problem, controlled vocabularies have been used in place of free text attributes, wherever possible. Several controlled vocabularies have been developed, including *interaction type*, *feature type*, *feature detection method*, *participant detection method*, and *interaction detection method* to describe specific aspects of both an interaction and the experimental methodology used to determine these (Fig. 1).

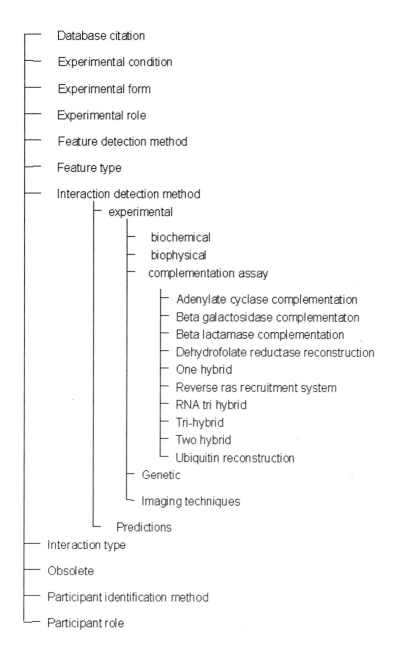

Database citation

Experimental condition

Experimental form

Experimental role

Feature detection method

Feature type

Interaction detection method

experimental

biochemical

biophysical

complementation assay

Adenylate cyclase complementation

Beta galactosidase complementaton

Beta lactamase complementation

Dehydrofolate reductase reconstruction

One hybrid

Reverse ras recruitment system

RNA tri hybrid

Tri-hybrid

Two hybrid

Ubiquitin reconstruction

Genetic

Imaging techniques

Predictions

Interaction type

Obsolete

Participant identification method

Participant role

part of both the *interaction detection method* and *feature detection method* controlled vocabularies.

Fig. 1 Controlled vocabularies to describe experimental methodologies developed for use with the PSI-MI XML interchange format

technologies that can be used to infer that two or more proteins form a molecular aggregate. This vocabulary of more than 80 terms has a hierarchical structure based on a limited number of high level terms that group similar methods and reflect commonly used classifications and technical distinctions. As one method may be a specialization of more than one technology, a term may have more than one parent. For example, the "colocalization by fluorescent probes cloning" method (MI:0021) is both a fluorescence technology (MI:0051) and an imaging technique (MI:0428) The *participant detection method* controlled vocabulary lists more than twenty methods commonly used to establish the identity of the interacting partners, for example peptide mass fingerprinting (MI:0082).

The controlled vocabularies described here are not static; they will be maintained and updated by the HUPO PSI workgroup to reflect new experimental methodologies, or requirements from the community, in a manner similar to the maintenance of the Gene Ontology. This will ensure consensus on the inclusion of new terms by the user community, a high degree of flexibility to define terms when they are needed, and the avoidance of vague or ambiguous categories such as "other methods". In accordance with the GO model, an editorial team has been appointed [7] and requests for new terms dealt with via a SourceForge tracker system. Anyone interested in becoming more directly involved in the process is directed to the mailing list psidev-vocab@lists.sf.net.

Major interaction data providers are currently establishing the "International Molecular-Interaction Exchange (IMEx)" collaboration, in which they will regularly exchange user-submitted data, using the HUPO-PSI data exchange format. This will operate on similar principles to the EMBL/GenBank/DDBJ collaboration, thus providing a synchronized, stable, reliable resource for molecular interaction data. The success of this approach, with several of the above databases, for example IntAct, DIP. MINT and HPRD providing some or all of their data in PSI-MI format and others planning to follow suit in the near future, has provided encouragement for the progress of the related HUPO-PSI sponsored projects which are at an earlier stage of development due to the more complex nature of the data which they have to deal with.

## 4. Mass Spectrometry and the Establishment of Controlled Vocabularies

The HUPO-PSI mass spectrometry work group (PSI-MS) is working to develop a common data repository for the deposition of mass spectrometry data generated by proteomics groups and data standards accepted by both the user community and by instrumentation manufacturers. To this end, the group has produced the mzData format, a vendor-independent representation of mass spectra, providing a unified format for data archiving, exchange, and search engine input [7]. It has been jointly developed by academic users, commercial users and instrument vendors, among them Eli Lilly, EBI, Bruker Biosciences, Shimadzu, MDS Sciex, Agilent, and Thermo Electron. Controlled vocabularies will be used throughout mzData., in particular for the description of source detection methods, instrument parameters and analysis techniques. It has been proposed that the current ASTM mass spectrometry standard data dictionary be adopted, and updated, for use as a controlled vocabulary within this model, with eventual ownership of this dictionary potentially passing to the American Society for Mass Spectrometry (ASMS) so that it could be used to support both the HUPO-PSI and the ASTM's raw data standardization efforts. Users will also have the ability to develop their own vocabularies to allow for the specific needs of individual laboratories and maximum flexibility in experimental design. Work is currently ongoing to create and expand these vocabularies, in line with the final development of a beta version of mzData, a final release of which is planned for the HUPO world congress in Beijing, October 2004.

The mzData model will also act as the mass spectrometry component of the MIAPE data model and top level processes such as sample identification will be better dealt with through the development of this model than independently by the mass spectrometry group. The design of specifications for a spectral analysis format (mzAnalysis) is underway, and also that of a common syntax for the identification of proteins and peptides (mzProtID), which must also have the ability to describe post-translational modifications.

## 5. General Proteomics Standards and the Establishment of Controlled Vocabularies

As already stated, it is intended that all efforts within the remit of the HUPO-PSI will be coordinated and united within a framework provided by the establishment of standards for the representation of a full proteomics experiment, the Global Proteomics Standards (GPS). Based on the PEDRo schema [18], this work group is tasked with developing the "Minimum Information About a Proteomics Experiment (MIAPE)" document analogous to the MIAME requirements for a micro-array experiment [19], and both an object model (PSI-OM) and XML format (PSI-ML) to fully represent a proteomics experiment. PSI-GPS will use the modules such as the more specific mzData format as components of a full experiment description, comprising sample preparation, analysis technologies, and results. To fully delineate these processes, controlled vocabularies are currently being written and appropriate terms will be contributed to the MGED Extended ontology under the "PSI" namespace. The MGED ontology is being written to support the micro-array object model, MAGE. The extended version adds further associations and classes to the core ontology which is intended to be stable and fully in synch with MAGE.

## 6. Summary

The design and use of controlled vocabularies to describe experimental data and enable the storage and exchange of proteomics data in a format that allows subsequent users a clear and comprehensive understanding of the conditions in which the experiments were performed, has only recently been tackled by workers in the field. The success of the HUPO-PSI MI format which uses a series of controlled vocabularies to describe molecular interactions, the features on a molecule responsible for such interactions and the experimental methods by which both interaction and features were determined suggests that ongoing work to describe detection of the proteome content of a sample by mass spectrometry will be equally successful. All these efforts should be seen in the wider framework of the GPS which is tackling more top level issues as sample description, acquisition and handling in conjunction with the MGED consortium, with the eventual aim of having a single combined ontology, with which to describe any experiment investigating the transcriptome and/or proteome content of a cell. This degree of cooperativity will ensure that these ontologies remain non-redundant and that the user community can access a single, common ontology suitable for describing complex experimental procedures.

It is always an issue, when launching a new standard, that this be seen as a fulfillment of a genuine need within the scientific community, rather than the bureaucratic imposition of unnecessary extra work without any perceived benefit. To this end, the HUPO-PSI proteomic standards are being written jointly by as wide a cross section of the perceived end users as can practicably be achieved, with consultation at all stages of the process being an absolute requirement. The strides made within the protein:protein interaction community as a result of the publication of these standards and ontologies only a few months ago, lead us to hope that extending this process to cover the wider field of experimental proteomics will be equally productive and be of great benefit to an increased understanding of the proteome across all species and cell types.

# References

1. M.A. Harris, J. Clark, A. Ireland, J. Lomax, M. Ashburner, R. Foulger, k. Eilbeck, S. Lewis, B. Marshall, C. Mungall, J. Richter, G.M. Rubin, J.A. Blake, C. Bult, M. Dolan, H. Drabkin, J.T. Eppig, D.P. Hill, L. Ni, M. Ringwald, R. Balakrishnan, J.M. Cherry, K.R. Christie, M.C. Costanzo, S.S. Dwight, S. Engel, D.G. Fisk, J.E. Hirschman, E.L. Hong, R.S. Nash, A. Sethuraman, C.L. Theesfeld, D. Botstein, K. Dolinski, B. Feierbach, T. Berardini, S. Mundodi, S.Y. Rhee, R. Apweiler, D. Barrell, E. Camon, E. Dimmer, V. Lee, R. Chisholm, P. Gaudet, W. Kibbe, R. Kishore, E.M. Schwarz, P. Sternberg, M. Gwinn, L. Hannick, J. Wortman, M. Berriman, V. Wood, N. de la Cruz, P. Tonellato, P. Jaiswal, T. Seigfried, R. White; Nucleic Acids Res, **32** D258 (2004).
2. R. Apweiler, A. Bairoch, C.H. Wu, W.C. Barker, B. Boeckmann, S. Ferro, E. Gasteiger, H. Huang, R. Lopez, M. Magrane, M.J. Martin, D.A. Natale, C. O'Donovan, N. Redaschi, L.S. Yeh. Nucleic Acids Res. **32**:D115 (2004)
3. S. Orchard, C. Taylor, W. Zhu, R. K. Julian, Jr, H. Hermjakob, R. Apweiler Expert Review in Proteomics (in press)
4. J. Whitfield Nature. **428**, 592. (2004)
5. P. Legrain , L. Selig .FEBS Lett. **480**, 32 (2000).
6. S. Orchard, H. Hermjakob, R. Apweiler. Proteomics, **3**, 1374 (2003)
7. S. Orchard, C. F.. Taylor, H. Hermjakob, Weimin-Zhu, R. K. Julian, Jr, R. Apweiler Proteomics, (in press).
8. H. Hermjakob, L. Montecchi-Palazzi, C. Lewington, S. Mudali, S. Kerrien, S. Orchard, M. Vingron, B. Roechert, P. Roepstorff, A. Valencia, H. Margalit, J. Armstrong, A. Bairoch, G. Cesareni, D. Sherman, R. Apweiler. Nucleic Acids Res., **32**, D452 (2004)
9. G.D. Bader, D. Betel, C.W.V. Hogue. Nucleic Acids. Res. **31**:248 (2003).
10. I. Xenarios, L. Salwinski, X.J. Duan, P. Higney, S.M. Kim, D. Eisenberg. Nucleic Acids. Res. **30**:303 (2002).

11. A. Zanzoni, L. Montecchi-Palazzi, M. Quondam, G. Ausiello, M. Helmer-Citterich, G. Cesareni.. FEBS Letts **513**: 135 (2002).
12. http://www.hybrigenics.fr
13. S. Peri, J.D. Navarro, T.Z. Kristiansen, R. Amanchy, V. Surendranath, B. Muthusamy, T.K. Gandhi, K.N. Chandrika, N. Deshpande, S. Suresh, B.P. Rashmi, K. Shanker, N. Padma, V. Niranjan, H.C. Harsha, N. Talreja, B.M. Vrushabendra, M.A. Ramya, A.J. Yatish, M. Joy, H.N. Shivashankar, M.P.Kavitha, M. Menezes, D.R. Choudhury, N. Ghosh, R. Saravana, S. Chandran, S. Mohan, C.K. Jonnalagadda, C.K. Prasad, C. Kumar-Sinha, K.S. Deshpande, A. Pandey. Nucleic Acids Res. **32** D497 (2004).
14. H.W. Mewes, C. Amid, R. Arnold. Nucleic Acids Res. **32** D41 (2004)
15. M. Hucka, A. Finney, H.M. Sauro, H. Bolouri, J.C. Doyle, H. Kitano, A. P. Arkin, B.J. Bornstein, D. Bray, A. Cornish-Bowden , A.A. Cuellar, S. Dronov, E.D. Gilles, M. Ginkel, V. Gor, I.I. Goryanin, W.J. Hedley, T.C. Hodgman, J.H. Hofmeyr, P.J. Hunter, N.S. Juty, J.L. Kasberger, A. Kremling, U. Kummer, N. Le Novere, L.M. Loew, D. Lucio, P. Mendes, E. Minch, E.D. Mjolsness , Y. Nakayama, M.R. Nelson, P.F. Nielsen, T. Sakurada, J.C. Schaff, B.E. Shapiro, T.S. Shimizu, H.D. Spence, J. Stelling, K. Takahashi, M. Tomita, J. Wagner, J. Wang;. Bioinformatics **19**:524-531 (2003)
16. H. Hermjakob, L. Montecchi-Palazzi, G. Bader, J. Wojcik.,L. Salwinski, A. Ceol, S. Moore, S. Orchard, U. Sarkans, C. von Mering, B. Roechert, S. Poux, E. Jung, H. Mersch, P. Kersey, M. Lappe, Y. Li, R. Zeng, D. Rana, M. Nikolski, H. Husi, C. Brun, K. Shanker, S.G. Grant, C. Sander, P. Bork, W. Zhu, A. Pandey, A. Brazma, B. Jacq, M. Vidal, D. Sherman, P. Legrain, G. Cesareni, I. Xenarios, D. Eisenberg, B. Steipe, C. Hogue, R. Apweiler. Nat Biotechnol. **22** 177 (2004)
17. J.S. Garavelli Proteomics. **4**, 1527 (2004).
18. C.F. Taylor, N.W. Paton, K.L.Garwood, P.D. Kirby, D.A. Stead, Z. Yin, E.W. Deutsch, L. Selway, J. Walker, I. Riba-Garcia, S. Mohammed, M.J. Deery, J.A. Howard, T. Dunkley, R. Aebersold, D.B. Kell, K.S. Lilley, P. Roepstorff, J.R. Yates 3rd, A. Brass, A.J. Brown, P. Cash, S.J. Gaskell, S.J. Hubbard, S.G. Oliver SG. Nat Biotechnol. 2003, **21**, 24 (2003)
19. A. Brazma, P. Hingamp, J. Quackenbush, G. Sherlock, P.Spellman, C. Stoeckert, J. Aach, W. Ansorge, C.A. Ball, H.C. Causton, T. Gaasterland, P. Glenisson, F.C. Holstege, I.F. Kim, V. Markowitz, J.C. Matese, H. Parkinson, A. Robinson, U. Sarkans, S. Schulze-Kremer, J. Stewart, R. Taylor, J. Vilo, M. Vingron M. Nat Genet., **29**, 365 (2001).

# A FLEXIBLE MEASURE OF CONTEXTUAL SIMILARITY FOR BIOMEDICAL TERMS[*]

## I. SPASIĆ[1], S. ANANIADOU[2]

[1]*Department of Chemistry, UMIST, Manchester, UK*
[2]*School of Computing, Science and Engineering, University of Salford, UK*
*E-mail: i.spasic@umist.ac.uk, s.ananiadou@salford.ac.uk*

We present a measure of contextual similarity for biomedical terms. The contextual features need to be explored, because newly coined terms are not explicitly described and efficiently stored in biomedical ontologies and their inner features (e.g. morphologic or orthographic) do not always provide sufficient information about the properties of the underlying concepts. The context of each term can be represented as a sequence of syntactic elements annotated with biomedical information retrieved from an ontology. The sequences of contextual elements may be matched approximately by edit distance defined as the minimal cost incurred by the changes (including insertion, deletion and replacement) needed to transform one sequence into the other. Our approach augments the traditional concept of edit distance by elements of linguistic and biomedical knowledge, which together provide flexible selection of contextual features and their comparison.

## 1. Introduction

Breakthrough advances in biotechnology have given rise to rapid production of biomedical data. New discoveries are being described in scientific papers (most often electronically available) with the intention of sharing the results with the scientific community. However, the rapid expansion of the bio-literature[a] makes it increasingly difficult to locate the right information at the right time. Clearly, for biomedical experts to experience the full benefits of electronically accessible literature, natural language processing (NLP) applications (such as information retrieval, information extraction,

---

[*]This work has been partially supported by the JISC-funded National Centre for Text Mining (NaCTeM), Manchester, UK.
[a]For example, the MEDLINE database (www.ncbi.nlm.nih.gov/PubMed) currently contains approximately 12 million references to biomedical articles, growing by more than 10,000 references weekly.

etc.) are becoming a necessity in order to facilitate navigation through huge volumes of biomedical texts.

Automatic extraction and retrieval of biomedical information subsumes identification of terms denoting biomedical concepts (such as compounds, genes, drugs, reactions, etc.), their properties and mutual relations from a corpus of relevant documents. Rule-based approaches to these problems cannot cope with an enormous and ever growing number of terms and the complex structure of terminologies[b]. Since rules would need to be defined for each NLP task and biomedical subdomain separately, manual rule engineering in such a broad and complex domain is hindered by inefficiency and inconsistency. Alternatively, a similarity measure could be used as a vehicle of machine learning approaches to a variety of NLP tasks, utilising the large body of biomedical texts as the training data. In this paper, we suggest a measure of contextual similarity between biomedical terms based on edit distance. The alignment between two contexts corresponding to their edit distance can be used to match terms occurring in similar contexts. This property can be further exploited to support tasks such as term classification and disambiguation, extraction of their relations, etc.

The remainder of the paper is organised as follows. In Section 2 we briefly overview edit distance. Section 3 introduces the SOLD measure, generally based on the idea of edit distance as a means of assessing contextual similarity of biomedical terms. Sections 4, 5 and 6 give details on the specific aspects of the SOLD measure, namely syntactic, ontology-driven and lexical components. Finally, in Section 7 we conclude the paper.

## 2. Background and Related Work

Edit distance (ED) has been widely used for approximate string matching, where the distance between identical strings equals zero and increases as the strings get more dissimilar with respect to the symbols they contain and the order in which they appear. ED is defined as the minimal cost incurred by the changes needed to transform one string into the other. These changes may include insertion or deletion of a single character, replacement of two characters in the two strings and transposition of two adjacent characters in a single string. The choice of edit operations and their costs influences the "meaning" of the corresponding approximate matching, and thus depends

---

[b]For example, UMLS (www.nlm.nih.gov/research/umls) currently contains over one million concepts named by 2.8 million terms, organised into a hierarchy of 135 classes and interconnected by 54 different relations.

on a specific application.

A most popular application area of ED is molecular biology, where it has been used to compare DNA and protein sequences in order to infer information about the common ancestry, functional equivalence, possible mutations, etc.[1] It has also been successfully utilised in NLP to deal with alternate spellings, misspellings, the use of upper- and lower-case letters, etc. Further, ED has been used in terminological processing for the recognition of term variants (namely, protein names) based on their *internal* properties focusing on orthographic features.[2] Our intention, however, is primarily to explore *contextual* properties of terms.

In this case, it is more convenient to apply ED at the *word level* rather than the *character level*. Namely, character-based ED does not cope well with permutations of words. For instance, judging by the "conventional" ED, *stone in kidney* is more similar to *stone in bladder* than *kidney stone*. Obviously, for some applications it is more useful to treat strings as sequences of words. For example, approximate string matching can be viewed as the problem of pairing up their words so as to minimise their ED.[3] Recently, ED has been applied at the word level[4] as support for extended phrase-based text search allowing different wordings and syntactic mistakes. In this approach, ED was simply applied to words as opposed to characters. We, however, enriched the basic ED approach with linguistic knowledge (relying on part-of-speech (POS) tagging and partial parsing) and domain-specific knowledge (using an ontology). In the following section, we point to the main developments in this direction.

## 3. Approximate Context Matching

ED usually relies on the exact matches between symbols unless "wild card" symbols are allowed. This is not suitable for words, because they are inflected. Also, term variation causes two terms not to match even when they are synonymous. We want to keep the main idea of ED to account for different orderings of words, but also to make it more flexible towards lexical variations. For example, two inflected word forms should match if both their lexical categories and their base forms are identical, e.g.: [c]

        `<tok><sur>`**better**`</sur><lem cat="`**adj**`"`>**good**`</lem></tok>`
        `<tok><sur>`**good**`</sur>`  `<lem cat="`**adj**`"`>**good**`</lem></tok>`

---

[c]In the given XML notation, elements `<tok>`, `<sur>` and `<lem>` stand for token, surface form and lemma respectively, while attribute `cat` corresponds to category.

When two terms are compared, information from an ontology may be utilised. If the terms match exactly or if they are identified as variants, the matching score should be the highest, slightly less if they are siblings in the "is-a" hierarchy, etc. For example, the following terms have been recognised as variants in UMLS and annotated as such in the corpus by mapping them to the same preferred term form:

```
<tok><sur>vitamin A</sur><lem cat="term">vitamin A</lem></tok>
<tok><sur>A vitamin</sur><lem cat="term">vitamin A</lem></tok>
<tok><sur>vitamin-A</sur><lem cat="term">vitamin A</lem></tok>
<tok><sur>retinol</sur>    <lem cat="term">vitamin A</lem></tok>
```

Similarly, classified terms can be compared through their classes, e.g. both *retinol* and *ascorbic acid* are mapped to the *Vitamin* class in UMLS, and therefore can be regarded similar:

```
<tok>
  <sur>retinol</sur>                    →        Vitamin
  <lem cat="term">vitamin A</lem>
</tok>
            ↑ similar ↓                 ←    ↑ identical ↓
<tok>
  <sur>ascorbic acid</sur>
  <lem cat="term">vitamin C</lem>       →        Vitamin
</tok>
```

When term classes are not identical, their superclasses can be compared analogously, e.g. the classes retrieved for terms *insulin* and *glycosidase* are respectively *Hormone* and *Enzyme*, which both descend from the *Biologically Active Substance* class, so the given terms can be regarded similar:

```
<tok>
  <sur>insulin</sur>                      →     Hormone
  <lem cat="term">insulin</lem>                      ↘
</tok>                                                     Biologically
            ↑ similar ↓                  ←                Active
<tok>                                                     Substance
  <sur>glycosidase</sur>                        ↗
  <lem cat="term">glycosidase</lem>      →     Enzyme
</tok>
```

When at least one of the compared terms is not classified, then lexical clues may indicate their semantic similarity. For example, suppose that *retinol* has been mapped to its preferred form *vitamin A* in the ontology and that *vitamin C* has only been identified in the corpus. Then the lexical similarity between the corresponding normalised forms *vitamin A* and *vitamin C* (e.g. measured by ED) can be used to reduce the cost of their replacement:

```
<tok>
    <sur>retinol</sur>                              →        Vitamin
    <lem cat="term">vitamin A</lem>
</tok>
        ↑ similar ↓    ←    ↑ lex. similar ↓        ←        ↑ ? ↓
<tok>
    <sur>vitamin C</sur>
    <lem cat="term">vitamin C</lem>        →        ?
</tok>
```

Apart from lexical and terminological clues, syntactic information can be utilised as well. For example, partial parsing can be applied to POS-tagged text to group subsequent words into basic syntactic structures (e.g. noun phrases). ED applied to chunks of words rather than individual words is "forced" to take into account the syntactic structure at the phrase level. By choosing to replace syntactic categories with similar properties at lower costs (e.g. nouns and pronouns), ED can also be used to compare the syntactic structure at the sentence level. Namely, the sentences receiving low ED values are the ones that can be transformed into one another using a small number of low-cost edit operations, implying that their overall syntactic structure is fairly isomorphic.

We suggested how the traditional concept of ED can be augmented by elements of linguistic and domain-specific knowledge. We continue to describe the specific solutions used to implement the SOLD (syntactic, ontology-driven, lexical distance) measure.

## 4. Syntactic Component

Let $X = (x_1, \ldots, x_m)$ and $Y = (y_1, \ldots, y_n)$ denote two sentences as the sequences of chunks (not individual words) denoted by $x_i$ $(1 \leq i \leq m)$ and $y_j$ $(1 \leq j \leq n)$. Their distance is defined as the minimal number of edit operations necessary to transform $X$ into $Y$. Figure 1 describes the computation of the SOLD measure using the standard dynamic programming approach,[5] where $\text{cost}(i, j)$ denotes ED between $(x_1, \ldots, x_i)$ and $(y_1, \ldots, y_j)$, IC and DC (where IC $\equiv$ DC) are the costs of inserting and deleting a given chunk, and RC is the cost of replacing two chunks (see Table 1[d]). Automatic optimisation of the cost function led to overfitting. Therefore, an appropriate cost function has been chosen empirically and supported by experiments conducted with equal weights, metric and non-metric cost functions. We

---

[d]The cost of replacement by the epsilon symbol represents the cost of inserting or deleting the other symbol.

describe the motivation behind the chosen cost function, which provided
satisfactory results.

$$
\begin{aligned}
&\text{cost}(0,0) = 0; \\
&\text{for } (i = 1;\ i \le m;\ i = i + 1) \quad \text{cost}(i,0) = \text{cost}(i-1,0) + \text{IC}(x_i); \\
&\text{for } (j = 1;\ j \le n;\ j = j + 1) \quad \text{cost}(0,j) = \text{cost}(0,j-1) + \text{DC}(y_j); \\
&\text{for } (i = 1;\ i \le m;\ i = i + 1) \\
&\text{for } (j = 1;\ j \le n;\ j = j + 1) \\
&\qquad \text{cost}(i,j) = \min \begin{cases} \text{cost}(i-1,j) & + \text{IC}(x_i) \\ \text{cost}(i,j-1) & + \text{DC}(y_j) \\ \text{cost}(i-1,j-1) & + \text{RC}(x_i,y_j) \end{cases} ; \\
&\text{sold}(X,Y) = \text{cost}(m,n);
\end{aligned}
$$

Figure 1.   Calculation of the SOLD distance.

The choice of specific costs is based on an assumption about the potential semantic content and importance of syntactic chunks. Deleting a term incurs the highest possible cost (1), since important domain-specific information is lost. High importance (0.9) is also given to verbs as they may represent domain-specific relations. Generally, noun phrases (NPs), together with verbs, carry "heavy" semantic load. This is emphasised even more in a sublanguage, because terms constitute a subclass of NPs. NPs other than terms are still semantically important, especially since they can be unrecognised terms, so they are assigned high cost (0.9). Further, pronouns are given high cost (0.85), because they can co-refer with terms (i.e. indirectly denote a domain-specific concept). Prepositions can model spatial, temporal and other types of relationships between terms, and for this reason they are relatively highly ranked (0.5). Similarly, adjectives and adverbs as potential modifiers of terms and domain-specific verbs, e.g.:

> ... the fragment of SMRT encoding amino acids 1192-1495, which **strongly**
> <u>interacts</u> with TRbeta, <u>interacts</u> **very weakly** with COUP-TFI ...

are given the same cost (0.5). Next ranked (0.4) are different forms of the verb *to be*, which can be used in a general sense, but can also model the "is-a" relationship between terms, e.g.:

> ... <u>acetylcysteine</u> **is a** <u>drug</u> usually used to reduce the thickness of mucus ...

so they are assigned a similar cost (0.4). Auxiliary verb phrases can be used to modify the meaning of domain-specific verbs and in that manner encode important semantic information, e.g.:

> ... the oestrogen receptor AF-2 antagonist hydroxytamoxifen **cannot** <u>promote</u>
> ER-TIF1 interaction ...

and they incur the same cost (0.4). A low cost (0.2) is given to conjunctions, whose main role is to support text cohesion and not to pass relevant domain-specific information. Other chunks are assigned zero cost, as they are regarded irrelevant. For example, linking phrases (e.g. *on the other hand*) guide a reader, but carry no explicit semantic content. Punctuation marks are used similarly to improve the readability, and are thus discarded. Determiners are ignored because of their insufficient semantic content and especially because they are not used consistently or even correctly.

Table 1.   The cost of edit operations for different chunks.

| Chunk | np | term | link | be | aux | adj | adv | cnj | det | prep | pron | pun | v | $\epsilon$ |
|---|---|---|---|---|---|---|---|---|---|---|---|---|---|---|
| np | 0.20 | 0.30 | 0.90 | 0.90 | 0.90 | 0.75 | 0.90 | 1.00 | 0.90 | 1.00 | 0.15 | 0.90 | 0.70 | 0.90 |
| term | 0.30 | 0.15 | 1.00 | 0.90 | 0.90 | 0.80 | 0.95 | 1.00 | 1.00 | 1.00 | 0.15 | 1.00 | 0.70 | 1.00 |
| link | 0.90 | 1.00 | 0.00 | 0.40 | 0.40 | 0.50 | 0.50 | 0.20 | 0.00 | 0.50 | 0.85 | 0.00 | 0.90 | 0.00 |
| be | 0.90 | 0.90 | 0.40 | 0.00 | 0.10 | 0.90 | 0.75 | 0.55 | 0.40 | 0.70 | 0.90 | 0.40 | 0.55 | 0.40 |
| aux | 0.90 | 0.90 | 0.40 | 0.10 | 0.00 | 0.90 | 0.75 | 0.55 | 0.40 | 0.70 | 0.90 | 0.40 | 0.55 | 0.40 |
| adj | 0.75 | 0.80 | 0.50 | 0.90 | 0.90 | 0.15 | 0.25 | 0.65 | 0.50 | 0.85 | 0.75 | 0.50 | 0.90 | 0.50 |
| adv | 0.90 | 0.95 | 0.50 | 0.75 | 0.75 | 0.25 | 0.15 | 0.65 | 0.50 | 0.85 | 0.90 | 0.50 | 0.80 | 0.50 |
| cnj | 1.00 | 1.00 | 0.20 | 0.55 | 0.55 | 0.65 | 0.65 | 0.00 | 0.20 | 0.55 | 1.00 | 0.20 | 0.95 | 0.20 |
| det | 0.90 | 1.00 | 0.00 | 0.40 | 0.40 | 0.50 | 0.50 | 0.20 | 0.00 | 0.50 | 0.85 | 0.00 | 0.90 | 0.00 |
| prep | 1.00 | 1.00 | 0.50 | 0.70 | 0.70 | 0.85 | 0.85 | 0.55 | 0.50 | 0.05 | 1.00 | 0.50 | 1.00 | 0.50 |
| pron | 0.15 | 0.15 | 0.85 | 0.90 | 0.90 | 0.75 | 0.90 | 1.00 | 0.85 | 1.00 | 0.05 | 0.85 | 0.70 | 0.90 |
| pun | 0.90 | 1.00 | 0.00 | 0.40 | 0.40 | 0.50 | 0.50 | 0.20 | 0.00 | 0.50 | 0.85 | 0.00 | 0.90 | 0.00 |
| v | 0.70 | 0.70 | 0.90 | 0.55 | 0.55 | 0.90 | 0.80 | 0.95 | 0.90 | 1.00 | 0.70 | 0.90 | 0.20 | 0.90 |
| $\epsilon$ | 0.90 | 1.00 | 0.00 | 0.40 | 0.40 | 0.50 | 0.50 | 0.20 | 0.00 | 0.50 | 0.90 | 0.00 | 0.90 | |

The costs of replacing two chunks depends on their types. Zero cost is used to make the chunks fully compatible (e.g. auxiliary verb phrases can freely interchange). Also, all chunks that are deleted with zero cost may freely replace one another. Generally, the replacement costs reflect the compatibility between the involved chunks. Therefore, low costs can be found along the main diagonal in Table 1 emphasising the highest compatibility between the same chunk types. An exception with this regard is the cost of replacing NPs and terms with pronouns (0.15), which can act as "wild cards" for these chunks. Note that the cost of replacing the same chunk types is not necessarily zero. Although compatible, they cannot always be freely replaced. This is used for high-content chunks (such as terms and NPs) in order to emphasise the importance of semantic information they encode and not only their syntactic function.

So far we have mostly relied on syntactic information acquired through POS tagging and partial parsing. We would like to incorporate more domain-specific knowledge into the SOLD measure in order to support semantic comparison. Since the ontology used incorporates hierarchies of terms and domain-specific verbs, the replacement costs can be fine-grained so as to reflect the semantic closeness of terms and verbs considered. The actual replacement cost involving such chunks depends on their content. In these cases, the replacement cost $r$ given in Table 1 is not fixed, but rather represents its upper limit. There are two basic principles used for the calculation of the replacement cost in such cases: a knowledge-rich approach based on domain-specific knowledge contained in the ontology and a knowledge-poor approach based on lexical similarity. In the following sections we discuss these two approaches to semantic comparison.

## 5. Ontology-Driven Component

Ontology-based replacement cost is used for terms and verbs contained in the ontology. Let us describe how the replacement cost is calculated for two classified terms. Figure 2 describes the computation of the replacement cost (RC) for two classified terms ($t_1$ and $t_2$) based on their similarity: the higher the similarity, the smaller the replacement cost. It is first checked if the terms are lexical variants, that is – if they are orthographic variants (differing in the use of hyphenation, lower and upper cases, spelling, etc.) or inflectional variants (differing in number – singular or plural), simply by checking if they are linked to the same term identifier in UMLS. Lexical variants are given the highest similarity value (1), since they identify the same concept and differ only in their textual realisation. For the same reason, semantic variants (i.e. synonyms) are given the same similarity score. It is checked if two terms are synonyms, by checking if they are mapped to the same concept identifier in UMLS. If they are not recognised as semantic variants, the class information is used for their further comparison. All semantic classes in UMLS are organised into a hierarchy, which can be used to quantify their similarity. The tree similarity (ts) between two classes ($C_1$ and $C_2$) is calculated according to the following formula:

$$ts(C_1, C_2) = \frac{2 \cdot \text{common}(C_1, C_2)}{\text{depth}(C_1) + \text{depth}(C_2)}$$

where common($C_1, C_2$) denotes the number of common classes in the paths leading from the root to the given classes, and depth($C$) is the number of classes in the path connecting the root and the given class. This formula is a derivative of Dice coefficient, where each ancestor class is treated as a separate feature. It was previously used to measure conceptual similarity in a hierarchically structured lexicon.[6] Other measures have been proposed and could be used as well. For example, Resnik[7] used a "probabilistic variation" of this model:

$$ts(C_1, C_2) = \frac{2 \cdot \log P(S(C_1, C_2))}{\log P(C_1) + \log P(C_2)}$$

where $S(C_1, C_2)$ is the deepest class that subsumes both classes, and $P(C)$ denotes the probability that a random object belongs to the given class. To be used with UMLS, this approach would require additional computation of these probabilities.

Further, since UMLS supports multiple classification, we estimate the similarity between two terms as the maximal similarity between their classes. Note that if two terms belong to the same class (among others if any), then their similarity reaches 1. In order to differentiate between compatible terms (i.e. non-synonymous terms belonging to the same class) and semantic variants (i.e. synonymous terms), we scale down the tree similarity by 10%. Finally, having calculated the similarity between two terms, it is converted to the corresponding distance and mapped to the interval $[0, r]$ (where $r = 0.15$ is the maximal replacement cost for two terms). The replacement cost for classified domain-specific verbs is calculated similarly.

| | | |
|---|---|---|
| if $t_1$ and $t_2$ are lexical variants, | then | $sim(t_1, t_2) = 1$, |
| else if $t_1$ and $t_2$ are synonyms, | then | $sim(t_1, t_2) = 1$, |
| else | | $sim(t_1, t_2) = 0.9 \cdot \max$, |
| | | (where max is the maximal value of $ts(C_1, C_2)$ |
| | | for all classes $C_1$ of $t_1$ and all classes $C_2$ of $t_2$) |
| $RC(t_1, t_2) = r \cdot (1 - sim(t_1, t_2))$; | | |

Figure 2.  Calculation of the variable replacement costs.

## 6. Lexical Component

The approach described in Section 5 applies only to *classified* terms or verbs. Currently, biomedical ontologies are inherently incomplete due to the fast-growing number of terms. Therefore, it would be useful to use clues other than the ones explicitly stated in the ontology in order to extend the semantic comparison to other terms and verbs. Since the syntactic clues are already being used when comparing term contexts by the SOLD distance, we opted for internal lexical properties of context constituents. Lexical comparison has been enabled for terms, NPs and verbs. We utilised the standard ED approach applied at the grapheme level. Similarly to Tsuruoka and Tsujii,[2] we differentiate between four types of graphemes (space and hyphen, digits, letters and all other graphemes) in order to determine the appropriate costs of edit operations (Table 2).

Table 2.   The cost of edit operations for different graphemes.

| Grapheme | " " or "–" | digit | letter | other | $\epsilon$ |
|---|---|---|---|---|---|
| " " or "–" | 0.05 | 1.00 | 1.00 | 1.00 | 0.50 |
| digit | 1.00 | 0.10 | 1.00 | 1.00 | 1.00 |
| letter | 1.00 | 1.00 | 0.90* | 1.00 | 1.00 |
| other | 1.00 | 1.00 | 1.00 | 0.05 | 0.50 |
| $\epsilon$ | 0.50 | 1.00 | 1.00 | 0.50 | |

The highest deletion costs are given to digits and letters as they convey more information compared to other graphemes. For example, spaces and hyphens basically serve to improve the readability of multi-word terms. In addition, they are not always used consistently and often cause orthographic variation by replacing each other or being omitted altogether (e.g. *EGR-1* vs. *EGR 1* vs. *EGR1*). Hence, these graphemes are assigned lower cost (0.5). The same cost is given to all other graphemes for similar reasons.

The replacement cost is generally chosen so as to "discourage" the replacement of graphemes of different types (e.g. digits and letters) by assigning the highest cost (1) to such operations. The replacement within the same type depends on the importance and similarity between the graphemes. Space and hyphen are regarded similar, thus are given a low cost (0.05). Similarly, digits can be interchanged at a relatively low cost (0.1). Letters are given high cost (0.9) since morphemes (as groups of letters), often in the form of neoclassical roots and affixes,[8] are used to encode important features of biomedical concepts. In order to make the cost function less case-sensitive, the cost of replacing the same letter differing only in case is obtained by subtracting the general replacement cost for letters from the highest possible cost: $1 - 0.9 = 0.1$. Note that we still maintain the case sensitivity. This may be important for some types of terms (e.g. case variants sometimes can be used to distinguish a gene from its protein[9]). ED between two chunks is used to adjust the cost of their replacement.

The lexical component adds to the robustness of the SOLD measure by comparing terms and verbs not covered by the ontology and, therefore, overlooked by the ontology-driven component. It also makes the SOLD distance approach robust with respect to spelling variations and typing errors occurring within semantically important chunks. Alternatively, word edit distance[3] or the MetaMap[10] program for the recognition of term variants can be used to support lexical comparison.

## 7. Discussion and Conclusions

We described the SOLD measure, which can be used to assess similarity between terms based on their contextual features. Compared to other measures of contextual similarity, our approach offers a significant degree of flexibility. For example, Nenadić et al.[11] relied on Dice coefficient using predefined lexico-syntactic patterns as contextual features. Such an approach lacks the necessary flexibility, since small syntactic variations may cause similar patterns not to match and to be accounted as different features. In our approach, the variability of a natural language has been accounted for at multiple levels. First, the choice of contextual features need not be rigidly predefined, as features may be matched, replaced or discarded as necessary through elastic matching, thus neutralising some types of syntactic variation. Further, lexical variability is partly neutralised by using an ontology to match different forms of terms and domain-specific verbs. In addition, lexical similarity is assessed by ED in order to compensate for incompleteness of the ontology.

We also compare our approach to that of Dagan et al.[12], who proposed a method for estimating word similarity from sparse data, the main assumption being that similar word co-occurrences should have similar mutual information. In our approach, sparsity of data can be partly compensated by non-exact matching driven by the ontology and lexical similarity. In other words, classes of terms (both lexical and semantic) are compared rather than individual terms, which means that individual frequencies are aggregated into collective frequencies of similar terms. In addition, syntactic knowledge and ED are used to generalise a rigid and knowledge-poor notion of co-occurrence into contextual similarity that takes into account not only the relative position and the frequency of co-occurrence, but also a wider context with respect to its syntactic structure and semantic content.

Word similarity approaches can be evaluated through the recognition of synonyms.[13] We generalise this idea to recognition of terms belonging to the same semantic classes in order to evaluate the SOLD measure. It has been fully implemented as part of a case-based reasoning system in which the similarity measure plays a key inferencing role.[14] The efficiency of comparison is improved through retrieval component developed to reduce the search space to potentially similar cases with respect to the SOLD measure, thus avoiding a brute-force nearest-neighbour approach and enhancing the scalability of the given measure. We tested our approach for functional classification of chemicals (13 UMLS classes) based on a training corpus of 2072 MEDLINE abstracts annotated with 18236 training, 2405 validation, 2838 testing and 30419 non-classified terms. The performance has been evaluated relative to three baseline methods (random, naive Bayes and rule-based classifier), which achieved 8.97%, 25.31% and 45.54% for

F-measure, while we obtained 58.52%.

We plan to use the SOLD measure to improve the flexibility of a rule-based information extraction (IE) system by identifying contexts similar to the ones to which the IE rules apply directly and to extract information of interest indirectly by the rules through alignment. In particular, we are interested in extracting information on protein-protein interactions. We believe that other methods developed for NLP tasks in biomedicine may similarly be facilitated by the use of the SOLD measure.

## References

1. A. Apostolico and R. Giancarlo. Sequence Alignment in Molecular Biology. In M. Farach-Colton et al. (Eds.), *DIMACS Special Year for Mathematical Support of Molecular Biology*, 47:85-116, 1999.
2. Y. Tsuruoka and J. Tsujii. Boosting Precision and Recall of Dictionary-Based Protein Name Recognition. In *ACL Workshop on NLP in Biomedicine*, Sapporo, Japan, 41-48, 2003.
3. J. French, A. Powell and E. Schulman. Applications of Approximate Word Matching in Information Retrieval. In *Int Conf on Knowledge and Information Management*, Los Angeles, USA, 1997.
4. G. Navarro, E. Silva de Moura, M. Neubert, N. Ziviani and R. Baeza-Yates. Adding Compression to Block Addressing Inverted Indexes. *Information Retrieval*, 3:49-77, 2000.
5. R. Wagner and M. Fischer. The String-to-String Correction Problem. *J of ACM*, 21(1):168-173, 1974.
6. Z. Wu, M. Stone Palmer. Verb Semantics and Lexical Selection. In *Annual Meeting of the ACL*, Las Cruces, USA, 133-138, 1994.
7. P. Resnik. Using Information Content to Evaluate Semantic Similarity in a Taxonomy. In *Int Conf on Artificial Intelligence*, Montreal, Canada, 448-453, 1995.
8. S. Ananiadou. A Methodology for Automatic Term Recognition. In *COLING*, Kyoto, Japan, 1034-1038, 1994.
9. M. Weeber, B. Schijvenaars, E. van Mulligen, B. Mons, R. Jelier, C. van der Eijk and J. Kors. Ambiguity of Human Gene Symbols in LocusLink and MEDLINE: Creating an Inventory and a Disambiguation Test Collection. In *AMIA Symposium*, 704-708, 2003.
10. A. Aronson. Effective Mapping of Biomedical Text to the UMLS Metathesaurus: the MetaMap Program. In *AMIA Symposium*, 17-21, 2001.
11. G. Nenadić, I. Spasić and S. Ananiadou. Mining Term Similarities from Corpora. *Terminology*, 10(1):55-80, 2004.
12. I. Dagan, S. Marcus and S. Markovitch. Contextual Word Similarity and Estimation from Sparse Data. *Computer, Speech and Language*, 9:123-152, 1995.
13. G. Grefenstette. *Exploration in Automatic Thesaurus Discovery*. 1994.
14. I. Spasić. *A Machine Learning Approach to Term Classification*. PhD Thesis. University of Salford, UK, 2004.

# UNDERSTANDING THE GLOBAL PROPERTIES OF FUNCTIONALLY-RELATED GENE NETWORKS USING THE GENE ONTOLOGY

L. TARI, C. BARAL AND P. DASGUPTA

*Dept. of Computer Science and Engineering, Arizona State University,*
*Brickyard Suite 501, 699 South Mill Avenue*
*Tempe, AZ 85287, USA*
*Email: {luis.tari,chitta,partha}@asu.edu*

The global behavior of interactions between genes can be investigated by forming the network of functionally-related genes using the annotations based on the Gene Ontology. We define two genes to be connected when the pair of genes is involved in the same biological process. There has been other work on the analysis of different kinds of cellular and metabolic networks, such as gene coexpression network, in which genes are paired when they are found to be coexpressed in the microarray experiments. We observe that our functionally-related gene networks among humans, fruit flies, worms and yeast exhibit the small-world property, but all except the network of worms show the existence of the scale-free property.

## 1. Introduction

Uncovering the underlying functions and interactions within a living cell is an important goal in the post-genomic era. Recent advances in technology, such as the development of microarrays and protein chips, allow biologists to study the functioning of the cell in many new ways. While it significantly speeds up the process of understanding bio-molecular interactions, modeling interactions of a cell in quantifiable terms is a major challenge for biologists. By modeling the interactions, the ultimate goal is to discover the fundamental properties that govern the behavior of a cell.

Work by Watts and Strogatz on the small world characteristic[1] and by Barabasi and Albert on the scale-free feature[2] of large, sparse and complex networks has been applied to various areas such as sociology and computer networks. The small-world phenomenon demonstrates the famous property of six degrees of separation between any two persons in the world[3]. The neural network of the worm *Caenorhabditis elegans*, the power grid of the western United States and the collaboration of film actors have been shown to exhibit small-world properties[4]. Some of the important outcomes of such small-world networks are the increase of signal-propagation speed, computational power and synchronizability. In biological domains, infectious diseases are found to be more easily spread in small-world networks than in regular networks[4].

209

Previously, complex networks have been thought to exhibit the property of classical random networks, in which the fundamental randomness of the model leads to the same number of edges in most nodes. Empirical studies on the structure of the World Wide Web[5] show that a few highly connected nodes dominate the structure. This phenomenon is known as scale-free. Scale-free networks are robust with respect to random attacks and component failures, since the chance of harming a highly connected node is low[6]. This is in contrast to a random network, in which the removal of several nodes can effectively disrupt the network. On the other hand, attacks on highly connected nodes in a scale-free network can catastrophically disrupt the network.

Most cellular functions are known to be carried out by groups of molecules within interacting functional modules[7]. It is essential to study interactions within a cell and the properties that govern the interactions. Recently, there has been a significant amount of interest in examining the universal property that oversees the complex molecular interactions between the cell components, by modeling the molecular interactions in the form of networks, equivalently known as graphs in mathematical terms. Numerous studies of various types of cellular and metabolic networks have shown the existence of small-world and scale-free properties. Some of these studies include the characterizations of physical interactions between protein-protein, protein-nucleic-acid, protein-metabolite molecule pairs[8,9,10]. Modeling of more complex functional interactions such as metabolites that are substrates or products in the same biochemical reaction[11,12], or chemical reactions that share at least one chemical component either as substrate or as product, also reveal such properties[13]. Further examples of small-world, scale-free organization includes the study of genetic-regulatory network such as protein domain interactions and coexpression of genes based on microarray data[14,15], in which coexpression of genes are paired to imply that the genes are involved in the same biological process.

The main focus of this work is on the characterization of the gene involvement in the same biological process in a large scale. Unlike the previous work[14,15] in which the study of the involvement of genes in the same biological process were based on coexpression of genes in microarray experiments, our work utilizes the Gene Ontology to study the global behavior of such networks.

The Gene Ontology is a hierarchy of controlled vocabulary that includes three independent ontologies for biological process, molecular function and cellular component. Standardized terms in the Gene Ontology describe roles of genes and gene products in any organism. Figure 1 illustrates the main terms in the biological process ontology. A gene product has one or more molecular functions, can be used in one or more biological processes, and can be associated with one or more cellular components[16]. As a way to share knowledge about functionalities of genes, the Gene Ontology itself does not contain gene products of any organisms. Rather, biologists annotate biological roles of gene products using the Gene Ontology, known as annotations.

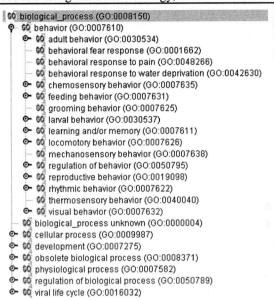

Figure 1 The hierarchy of main terms in the biological process ontology, with the term "behavior" expanded to show its children. Screenshot was taken from the TGen GOBrowser (to be published.)

Our model is a network composed of genes or proteins as nodes and an edge exists between two nodes if they are involved in the same biological process[a]. A biological process is defined as a biological objective to which the gene or gene product contributes[16]. We selected four evolutionarily conserved organisms: *Homo sapiens, Drosophila melanogaster, Caenorhabditis elegans* and *Saccharomyces cerevisiae* to construct functionally-related gene networks from the Gene-Ontology-based annotations and study the global behavior that governs such network.

The remainder of the work is organized as follows. Section 2 provides the descriptions and definitions of the formal measures of the small-world and scale-free properties. The process of constructing our functionally-related gene networks is described in Section 3. Section 4 describes the existence of small-world and scale-free properties in our networks, and section 5 discusses some of the possible implications of such properties in the networks.

---

[a] Currently the generality or specificity of the Gene Ontology terms is not taken into consideration.

## 2. Preliminaries

The small-world property can be characterized by two statistical quantities: clustering coefficient $C$ and characteristic path length $L$[1]. $L$ is the average minimal distance between any two nodes in the network, while $C$ is a measure of how clustered a graph is, which implies an average of interconnectivity among the neighbors of each node. More formally,

$$C = \frac{\sum_{i=1}^{n} C_i}{n} \text{ and } C_i = \frac{e_i}{k_i \times (k_i - 1)/2} \tag{1}$$

if node $i$ has $k_i$ immediate neighbors with $e_i$ number of edges between $i$'s neighbors in a graph of $n$ nodes. It is easy to see that a fully connected graph has a clustering coefficient of 1. If a given node $j$ has no neighbors or one neighbor, then we define $C_j = 1$. Regular networks have large $C$ and $L$ grows linearly with $n$, while random networks have small $C$ and $L$ only grows logarithmically with $n$[17]. In other words, regular networks have relatively large $L$, while random networks have relatively small $L$. By having the same configuration (i.e. the same number of nodes and edges) but with different probability $p$ of rewiring edges, a collection of graphs between a regular network ($p=0$) and a random network ($p=1$) can be generated. Such random rewiring procedure shows that for intermediate values of $p$, the graph is a small-world network[4]. This phenomenon implies that small-world networks fall in between the two; small-world networks are highly clustered like regular networks, while the characteristic path length is as small as random networks[4]. With the relation among the three kinds of networks, showing a network exhibit the small-world property requires the comparison of the actual configuration of the network with the random configuration of itself. For random networks, the two quantities can be computed as

$$C \approx \bar{k}/(N-1), L \approx \ln N / \ln \bar{k} \tag{2}$$

where $N$ is the number of nodes in a network and $\bar{k}$ is the average number of edges per node[18].

The scale-free property is defined by an algebraic behavior in the probability of degree distribution $P(k)$, i.e. the probability that a selected node has exactly $k$ edges. Scale-free networks are networks that have a degree distribution approximated as the power law, $P(k) \sim k^{-\gamma}$, where $k$ is the number of edges and $\gamma$ is the degree exponent[10]. The existence of the scale-free property in a network implies that there can be a few nodes with a significantly larger number of edges than the typical nodes.

## 3. Construction of functionally-related gene networks

The Gene Ontology is composed of three independent ontologies: molecular function, biological process and cellular component. Our functionally-related gene networks are constructed from annotations based on the biological process ontology of the Gene Ontology. The networks are composed of genes or proteins as nodes and two nodes are connected if they are involved in the same biological process based on the annotations. Figure 2 illustrates the idea.

The annotations that we used to construct the networks are curated by various highly recognized organizations and institutes. The annotation of *Homo sapiens* is obtained from the collection at the European Bioinformatics Institute, while the annotation of *Drosophila melanogaster* is from the FlyBase organization. The annotation for *Caenorhabditis elegans* is obtained from the WormBase organization, while the annotation for *Saccharomyces cerevisiae* is from the collection at Stanford University. In all four cases, the actual files used were the annotation files lodged with the Gene Ontology Consortium by the four currating organizations.

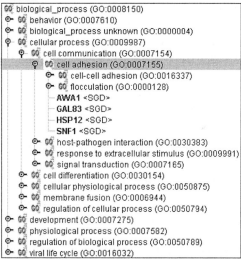

Figure 2 The genes products for yeast genes AWA1, GAL83, HSP12, SNF1 are all annotated as being involved in the biological process of cell adhesion. Pairs of the four genes are linked in the functionally-related gene network for yeast. Screenshot was taken from the TGen GOBrowser (to be published.)

Annotations of the four organisms were first preprocessed to remove genes that are mapped to unknown and obsolete Gene Ontology terms. Unknowns, referred to as the term "biological process unknown" in the biological process ontology, are used when annotation of gene products whose function are not known or cannot be inferred. Obsolete Gene Ontology terms are terms that have been removed from the active biological process ontology[19]. The Gene Ontology

Consortium defines a set of evidence codes to support the functional assignments of gene products. As part of the annotation process, curators are required to provide an evidence code when assigning a Gene Ontology term to a gene product. Reliability of annotations varies with different evidences. To further increase the reliability of our network, gene product annotations that are inferred from electronic annotation (IEA) were removed from our network. The evidence IEA is used when no curator has checked the annotation to verify its accuracy, and thus has the lowest quality among all evidences[20]. The resulting networks are composed of 7512 proteins for humans, 4641 genes for fruit flies, 3254 proteins for worms and 4660 genes for yeast.

## 4. Results

Our results show that the functionally-related gene networks of the four organisms exhibit the small-world property. All networks except the network for worms also demonstrate the scale-free property.

We first present the results regarding the existence of small-world property of the networks. As shown in Table 1 (referred to as the actual configuration of the networks), the clustering coefficients $C$, computed by equation 1, of the four networks are very high, while the characteristic path lengths $L$ are surprisingly quite small. The low characteristic path lengths in the actual configuration are related to a high number of edges for each node. In particular, the human functionally-related gene network has an average number of 276.68 edges for each node and on average a node can be reached by another node within 2.58 links. To examine the existence of small-world property of the networks, the clustering coefficients $C$ and characteristic path lengths $L$ of the random configuration of the networks with the same parameters $N$ and $\bar{k}$ were approximated by equation $2^{18}$, as shown in Table 2. The results show that the networks with actual configuration have much higher clustering coefficients $C$, while the characteristic path lengths $L$ are about the same as the random configurations. Table 3 describes the minimal paths between any two nodes among the four networks in the actual configuration, showing that in the worst case there can be 8 degrees of separation between two nodes in the human network, but with a very low probability of $6.38 \times 10^{-7}$. These results confirm the networks are highly clustered but with short characteristic path lengths. In other words, the functionally-related gene networks of the four organisms are highly clustered and at the same time have small path lengths, which coincide with the property of small-world network.

As for the scale-free property, our results shown in figure 3 illustrates convincingly that the functionally-related gene network of humans, fruit flies and yeast follow a power-law distribution. Figures 4, 5, 6, 7 show the exact degree distribution for each of the organism. However, in the case of worms, it does not seem to follow a power-law distribution. The property of the power-law distribution shows that the functionally-related gene network of humans, fruit flies and yeast can be modeled by scale-free networks, while we cannot make the same observation for worms.

Table 1 Results for the functionally-related gene network constructed from the Gene-Ontology-based annotations (actual configuration). $N$ is the total number of nodes (genes or proteins), $\bar{k}$ is the average number of edges per node, $C$ is the clustering coefficient and $L$ is the average shortest path.

|  | $N$ | $\bar{k}$ | $C$ | $L$ |
|---|---|---|---|---|
| Humans | 7512 | 276.68 | 0.87 | 2.58 |
| Fruit Flies | 4641 | 100.22 | 0.87 | 2.90 |
| Worms | 3254 | 1573.85 | 0.87 | 1.55 |
| Yeast | 4660 | 73.44 | 0.88 | 3.51 |

Table 2 Results for the functionally-related gene networks (random configuration) with the same parameters $N$ and $\bar{k}$ as in Table 1.

|  | $N$ | $\bar{k}$ | $C$ | $L$ |
|---|---|---|---|---|
| Humans | 7512 | 276.68 | 0.037 | 1.59 |
| Fruit Flies | 4641 | 100.22 | 0.022 | 1.83 |
| Worms | 3254 | 1573.85 | 0.48 | 1.10 |
| Yeast | 4660 | 73.44 | 0.016 | 1.97 |

Table 3 The length distribution of the minimal paths between two nodes of length $L_n$ among the four organisms. No path exists between two nodes if $n = 0$.

| $n$ | $L_n$ (Humans) | $L_n$ (Fruit Flies) | $L_n$ (Yeast) | $L_n$ (Worms) |
|---|---|---|---|---|
| 0 | 0.0834 | 0.1840 | 0.2004 | 0.0129 |
| 1 | 0.0368 | 0.0216 | 0.0158 | 0.4838 |
| 2 | 0.3958 | 0.2140 | 0.0929 | 0.4698 |
| 3 | 0.4100 | 0.4297 | 0.3032 | 0.0321 |
| 4 | 0.0676 | 0.1295 | 0.2721 | 0.0013 |
| 5 | 0.0058 | 0.0186 | 0.0923 | $1.927 \times 10^{-5}$ |
| 6 | $5.24 \times 10^{-4}$ | 0.0022 | 0.0203 | - |
| 7 | $2.3 \times 10^{-5}$ | 0.00036 | 0.0028 | - |
| 8 | $6.38 \times 10^{-7}$ | $2.158 \times 10^{-5}$ | $3.002 \times 10^{-4}$ | - |
| 9 | - | $5.57 \times 10^{-7}$ | $2.34 \times 10^{-5}$ | - |
| 10 | - | - | $1.29 \times 10^{-6}$ | - |

Table 4 Reachability of the network. *U* is the percentage of unconnected pairs in the networks, *L* is the average shortest path and *R* is the percentage of node pairs that can be reached within $\lceil L \rceil$ (ceiling of *L*) number of links.

|  | U | L | R |
|---|---|---|---|
| Humans | 8.34% | 2.58 | 91.93% |
| Fruit Flies | 5.1% | 2.90 | 94.51% |
| Worms | 1.29% | 1.55 | 96.61% |
| Yeast | 20.03% | 3.51 | 85.53% |

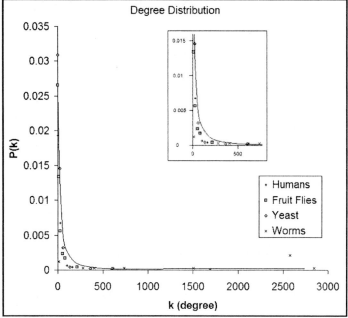

Figure 3 - Alegbraic scaling behavior of *P(k)* for humans, fruit flies and yeast, but not worms. *P(k)* is the probability that a selected node has *k* number of edges. The inset shows a clearer view of the curve.

We also examine the reachability of the four networks. Table 4 shows that all four networks have surprisingly small path lengths. This behavior can be explained with the presence of highly connected nodes in the networks, as illustrated in figure 3. In fact, less than 8% of the proteins in the human network can be reached by more than 3 proteins from any given protein. Similarly, the other 3 networks also exhibit such close connectivity between any two nodes. On the other hand, we also observe that there are nodes that cannot be reached by another given node in the networks. Specifically, out of all possible pairings in the human network, 8.34% of the pairs cannot be reached by each other. Among all of the four networks, the yeast network has the most number of unconnected pairs – about 20% of all pairs. The existence of unconnected pairs can be explained by the fact that not all functions of genes for each of the four organisms have been fully discovered.

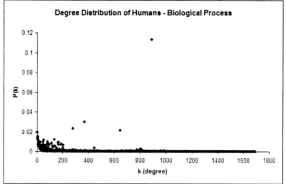

Figure 4 The exact degree distribution of the functionally-related gene network of Humans, where $P(k)$ refers to the probability that a selected node has $k$ edges.

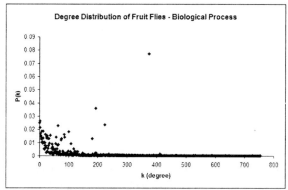

Figure 5 The exact degree distribution of the functionally-related gene network of Fruit Flies, where $P(k)$ refers to the probability that a selected node has $k$ edges.

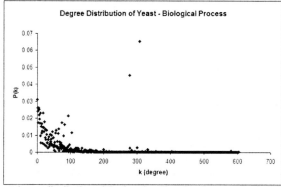

Figure 6 The exact degree distribution of the functionally-related gene network of Yeast, where $P(k)$ refers to the probability that a selected node has $k$ edges.

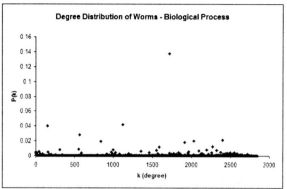

Figure 7 The exact degree distribution of the functionally-related gene network of Worms, where $P(k)$ refers to the probability that a selected node has $k$ edges.

## 5. Discussion

There have been various studies on the global property that governs the behavior of different aspects of metabolic and cellular networks. Our work differs from the others as we focus on a different perspective of biological network, specifically gene products that are involved in the same biological process. Our work is most closely related to the study of the gene coexpression network, in which coexpressed genes in the microarray experiments are connected to form the network. Coexpressed genes may imply that the genes are functionally related, i.e. genes are involved in the same biological process. However, coexpression of genes depends on the threshold of coexpression correlation and thus has an effect of the size and connectivity of such network. The implication of coexpression of genes to be functionally related genes can arguably be an assumption. Our use of the Gene-Ontology based annotations is independent of such assumption. In addition, our method does not have dependence on threshold values and experimental bias in the microarray data. It is inevitable that our method also introduces some potential bias by utilizing the annotations to construct our networks. Annotations are curated based on different evidences such as direct assay, sequence similarity and expression pattern, in which each has its own experimental bias. However such bias should be restrained to the minimum, as there are strict guidelines on the approval of the annotations, and the Gene-Ontology based annotations are widely accepted by the biomedical community. Among all of the evidence codes used for the annotations, the evidence code "inferred from electronic annotation" is applied to annotations that are yet to be verified for their accuracy by the curators. Such annotations are removed to ensure high quality and reliability of our networks.

Our work also goes in line with the common application of the Gene Ontology – interpreting microarray data from a biological point of view[21,22]. Microarray experiments allow biologists to find a set of differentially expressed genes between two or more conditions being studied, for instance among tissues treated with drugs and untreated tissues. Identifying which genes are differentially expressed is important, but it is also essential to interpret the biological roles of these genes. With the Gene Ontology, biologists can acquire a list of functionally-related genes from microarray experiments.

Our results of functionally-related gene networks are consistent with other studies of metabolic and cellular networks. We find that the network of the four organisms – human, fruit flies, yeast and worms have the property of small-world. Due to the fact that the studies of functionalities of genes in the organisms have not been completed, annotations are updated periodically. Even as the annotations evolve, the conclusion of the existence of small-world property still holds as the characteristic path lengths $L$ in small-world networks grow only logarithmically with the number of nodes[4]. In other words, as more gene products are added to the annotations, $L$ would not be changed by much. Other than the network of worms, all of them also exhibit the scale-free property. Such findings can be of huge implication for the evaluation of newly derived gene product interactions and the practice of medicine. As described in work[23] by Goldberg and Roth, a potential application is to exploit the neighborhood cohesiveness derived from the small-world property of our network to define measures of confidence. Such confidence can be applied to evaluate gene product interactions that are inferred from new data. Another possible scenario could be on searching for new targets for antibiotics, a pharmacologist can utilize the functionally-related gene network to find gene products that are involved in bacterial protein synthesis[b] and other known involvement of biological process in such gene products. Because of the small-world and scale-free properties of the networks, the number of genes that a pharmacologist needs to consider can be significantly reduced.

## Acknowledgement

We would like to acknowledge John Pearson at Translational Genomics Research Institute (TGen) for introducing the Gene Ontology and his valuable comments on this paper. Luis Tari is grateful for the support of Edward Suh at TGen. Chitta Baral acknowledges the support of NSF through grant number 0412000. We also appreciate the helpful comments provided by the reviewers of this paper.

---

[b] Example taken from the Gene Ontology website.

**References**

1  D. Watts, Networks, dynamics and the small world phenomenon, *American Journal of Sociology.* **105(2)**, 493-527 (1999).
2  A. Barabási, R. Albert, Emergence of scaling in random networks, *Science.* **286**, 509-512 (1999).
3  D. Watts, Six degrees of interconnection, *Wired Magazine.* **136** (2003).
4  D. Watts, S. Strogatz, Collective dynamics of 'small-world' networks, *Nature.* **393**, 440-442 (1998).
5  R. Albert, H. Jeong, A. Barabasi, Diameter of the World-Wide Web, *Nature.* **400**, 130-131 (1999).
6  S. Bornholdt, H. Schuster, Handbook of Graphs and Networks: from Biological Nets to the Internet and WWW, *Oxford University Press* (2003).
7  L. Hartwell, J. Hopfield, S. Leibler, A. Murray. From molecular to modular cell Biology. *Nature.* **402** supplement. 6761, C47-C52 (1999).
8  H. Jeong, S. Mason, A. Barabasi, Z. Oltva. Lethality and centrality in protein networks. *Nature.* **411**, 41-42 (2001).
9  A. Wagner, The yeast protein interaction network evolves rapidly and contains few redundant duplicate genes. *Mol. Bio. Evol.* **18**, 1283-1292 (2001).
10  A. Barabási, Z. Oltvai, Network biology: understanding the cell's functional organization. *Nature Reviews Genetics.* **5**, 101 -113 (2004).
11  H. Jeong, B. Tombor, R. Albert, Z. Ottvai, A. Barabasi. The large scale of metabolic networks. *Nature,* **407**, 651-654 (2000).
12  D. Fell, A. Wagner, The small world of metabolism. *Nat Biotechnology.* **18,**112-122 (2000).
13  U. Alon, M. Surette, N. Barkai, S. Leibler, Robustness in bacterial chemotaxis. *Nature (London).* **397,** 168–171 (1999).
14  H. Agrawal, Extreme self-organization in networks constructed from gene expression data. *Phys. Rev. Lett.* **89**, 268702 (2002).
15  V. van Noort, B. Snel, M. Huynen, The yeast coexpression network has a small-world, scale-free architecture and can be explained by a simple model. *EMBO Reports.* **5,** 280-284 (2004).
16  M. Ashburner, C. Ball, J. Blake, D. Botstein, et al., Gene Ontology: tool for the unification of biology. *Nature Genetics.* **25**, 25 – 29 (2000).
17  B. Bollobas, Random Graphs. *Academic Press*, London (1985).
18  D. Watts, Small Worlds. *Princeton University Press*, Princeton (1999).
19  http://www.geneontology.org/ontology/GO.defs Gene Ontology definitions
20  http://www.geneontology.org/GO.evidence.html GO evidence codes
21  S. Draghici, P. Khatri, R. Martins, C. Ostermeier, S. Krawetz, Global functional profiling of gene expression. *Genomics.* **81(2)**, 98-104 (2003).
22  B. Zeeberg et al., GoMiner: A Resource for Biological Interpretation of Genomic and Proteomic Data. *Genome Biology.* **4(4)**, R28 (2003).
23  D. Goldberg, F. Roth, Assessing experimentally derived interactions in a small world. *Proc. of the Nat. Acad. of Sciences.* **100(8)**, 4372-4376 (2003).

# IDENTIFICATION OF FUNCTIONAL MODULES IN PROTEIN COMPLEXES VIA HYPERCLIQUE PATTERN DISCOVERY

HUI XIONG$^{a,b,*}$ , XIAOFENG HE$^{b,*}$ , CHRIS DING$^{b}$, YA ZHANG$^{b,c}$, VIPIN KUMAR$^{a}$, STEPHEN R. HOLBROOK$^{b}$

$^{a}$ *Computer Science & Engineering, University of Minnesota, MN, USA*
*E-mail: {huix, kumar}@cs.umn.edu*

$^{b}$ *Computational Research Division and Physical Biosciences Division*
*Lawrence Berkeley National Laboratory, Berkeley, CA, USA*
*E-mail: {xhe, chqding, srholbrook}@lbl.gov*

$^{c}$ *Information Sciences and Technology, Penn State University, PA, USA*
*E-mail: yzhang@ist.psu.edu*

Proteins usually do not act isolated in a cell but function within complicated cellular pathways, interacting with other proteins either in pairs or as components of larger complexes. While many protein complexes have been identified by large-scale experimental studies [7,8], due to a large number of false-positive interactions existing in current protein complexes [10], it is still difficult to obtain an accurate understanding of functional modules, which encompass groups of proteins involved in common elementary biological function. In this paper, we present a hyperclique pattern discovery approach for extracting functional modules (hyperclique patterns) from protein complexes. A hyperclique pattern is a type of association pattern containing proteins that are *highly affiliated* with each other. The analysis of hyperclique patterns shows that proteins within the same pattern tend to present in the protein complex together. Also, statistically significant annotations of proteins in a pattern using the Gene Ontology suggest that proteins within the same hyperclique pattern more likely perform the same function and participate in the same biological process. More interestingly, the 3-D structural view of proteins within a hyperclique pattern reveals that these proteins physically interact with each other. In addition, we show that several hyperclique patterns corresponding to different functions can participate in the same protein complex as independent modules. Finally, we demonstrate that a hyperclique pattern can be involved in different complexes performing different higher-order biological functions, although the pattern corresponds to a specific elementary biological function.

## 1   Introduction

Complex cellular processes are modular and are accomplished by proteins in complex multi-protein assemblies. Often these multi-protein complexes act as highly efficient protein machines and perform activities related to complex

---

*These authors contributed equally to the work reported in this paper

biological phenomena, such as DNA replication, transcription, metabolism, and signal transduction. A variety of experimental and computational approaches have been employed to deduce the constituents of protein macromolecular complexes. Experimental approaches such as the yeast two-hybrid genetic screen [14,9] yield binary interaction data while more recent large-scale methods [7,8] combine tagged "bait" proteins and protein-complex purification schemes with mass spectrometric measurements to identify protein complexes that contain three or more components.

While proteomic studies [7,8] have generated large amount of interesting protein complex data, much remains to be learned before we have a comprehensive knowledge of functional modules - groups of proteins involved in common elementary biological function. Along this line, an important issue is the effective extraction of functional modules. Previous research on this topic can be grouped into two approaches. One approach is targeted on extraction of densely connected subgraphs from the protein interaction network, such as fully connected subgraphs (cliques)[13] and almost fully connected subgraphs ($k$-cores)[2]. However, algorithms for finding cliques and $k$-core are typically quite expensive. Another approach for detection of functional modules is through clustering analysis [5,12], which divide proteins into groups (clusters) in the way such that similar proteins are in the same cluster and dissimilar proteins are in different clusters.

In this paper, we present a hyperclique pattern discovery approach for identifying functional modules (hyperclique patterns) from protein complex data. A hyperclique pattern is a type of association pattern containing proteins that are *highly affiliated* with each other; that is, every pair of proteins within a hyperclique pattern is guaranteed to have the cosine similarity (uncentered Pearson correlation coefficient [†]) above certain level. As a result, our method is more robust than related approaches in the presence of large number of false-positive protein interactions. Indeed, a significant number of false-positive protein interactions are present in current experimentally identified protein complexes. Gavin *et al.* [7] estimate that 30% of the protein interactions they detect may be spurious, as inferred from duplicate analyses of 13 purified protein complexes. Finally, please note that clustering analysis finds related proteins with a global constraint, while hyperclique patterns capture relationships among proteins on a local level and thus are more compact representations of proteins.

Hyperclique pattern discovery is especially effective on protein complex data, because protein complex data can be viewed as a bipartite graph[5] (a ma-

---

[†]When computing Pearson correlation coefficient, the data mean is not subtracted.

trix in which rows represent protein complex and column represents proteins). In contrast, previous approaches are usually based on a graph of pairwise similarities. A bipartite graph representation of protein complexes allows us to efficiently compute hyperclique patterns, much faster than finding cliques or k-cores in a graph.

The analysis of discovered hyperclique patterns from protein complexes using the Gene Ontology suggests that proteins within the same hyperclique pattern more likely perform the same function and participate in the same biological process. For example, all proteins of an identified pattern {Pre2, Pre4, Pre5, Pre6, Pre8, Pre9, Pup3, Scl1} corresponds to the same function annotation "*endopeptidase activity*" and the 3-D structural view of these proteins reveals that they physically interact with each other. Furthermore, we show that several hyperclique patterns with different functions can participate in the same protein complex as independent modules. Finally, we demonstrate that a hyperclique pattern can be involved in different protein complexes performing different higher-order biological functions, although the pattern corresponds to a specific biological function.

## 2 Hyperclique Pattern Discovery

In this section, we describe the concept of hyperclique patterns [15] after first introducing the concept on which it is based: the association rule [1].

### 2.1 Association Rules

We present the concept of association rules [1] within the context of biology. Let $P = \{p_1, p_2, \ldots, p_n\}$ be a set of proteins and $C = \{c_1, c_2, \ldots, c_l\}$ be the set of protein complexes, where each complex $c_i$ is a set of proteins and $c_i \subseteq P$. A pattern is a set of proteins $X \subseteq P$, and the **support** of $X$, $supp(X)$, is the fraction of protein complexes containing $X$. For example, in Table 1, the support of the pattern $\{p_3, p_4\}$ is $3/5 = 60\%$, since three protein complexes (c2, c3, c4) contain both $p_3$ and $p_4$.

An association rule is of the form $X \to Y$, which means the presence of pattern $X$ implies the presence of pattern $Y$ in the same protein complex, where $X \subseteq P$, $Y \subseteq P$, and $X \cap Y = \phi$. The **confidence** of the association rule $X \to Y$ is written as $conf(X \to Y)$ and is defined as $conf(X \to Y) = supp(X \cup Y)/supp(X)$. For instance, for protein complex data shown in Table 1, the confidence of the association rule $\{p_3\} \to \{p_4\}$ is $conf(\{p_3\} \to \{p_4\}) = supp(\{p_3, p_4\})/supp(\{p_3\}) = 60\% / 80\% = 75\%$. In biology domain, there are many interesting patterns occuring at low levels of support, such as the ones

Table 1. A Sample Protein Complex Data Set.

| Protein Complex | Proteins |
|---|---|
| c1 | $p_1, p_2$ |
| c2 | $p_1, p_3, p_4, p_5$ |
| c3 | $p_2, p_3, p_4, p_6$ |
| c4 | $p_1, p_2, p_3, p_4$ |
| c5 | $p_1, p_2, p_3, p_6$ |

identified in this paper. However, existing association-rule mining algorithms often have difficulties in finding patterns at low levels of support. Also, many patterns discovered by association-rule mining algorithms contain proteins which are poorly correlated with each other.

## 2.2  Hyperclique Patterns

A hyperclique pattern is a new type of association pattern that contains proteins that are *highly affiliated* with each other; that is, every pair of proteins within a pattern is guaranteed to have the cosine similarity (uncentered Pearson correlation coefficient) above a certain level. Indeed, the presence of a protein in one protein complex strongly implies the presence of every other protein that belongs to the same hyperclique pattern. The h-confidence measure is specifically designed to capture the strength of this association.

**Definition 2.1** *The* **h-confidence** *of a pattern* $X = \{p_1, p_2, \cdots, p_m\}$, *denoted as* $hconf(X)$, *is a measure that reflects the overall affinity among proteins within the pattern. This measure is defined as* $min(conf(\{p_1\} \rightarrow \{p_2, \ldots, p_m\}), conf(\{p_2\} \rightarrow \{p_1, p_3, \ldots, p_m\}), \ldots, conf(\{p_m\} \rightarrow \{p_1, \ldots, p_{m-1}\}))$, *where conf is the confidence of association rule as given above.*

**Example 2.1** *For the sample protein complex data set shown in Table 1, let us consider a pattern* $X = \{p_2, p_3, p_4\}$. *We have* $supp(\{p_2\}) = 80\%$, $supp(\{p_3\}) = 80\%$, $supp(\{p_4\}) = 60\%$, *and* $supp(\{p_2, p_3, p_4\}) = 40\%$. *Then,*

$$conf(\{p_2\} \rightarrow \{p_3, p_4\}) = supp(\{p_2, p_3, p_4\})/supp(\{p_2\}) = 50\%$$
$$conf(\{p_3\} \rightarrow \{p_2, p_4\}) = supp(\{p_2, p_3, p_4\})/supp(\{p_3\}) = 50\%$$
$$conf(\{p_4\} \rightarrow \{p_2, p_3\}) = supp(\{p_2, p_3, p_4\})/supp(\{p_4\}) = 66.7\%$$

*Therefore,* $hconf(X) = min(conf(\{p_2\} \rightarrow \{p_3, p_4\}), conf(\{p_3\} \rightarrow \{p_2, p_4\}), conf(\{p_4\} \rightarrow \{p_2, p_3\})) = 50\%$.

**Definition 2.2** *A pattern X is a* **hyperclique pattern** *if* $hconf(X) \geq h_c$, *where* $h_c$ *is a user-specified minimum h-confidence threshold. A hyperclique pattern is a* **maximal hyperclique pattern** *if no superset of this pattern is also a hyperclique pattern.*

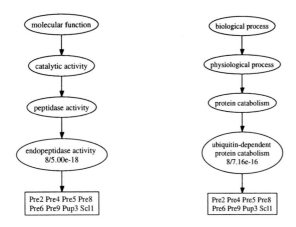

Figure 1. Gene Ontology annotations of pattern {Pre2, Pre4, Pre5, Pre6, Pre8, Pre9, Pup3, Scl1}. Figure on the left shows subgraph of the function annotation. Figure on the right shows subgraph of process annotation. Proteins are shown in square box and significant nodes are labeled with the number of proteins annotated directly or indirectly to that term and the p-value for the term.

Table 2. Examples of Hyperclique Patterns from Yeast Protein Complex Data.

| Yeast Protein Complex Data | | |
|---|---|---|
| Hyperclique patterns | Supp | Hconf |
| {Cus1, Msl1, Prp3, Prp9, Sme1, Smx2, Smx2, Smx3, Yhc1} | 1.25% | 100% |
| {Pre2, Pre4, Pre5, Pre6, Pre8, Pre9, Pup3, Scl1} | 1.7% | 66.7% |
| {Cwc2, Ecm2, Hsh155, Prp19, Prp21, Snt309} | 1.7% | 100% |
| {Emg1, Imp3, Imp4, Kre31, Mpp10, Nop14, Sof1, Utp15, Noc4} | 1.25% | 100% |

Let us consider the sample protein complex data in Table 1. For the h-confidence threshold 0.5, the pattern $\{p_2, p_3, p_4\}$ is a hyperclique pattern. Furthermore, since no superset of this pattern is a hyperclique pattern at the threshold 0.5, this pattern is also a maximal hyperclique pattern.

Table 2 shows some hyperclique patterns identified from a yeast protein complex data set [7]. One hyperclique pattern in that table is {Pre2, Pre4, Pre5, Pre6, Pre8, Pre9, Pup3, Scl1}. Figure 1 shows the function and process subgraphs of the Gene Ontology corresponding to this pattern. One observation is that all proteins in this pattern perform the same biological function, *endopeptidase activity*. Also, all proteins in the pattern involve in the same biological process, *ubiquitin-dependent protein catabolism*. In other words, proteins in the same hyperclique pattern are highly-affiliated with each other. Indeed, the following Theorem 2.1 guarantees if a hyperclique pattern has an h-confidence value above the h-confidence threshold, $h_c$, then every pair of proteins within the pattern must have a cosine similarity (uncentered Pearson correlation coefficient) greater than or equal to $h_c$.

**Theorem 2.1** *Given a hyperclique pattern* $X = \{p_1, p_2, \ldots, p_m\}$ *at the h-confidence threshold* $h_c$, *for two proteins* $p_l$ *and* $p_k$ *such that* $\{p_l, p_k\} \subset X$, *we have* $cosinesim(p_l, p_k) \geq h_c$, *where* $cosinesim(p_l, p_k) = \dfrac{supp(\{p_l, p_k\})}{\sqrt{supp(\{p_l\})supp(\{p_k\})}}$, *which is the cosine similarity between* $p_l$ *and* $p_k$.

### 2.3 Computation Algorithm

In a nutshell, the process of searching hyperclique patterns can be viewed as the generation of a level-wise pattern tree. Every level of the tree contains patterns with the same number of proteins. If the level is increased by one, the pattern size (number of proteins) is also increased by one. Every pattern has a branch (sub-tree) which contains all the superset of this pattern. Our algorithm for finding hyperclique patterns is breath-first. We first check all the patterns at the first level. If a pattern is not satisfied with the user-specifed support and h-confidence thresholds, the whole branch corresponding to this pattern can be pruned without further checking. This is due to the the anti-monotone property of support and h-confidence measures. Consider the h-confidence measure, the anti-monotone property guarantees that the h-confidence value of a pattern is greater than or equal to that of any superset of this pattern. Following this manner, the pattern tree is growing level-by-level until all the patterns have been generated. This algorithm is very efficient for handling large-scale datasets [15].

## 3 Protein Complex Data and Analysis Tools

**Protein Complex Data:** Two datasets [7,8] summarizing large-scale experimental studies of multi-protein complexes are available for the yeast *Saccharomyces Cerevisiae*. Coupling different purification (immunoprecipitation and tandem affinity purification (TAP)) and labeling schemes with mass spectrometry (MS), both studies used bait proteins to identify physiologically intact protein complexes. Independent research [4,11] showed that the TAP-MS dataset by Gavin, *et al.* [7] has a relatively better accuracy for predicting protein functions, therefore we take this dataset to illustrate our method. In this TAP-MS dataset, there are a total of 1,440 distinct proteins within 232 multi-protein complexes, and the data format is illustrated in Table 1.

**Analysis Tools:** The Gene Ontology (http://www.geneontology.org) was used to annotate the proteins of hyperclique patterns identified in the TAP-MS dataset. A graph drawing package GraphViz (http://www.research.att.com/sw/ tools/graphviz/) was used to produce the graph representation of the annotation. The functional description of each protein (if available) was obtained from the Saccharomyces Genome Database (SGD)[6]. The 3-D structure information of yeast proteins was obtained from the Protein Data Bank (PDB) (http://www.rcsb.org/pdb), and PyMOL[3] was used for visualizing the 3-D structure of proteins within a hyperclique pattern.

## 4   Analysis of Hyperclique Pattern using Gene Ontology

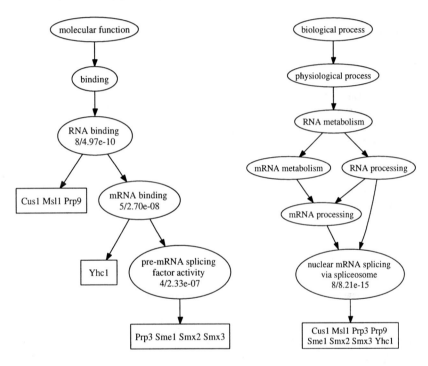

Figure 2. The Gene Ontology annotations of pattern {Cus1, Msl1, Prp3, Prp9, Sme1, Smx2, Smx3, Yhc1}. Figure on the left shows subgraph of function annotation of the pattern. Figure on the right shows subgraph of process annotation. Proteins are listed in square box. Significant nodes are labeled with the number of proteins annotated directly or indirectly to that term and the p-value for the term.

Setting a support threshold to be 0 and an h-confidence threshold to be 0.6, we obtained 60 maximal hyperclique patterns. Limited by space, we analyze some of the patterns obtained. Detailed results are available at our project web site [‡] .

The proteins within the same hyperclique pattern have strong association with each other. To investigate this, we analyze the annotations of the patterns using the terms from the Gene Ontology. Figure 2 shows the subgraphs of the Gene Ontology corresponding to pattern {Cus1, Msl1, Prp3, Prp9, Sme1, Smx2, Smx3, Yhc1}. The left subgraph in the figure is the molecular function annotation of the proteins in the pattern. Note that all 8 proteins from this pattern are annotated to the term *RNA binding* with p-value 4.97e-10. The p-value is calculated as the probability that $n$ or more proteins would be assigned to that term if proteins from the entire genome are randomly assigned to that pattern. The smaller the p-value, the more significant the annotation. Among the pattern, 4 proteins {Prp3, Sme1, Smx2, Smx3} are annotated to a more specific term *pre-mRNA splicing factor activity* with p-value 2.33e-07. The annotation of these proteins confirms that each pattern form a module performing specific function. The right subgraph in Figure 2 shows the biological process this pattern is involved. The proteins are annotated to the term *nuclear mRNA splicing via spliceosome* with p-value 8.21e-15 which is statistically significant.

Table 3. The Hyperclique pattern {Pre2, Pre4, Pre5, Pre8, Pup3, Pre6, Pre9, Scl1} contained in four protein complexes. All proteins in the pattern are in bold.

| CID | Protein Complexes | Function Category |
|-----|-------------------|-------------------|
| 106 | Blm3 Dam1 Dbp9 Ecm29 Est3 Gfa1 Ino4 Kap95 Lys12 Mds3 Nud1 Pda1 Pdb1 Pre10 **Pre2** Pre3 **Pre4 Pre5 Pre6 Pre8 Pre9** Pse1 **Pup3** Rgr1 Rpt3 Rpt5 **Scl1** Spa2 Srp1 Ulp1 YFL006W YGR081C YMR310C YPL012W Yra1 | Protein Synthesis and Turnover |
| 148 | Cdc6 Ecm29 Gfa1 Mlh2 Nas6 Pgk1 Pre1 **Pre2** Pre3 **Pre4 Pre5 Pre6** Pre7 **Pre8 Pre9 Pup3** Rpn10 Rpn11 Rpn12 Rpn13 Rpn3 Rpn5 Rpn6 Rpn7 Rpn8 Rpn9 Rpt1 Rpt2 Rpt3 Rpt4 Rpt5 Rpt6 **Scl1** Ubp6 | Protein Synthesis and Turnover |
| 157 | Blm3 Cdc6 Ecm29 Mlh2 Pgk1 Pre1 Pre10 **Pre2** Pre3 **Pre4 Pre5 Pre6** Pre7 **Pre8 Pre9 Pup3** Rgr1 Rpn10 Rpn11 Rpn12 Rpn13 Rpn3 Rpn5 Rpn6 Rpn7 Rpn8 Rpn9 Rpt1 Rpt2 Rpt3 Rpt4 Rpt5 Rpt6 **Scl1** Ubp6 YFL006W | Protein Synthesis and Turnover |
| 151 | Blm3 Cdc55 Cin1 Erg13 Hhf2 Hos2 Iml1 Kap95 Kel1 Lte1 Myo5 Pfk1 Pph21 Pph22 Pre1 Pre10 **Pre2 Pre4 Pre5 Pre6** Pre7 **Pre8 Pre9** Pup1 Pup2 **Pup3** Rrd2 Rts1 **Scl1** Sif2 Srp1 Tdh2 Tdh3 Tef4 Tpd3 YBL104C YCR033W YGL245W YGR161C YIL112W YKR029C Yef3 Yor1 Yra1 Zds1 Zds2 | Signalling |

---

[‡] http://www.cs.umn.edu/~huix/pfm/pfm.html

Figure 3. The 3-D structure of the yeast proteasome including all proteins in the hyperclique pattern {Pre2, Pre4, Pre5, Pre8, Pup3, Pre6, Pre9, Scl1}. In the figure, (a), (b), and (c) show 3-D structures of proteins only in the pattern. (Pre2(blue), Pre4(red), Pre5(yellow), Pre6(brown), Pre8(green), Pre9(pink), Pup3(magenta), Scl1(cyan)). In contrast, (d) shows 3-D structures of all proteins in the proteasome complex.

## 5   Hyperclique Patterns as Functional Modules

Gene Ontology annotations reveal that proteins in the same hyperclique pattern tend to perform a common function and be involved in the same biological process. In this subsection, we describe the role of hyperclique patterns as functional modules.

Consider the hyperclique pattern {Pre2, Pre4, Pre5, Pre6, Pre8,

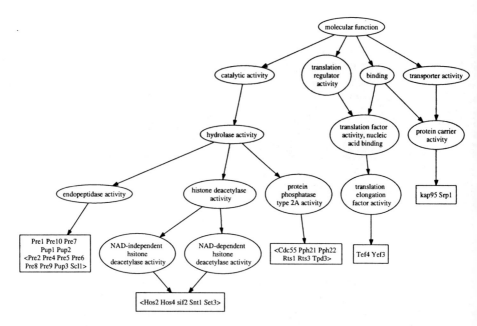

Figure 4. Subgraph of the Gene Ontology (function) corresponding to the protein complex 151. Proteins within a pair of < > form a hyperclique pattern.

Pre9, Pup3, Scl1}. All proteins in this pattern are components of the proteasome complex which destroys the proteins no longer in use or the failures of the translation products. The proteasome constitutes nearly 1% of cellular proteins. The 3-D structure of this complex (PDB ID: 1fnt) is available in the Protein Data Bank (PDB). Figure 3 (d) shows the 3-D structure of all proteins in the proteasome. Figures 3 (a), (b), and (c) show the 3-D structure of all proteins in the hyperclique pattern from different view angles. As can be seen, proteins in the pattern have physical interactions with each other. This is compelling physical evidence implying that proteins in the same hyperclique pattern tend to physically interact together to form a compact structure and perform a common molecular function. Figure 1 illustrates the molecular function and biological process of this pattern. It is also interesting to observe that this hyperclique pattern is contained in four protein complexes in the TAP-MS data set, as shown in Table 3. According to Gavin's function category, these four protein complexes belong to two different function categories: *protein synthesis and turnover* and *signalling*. In other words, this hyperclique pattern acts as a functional module participating in protein complexes which perform different high-order functions.

Furthermore, we observed that three identified hyperclique patterns are contained in the protein complex 151 (refer to Table 3). Figure 4 shows the subgraph of the Gene Ontology (function) corresponding to the protein complex 151. As can be seen, three hyperclique patterns correspond to three different functions: the pattern {Pre2, Pre4, Pre5, Pre6, Pre8, Pre9, Pup3, Scl1} corresponds to the function *endopeptiase activity*, the pattern {Hos2, Hos4, Sif2, Snt1, Set3} corresponds to the function *histone deacetylase activity*, and the pattern {Cdc55, Php21, Php22, Rts1, Rts3, Tpd3} corresponds to the function *Protein phosphatase type 2A activity*. This indicates that hyperclique patterns can serve as different functional modules to participate in a common protein complex.

## 6 Discussion

In this paper, we describe a hyperclique pattern discovery approach to identify functional modules in protein complex data. The tight threshold in the definition of hyperclique patterns ensures the strong associations among the proteins in the same functional module. Analysis using the Gene Ontology indicates that the computationally discovered hyperclique patterns are biologically significant. Our approach can not only effectively identify the basic functional modules in protein complexes, but also is robust in the presence of large number of false-positive protein interactions, due to the strong associations among the constituent proteins.

Our work discovered several interesting protein functional modules. For example, one discovered protein functional module {Pre2, Pre4, Pre5, Pre8, Pup3, Pre6, Pre9, Scl1} focuses on the function *endopeptidase activity* by the Gene Ontology. This hyperclique pattern with specific function is also found to exist in four experimentally determined protein complexes performing different higher level biological functions.

**Acknowledgments**

This work was supported by U.S. Department of Energy, Office of Science, Office of Laboratory Policy and Infrastructure, through an LBNL LDRD, under contract DE-AC03-76SF00098. This work was also supported by NSF grant # IIS-0308264, NSF ITR #0325949, and by Army High Performance Computing Research Center under the auspices of the Department of the Army, Army Research Laboratory cooperative agreement number DAAD19-01-2-0014. The content of this work does not necessarily reflect the position or policy of the government and no official endorsement should be inferred.

## References

1. R. Agrawal, T. Imielinski, and A. Swami. Mining association rules between sets of items in large databases., In *ACM SIGMOD*, 1993.
2. G.D. Bader and C.W. Hogue. Analyzing yeast protein-protein interaction data obtained from different sources. *Nature Biotechnology*, 2002.
3. W.L. Delano. *The PyMOL User's Manual.* DeLano Scientific, 2002.
4. M. Deng, F. Sun, and T. Chen. Assessment of the reliability of protein-protein interactions and protein function prediction. *Pac Symp Biocomput*, pages 140–151, 2003.
5. C. Ding, X. He, R.F. Meraz, and S.R. Holbrook. A unified representation of multi-protein complex data for modeling interaction networks. *Proteins: Structure, Function, and Genetics, to appear*, 2004.
6. S. S. Dwight, M. A. Dolinski, and K. Ball et al. Saccharomyces genome database (sgd) provides secondary gene annotation using the gene ontology (go). *Nucleic Acids Research*, 2002.
7. A. Gavin et al. Functional organization of the yeast proteome by systematic analysis of protein complexes. *Nature*, 415:141–147, 2002.
8. Y. Ho et al. Systematic identification of protein complexes in saccharomyces cerevisiae by mass spectrometry. *Nature*, 415:180–183, 2002.
9. T. Ito, T. Chiba, R. Ozawa, M. Yoshida, M. Hattori, and Y. Sakaki. A comprehensive two hybrid analysis to explore the yeast protein interaction. in *Proc. Natl. Acad. Sci.*, 98(8):4569–4574, 2001.
10. A. Kumar and M. Snyder. Protein complexes take the bait. *Nature*, 415:123–124, 2002.
11. C. Mering, R. Krause, B. Snel, M. Cornell, S.G. Oliver, S. Fields, and P. Bork. Comparative assessment of large-scale data sets of protein-protein interactions. *Nature*, 417, 2002.
12. J.B. Pereira-Leal, A.J. Enright, and C.A. Ouzounis. Detection of functional modules from protein interaction networks. *PROTEINS: Structure, Function, and Bioinformatics*, 54:49–57, 2004.
13. V. Spirin and L.A. Mirny. Protein complexes and functional modules in molecular networks. *Proceedings of the National Academy of Sciences*, 100(21):12123–12128, October 2003.
14. P. Uetz, L. Cagney, and G. Mansfield et al. A comprehensive analysis of protein-protein interactions in *saccharomyces cerevisiae*. *Nature*, 403:623–627, 2000.
15. H. Xiong, P. Tan, and V. Kumar. Mining strong affinity association patterns in data sets with skewed support distribution. In *Proc. of the IEEE International Conference on Data Mining*, pages 387–394, 2003.

# MULTI-ASPECT GENE RELATION ANALYSIS

## H. YAMAKAWA, K. MARUHASHI, Y. NAKAO

*Fujitsu Laboratories Ltd., 1-1, Kamikodanaka 4-chome,*
*Nakahara-ku, Kawasaki, Kanagawa,211-8588, Japan*

## M. YAMAGUCHI

*Fujitsu Limited, 9-3, Nakase 1-chome, Mihama-ku, Chiba,*
*Chiba, 261-8588, Japan*

Recent progress in high-throughput screening technologies has led to the production of massive amounts data that we can use to understand biological systems. To interpret this data, biologists often need to analyze the characteristics of a set of genes by using Gene Ontology (GO) annotation. We are proposing a novel method for assisting such an analysis. Given a set of genes, the method automatically extracts several analyzing aspects in terms of a subset of genes that are attached to some related GO terms. It then creates a gene-attribute bipartite graph that highlights the aspect selected by the user according to his/her interests. We describe this method in detail and report on an experiment where the proposed method is applied to the analysis of rat kidney expression data.

## 1 Introduction

The DNA microarray is an effective tool for monitoring and profiling gene expression patterns. It can measure the expression levels of thousands of genes simultaneously and provide a set of expression patterns for a given list of genes. Biologists analyze gene expression patterns to determine their biological meanings (e.g., interactions between specific genes, dependencies between changes in gene expressions, and patient's responses to treatment). In the bioinformatics field, many methods, including a kind of data mining, have been applied to assisting such analysis[1]. For example, Kennedy et al. used a clustering method to assist in microarray dataset analysis. They extracted the gene list by preprocessing the gene expression data. They then applied a clustering method to the gene list and then presented the resulting gene clusters together with meaningful descriptions using the functional information obtained from the Gene Ontology (GO)[2].

The GO is a vocabulary that describes the attributes of genes (for example, their biological functions). Each term in the vocabulary, called a GO term, represents a possible attribute value that is possessed by a gene. The GO has a hierarchical structure, i.e., GO terms are connected by *is-a* relations and construct a directed acyclic graph. The GO Consortium is currently creating three standard gene ontologies that will describe the associated biological processes, cellular components, and molecular functions for genes and their products (RNA or protein products encoded by genes). Many biological resources, including LocusLink (http://www.ncbi.nlm.nih.gov/LocusLink), use the GO terms to annotate gene properties.

Because there are many kinds of GO terms, attaching GO annotations to a gene list produces high-dimensional data. In fact, computer-aided analysis methods are required, such as clustering, principal component analysis (PCA), and self-organizing maps (SOM). Many of these conventional methods are helpful in understanding the overall characteristics of a gene list (e.g., the major gene relationships found within that gene list). For example, as shown by Kennedy et al.[3], a dendrogram display of hierarchical clustering with GO terms can be used to illustrate the overall structure of the major characteristics of a gene list.

There is a case, however, in which the overall structure of the gene list characteristics is not appropriate. Because high-dimensional data can contain many aspects, an overall structure is incapable of illustrating all of these aspects, such that the user may miss some important aspects relating to his or her interests. For example, consider a case in which collagen activity is observed in a biological phenomenon related to osteoblasts and osteoclasts. One possible analysis aspect is the collagen metabolism. A biologist whose main interest resides in the metabolism process will want to understand the collagen biosynthetic pathway from the viewpoint of how the collagen biosynthesis pathway interacts with other metabolic pathways, or the requirements for collagen biosynthesis. In this case, it is important to be able to distinguish between the intracellular and the extracellular phenomena, as well as distinguish the metabolism phenomena from other phenomena (see "The extracellular matrix of animals", pp 971–995, in [4]). Another possible aspect is animal development. A biologist whose main interest is in the developmental program of animals will want to identify the types of collagen-related developmental processes that occur. In this case, it is appropriate to classify the active genes based on their relationship to the development processes (e.g. the formation of an extracellular matrix, arrangement of the cytoskeleton) (see "Fibroblasts and Their Transformations: The Connective-Tissue Cell Family", pp.1179–1187, in [4]).

In response to the above demands, we have developed a tool for supporting gene relationship analysis, called Genesphere Connection Miner (Cminer) [5], which creates a gene-term bipartite graph that can be focused on the user's interests. The contents of the display can be changed according to the user' s settings, as specified by a list of GO terms. For example, Wagatsuma et al. reported that they had obtained results related their interest by using this tool to analyze the temporal expression data for a hepatitis model rat (about 500 genes)[6]. There is a problem with this technique, however, in that it takes a long time to create an appropriate list of GO terms by hand. To solve this problem, this paper proposes a method for automatically extracting the analysis aspects. Given a gene list, the method automatically extracts several aspects in terms of a subset of genes attached with related GO terms. The user can easily specify his/her interest by selecting one appropriate aspect from those that are available. In Chapter 2, we describe the aspect extraction method in detail. In Chapter 3, we report on an experiment in which the proposed method is applied to the analysis of rat kidney expression data.

## 2   Multi-aspect gene relation analysis system

We are proposing a multi-aspect gene relation analysis system, which outputs multiple aspects about a given gene list. For an aspect selected by the user, this system displays a bipartite graph consisting of gene symbols and GO terms. This system uses the Gene Ontology (GO) that is a vocabulary used to describe the attributes of genes, LocusLink that is a gene database in which genes are annotated with GO terms, and HomoloGene that is used to provide orthologous information.

A unique feature of this system is that it automatically extracts multiple aspects for analyzing a gene list by using a conceptual clustering technique called ETMIC situation decomposition (E-SD) [7,8]. The E-SD method simultaneously selects a gene subset and a GO term subset as an aspect. To date, however, the E-SD method has faced two issues related to this function. One involves the comparison of GO terms of different abstraction levels. The second relates to the fact that only a few combinations of genes can be compared because many genes have no GO term annotations [a]. Our new system overcomes the first problem by: [a] a GO term is inferred by using a transitive relationship. Then, to overcome the latter problem: [b] a GO term is extended by using orthologous information, [c] a GO term is summarized by using weighted singular value decomposition (SVD) [b]

To realize these countermeasures, our new system is composed of the seven subprocesses described below (see Figure 1).

(1) Inference of GO terms using transitive relationships [a]
(2) Logarithmic probability weighted SVD [c]
(3) Additions of terms of orthologous genes for each gene[b]

---

[a] Actually, over 20% of genes of human are annotated with GO term, but only 4% of genes of rat are annotated.

[b] Although these countermeasures may introduce noisy information, the advantage is expected to outweigh the disadvantages in such a case of very poor gene annotations.

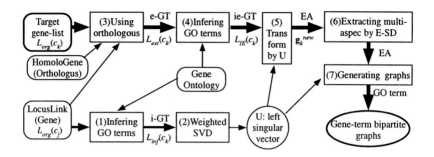

Figure 1: Multi-aspect gene relationship analysis system.
Rounded-rectangles indicate input data, rectangles are processing, and circles are output.
Bold arrows indicate the gene list flow for which attributes are added. i-GT: inferred GO
term, e-GT: extended GO term, ie-GT: inferred extended GO term, EA: eigen-attributes.

 

(4) Inference of GO terms using transitive relationships [a]
(5) Transformation of GO terms to eigen-attributes.[c]
(6) Extraction of multi-aspects by E-SD method
(7) Generation of gene-term bipartite graph for each aspect [c]

 

The main process path for the target gene list runs from steps (3) to (7),
with the multi-aspect being extracted in (6). The process path from (1) to (2)
generates a left singular vector using all the available genes. To collect terms
that have a similar appearance distribution, the vector transforms the target
gene list from GO terms to eigen-attributes in (5).

## 2.1 Inferring GO terms using transitive relations

LocusLink contains a GO term list $L_{org}(c_j)$ for each gene $c_j$. Each term list
$L_{org}(c_j)$ in LocusLink is converted to an inferred GO term list $L_{inf}(c_j)$. In
practice, all of the ancestor terms for each GO term are added to the new term
list $L_{inf}(c_j)$, using the transitive relationship of the GO.

## 2.2 Logarithmic probability weighted SVD

To gather up those GO terms having a similar distribution, a gene representation is transformed into eigen-attributes. This transformation subprocess (5)
uses a left singular vector $U$, calculated by the SVD method in subprocess (2).

## Logarithmic probability weighting:

When we give all the GO terms an equal weighting, the SVD process tends to select those terms in the higher layers of the GO hierarchy, which has very little meaning. This is because the SVD process tends to select those GO terms associated with many genes. In addition, those terms in a higher layer tend to associate with many genes (e.g., "binding" in the molecular function ontology).

To overcome this problem, each GO term $i$ is weighted with a logarithmic probability weight $W_i$ which highlights a moderate abstraction level (depth) in the GO hierarchy. $W_i$ is calculated using the following formula.

$$W_i = -\log(\frac{m_i}{m}) \tag{1}$$

Where $m$ is the total number of genes, and $m_i$ is the number of genes with term $i$. For instance, the topmost concept in a hierarchy is annotated by all the genes, its weighting becomes zero $W_i = 0$, and so is ignored. On the other hand, a special narrower term is emphasized.

The idea of logarithmic probability weighting relates to the formula of similarities in the layered structure as proposed by Resnik, Lin, and Jian [9].

## Singular value decomposition (SVD):

In preparation for the SVD process, the inferred terms $L_{inf}(c_j)$ for all the genes are converted to a matrix. Firstly, each inferred term list $L_{inf}(c_j)$ $(j \in [1, m])$ is converted into a GO term vector $\mathbf{g}_j$ of length $n$. Here, $n$ is the number of GO terms used in all the genes. An element registered in the inferred term list $L_{inf}(c_j)$ is set to $W_i$ while others are set to 0. Secondly, all the GO term vectors $\mathbf{g}_j$ $(\forall j \in [1, m])$ are collected into a matrix $G$ with $n$ columns (GO terms) and $m$ rows (genes). SVD decomposes matrix $G$, as follows.

$$G = USD^T \tag{2}$$

$U$ and $D$ are a unitary matrix that satisfies $U^T U = I_n$ and $D^T D = I_m$ respectively. The column vector of $U$ is called the left singular vector. The column vector of $D$ is called the right singular vector.

### 2.3   Adding terms of the orthologous genes to the target gene list

Each gene $c_k$ in the target gene list $\zeta$ is annotated by the GO term list $L_{org}(c_k)$. To compensate for insufficient GO term annotation in the target gene list $L_{org}(c_k)$, GO terms are added using orthologous information. Each target gene $c_k$ of rat has corresponding orthologous genes (human and mouse). GO terms that belong to orthologous genes and rat genes $c_k$ are added to a new GO term list $L_{ext}(c_k)$. This is called the extended GO term list for gene $c_k$. Orthologous information for three species (human, mouse, and rat) is obtained from HomoloGene. This is in the form of a 13952-row (entry) by 3-column (race) LocusID matrix.

### 2.4 Infering GO terms using transitive relations

In the same way as in subprocess (1), we obtain an extended inferred term list $L_{IE}(c_k)$ from term list $L_{ext}(c_k)$ for each gene $c_k$ in target gene list $\zeta$.

### 2.5 Transforming GO terms to eigen-attributes

As part of the preparations for the E-SD process, the genes' representations are transformed into eigen-attribute set $\alpha$ from inferred extended term lists $L_{IE}(c_k)$. The set $\alpha$ consists of the top 20 eigen-attributes acquired by the SVD process. Each term list $L_{IE}(c_k)$ for one gene is converted into a GO term vector $\mathbf{g}_k^{IE}$ of length $n$. Here, $n$ is the number of GO terms used in all the genes. An element registered in a term list $L_{IE}(c_k)$ is set to $W_i$ and others are set to 0. Here $W_i$ is the weighting for the $i$-th GO term.

Each gene vector $\mathbf{g}_k^{new}$ which is described by eigen-attribute set $\alpha$ is transformed from $\mathbf{g}_k^{IE}$ using left singular vector $U$.

$$\mathbf{g}_k^{new} = U^T \mathbf{g}_k^{IE} \tag{3}$$

### 2.6 Extracting multi-aspects with the E-SD method

The ETMIC situation decomposition (E-SD) method [7,8] is used to extract multiple aspects from the target gene list. The target gene list $\zeta$ is a collection of vectors $\mathbf{g}_k^{new}$ described by eigen-attribute set $\alpha$. This list is described as the large square in the left-hand part of Figure 2. Each extracted aspect $J = \{A, C\}$ is a combination of subset $A$ of eigen-attribute set $\alpha$ (horizontal axis in Fig. 2), and subset $C$ of the gene sets $\zeta$ (vertical axis in Fig.2).

The E-SD algorithm selects some useful multi-aspects from the enormous combination of subsets $A$ and $C$. This process is based on searching for the local maximum point of an ETMIC criterion to change the gene selection $C$ in every partial space $A$ [8]. The ETMIC criterion that evaluates each aspect $J = \{A, C\}$ is as follows.

$$E(A, C) = n_C \left( \min_i \left( I_{X_A^{-i}; X_i}(C) \right) - \max_j \left( I_{X_A; X_j}(C) \right) \right) \tag{4}$$

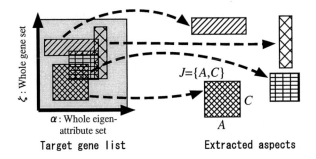

Figure 2: Composition of aspect decomposition process
Because of limitations of the figure, the selections of gene
subsets and eigen-attribute subsets are drawn as a continuous area.

Here, $n_C$ is the selected number of genes. $I_{X_A^{-i};X_i}(C)$ and $I_{X_A;X_j}(C)$ are mutual information on each of two partial spaces for gene subset $C$. $X_A^{-i}$ is a partial space for eigen-attribute subset $i$ to eigen-attribute subset $A - a_i$. $X_i$ is a partial space for eigen-attribute $a_i$. $X_A$ is a partial space for eigen-attribute subset $A$. $X_j$ is a partial space for eigen-attribute $j$.

### 2.7 Generation of gene-term bipartite graph for each aspect

For each aspect, genes and GO terms should be selected to draw the gene-term bipartite graph, which is helpful in understanding the gene relations. The members of gene subset $C$ are displayed on this graph. Unfortunately, it is difficult to select the GO terms to be displayed on this graph, because the eigen-attribute subset $A$ corresponds to a weighted selection of GO terms.

The norm of the vector $U$ for $i$-th GO term $|U_i| = \sqrt{\sum_{a \in A} U_{ai}^2}$ is used for selecting the significant GO terms, where $U_{ai}$ is the $a$-th eigen-attribute and the $i$-th GO term element of matrix $U$. GO terms are selected so that every gene connects to at least one GO term. This process selects one GO term from the inferred extended term list $L_{IE}(c_k)$ of each gene. This selection for each gene $c_k$ is based on the equation using the maximum norm of $U$, as follows.

$$i_{max}(c_j) = \arg \max_{i \in L_{IE}(c_j)} |U_i| \tag{5}$$

Finally, each aspect is drawn by a gene-term bipartite graph, includes the gene subset $C$ and the GO term subset indicated by $i_{max}(c_j)$.

## 3  Experiment: Analysis of gene-list of rat embryonic kidney

In this section, we describe an experiment in which the proposed method is applied to the analysis of a gene list obtained through microarray expression profiling.

### 3.1  Experimental setup: Analyzed gene list

Stuart et al.[10] analyzed gene expression patterns during kidney organogenesis using DNA array technology and, as a result, classified 8,740 genes into five discrete clusters based on the temporal patterns of their expression levels. We used the second cluster (group 2), whose expression pattern peaks with mid-nephrogenesis, as an example gene list for our experiment. The following lists the symbols (bold font) and the names of the genes used for our experiment.

> **AGR**: Agrin, **Calm1**: Calmodulin 1 (phosphorylase kinase, delta), **Col1a1**: collagen, type 1, alpha 1, **Dcn**: decorin, **ENP1MR**: Epithelial membrane protein 1, **Erbb2**: Avian erythroblastosis viral (v-erb-B2) oncogene homologue 2 **Erp29**: endoplasmic retuclum protein 29, **Galr3**: galanin receptor 3, **ID125A**: Inhibitor of DNA binding 1, helix-loop-helix protein, **ILGF-BPA**: Insulin-like growth factor binding protein 2, **Lamc1**: laminin, gamma 1, **Lbp**: lipopolysaccharide binding protein, **Mmp2**: matrix metalloproteinase 2, **Mtap6**: microtubule-associated protein 6, **Nkaa1b**: ATPase, Na+K+ transporting, alpha 1, **Npr1**: natriuretic peptide receptor 1, **Phb**: Prohibitin, **Pkcb**: protein kinase C, beta 1, **Pmp22**: peripheral myelin protein 22, **SO-MATO**: somatostatin receptor 5, **Serpina1**: serine (or cysteine) proteinase inhibitor, clade A, member 1, **Sm22**: Transgelin (Smooth muscle 22 protein), **Sparc**: Secreted acidic cystein-rich glycoprotein (osteonectin), **Ucp2**: Uncoupling protein 2, mitochondrial.

These 24 genes were derived from the group 2 cluster (containing 168 accession numbers of GenBank genes)$^c$ as follows. Every gene in the cluster was converted into the LocusID or removed if a LocusID could not be found. For each of the 66 genes successfully converted as a result of that process, GO terms were retrieved from the LocusLink database. Finally, we constructed a GO term list $L_{org}(c_k)$ for each gene $c_k$ that has one or more associated GO terms.

---

$^c$The list of genes was acquired from the Kidney Development Gene Expression Database (*http://organogenesis.ucsd.edu/*).

Table 1: Extracted aspects list of top five.

| rank | ETMIC score | eigen-attributes | # genes | | # different genes | | | | |
|------|-------------|------------------|---------|-----|-----|-----|-----|-----|-----|
| | | | | | $\mathcal{A}$ | $\mathcal{B}$ | $\mathcal{C}$ | $\mathcal{D}$ | $\mathcal{E}$ |
| 1 | 8.21 | 4 6 | 18 | $\mathcal{A}$ | 0 | 7 | 11 | 8 | 8 |
| 2 | 8.02 | 9 13 | 17 | $\mathcal{B}$ | 7 | 0 | 12 | 11 | 9 |
| 3 | 7.45 | 15 18 | 13 | $\mathcal{C}$ | 11 | 12 | 0 | 15 | 11 |
| 4 | 7.25 | 10 13 | 14 | $\mathcal{D}$ | 8 | 11 | 15 | 0 | 14 |
| 5 | 6.93 | 10 11 | 14 | $\mathcal{E}$ | 8 | 9 | 11 | 14 | 0 |

### 3.2 Experimental setup: Compressing GO term annotation

For the E-SD subprocess (6), we constructed twenty attributes, that is, eigen-attributes, which express the major features of all the GO term annotations in the LocusLink database (see Fig.1). These twenty eigen-attributes $A$ were calculated as follows. Firstly, SVD compressed all the GO term annotations obtained from the LocusLink database into a twenty-dimensional feature space by means of subprocesses (1) and (2). In this process, we used 13,557 $(=n)$ effective entries (i.e., entries with GO annotations) among all the 24,489 LocusLink entries (human, mouse and rat) and 5,325 $(=m)$ effective GO terms (i.e., GO terms found in the GO annotations of LocusLink entries). Then, to input the E-SD method, each eigen-attribute was digitized to a nominal variable with five domains.

### 3.3 Experimental results: List of extracted aspects

Table 3.3 summarizes the top five aspects $(\mathcal{A}, \mathcal{B}, \mathcal{C}, \mathcal{D}, \mathcal{E})$ that the proposed method extracted from the gene list mentioned above.

This table shows that these five aspects have different features. For example, the first aspect $\mathcal{A}$ has 18 genes in a two-dimensional feature space that is spanned by the 4th and 6th axes of the compressed GO annotation, while the second aspect $\mathcal{B}$ shows 17 genes in another feature space spanned by the 9th and 13th axes. They are different not only in the axes spanning the feature space but also in the focused genes, i.e., 7 genes out of 18 and 19 are different, as shown in the right-hand part of this table. This fact indicates a feature of the E-SD process in that it selects genes in such a manner that the selected genes exhibit a relatively simple scattering pattern in the selected feature space. In other words, the selected genes are expected to construct clusters having a relatively simple structure. As a result, it should be easy to interpret each aspect by analyzing the GO annotations associated with some of the clusters of the selected genes.

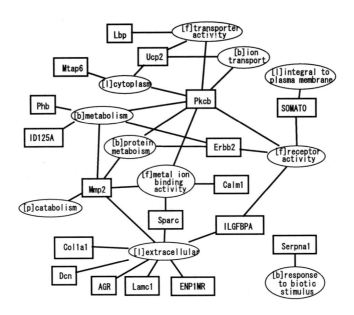

Figure 3: Gene-term bipartite graph for first aspect ($\mathcal{A}$)
Rectangles represent genes. Ovals represent GO terms. [p]=biological_process,
[f]=moleculer_function,[l]cellular_componet.eigen-attribute=(4 6) # genes=18.

### 3.4 Experimental results: Investigation of each aspect

**Gene-term bipartite graph for first aspect ($\mathcal{A}$) :** As shown in Figure 3, the largest gene cluster in the first aspect $\mathcal{A}$ is *extracellular*. In comparison with the analysis by Stuart et al.[10], the genes of *extracellular* exhibit such a good concordance that they contain all the five representative genes (AGR, Col1a1, Dcn, Sparc, and Mmp2) listed by them Stuart et al. [d] This aspect also presents a contrast between the *extracellular* and *cytoplasm* that are linked through genes related to *metabolism process* and/or *metal ion binding activity* (Pkcb, Mmp2 and Sparc). This could be a good indication for identifying some midnephrogenesis-specific metabolic processes. In addition, it should be noted that gene selection by the E-SD process makes the graph simpler and thus makes the aforementioned focus of the graph clearer. For instance, the graph would be more complicated if the Nkaa1b gene, which connects five GO

---

[d]Stuart et al.[10] listed ten representative genes of the group 2 cluster. Among them, five genes were used here because no GO terms could be retrieved for the other genes.

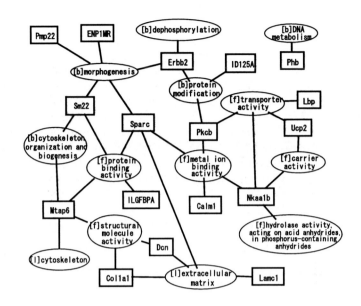

Figure 4: Gene-term bipartite graph for second aspect ($\mathcal{B}$)
Rectangles represent genes. Ovals represent GO terms. [p]=biological_process,
[f]=moleculer_function,[l]cellular_componet.eigen-attribute=(9 13) # genes=17.

terms[e], had not been dropped by the E-SD process.

**Gene-term bipartite graph for second aspect ($\mathcal{B}$):** As shown in Figure 4, the largest gene cluster in the second aspect $\mathcal{B}$ is *morphogenesis*. This aspect also focuses on two cellular components, namely, *cytoskeleton* and *extracellular matrix*. These suggestions for this aspect are similar to the statements made by Stuart et al.[10] in that group 2 was most notable for genes of the extracellular matrix as well as morphogenetic genes.

## 4 Conclusion

In this paper, we have addressed the need for a method for the multi-aspect analysis of biological data and proposed a novel method for assisting in this kind of analysis. The unique feature of our method is that it automatically

---

[e]In Fig. 3, five terms (*metabolism, catabolism, transporter activity, metal ion binding activity,* and *integral to plasma membrane*) are associated to Nkaa1b.

extracts multiple aspects for analyzing a gene list by using E-SD (a type of conceptual clustering). We conducted an experiment in which the method was applied to the analysis of rat kidney expression data. In this, our method successfully extracted different analyzing aspects, each of which consisted of relatively few genes and GO terms in a fairly simple structure. This suggests that the analyzing aspects identified by our method can be helpful in examining biological data from a range of viewpoints. For example, the user might mine some interesting viewpoints that he/she had not been aware of previously.

We are currently trying to apply our method to a larger set of genes. A preliminary experiment conducted for about thousand or more genes suggested that our method can extract different analyzing aspects but that extracted aspects tend to consist of too many genes and GO terms (e.g., 500 genes and 30 GO terms) for researchers to quickly find the major characteristics of each aspect. With regard to this point, one of the future issues is to develop a function for summarizing each aspect as well as a function for helping detailed analysis of each aspect. These functions could be realized by using some conventional methods for producing the overall characteristics of a gene list.

## References

1. J. Han. How can data mining help bio-data analysis? *Proc. 2nd Workshop on Data Mining in Bioinformatics (BIOKDD02)*, 1–2, 2002.
2. M. Ashburner *et al.* Gene ontology: tool for the unification of biology. *Nature Genet.*, 25:25–29, 2000.
3. P. J. Kennedy *et al.* Extracting and explaining biological knowledge in microarray data. *Proc. PAKDD 2004*, 699–703, 2004.
4. M. Robertson, ed. *Molecular biology of the cell.* Garland, 3 ed., 1994.
5. FUJITSU LIMITED. Biological information mining tool GeneSphere. http://www.fqspl.com.pl/life_science/xminer/GeneSphere.pdf.
6. H. Wagatsuma *et al.* A method of gene expression profiling using Xminer software (in japanese). *Proc. 26th Annual Meetings of the Molecular Biology Society of Japan*, 1PC–146 (154), 2003.
7. H. Yamakawa. Proposing matchability criterion for situation decomposition. *Proc. Int. 1998 Conf. on Neural Information Processing*, 3, 514–517, 1998.
8. H. Yamakawa *et al.* Concept acquisition and reasoning process in a card classification task by situation decomposition using ETMIC criterion. *Cognitive Studies*, 11(2):143–154, 2004.
9. P.W. Lord *et al.* Semantic similarity measures as tools for exploring the gene ontology. *Proc. Pacific Symposium on Biocomputing*, vol. 8, 601–612, 2003.
10. R. O. Stuart *et al.* Changes in global gene expression patterns during development and maturation of the rat kidney. *Proc. Natl, Acad. Sci.*, 98(10):5649–5654, 2001.

# COMPUTATIONAL APPROACHES FOR PHARMACOGENOMICS

MARYLYN D RITCHIE

*Center for Human Genetics Research, Department of Molecular Physiology &
Biophysics, Vanderbilt University, 519 Light Hall
Nashville, TN 37232 USA*

MICHELLE W CARILLO

*Stanford University, 251 Campus Drive. MSOB X-215
Stanford, CA 94305-5479, USA*

RUSSELL A WILKE

*Personalized Medicine Research Center, Marshfield Clinic Research Foundation, 1000
N Oak Avenue, Marshfield, WI 54449 USA*

Pharmacogenomics is a fascinating, emerging area of biomedical research. As implied by the name, the field lies at the intersection of pharmacology and genomics. Corresponding research efforts are therefore necessarily multidisciplinary. There is evidence that an individual's response to drug treatment can be explained, in part, by their genetic variation in certain areas of the genome. Environmental factors including diet, age, and lifestyle can influence a person's response to medications. However, environmental factors need to be studied in the context of an individual's genetic makeup, in order to assign relative impact. Pharmacogenomics holds promise for the development of personalized medicine wherein individual drug prescriptions will be adapted to each person's genetic signature.

Pharmacogenomics combines laboratory-based pharmaceutical sciences such as physiology, biochemistry, and molecular genetics with population biology. With the sequencing of the human genome near completion, the ability to detect genetic variations associated with drug response has become increasingly feasible. Molecular technology has advanced, and progress appears to be limited primarily by issues related to cost and throughput. As these issues are rapidly addressed, there is a growing need for the statistical and computational capacity to store and manage the data, as well as interpret the wealth of resulting information.

Many research endeavors are underway to deal with the statistical and computational challenges being faced in pharmacogenomics. PharmGKB is a National Institutes of Health (NIH) funded research effort to build a centralized repository for genetic and clinical information on all individuals participating in pharmacogenomics studies. This database, built by Stanford University, is now a publicly available Internet tool. In conjunction with the PharmGKB, the NIH also developed the Pharmacogenomics Research Network (PGRN), which is a national collaborative research consortium. Members of the PGRN have various projects to explore tools for analyses of the data generated. Some groups are applying traditional biostatistics approaches such as logistic regression, linear discriminant analysis, and classification and regression trees (CART). Others are utilizing computer science approaches such as support vector machines and neural networks. Still others are developing novel statistical methods such as new cluster analysis algorithms and data reduction algorithms such as the Multifactor Dimensionality Reduction method. The number of statistical and computational approaches continues to expand with the challenges facing the pharmacogenomics community.

The goal of The Pacific Symposium on Biocomputing is to explore current research in the theory and application of computational methods as they apply to problems of biological significance. The significance of pharmacogenomics is clear. Many population geneticists anticipate that this will be the first discipline wherein functional genomics translates into clinical application on a large scale. As such, PSB represents an ideal forum to further the analytical and computational approaches associated with such an endeavor. The biological and chemical technology is advancing, and a merging of pharmacogenomics with biocomputing is inevitable.

We solicited papers related to computational approaches for pharmacogenomics. This includes database design and implementation, data sharing among pharmacogenomics centers, statistical analysis, statistical and computational method development, and real data applications. We received a number of excellent submissions, which made decisions difficult. Due to space and time constraints, we selected six papers for publication, four of which will be presented orally at the conference. These papers cover the breadth of topics we had solicited for in the original call for papers.

Sirohi and Peissig describe their methods for using various drug lexicons to automatically extract patient prescription information from their electronic medical records. Ferrin et al. describe DASH, a software framework for the access, maintenance, curation, and sharing of data amongst collaborators of a large scale pharmacogenomics project. This includes the design of the system as well as an application to a prototype problem. Khatri et al. describe a bioinformatics approach to identify genomic fingerprints for a target organism. This is especially important for identifying biological agents that might be used in terrorism. Peccoud and Vander Velden use mathematical models of molecular interaction networks for genotype to phenotype maps. Gui and Li describe a Threshold Gradient Decent method for the Cox model to select genes relevant to patient survival. Reif et al. explain the Exploratory Visual Analysis software and database for the visual exploration of statistical results and annotation information to extract new information from experimental results. These six papers demonstrate the breadth of computational technologies being developed in pharmacogenomics.

# IDENTIFICATION OF GENOMIC SIGNATURES FOR THE DESIGN OF ASSAYS FOR THE DETECTION AND MONITORING OF ANTHRAX THREATS

SORIN DRAGHICI[1,†], PURVESH KHATRI[1,2,†], YANHONG LIU[4], KITTY J CHASE[3], ELIZABETH A BODE[3], DAVID A KULESH[3], LEONARD P WASIELOSKI[3], DAVID A NORWOOD[3], JAQUES REIFMAN[2]

[1] *Dept. of Computer Science, Wayne State University, Detroit, MI 48202*
[2] *Bioinformatics Cell, Telemedicine and Advanced Technology Research Center, US Army Medical Research and Materiel Command, Ft. Detrick, MD 21701*
[3] *Diagnostic Systems Division, US Army Medical Research Institute of Infectitious Diseases, Ft. Detrick, MD 21701*
[4] *US Dept. of Agriculture, Agricultural Research Service, Eastern Regional Research Center, Wyndmoor, PA 19038*

Sequences that are present in a given species or strain while absent from or different in any other organisms can be used to distinguish the target organism from other related or un-related species. Such DNA signatures are particularly important for the identification of genetic source of drug resistance of a strain or for the detection of organisms that can be used as biological agents in warfare or terrorism. Most approaches used to find DNA signatures are laboratory based, require a great deal of effort and can only distinguish between two organisms at a time. We propose a more efficient and cost-effective bioinformatics approach that allows identification of genomic fingerprints for a target organism. We validated our approach using a custom microarray, using sequences identified as DNA fingerprints of *Bacillus anthracis*. Hybridization results showed that the sequences found using our algorithm were truly unique to *B. anthracis* and were able to distinguish *B. anthracis* from its close relatives *B. cereus* and *B. thuringiensis*.

## 1. Introduction

The area of organism identification using DNA sequences has many applications in various life science areas. However, there are also many challenges. For instance, sheep pox and goat pox viruses are so closely related that they cannot be distinguished using clinical signs, pathogenesis or seroreactivity.[30] Furthermore, both cross-infectivity and cross-resistance have

---

† These authors should be considered joint first authors.

been reported[38] to the point that the two were thought to be caused by a single viral species. However, genetic analysis demonstrated that sheep pox and goat pox are actually caused by two related, but genetically distinct viruses. Furthermore, the identification of a few base pair differences in the sequence coding for the P32 protein allowed the design of a polymerase chain reaction (PCR) restriction fragment length polymorphism (PCR RFLP) assay able to distinguish between the two species. This assay involves a PCR amplification with a common primer, followed by a digestion with a *Hinf I* restriction enzyme that produces fragments of different sizes allowing the identification of the two species.

The issue of distinguishing between different species is somewhat academic if the two species exhibit both cross-infectivity and, most importantly, allow passive cross-protection as the sheep pox and goat pox do.[37] However, this is not always the case. Genes that are present in certain isolates of a given bacterial species and are substantially different or absent from others can determine important strain-specific traits such as drug resistance[13] and virulence.[51] As an example, *B. anthracis*, *B. cereus*, and *B. thuringiensis* are genetically so close that it has been proposed to consider them a single species.[27] At the same time, these bacteria are very different on a phenotypic level. *B. cereus* is a frequent food contaminant but only a mild opportunistic human pathogen;[16,28] *B. thuringiensis* is actually a useful bacterium being used as a pesticide[46] while *B. anthracis* is a virulent pathogen for mammals that has been used as a bio-terror and biological warfare agent.[12,53]

In such cases, the identification of an organism-specific DNA sequence gains an increased importance. Even if such sequences are not functionally active, they can still be extremely useful if used as genetic fingerprints. DNA sequences that are present in a given species while absent from any other organisms can be used to distinguish the target organism from other related or un-related species. If such genetic fingerprints were available for organisms that can be potentially used as biological or terrorist weapons, the task of rapid threat identification, characterization, and selection of appropriate medical countermeasures could be immensely facilitated. Genetic fingerprints can also aid identification of genetic source of drug resistance of a strain,[17] which can be useful to drug developers in pharmacogenomics.

## 2. Existing work

The existing work in the areas of organism identification using DNA signatures can be divided into two different categories. One approach uses a laboratory assay to identify the organism. Techniques used include amplified fragment length polymorphism (AFLP),[44,45] suppression subtractive hybridization (SSH)[3] and custom DNA microarrays.[36] A second approach uses a purely bioinformatics analysis of the characteristics of the genomes of various species and extracts those features that are characteristic to individual species.

The laboratory based approach does not necessarily require information about the entire genomes involved and is better suited for the development of assays for monitoring and identification of biological threats. For instance, SSH, a PCR-based DNA subtraction method, allows identification of genomic sequence differences in a "tester" DNA relative to a "driver" DNA. AFLP relies on the analysis of a fluorescence based signal proportional to the size of various DNA fragments.[49] SSH and AFLP have been successfully used to identify genomic sequence differences between various strains or species of bacteria.[4,5,10,31,44] The major drawback of this approach is that it permits identification of genomic differences only between two organisms. For instance, in order to differentiate two species, one needs to use an SSH assay to compare each strain of one species with each strain of the other species.[44] Clearly, this approach cannot be used to provide a genomic signature that would differentiate a given organism from all others.

The *in silico* approach to identifying genomic signatures is usually based on an analysis of the entire genomes involved and aims at extracting features such as species-specific codon usage.[1,2,23,32–34,52] While this type of genomic signature can be informative about the given organisms and the relationships among them, it may not be directly usable for detection and monitoring purposes.

Comparative sequence analysis has also been useful in detecting intronic and intergenic regions[25,40] as well as uncovering novel repeated structures.[18,26] Several genome scale alignment tools are available: MUMmer,[14,15,39] AVID,[11] MGA,[29] WABA,[35] and GLASS[7] among others. Tax-Plot[22] provides visual representation of protein homologs in microbial and eukaryotic genomes. Most of these pair-wise[a] alignment tools assume that the input genomes are closely related. Therefore, there will be a mapping

---

[a]MGA is a multiple alignment tool but the alignment is still computed pair-wise.

of large subsequences between the two input genomes. In turn, they assume that these large subsequences, appearing in the same order in the closely related genomes, are very likely to be part of the final alignment. These regions are used as anchors for the alignment of the input genomes.

In general, anchor-based genome alignment programs first create a suffix tree from the two input genomes. A suffix tree is a compact representation of all suffixes in the input string.[41,54] A suffix of a string is a substring starting at any position in the string and extending up to the end of the string. Next, the suffix tree is searched for sequences that appear in both input genomes. These exact matching subsequences are known as maximal exact matches (MEMs). The anchors are chosen from these MEMs. Different programs apply different criteria for the selection of anchors. For instance, MUMmer uses the longest increasing subsequence (LIS)[24] for the selection of anchors.[14] MUMmer allows the selection of overlapping anchors whereas AVID and MGA only select non-overlapping anchors. Since MGA allows alignment of more than two genomes, it only selects MEMs that are present in all of the input genomes. AVID first finds the length of the longest MEM and discards all the MEMs that are less than half the length of the longest MEM. After selecting the anchors, MUMmer employs a variant of the Smith-Waterman algorithm[47] to close the gaps between the anchors. MGA and AVID close the gaps by recursively creating suffix trees for the non-anchored parts of the input genomes and hence, gradually reducing the gap sizes. Once the gaps are smaller than a threshold, MGA and AVID close them using the ClustalW[48] and Needleman-Wunsch algorithms,[42] respectively.

These large number of tools are all geared towards finding large-scale *similarities* between two or more genomes. Our focus here is different. While these algorithms were developed to find sequence similarities, our goal is to find sequence dissimilarities. These two problems are related but not reciprocal. Simply put, one cannot just take the complement of the sequences found in a similarity search and use them as genomic signatures. The main reason is related to the fact that a search aiming to find similarity will sometimes discard entire blocks after only a summary inspection because they are not sufficiently similar to the target sequence. On the other hand, a search aiming to find dissimilarities, i.e., unique signatures, has to actually focus on exactly those areas that are discarded without extensive analysis during the similarity search.

Here, we propose an algorithm for finding genomic fingerprints that distinguish an organism from all other organisms with known genomes.

As the number of sequenced organisms increases, this approach has the potential to substitute existing laboratory based approaches such as AFLP and SSH.

In this paper, we used this approach to find a genetic signature for *B. anthracis*. Identification of genomic regions unique to *B. anthracis* can provide clues to its genetic relationship to other highly similar organisms. Related work for the detection of *B. anthracis* used plasmid-encoded toxin genes for rapid DNA-based assays.[8] However, these failed to detect non-plasmid containing strains of *B. anthracis* isolated from the environment.[50] Also, there have been efforts to design real-time PCR assays. However, these assays only targeted a single locus and they yielded false-positive results with some strains of *B. cereus*.[20,43]

## 3. Analysis methods

Our goal is to find unique DNA sub-sequences for a given target genome across all available known genomes. An obvious approach is to compare (i.e., align) the genome of our interest against all available known genomes. These alignments will reveal the parts of the target genome that do not align with any other genome (i.e., are unique to the target genome). However, this seemingly simple approach is computationally very expensive. The GenBank database at NCBI contains nucleotide sequences from more than 140,000 organisms.[9] The length of these genomes vary from a few thousand base pairs to a few billion base pairs. Aligning the input genome with each of these genomes is computationally unfeasible.

The amount of computation can be considerably reduced by using the phylogenetic background of the target. Today biologists agree that various organisms have evolved from common ancestors. During evolution, functional genomic elements are conserved. Hence, two closely related genomes are expected to have many matching subsequences. If a subsequence that distinguishes the target from all organisms exists, this subsequence will also distinguish the target from its closest relative. Hence, a good initial set of potential genomic signatures can be obtained by comparing the target only with its closest relative and by retaining only those sequences that are different. Subsequently, each of these potential signatures is compared with all other known genomes. This approach drastically reduces both the number of comparisons required as well as the length of sequences to be compared (from a few million to a few thousand base pairs, at most).

In order to find the exact matching sequences between the target and its closest relative, we start by using their concatenated sequences to create a suffix tree. We then use a suffix tree search algorithm as the one employed in MUMmer to find the exact matching sequences in both genomes. Since our goal is to determine a set of relatively short sequences to be used on a microarray type assay, we have to search both the forward and the reverse strands. Any sequences that match between the two organisms are removed from further consideration. The result is a set of short segments of the target genome that can be considered potential signatures. These are then compared with all sequences in the blast-$nt$[21] database from NCBI.[6] We consider a sequence is unique for the target genome if it does not align to any sequence from any other organism with an expected value ($E$-value) less than a threshold of 0.01. Fig. 1 provides an overview of this approach.

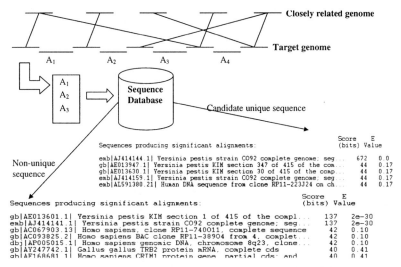

Figure 1. The genomic fingerprinting approach. Two genomes are searched for exact matching subsequences (MEMs). The MEMs are removed from the target genome and the remaining segments of the target genome ($A_1$, $A_2$, ..., $A_n$) are searched against the $nt$ database. If the length of a segment is less than the user specified length, it is discarded and not searched in the $nt$ database. As shown, if a sequence does not align with any sequence from another organism with $E$ value less than the specified threshold it is considered as a sequence unique to the target genome.

## 4. Results and discussion

In order to validate our approach, we designed a custom microarray using sequences identified as genomic fingerprints for *B. anthracis*. This array was then hybridized with *B. anthracis* and *B. cereus*.

In order to find a genomic signature for *B. anthracis* we proceeded as follows. We searched the *B. anthracis str. Ames* genome (GenBank contig accession number NC_03997) for subsequences of 30 base pairs or more matching anywhere (direct and reverse strand) with sequences from the genome of *B. cereus ATCC 14579* (GenBank contig accession number NC_004722). We chose *B. cereus ATCC 14579* genome as a closely related genome because it is considered to be a good representative of the *B. cereus* family.[19] Then, we removed all of matching sequences from the *B. anthracis* genome. This step produced over 6,000 sequences of length 50 or more. These sequences were then searched against the *nt* database using *blastn*. The sequences in the BLAST output that were not found in any other organism with $E$ value less than 0.01 were retrieved and considered part of the genomic fingerprints of *B. anthracis*. There were 140 such sequences. Note that this analysis stage also removed sequences that matched the genomes of other close relatives of *B. anthracis*, such as *B. thuringiensis*, without ever directly comparing them. These 140 target sequences were provided to CombiMatrix (Mukilteo, WA) for the design of a custom microarray. CombiMatrix designed 2 probes for 80 target sequences and 1 probe for 22 target sequences (for a total of 182 probes for 102 target sequences) with melting temperature in the range of 70°C to 75°C and a length of 35 base pairs or more. Probes of the required length and melting temperatures could not be identified for the remaining 38 target sequences. The microarray was designed with three replicates of each of the 182 probes.

The custom microarray was then hybridized with samples of *B. anthracis* and *B. cereus*. The hybridization results showed that 18 probes only hybridized to the *B. anthracis* sequences indicating that they were true genomic fingerprints of *B. anthracis*. Table 1 provides the positions of the sequences on *B. anthracis* genome that were found to be unique in the microarray experiment.

Surprisingly, many of the initial 182 probes also hybridized with *B. cereus*. We further searched these cross-hybridizing probes against the blast-*nt* database. For the probes that hybridized to *B. cereus* the results of this comparison showed that although the target sequences of those probes are only present in *B. anthracis*, the part of the target sequence on which the probes were designed was not unique to *B. anthracis* and is present in other genomes. This shows that the probe design stage lost some specificity due to its unique added requirements: melting temperatures in a very narrow range, limited lengths, etc. In all cases, although the initial, longer sequence was unique across the blast-*nt* database, by selecting a shorter subsequence, the probe became unspecific. Hence, another BLAST search is recommended before printing the assay, to check whether the subsequences selected as probes continue to be good signatures for the target organism.

Table 1. The following 18 probes identify 17 unique sequences of *B. anthracis (Ames)*. The first and second columns indicate the start and end, respectively, of the target sequences from *B. anthracis*. The third and the fourth column are the start and end positions, respectively, on the corresponding target sequences for which probes were designed.

| Sequence start | Sequence end | Probe start | Probe end |
|---:|---:|---:|---:|
| 175,231 | 175,455 | 6 | 44 |
| 175,567 | 175,677 | 36 | 71 |
| 488,976 | 489,620 | 130 | 166 |
| 945,569 | 946,596 | 151 | 190 |
| 1,629,522 | 1,630,538 | 489 | 523 |
| 1,629,522 | 1,630,538 | 529 | 568 |
| 1,845,001 | 1,845,363 | 111 | 145 |
| 2,021,535 | 2,022,919 | 491 | 529 |
| 2,098,619 | 2,099,274 | 591 | 625 |
| 2,783,190 | 2,783,405 | 17 | 54 |
| 2,918,788 | 2,920,251 | 977 | 1013 |
| 3,037,856 | 3,038,113 | 115 | 152 |
| 3,524,649 | 3,524,731 | 17 | 55 |
| 3,808,069 | 3,809,046 | 797 | 834 |
| 3,821,617 | 3,822,163 | 449 | 483 |
| 4,374,364 | 4,375,478 | 227 | 311 |
| 4,375,581 | 4,376,123 | 149 | 186 |
| 4,933,405 | 4,933,482 | 9 | 43 |

## 5. Conclusion

DNA sequences that are present in a given species or strain while absent from any other organism can be used to distinguish the target organism from other related or un-related species. The identification of such DNA signatures is particularly important for organisms that may be potentially used as biological warfare agents or terrorism threats.

Most approaches used to identify DNA signatures are laboratory based and require a significant effort and time. A bioinformatics approach can provide results faster and more efficiently. However, most tools built for genome comparisons only allow alignment of two genomes at a time. Using this approach to find unique DNA signatures across all known organisms is unfeasible. In addition, all existing tools are limited to finding the similarity between two genomes. In contrast, looking for DNA signatures requires the development of tools that identify sequence dissimilarities. In this paper, we describe an approach to find the DNA fingerprints of an organism. We used this approach to find a set of unique sequences for *B. anthracis* which were then used to design probes for a DNA microarray. The hybridization results revealed that a subset of these probes were truly unique to *B. anthracis* and were able to distinguish between *B. anthracis* and *B. cereus*, which is a close genetic relative.

## Acknowledgements

This work was supported by the research area directorates of the US Army Medical Research and Materiel Command and the Defense Threat Reduction Agency. The first two authors are also supported by: NSF DBI-0234806, NIH 1S10 RR017857-01, MLSC MEDC-538 and MEDC GR-352, NIH 1R21 CA10074001, 1R21 EB00990-01 and 1R01 NS045207-01.

## References

1. T. Abe, S. Kanaya, M. Kinouchi, Y. Ichiba, T. Kozuki, and T. Ikemura. A novel bioinformatic strategy for unveiling hidden genome signatures of eukaryotes: Self-organizing map of oligonucleotide frequency. *Genome Informatics*, (13):12–20, 2002.

2. T. Abe, S. Kanaya, M. Kinouchi, Y. Ichiba, T. Kozuki, and T. Ikemura. Informatics for unveiling hidden genome signatures. *Genome Research*, 13(4):693–702, 2003.

3. P. G. Agron, M. Macht, L. Radnedge, E. W. Skowronski, W. Miller, and G. L. Andersen. Use of subtractive hybridization for comprehensive surveys of prokaryotic genome differences. *FEMS Microbiology Letters*, 211(2):175–182, Jun 2002.

4. I. Ahmed, G. Manning, T. Wassenaar, S. Cawthraw, and D. Newell. Identification of genetic differences between two *Campylobacter jejuni* strains with different colonization potentials. *Microbiology*, 148:1203–1212, 2002.

5. N. Akopyants, A. Fradkov, L. Diatchenko, and et. al. PCR-based subtractive hybridization and differences in gene content among strains of *Helicobacter pylori*. *Proc. Natl. Acad. Sci.*, 95:13108–13113, 1998.

6. S. Altschul, T. Madden, A. Schaffer, J. Zhang, Z. Zhang, M. W., and D. Lipman. Gapped BLAST and PSI-BLAST: a new generation of protein database search programs. *Nucleic Acids Research*, 25(17):3389–3402, Sept 1997.

7. S. Batzoglou, L. Pachter, J. Mesirov, B. Berger, and E. Lander. Human and

mouse gene structure: comparative analysis and application to exon prediction. *Genome Research*, 10:950–958, 2000.

8. C. Bell, J. Uhl, T. Hadfield, J. David, R. Meyer, T. Smith, and F. Cockerill III. Detection of *Bacillus anthracis* dna by light-cycle pcr. *J. Clin. Microbiol.*, 40:2897–2902, 2002.

9. D. Benson, I. Karsch-Mizrachi, D. Lipman, J. Ostel, and D. Wheeler. GenBank: update. *Nucleic Acids Research*, 32(1):D23–D26, January 2004.

10. M. Bogush, T. Velikodvorskaya, Y. Lebedev, and et. al. Identification and localization of differences between *Escherichia coli* and *Salmonella typhimurium* genomes by suppressive sutractive hybridization. *Mol Gen Genet*, 262:721–729, 1999.

11. N. Bray, I. Dubchak, and L. Pachter. AVID: A global alignment program. *Genome Research*, 13(1):97–102, January 2003.

12. E. Check. Bioshield defence programme set to fund anthrax vaccine. *Nature*, 429(6987):4, May 2004.

13. J. Davies. Inactivation of antibiotics and the dissemination of resistance genes. *Science*, 264(5157):375–382, Apr 1994.

14. A. Delcher, S. Kasif, R. Fleischmann, J. Peterson, O. White, and L. Salzberg. Alignment of whole genomes. *Nucleic Acids Research*, 27(11):2369–237, 1999.

15. A. Delcher, A. Phillippy, J. Carlton, and S. Salzberg. Fast algorithms for large-scale genome alignment and comparison. *Nucleic Acids Research*, 30(11):2478–2483, 2002.

16. F. Drobniewski. *Bacillus cereus* and related species. *Clin. Microbiol. Rev.*, 6:324–338, 1993.

17. S. Drăghici and B. Potter. Predicting HIV drug resistance with neural networks. *Bioinformatics*, 19(1):98–107, January 2003.

18. I. Dunham, N. Shimuzu, B. Roe, S. Chissoe, and et. al. The DNA sequences of chromosome 22. *Nature*, 402:489–495, 1999.

19. K. Dwyer, J. Lamonica, J. Schumacher, L. Williams, J. Bishara, A. Lewandowski, R. Redkar, G. Patra, and D. V.G. Identification of bacillus anthracis specific chromosomal sequences by suppressive subtractive hybridization. *BMC Genomics*, 5(1):15, Feb 2004.

20. H. Ellerbrok, H. Nattermann, M. Ozel, L. Beutin, B. Appel, and G. Pauli. Rapid and sensitive identification of pathogenic and apathogenic *Bacillus anthracis* by real-time pcr. *FEMS Microbiol. Lett.*, 214:51–59, 2002.

21. N. C. for Biotechnology Information. Blast nucleotide database. ftp://ftp.ncbi.nih.gov/blast/db/.

22. N. C. for Biotechnology Information. Taxplot. http://www.ncbi.nlm.nih.gov/sutils/taxik2.cgi?

23. R. Grantham, C. Gautier, M. Gouy, R. Mercier, and A. Pave. Codon catalog usage and the genome hypothesis. *Nucleic Acids Research*, 8:r49–r62, 1980.

24. D. Gusfield. *Algorithms on Strings, Trees and Sequences: Computer Science and Computational Biology*. Cambridge University Press, New York, 1997.

25. R. Hardison, J. Oeltjen, and W. Miller. Long human-mouse sequence alignments reveal regulatory elements: a reason to sequence the mouse genome. *Genome research*, 7:959–966, 1997.

26. M. Hattori, A. Fujiyama, T. Taylor, H. Watanabe, and et. al. The DNA sequence of human chromosome 21. *Nature*, 405:311–319, 2000.
27. E. Helgason, D. Caugant, I. Olsen, and A. Kolsto. Genetic structure of population of *Bacillus cereus* and *B. thuringiensis* isolates associated with periodontitis and other human infections. *J. Clin. Microbiol.*, 38:1615–1622, 2000.
28. O. I. K. A. Helgason E, Caugant DA. Genetic structure of population of bacillus cereus and b-thuringiensis isolates associated with periodontitis and other human infections. *Journal Of Clinical Microbiology*, 38(4):1615–1622, Apr 2000.
29. M. Hohl, S. Kurtz, and E. Ohlebusch. Efficient multiple genome alignment. *Bioinformatics*, 18(Suppl. 1):S312–S320, 2002.
30. M. Hosamani, B. Mondal, P. A. Tembhurne, S. K. Bandyopadhyay, R. K. Singh, and T. J. Rasool. Differentiation of sheep pox and goat poxviruses by sequence analysis and pcr-rflp of p32 gene. *Virus Genes*, 29(1):73–80, Aug 2004.
31. B. Janke, U. Dobrindt, J. Hacker, and G. Blum-Oehler. A subtractive hybridization analysis of genomic differences between the uropathogenic *E. coli* strain 536 and the *E. coli* k-12 strain mg1655. *FEMS Microbial Lett.*, 199:61–66, 2001.
32. S. Kanaya, M. Kinouchi, T. Abe, Y. kudo, Y. Yamada, T. Nishi, H. Mori, and T. Ikemura. Analysis of codon usage diversity of bacterial genes with a self-organizing map (som): Characterization of horizontally transferred genes with emphasis on the e. coli o157 genome. *Gene*, 276:89–99, 2001.
33. S. Kanaya, Y. Kudo, T. Abe, T. Okazaki, D. Carlos, and T. Ikemura. Gene classification by self-organizing mapping of codon usage in bacteria with completely sequences genome. *Genome Informatics*, 9:369–371, 1998.
34. S. Kanaya, Y. Kudo, Y. Nakamura, and T. Ikemura. Detection of genes in escherichia coli sequences determined by genome projects and prediction of protein production levels, based on multivariate diversity in codon usage. *CABIOS*, 12:213–225, 1996.
35. W. Kent and A. Zahler. Conservation, regulation, synteny and introns in large-scale *C. briggsae-C. elegans* genomic alignment. *Genome research*, 10:1115–1125, 2000.
36. M. Kingsley, T. Straub, D. Call, D. Daly, S. Wunschel, and D. Chandler. Fingerprinting closely related xanthomonas pathovars with random nonamer oligonucleotide microarrays. *Appl. Environ. Microbiol.*, 68:6361–6370, 2002.
37. R. P. Kitching. Passive protection of sheep against capripoxvirus. *Res Vet Sci*, 41(2):247–250, Sep 1986.
38. R. P. Kitching and W. P. Taylor. Clinical and antigenic relationship between isolates of sheep and goat pox viruses. *Trop Anim Health Prod*, 17(2):64–74, May 1985.
39. S. Kurtz, A. Phillippy, A. Delcher, M. Smoot, M. Shumway, C. Antonescu, and S. Salzberg. Versatile and open software for comparing large genomes. *Genome biology*, 5:R12, 2004.
40. G. Loots, R. Locksley, C. Blankespoor, Z. Wang, W. Miller, E. Rubin, and

K. Frazer. Identification of a coordinate regulator of interleukins 4, 13 and 5 by cross-species sequence comparisons. *Science*, 288:136–140, 2000.

41. E. McCreight. A space-economical suffix tree construction algorithm. *Journal of the ACM*, 23(2):262–272, 1976.

42. S. Needleman and C. Wunsch. A general method applicable to the search for similarities in the amino acid sequence of two proteins. *J. Mol. Biol.*, 48(3):443–453, 1970.

43. Y. Qi, G. Patra, X. Liang, L. Williams, S. Rose, R. Redkar, and V. DelVecchio. Utilization of the *rpoB* gene as a specific chromosomal marker for real-time pcr detection of *Bacillus anthracis*. *Appl. Environ. Microbiol.*, 67:3720–3727, 2001.

44. L. Radnedge, P. G. Agron, K. Hill, P. Jackson, L. Ticknor, P. Keim, and A. G. L. Genome differences that distinguish bacillus anthracis from bacillus cereus and bacillus thuringiensis. *Applied And Environmental Microbiology*, 69(5):2755–2764, May 2003.

45. L. Radnedge, S. Gamez-Chin, P. McCready, P. Worsham, and G. Andersen. Identification of nucleotide sequences for the specific and rapid detection of yersinia pestis. *Applied And Environmental Microbiology*, 67(8):3759–3762, Aug 2001.

46. E. Schnepf, N. Crickmore, J. Van Rie, D. Lereclus, J. Baum, J. Feitelson, D. R. Zeigler, and D. H. Dean. Bacillus thuringiensis and its pesticidal crystal proteins. *Microbiology And Molecular Biology Reviews*, 62(3):775–806, Sep 1998.

47. T. Smith and M. Waterman. Identification of common molecular subsequences. *J. Molecular Biology*, 147(1):195–197, 1981.

48. J. Thompson, D. Higgins, and T. Gibson. CLUSTALW: improving the sensitivity of progressive multiple sequence alignment through sequence weighting, position specific gap penalties and weight matrix choice. *Nucleic Acids Research*, 22:4673–4680, 1994.

49. L. Ticknor, A. Kolsto, K. Hill, P. Keim, M. Laker, M. Tonks, and P. Jackson. Fluorescent amplified fragment length polymorphism analysis of norwegian bacillus cereus and bacillus thuringiensis soil isolates. *Appl Environ Microbiol.*, 67(10):4863–4873, 2001.

50. P. Turnbull, R. Hutson, M. Ward, M. Jones, C. Quinn, N. Finnie, C. Duggleby, J. Kramer, and J. Melling. *Bacillus anthracis* but not always anthrax. *J. Appl. Bacteriol.*, 72:21–28, 1992.

51. M. K. Waldor and M. J. J. Lysogenic conversion by a filamentous phage encoding cholera toxin. *Science*, 272(5270):1910–1914, Jun 1996.

52. H. Wang, J. Badger, P. Kearney, and M. Li. Analysis of codon usage patterns of bacterial genomes using the self-organizing map. *Molecular Biology and Evolution*, 18:792–800, 2001.

53. G. Webb. A silent bomb: the risk of anthrax as a weapon of mass destruction. *Proceedings of the National Academy of Sciences USA*, 100(7):4346–4351, 2003.

54. P. Weiner. Linear pattern matching algorithms. In *Proc. 14th IEEE Symp. Switching & Automata Theory*, pages 1–11, 1973.

# ENHANCING DATA SHARING IN COLLABORATIVE RESEARCH PROJECTS WITH DASH

THOMAS E. FERRIN, CONRAD C. HUANG, DANIEL M. GREENBLATT,
DOUG STRYKE, KATHLEEN M. GIACOMINI, AND JOHN H. MORRIS

*Departments of Pharmaceutical Chemistry and Biopharmaceutical Sciences University
of California San Francisco, 600 16th Avenue,
San Francisco, CA 94143, USA*

We describe a software framework, called DASH, that enables the facile access, maintenance, curation and sharing of computational biology data among collaborating research scientists. The DASH event-based framework enables members of team-based research projects to describe the multistep computational processing pipelines frequently required to generate data for sharing, monitors multiple distributed data stores for changes, and will then automatically invoke the appropriate processing pipeline(s). These pipelines can be used to communicate the results of data analyses to collaborators using mechanisms such as Web Services. We describe the overall design of the DASH system and the application of a simple DASH prototype to a collaborative pharmacogenomics research project involving several dozen researchers located at several different sites—the UCSF Pharmacogenetics of Membrane Transporters project.

## 1. Introduction

In a collaborative, team-based research project, each group must be able to share results with others and access data generated by others. Traditionally, this has been done by exchanging data via electronic mail or file transfer. While a more facile approach is to use a shared database, this often incurs the challenge of properly maintaining data integrity in the presence of updates by multiple researchers. For example, in a database used for computational biology, adding and altering data may require the invocation of additional computational protocols that automatically update all related—and especially derived— information. When performed manually, this tedious curation process can act as a deterrent, limiting either the number of participants or the growth of the database, or even preventing a collaborative project from being realized in the first place. While several technological solutions to support collaboration, such as workflow and data integration, already exist, no single solution addresses the needs of collaborative computational biology without a significant expenditure of money or personnel. The goal of the project we describe here is to create a system that enables the facile sharing of data, and is specifically targeted at small- to medium-sized collaborative computational biology projects. We believe this represents a very important class of collaborative science projects, as was discussed at the "Models of Team Science" session [1] at last year's BECON 2003 Symposium on Catalyzing Team Science [2]. "Team science" and the formation of integrated research networks are also common themes within the NIH Roadmap Initiatives [3, 4], the success of which depends crucially upon the sharing of research data. This fundamental need to efficiently and effectively share research data provides the motivation for the DASH project.

## 2. User Requirements

The core of collaborative science is the exchange of data among cooperating research groups. However, before any sharing can occur, participants must first agree on *what* data will be exchanged (e.g., experimental data, analysis results) and *how* exchanges will happen (e.g., data format, transfer media). When only a few data sets must be shared, data preparation may be done manually. However, as the number and types of data sets increase, the shortcomings of manual preparation, such as human error during processing and dependence on vigilant monitoring of available data, can become serious hindrances to the collaboration. Timely sharing of data becomes even more difficult if complex and time-consuming manual manipulation of data is needed. For small- to medium-sized academic laboratories, data preparation for collaborations can prove to be quite challenging due to limited funding and staffing. Tools are needed to help streamline data preparation and sharing by addressing the requirements listed in Table 1.

Table 1. User requirements for managing data exchange with collaborators.

| | |
|---|---|
| R1 | *Document data exchange protocol.* Having protocol documentation will ease transitions such as staff or student turnover, which is a particularly difficult problem for a small group where the person leaving may also be the single person who handles all data preparation. |
| R2 | *Facilitate changes in protocol.* In a lengthy collaboration, new types of data may be acquired or analyzed over the course of the collaboration. As the data domain evolves, so must the data exchange protocols. |
| R3 | *Support multiple collaborations.* Multi-group collaborations are becoming more common as larger and/or interdisciplinary scientific projects are being tackled. While the project groups share the same goals, they may not share the same research tools. As part of a multi-group collaboration, a lab must be able to prepare data in multiple formats to fit the needs of multiple collaborators. |
| R4 | *Automate data manipulation.* Automation removes the burden of repetitive activities from users and helps minimize human errors. In addition, scaling up is much more feasible for an automated system than a manual one. When preparing data for collaborations, automated data manipulation protocols can also be used to facilitate internal data processing. Having a single mechanism for initiating automated processes can simplify overall data management for users. |
| R5 | *Control data access.* Data sharing must not usurp the data owners' ability to control how data is published. In particular, sensitive data, e.g., patient information, are often used in clinical research but must not be shared with all collaborators without careful consideration and adherence to applicable regulations or restrictions. This implies that data owners are not forced to store their data in a centralized repository or in a prescribed format. |

## 3. The PMT Project

The UCSF Pharmacogenetics of Membrane Transporters project (PMT, http://pharmacogenetics.ucsf.edu) [5] provides an excellent example of the type of collaborative science project that can benefit from the DASH infrastructure. The goal of the PMT project is to understand the genetic basis for variation in drug response for drugs that interact with membrane transport proteins. Membrane transporters, a major determinant of pharmacokinetics, are of great pharmacological importance.

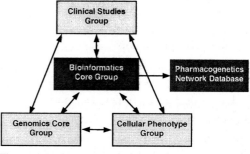

Figure 1. Components of the Pharmacogenetics of Membrane Transporters project.

The UCSF PMT project, begun in April 2000, involves more than 50 researchers from diverse disciplines, distributed across 19 labs at UCSF, UCLA, San Francisco General Hospital, and Kaiser Foundation Hospitals. These investigators are systematically identifying sequence variants in transporters and determining the functional significance of these variants through evaluation of relevant cellular and clinical phenotypes. Experimental results are deposited in the Pharmacogenetics Research Network and Knowledge Base (PharmGKB, http://www.pharmgkb.org) [6], hosted at Stanford University. The PMT is organized into four major components:

- Genomics Core Group (GCG), which is sequencing the DNA of 200 genes from several sample sets of more than 250 individuals;
- Cellular Phenotype Group (CPG), which determines the pharmacological effects of single nucleotide polymorphisms (SNPs) in cellular assays;
- Clinical Studies Group (CSG), which tests drug response in volunteer human subjects with known genotypes; and
- Bioinformatics Core Group (BCG), which performs data analyses, provides computing infrastructure to facilitate information exchange, and exports results to the PharmGKB.

Figure 1 shows the flow of data within the PMT. GCG-BCG data exchanges include trace files from DNA sequencers, per-sample single nucleotide polymorphism sites and variants. Data from GCG to BCG are uploaded to a shared network file system in Common Assembly File and

Standard Chromatogram File formats. BCG analyzes the uploaded experimental data and makes the results available through the PMT intranet Web site (password protected), either as Hypertext Markup Language (HTML) pages or as a tab-separated-values plain text file suitable for importing into spreadsheets and databases. With over 50 sequencing experiments completed and 150 proposed for the next five years, the timely analysis of data sets is critical to the project. In addition to standard analyses such as Hardy-Weinberg equilibrium of SNPs, other more speculative analyses are constantly being proposed and tested. Leabman has recently published some of these "data mining" analyses [7]. While the BCG has kept pace with the current volume of PMT data primarily through use of manual approaches to data curation, automation tools are needed to handle the planned increase in volume of data, as well as new hypotheses-testing analyses, for the next five years.

## 4. Existing Technologies

Tools exist to address some of the requirements listed in Table 1. The four most relevant technologies are workflow management, data integration, distributed resource management, and event-based technologies.

*Workflow management* is a protocol-centric paradigm for controlling activities within a system; data is often treated as auxiliary information attached to process instances. There are many commercial [8-10] and open-source [11-13] workflow management systems. Most provide central management of workflows, where an analysis produces the initial workflow, which is then revised under strict access control. However, the classic workflow paradigm does not fit well in a collaborative science environment, where many researchers need to be able to introduce new data and activities into a process definition. Centralized control over workflow modification would introduce unacceptable overhead (i.e., does not address requirement R2 in Table 1). Additionally, many commercial workflow systems are devoted to streamlining the execution of a series of manual activities. However, in a research environment workflows are often data-driven or data-triggered, and updates can be handled by automated activities rather than manually. With their focus on manual activities, traditional workflow systems often do not handle requirement R4 gracefully.

We have investigated both commercial and open-source workflow solutions. The system that most nearly satisfies the requirements in Table 1 is myGrid [11]. myGrid emphasizes the large-scale, geographically distributed e-science environment, and is necessarily complex. The subset of myGrid's functionality most pertinent to our requirements is limited to three of these components: an information repository, workflow enactor, and notification service. However, there does not appear to be the level of functional integration between these components necessary to support the kind of features outlined in our requirements. While myGrid seems like a promising technology, it appears to be overly complex for use in simple collaborations.

*Data integration* is data-centric; processing activities are not explicitly included other than as clients that access the integrated data views. Thus, requirements R1 and R2 are often not well addressed by data integration products. It is also unclear whether the benefits of data integration outweigh its cost. The volume of exchanged data in a research collaboration is typically low compared with enterprise-level data stores. Frequently, a simple data transfer and processing strategy is sufficient and does not incur the overhead of creating and managing an integrated data view between collaborators.

*Distributed Resource Management* solutions, or DRMs, include workload balancers and batch management systems like OpenPBS [14] and Platform Computing's Load Sharing Facility (LSF) [15], as well as grid solutions such as the Globus Grid Toolkit [16]. While these systems are very useful for the utilization of computational resources, they do not provide the capability to define pipelines based on data availability or modification. These systems could be used as part of a pipeline to distribute the computational task across multiple nodes, but do not themselves meet the requirements discussed above.

*Event-based technologies* are commonly used in conjunction with user interfaces and user-oriented systems [17, 18], but event-based approaches have also been used for distributed systems [19-21]. All of these systems function in a similar manner: events are generated in response to some action and are processed by an event handler for that particular event. Event handlers may generate additional events or might update files, database tables, or a user's display. Event-based approaches provide a firm foundation for building computational pipelines by linking together various events, but with the exception of Metis [22] have not been widely utilized for that purpose. Many of the requirements defined in Table 1 are very data-centric, and it may not be apparent to users how these can be met by the finer granularity event-based approach. However, it should be noted that this approach offers a great deal of promise due to its flexibility and adaptability.

While none of these approaches independently addresses all the requirements in Table 1, based on our analysis we felt that designing DASH using an event-based model offered the best tradeoff between functionality and usability. Our approach to ameliorate the complexities of the event model was to layer a data flow representation on top of the more granular event model.

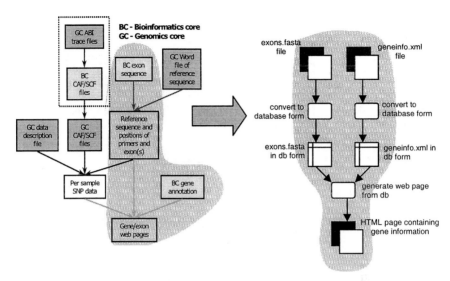

Figure 2. Illustration of the pipeline constructed to automate data processing within the Bioinformatics Core component of the PMT project. The shaded section on the left represents the portion of PMT data processing encapsulated by the data flow diagram shown on the right. This data flow is then used to automate the generation of web pages from source data.

## 5. Workflow within the PMT

The diagram on the left in Figure 2 is a high-level representation of the workflow between two of the components in the PMT project. (See reference [5] for a detailed description.) The PMT, with its complex interactions between heterogeneous, distributed resources, well represents the many small- to medium-sized collaborative science projects for which we have designed DASH. In Figure 2, we identify a subsection (shaded) of the PMT processing pipeline from which we isolated a simple data processing activity that generates web pages from data files. From this existing process we extracted a simple, two-staged pipeline. The image on the right illustrates this pipeline utilizing a standard set of data flow diagram symbols adapted from the Gane and Sarson method of process notation [23]. Using this PMT pipeline as an example, we implemented a simple proof-of-concept prototype of DASH that monitors a file or database table and triggers a protocol to run whenever the associated data source changes. As depicted in Figure 2, this two-staged pipeline creates a web page containing gene information anytime a new *exons.fasta* or *geneinfo.xml* file appears.

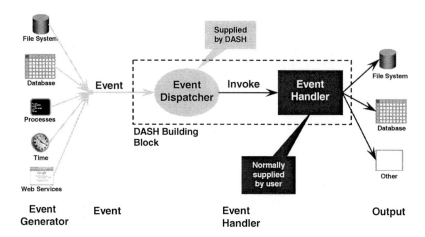

Figure 3. Overview of components in DASH's event model.

## 6. The DASH Event Model

Science can be viewed as an event-driven process; the development of a new hypothesis, availability of new data in a public database, or generation of experimental results are all examples of important events in the scientific research environment. Some of these events originate from human actors such as fellow researchers, while others occur as the result of automated processes. Each one of these events may warrant any number of follow-up actions: experiments may need to be designed to test the hypothesis, automated protocols could be invoked to process the novel data, or the laboratory results could be compared against previous experiments. Note that any of these activities could, in turn, generate more events that will require additional processing, and so on. In general terms, this process consists of three core concepts: actions that generate events, the events themselves and attached data, and handlers that process events.

Even the simple pipeline shown in Figure 2 evokes some of the potential complexity of automated data monitoring and processing. The two input branches of the prototype pipeline run independently, each updating its own set of database tables. An update to either table triggers the web page generation protocol with sufficient data to produce a web page. The two inputs can be thought of as having a Boolean OR relationship. Within the PMT, this is not always the case; some data sources have a Boolean AND relationship. That is, a step in the processing pipeline requires updated data from two or more data sources before it can run. Therefore, a useful data automation system must also handle a group of data sources having an AND relationship. These relationships suggest the concept of data groups as opposed to individual sources. Taken to

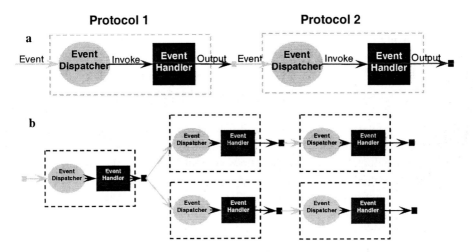

Figure 4. (a) Building blocks of DASH's event model; (b) Several building blocks linked together form a complex event processing pipeline.

the next level, this analysis implies the requirement of nested groups to support arbitrarily complex data processing scenarios.

Figure 3 shows an overview of all of the components in the DASH event model. The Event Dispatcher is a component of DASH that monitors Event Generators for Events and subsequently dispatches these Events to their respective handlers. The dotted line labeled "DASH Building Block" encompasses one processing unit of the DASH event management architecture.

Figure 4a shows two event processing units. The role of an individual event-processing unit is simple; an event arrives from DASH's event dispatcher, and the handler is invoked. This unit need not have any knowledge of where the event came from, which other handlers may process the same event, or what types of events may be generated as the result of the event handler invocation. This restricted local view keeps the conceptual model simple, while providing a high level of flexibility and applicability to known and unforeseen problem domains.

The real power in this model comes from associating these simple event-processing units. Associations between individual units are implicitly established when events generated from the action of one unit's event handler are dispatched to the event handler of the second. DASH provides a registration interface that allows users to specify what types of events each Event Generator is capable of producing, as well as which Events should be dispatched to which Handlers. The linking of processing units can result in complex relationships between multiple event generators and handlers, as illustrated by the branching pipeline in Figure 4b.

## 7. The DASH Data Flow Layer

While events and handlers provide a sound foundation upon which to build, from the researcher's perspective it is advantageous to view the interactions in terms of the more familiar concepts of data and processing protocols; these can be represented using data flow notation. Figure 5a introduces a new view of the DASH building block using data flow notation. Conceptually similar to the event model building block introduced in Figure 3, this construct represents the functional unit in the context of data flow. D1 represents a data store and P1 a protocol that is invoked in response to changes in that data store. Within this simple system, there is only one execution pattern: D1 is altered and P1 is invoked to process the changed data.

These building blocks can be aggregated to form complex pipelines of data and processing protocols, as shown in Figure 5b. While representing relationships in terms of data and protocols (as opposed to events, generators, and handlers) is more suitable for a data sharing application, the underlying implementation still uses the event model to propagate changes throughout the system. A change in data store E1 generates a data change event. Protocols P1 and P2 are registered as handlers for data change events in E1, and are invoked by the event dispatcher to process the changed data. Similarly, the actions taken by protocols P1 and P2 (e.g. writing information into a database table) could result in the generation of further data change events from data stores D1 and D2, respectively. This process of event generation and event consumption thus

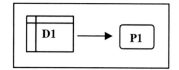

Figure 5a.  Data flow building block.

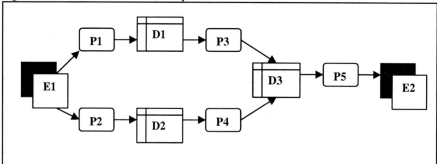

Figure 5b.  Complex data flow diagram using building blocks.

drives the propagation of data updates throughout a processing pipeline.

In the context of data flow, it is sometimes advantageous to process events in an order different from that in which they are generated. Reordering event processing enables DASH to maintain data integrity while optimizing use of resources such as data access and processor time. The end result of this event-processing model is to propagate all data updates throughout the system in the most efficient manner possible while still maintaining an internally consistent set of data. Events can be combined into a series which can then be optimized to allow for the efficient processing of large volumes of data by running protocols in parallel whenever possible.

Our DASH infrastructure, consisting of an underlying event model that is optimized for efficiency and ease of use in a data-sharing context, addresses several of the requirements given in Table 1. Using a preexisting body of distributed data stores (R5) and heterogeneous processing protocols, DASH can automate the propagation of updates through the system in order to provide a consistent set of data for use in subsequent processing steps (R4), or for consumption by collaborators or end-users (R3).

## 8. DASH Applied to the PMT

Figure 6 shows a page from the PMT Website generated using an expanded version of the prototype described in section 5 above. The text accompanying the arrows gives the source files for the data. The data used to generate this single Web page comes from at least five different files—some of them in XML format, others in fasta format, and still others in plain text. Changes in any of the source files will result in the automatic regeneration of the Web page by DASH.

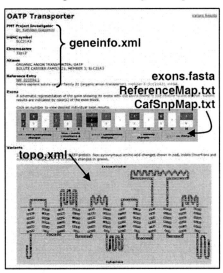

## 9. Current Status

DASH is currently in the early development stage. A simple prototype has been implemented and applied to the PMT project to obtain the results described in this

Figure 6. Sample PMT web page containing data about the exons of a transporter gene and showing the data files used to generate the HTML representation.

manuscript, and further implementation work is ongoing. As DASH becomes more functional, our intent is to make the software available at no cost and in documented source code form so that other research groups can directly benefit from our work. Further information on DASH is available from http://www.cgl.ucsf.edu/Research/DASH/.

## 10. Future Work

There are two areas that we will be focusing on for our future implementations of DASH. The first of these is the development of a Web-based user interface that will allow researchers to create, modify, and monitor DASH processing pipelines. By enabling researchers to graphically manipulate active processing pipeline components, we can address requirement R2. The ability to discover, display, and print relationships among the processing pipelines of collaborating researchers allows us to address requirement R1.

The second area for future work is the extension of DASH to cross organizational boundaries. This often requires that the tools and libraries support some form of distributed computing model. For DASH, distributed computing will be supported through Web Services [24] interfaces. The use of Web Services allows us to leverage the security mechanisms already supported through web services technologies and provides a general interface between two DASH instances running in different computing environments. Web Services will also be used to support communication between DASH and other related systems that support a Web Services interface. DASH will act as a Web Services endpoint as well as a Web Services client. One additional use of Web Services will be to export administrative information for the discovery and status of DASH pipelines (bounded, of course, by security restrictions). This will be used to present a broader view of processing pipelines across multiple computing environments (addresses requirements R3 and R5).

## Acknowledgments

The UCSF Pharmacogenetics of Membrane Transporters project is sponsored by NIH grant U01 GM61390. Support for the DASH project comes from NIH P41-RR01081.

## References

1. Models of Team Science (www.becon1.nih.gov/symposia_2003/BECON2003_sessionIV.ppt).
2. BECON 2003 Symposium on Catalyzing Team Science (www.becon1.nih.gov/symposium2003.htm).

3. Zerhouni, E., *The NIH Roadmap*. Science, 2003. **203**: p. 63-64,72.
4. NIH Roadmap Initiatives (http://nihroadmap.nih.gov/initiatives.asp).
5. Stryke, D., et al. *SNP analysis and presentation in the Pharmacogenetics of Membrane Transporters Project*. in *Pacific Symposium Biocomputing*. 2003: World Scientific.
6. Hewett, M., et al., *PharmGKB: the Pharmacogenetics Knowledge Base*. Nucleic Acids Res., 2002. **30**(1): p. 163-165.
7. Leabman, M.K., et al., *Natural variation in human membrane transporter genes reveals evolutionary and functional constraints*. Proc Natl Acad Sci U S A, 2003. **100**(10): p. 5896-901.
8. IBM Lotus Workflow (http://www.lotus.com/).
9. Oracle Workflow 11i (http://www.oracle.com/).
10. BEA Weblogic Integration (http://www.bea.com/framework.jsp).
11. Stevens, R.D., A.J. Robinson, and C.A. Goble, *myGrid: personalised bioinformatics on the information grid*. Bioinformatics, 2003. **19 Suppl 1**: p. I302-I304.
12. Oinn, T., et al., *Taverna: a tool for the composition and enactment of bioinformatics workflows*. Bioinformatics, 2004.
13. Vivtek Inc., *wftk (Workflow Toolkit)*. 2003.
14. Portable Batch System (http://www.openpbs.org).
15. Platform LSF (http://www.platform.com/products/LSF/).
16. Foster, I. and C. Kesselman, *Globus: A Metacomputing Infrastructure Toolkit*. International Journal of Supercomputer Applications, 1997. **11**(2): p. 115-128.
17. Document Object Model (DOM) Level 2 Events Specification (http://www.w3.org/TR/2000/REC-DOM-Level-2-Events-20001113/).
18. Jacob, R.J.K., L. Deligiannidis, and S. Morrison, *A software model and specification language for non-WIMP user interfaces*. ACM Transactions on Computer-Human Interaction (TOCHI), 1999. **6**(1).
19. Carzaniga, A., D.S. Rosenblum, and A.L. Wolf, *Design and Evaluation of a Wide-Area Event Notification Service*. ACM Transactions on Computer Systems, 2001. **19**(3): p. 332-383.
20. Cugola, G., E. Di Nitto, and A. Fuggetta. *Exploiting an event-based infrastructure to develop complex distributed systems*. in *20th international conference on Software engineering*. 1998.
21. Ben-Shaul, I.Z. and G.E. Kaiser. *A paradigm for decentralized process modeling and its realization in the Oz environment*. in *16th international conference on Software engineering*. 1994.
22. Anderson, K.M., et al. *Metis: Lightweight, Flexible, and Web-based Workflow Services for Digital Libraries*. in *2003 Joint Conference on Digital Libraries*. 2003. Houston, Texas: IEEE.
23. Gane, C.P. and T. Sarson, *Structured Systems Analysis: Tools and Techniques*. 1979: Prentice Hall. 241.
24. Web Services Activity (http://www.w3.org/2002/ws/).

# THRESHOLD GRADIENT DESCENT METHOD FOR CENSORED DATA REGRESSION WITH APPLICATIONS IN PHARMACOGENOMICS*

J. GUI AND H. LI

*Department of Statistics and Rowe Program in Genetics
University of California, Davis, CA 95616, USA
E-mail: jgui@ucdavis.edu, hli@ucdavis.edu*

An important area of research in pharmacogenomics is to relate high-dimensional genetic or genomic data to various clinical phenotypes of patients. Due to large variability in time to certain clinical event among patients, studying possibly censored survival phenotypes can be more informative than treating the phenotypes as categorical variables. In this paper, we develop a threshold gradient descent (TGD) method for the Cox model to select genes that are relevant to patients' survival and to build a predictive model for the risk of a future patient. The computational difficulty associated with the estimation in the high-dimensional and low-sample size settings can be efficiently solved by the gradient descent iterations. Results from application to real data set on predicting survival after chemotherapy for patients with diffuse large B-cell lymphoma demonstrate that the proposed method can be used for identifying important genes that are related to time to death due to cancer and for building a parsimonious model for predicting the survival of future patients. The TGD based Cox regression gives better predictive performance than the $L_2$ penalized regression and can select more relevant genes than the $L_1$ penalized regression.

## 1. Introduction

With the sequencing of the human genome near completion and with the development of high-throughput technologies, we are now able to obtain information about an individual's entire genome or the entire genomic profiles of a tumor. Very high-dimensional genetic and genomic data are being generated in pharmaceutical industries and in biomedical and clinical research. Examples of high-throughput data include whole genome wide SNP data, microarray-based gene expression data and proteomic data. Traditional environmental risk factors such as diet, age and lifestyle can influ-

---

*This work is supported by NIH grantES09911

ence a person's response to treatments, it is believed that understanding an individual's genetic makeup or individual tumor's genomic profiles would provide the key for explaining such variation and for creating personalized drugs with greater efficacy and safety. For example, with the DNA microarray technology, one can simultaneously measure expression profiles for thousands of genes in cancer tissues, which offers the possibility of a powerful, genome-wide approach to the genetic basis of different types of tumors. Recent studies [1,2] have demonstrated great succuss in predicting cancer class using the gene expression data. Different classes of cancer may correspond to different clinical outcomes of a given treatment. In addition, studies also demonstrated that additional predictive power can be obtained by incorporating genomic information in addition to the traditional predictive factors such as tumor grades, sizes and stages [2].

In pharmacogenomics, an important area of research is to relate the high-dimensional genetic, genomic or proteomic data to various phenotypes such as continuous drug response levels, response to treatment which can be categorical or censored clinical outcomes such as time to cancer recurrence or death after treatment. Due to large variability in time to cancer recurrence among cancer patients, studying possibly censored survival phenotypes can be more informative than treating the phenotypes as binary or categorical variables. Since the follow up time is limited, some patients' exact survival time can't be measured. For those patients, we only have their right censored survival time. The emphasis of this paper is to develop methods for predicting patient's time to clinical event using high-dimensional genetic or genomic data.

The most popular method in regression analysis for censored survival data is the Cox regression model [3]. However, due to the very high dimensional space of the predictors, e.g., the genes with expression levels measured by microarray experiments, the standard maximum Cox partial likelihood method cannot be applied directly to obtain parameter estimates. There are mainly two solutions to this problem. One approach is based on dimension deduction such as singular value decomposition (SVD) or the partial Cox egression (PCR) [4]. The other approach is to use the penalized partial likelihood. This includes $L_2$ and $L_1$ penalization. Li and Luan [5] was the first to investigate the $L_2$ penalized estimation of the Cox model in the high-dimensional low-sample size settings and applied their method to relate the gene expression profile to survival data. As pointed out by Li and Luan [5], one limitation of $L_2$ penalization is that it uses all the genes in the prediction and does not provide a way of selecting relevant genes.

An alternative is to use the $L_1$ penalized estimation, which was proposed by Tibshirani [6] and was called the least absolute shrinkage and selection operator (Lasso). Using newly developed least angle regression (LARS) by Efron et al. [7], Gui and Li [8] proposed an efficient way to estimate $L_1$ penalized Cox regression model, which was called the LARS-Lasso procedure. One limitation of the LARS-Lasso procedure is that the number of genes selected cannot be greater than the sample size. In addition, if there is a group of variables among which the pair-wise correlations are very high, the LARS-Lasso procedure tends to select only one variable from the group. This could be a potential limitation when the goal is to select all important genetic or genomic features which are related to the risk of a clinical event.

Friedman and Popescu [9] have recently proposed a stepwise optimization method called threshold gradient descent (TGD) and have demonstrated it application in linear regression and classification problems. They showed that with different threshold value, TGD can approximate the estimates of partial least square, ridge regression, Lasso and LARS. The TGD based methods provide a data-driven approach for selecting the penalty function. Friedman and Popescu [9] further demonstrated that with small threshold, the TGD method approximates the PLS or ridge estimates and has better predictive power for data simulated with small variability in true coefficients. On the other hand, the TGD with large threshold value can approximate the Lasso or LARS estimates, which provide better predictive performance when there is high variability of the coefficients. In this paper, we extend the TGD method to censored survival data and to the Cox regression model. We demonstrate the method by analyzing a real data set of diffuse large B-cell lymphoma (DLBCL) survival times and gene expression levels [2]. Finally, we give a brief discussion of the methods and conclusions.

## 2. Statistical Models and Methods

### 2.1. *Cox proportional hazards model*

Suppose that we have a sample size of $n$ from which to estimate the relationship between the survival time and the genetic/genomic profiles such as the gene expression levels $X_1, \cdots, X_p$ of $p$ genes. Due to censoring, for $i = 1, \cdots, n$, the $i$th datum in the sample is denoted by $(t_i, \delta_i, x_{i1}, x_{i2}, \cdots, x_{ip})$, where $\delta_i$ is the censoring indicator and $t_i$ is the survival time if $\delta_i = 1$ or censoring time if $\delta_i = 0$, and $x_i = \{x_{i1}, x_{i2}, \cdots, x_{ip}\}'$ is the vector of the genetic/genomic profiles of $p$ genes for the $i$th sample. Our aim is to build the following Cox regression model for the hazard of cancer recurrence or death at time $t$

$$\lambda(t) = \lambda_0(t) \exp(\beta_1 X_1 + \beta_2 X_2 + \cdots + \beta_p X_p)$$
$$= \lambda_0(t) \exp(\beta' X), \tag{1}$$

where $\lambda_0(t)$ is an unspecified baseline hazard function, $\beta = \{\beta_1, \cdots, \beta_p\}$ is the vector of the regression coefficients, and $X = \{X_1, \cdots, X_p\}$ is the vector of genetic/genomic profiles with the corresponding sample values of $x_i = \{x_{i1}, \cdots, x_{ip}\}$ for the $i$th sample. We define $f(X) = \beta' X$ to be the linear risk score function.

Based on the available sample data, the Cox's partial likelihood [3] can be written as

$$PL(\beta) = \prod_{k \in D} \frac{\exp(\beta' x_k)}{\sum_{j \in R_k} \exp(\beta' x_j)},$$

where $D$ is the set of indices of the events (e.g., deaths) and $R_k$ denotes the set of indices of the individuals at risk at time $t_k - 0$. Our goal is to find the coefficient vector $\beta$ which minimizes the negative log partial likelihood function. However, in the settings when $p >> n$, there is no unique solution to this optimization problem. In addition, one expects the high variability of the estimates based on different random samples drawn from the population distribution. A common remedy is to regularize this optimization problem by addition a penalty $\lambda P(\beta)$ to the negative partial likelihood, i.e.,

$$\hat{\beta}(\lambda) = \operatorname{argmin}_\beta \{-\log PL(\beta) + \lambda P(\beta)\}. \tag{2}$$

Here the negative log partial likelihood is treated as a loss function $l(\beta)$ which we want to minimize. The most popular penalties include the $L_2$ penalty where $P(\beta) = \sum \beta_j^2$ and the $L_1$ penalty where $P(\beta) = \sum |\beta_j|$.

## 2.2. The threshold gradient descent algorithm

As observed in Friedman and Popescu [9], for a given penalty function $P(\beta)$, the procedure represented by (2) produces a family of estimates, $\hat{\beta}(\lambda)$, each is indexed by a particular value of the tuning parameter $\lambda$. This family lies on a one-dimensional path of finite length in the $p$ dimensional space of all joint parameter values $\beta$. Model selection procedure such as cross-validation (see next section) can be use for selecting a point (i.e., a $\lambda$ parameter) on that path. Different penalty $P(\beta)$ therefore corresponds to different path. To account for different possible true values of $\beta$, Friedman and Popescu [9] further suggested a gradient directed path finding algorithm for estimating $\beta$. Specifically, let $l(\beta) = -\log PL(\beta)$, $\eta = X\beta$, and define

$$\mu_i = -\partial l/\partial \eta_i = \delta_i - \exp(\eta_i) \sum_{k \in C_i} \frac{d_k}{\sum_{j \in R_k} \exp(\eta_j)},$$

where $C_i = \{k : i \in R_k\}$ denotes the risk sets containing individual $i$ and $d_k$ is the number of events at time $t_k$. Then $\mu = (\mu_1, \cdots, \mu_n)$ is the negative gradient of the loss function with respective to $\{\eta_1, \cdots, \eta_n\}$. The negative gradient with respective to $\beta$ is therefore $g = -\partial l/\partial \beta = X\mu$. Starting from $\hat{\beta} = 0$, the gradient directed paths can be updated as

$$\hat{\beta}(\nu + \Delta\nu) = \hat{\beta}(\nu) + \Delta\nu h(\nu),$$

where $\Delta\nu > 0$ is an infinitesimal increment and $h(\nu)$ is the direction in the parameter space tangent to the path evaluated at $\hat{\beta}(\nu)$. This tangent vector at each step represents a descent direction. In order to direct the path towards parameter points with diverse values, Friedman and Popescu [9] suggested to define $h(\nu)$ as

$$h(\nu) = \{h_j(\nu)\}_1^p = \{f_j(\nu) \cdot g_j(\nu)\}_1^p,$$

where

$$f_j(\nu) = I[|g_j(\nu)| \geq \tau \cdot max_{1 \leq k \leq p}|g_k(\nu)|],$$

where $I[.]$ is an indicator function, and $0 \leq \tau \leq 1$ is a threshold parameter that regulates the diversity of the values of $f_j(\nu)$; larger values of $\tau$ lead to more diversity [9]. $g(\nu)$ is the negative gradient evaluated at $\beta = \hat{\beta}(\nu)$.

For any threshold value $0 \leq \tau \leq 1$, the threshold gradient decent path finding algorithm for the Cox model involves the following five steps,

(1) Set $\beta(0) = 0$, $\nu = 0$.
(2) Calculate $\eta$, $\mu$, $g(\nu) = -\partial l/\partial \beta$ for the current $\beta$.
(3) Calculate $f_j(\nu) = I[|g_j(\nu)| \geq \tau \cdot max_{1 \leq k \leq p}|g_k(\nu)|]$
(4) Update $\beta(\nu + \Delta\nu) = \beta(\nu) + \Delta\nu \cdot g(\nu) \cdot f(\nu)$, $\nu = \nu + \Delta\nu$.
(5) Repeat 2-4. Cross validation (see next section) is then employed to determine a point on the path $(\nu)$ and to terminate the iterations.

### 2.3. Model selection through the cross validated partial likelihood

For a given gradient threshold $\tau$, to determine the value of the tuning parameter $\nu$ in the final model, one can choose $\nu$ which minimizes the cross-validated partial likelihood (CVPL), which is defined as

$$CVPL(\nu) = -\frac{1}{n}\sum_{i=1}^{n}[l(\hat{f}^{(-i)}(\nu)) - l^{(-i)}(\hat{f}^{(-i)}(\nu))],$$

where $\hat{f}^{(-i)}(\nu)$ is the estimate of the score function based on the threshold gradient descent with tuning parameter $\nu$ from the data without the $i$th subject. The terms $l(f)$ and $l^{(-i)}(f)$ are the log partial likelihoods with all the subjects and without the $i$th subject, respectively. The optimal value of $\nu$ is chosen to maximize the sum of the contributions of each subject to the log partial likelihood. When the threshold $\tau$ is unknown, we can perform a two-dimensional parameter search using CVPL for $\tau$ and $\nu$ simultaneously.

## 3. Application to prediction of survival time of patients with DLBCL

To demonstrate the utility of the TGD based Cox regression in relating genomic data to censored survival phenotypes, we re-analyzed a recently published data set of DLBCL by Rosenwald et al. [2]. This data set includes a total of 240 patients with DLBCL, including 138 patient deaths during the followups with median death time of 2.8 years. Rosenwald et al. divided the 240 patients into a training set of 160 patients and a validation set or test set of 80 patients and built a multivariate Cox model. The variables in the Cox model included the average gene expression levels of smaller sets of genes in four different gene expression signatures together with the gene expression level of BMP6. It should be noted that in order to select the gene expression signatures, they performed a hierarchical clustering analysis for genes across all the samples (including both test and training samples). In order to compare our results with those in Rosenwald et al., we used the same training and test data sets in our analysis. In this data set, the gene expression measurements of 7,399 genes (not unique since many genes were spotted multiple times on the arrays) are available for analysis.

Table 1.   Number of features selected for different threshold value $\tau$

| Threshold value ($\tau$) | 0.0 | 0.2 | 0.4 | 0.6 | 0.8 | 1.0 |
|---|---|---|---|---|---|---|
| Number of non-zero coefficients | 7399 | 1171 | 464 | 140 | 43 | 4 |

### 3.1. Effects of the threshold value $\tau$

Table 1 shows for several threshold $\tau$ values, the number of coefficients estimated to be non zero for the training set of 160 patients. For each threshold value $\tau$, we obtain the optimal predictive point along the path using a 10-fold CVPL. As expected, larger values of $\tau$ give rise to fewer non-zero coefficient values, with only 4 out of 7399 features having any influence on survival for $\tau = 1.0$. Figure 1 shows the coefficient values obtained for the DLBCL training data using threshold gradient descent with threshold values $\tau = 0, 0.4, 0.6$ and 1, sorted by the gene order in the original data set of Rosenwald *et al.* All solutions produce relatively large absolute coefficient values in similar regions, with larger values of $\tau$ selecting fewer non-zero values within each one. In addition, we clearly observed that smaller $\tau$ resulted in smaller absolute coefficients of all the genes, and larger $\tau$ resulted in very different estimates of the coefficients among all the genes.

### 3.2. *Predictive performance*

In order to assess how well the model predicts the outcome, we employ the idea of time dependent receiver-operator characteristics (ROC) curve for censored data and area under the curve (AUC) as our criteria. These methods were recently developed by Heagerty *et al.* [10] in the context of the medical diagnosis and were proposed as criteria for censored data regression with microarray gene expression data [5,4]. Note that larger AUC at time $t$ based on a score function $f(X)$ indicates better predictability of time to event at time $t$ as measured by sensitivity and specificity evaluated at time $t$. Figure 2 (a) shows the areas under the ROC curves for different threshold values $\tau$ based on 10-fold cross-validation for the training data set. This plot suggested that the model with $\tau = 0.4, 0.6$ and 0.8 gave the best predictive results as measured by the areas under the curves. Since the model with $\tau = 0.8$ includes fewer genes in the model, one should choose this model for future prediction. Figure 2 (b) shows the areas under the ROC curves based on the predicted scores for the patients in the testing data set. This plot indicates that $\tau = 0.8$ and $\tau = 1.0$ gave almost the same predictive performance, which implies that the DLBCL data set encourages variability among the coefficients of different genes. The AUCs are between 0.66 and 0.68 in the first 10 years of followups, indicating a reasonable predictive performance. In contrast, smaller values of $\tau$, which correspond approximately to $L_2$ penalized estimation, gave much lower values of AUCs, indicating worse predictive performance.

To further examine whether clinically relevant groups can be identified by the model, we divided the patients in the test data into two groups based

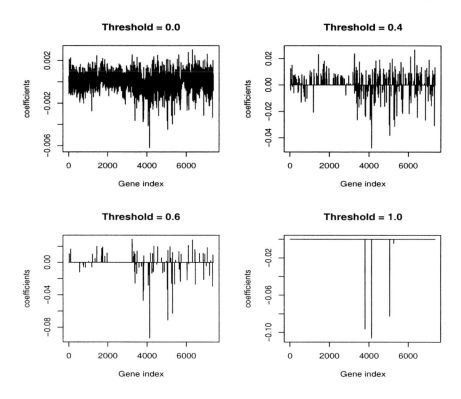

Figure 1. Coefficient values obtained for the DLBCL data using threshold gradient descent with threshold values $\tau = 0, 0.4, 0.6$ and 1, sorted by the gene order in the original data set of Rosenwald *et al.* All solutions produce relatively large absolute coefficient values in similar regions, with larger values of $\tau$ selecting fewer non-zero values within each one.

on their estimated risk scores $\beta' X$ in the Cox model (1) using the mean score as a cutoff value. Figure 3 shows the Kaplan-Meier curves for the two groups of patients, showing very significant difference ($p$-value=0.0004) in overall survival between the high risk group (36 patients) and low risk group (44 patients). Similar analysis was done using $L_2$ penalization, less significant difference was observed ($p$-value=0.003).

### 3.3. *Genes identified*

As shown previously, cross-validation analysis for the training data set suggested that threshold value $\tau = 0.8$ should result in better predictive perfor-

Table 2. Genes that were identified to be related to the risk of death when $\tau = 0.8$.

| Gene ID | Group | Description |
|---------|-------|-------------|
| AA286871 | | tumor necrosis factor receptor superfamily, member 17 |
| AI361763 | | UDP-Gal:betaGlcNAc beta 1,4- galactosyltransferase, polypeptide 2 |
| AA213564 | PS | ribosomal protein S21 |
| AA443696 | PS | ribosomal protein S21 |
| AA714637 | PS | ribosomal protein S12 |
| AA714637 | PS | ribosomal protein S12 |
| AA837360 | | proapoptotic caspase adaptor protein |
| AA760674 | | COX15 homolog, cytochrome c oxidase assembly protein (yeast) |
| AI087048 | | interferon regulatory factor 4 |
| W72411 | | tumor protein p63 |
| H92332 | MHC | major histocompatibility complex, class II, DQ alpha 1 |
| AA411017 | MHC | major histocompatibility complex, class II, DQ alpha 1 |
| AA159668 | MHC | major histocompatibility complex, class II, DQ alpha 1 |
| AA411017 | MHC | major histocompatibility complex, class II, DQ alpha 1 |
| AA729055 | MHC | major histocompatibility complex, class II, DR alpha |
| AA032179 | MHC | major histocompatibility complex, class II, DR beta 5 |
| AA714513 | MHC | major histocompatibility complex, class II, DR beta 5 |
| AA729003 | | T-cell leukemia/lymphoma 1A |
| R97095 | | T-cell leukemia/lymphoma 1A |
| AA480985 | GCB | Weakly similar to germinal center expressed transcript |
| AA805575 | GCB | Weakly similar to A47224 thyroxine-binding globulin precursor |
| AA278822 | | Fc receptor-like protein 1 |
| AA485725 | | immunoglobulin kappa constant |
| AA487453 | PS | GRO2 oncogene |
| AA598653 | LNS | osteoblast specific factor 2 (fasciclin I-like) |
| LC-29222 | LNS | |
| AA495985 | LNS | small inducible cytokine subfamily A (Cys-Cys), member 18 |
| H98765 | LNS | cytochrome P450, subfamily XXVIIA, polypeptide 1 |
| AA579913 | LNS | leukocyte immunoglobulin-like receptor, subfamily B, member 1 |
| R62612 | LNS | fibronectin 1 |
| X14420 | LNS | collagen, type III, alpha 1 |
| W87899 | | aryl hydrocarbon receptor |
| AI370252 | | T cell receptor beta locus |
| AA147638 | | T cell receptor beta locus |
| AA147638 | | T cell receptor beta locus |
| N29376 | | myeloid cell nuclear differentiation antigen |
| AA833786 | | hemoglobin, alpha 2 |
| W63749 | | B-cell CLL/lymphoma 2 |
| AA495936 | | microsomal glutathione S-transferase |
| H44867 | | mal, T-cell differentiation protein |
| AA292532 | | regulator of G-protein signalling 16 |
| AA804793 | | ESTs |
| AA243583 | | KIAA0084 protein |

mance. For $\tau = 0.8$, 43 non-unique genes have non-zero coefficients. Table 2 lists the gene IDs and their descriptions for these 43 genes. Note some genes appear more than once due to replicates on arrays. These genes are

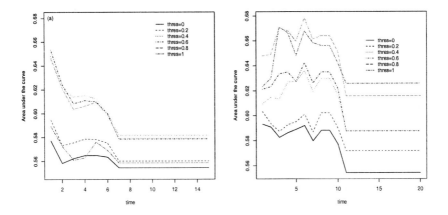

Figure 2. ROC curves based on (a) the 10-fold cross validated scores based on the training data set and (b) the estimated scores for the 80 patients in the testing data set for different gradient threshold values $\tau$.

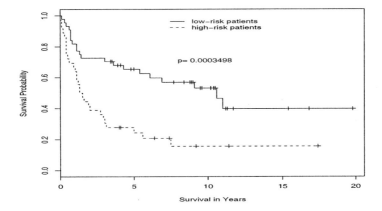

Figure 3. The Kaplan-Meier curves for the high and low risk groups defined by the estimated scores for the 80 patients in the test data set.

related to the risk of death among the DLBCL patients. It is interesting to note that many of these 43 genes belong to the four signature groups defined by Rowsenald *et al.* using clustering analysis of genes. The four groups are MHC class II (MHC), Proliferation signature (PF), Lymph node signature (LNS) and Germinal center B-cell signature (GCB). The genes in these groups were shown to be related to the risk of death for the DLBCL patients. The estimated coefficients for the genes in MHC, GCB and LNS signature groups were all negative except for AA495985, indicating that high expression levels of these genes reduce the risk of death among the patients with DLBCL. All the genes in the PF group have positive coefficients. This agrees with what Rosenwald *et al.* (2002) has found. In addition, since many genes were spotted several times on the arrays, it is interesting that these genes were all selected as predictors. Other genes which do not belong to the four signature groups include T-cell receptor beta locus, T-cell leukemia/lymphoma 1A and T-cell differentiation protein.

As a comparison, when $\tau = 1.0$, only four genes (H92332, LC_29222, AA480985, H98765) were selected by the model. These four genes belong to three of the four signature groups and have large absolute coefficients. This is in direct contrast with the genes selected by the TGD Cox regression with $\tau = 0.8$, where genes which are highly correlated were also selected.

## 4. Discussion and Conclusions

In pharmacogenomics, one important problem is to predict patient's time to cancer relapse or time to death due to cancer after treatment using genomic profiles of the cancerous cells prior to the treatment. Powerful statistical methods for such prediction allow for high-dimensional genomic data such as microarray gene expression data to be used most efficiently. In this paper, we have extended the threshold gradient descent based method [9] for censored survival data in order to identify important predictive genes for survival using high throughput genetic or genomic data. Since the risk of cancer recurrence or death due to cancer may result from the interplay between many genes, methods which can utilize data of many genes, as in the case of our proposed method, are expected to show better performance in predicting risk. We have demonstrated the applicability of our methods by analyzing time to death of the diffuse large B-cell lymphoma patients and obtained satisfactory results, as evaluated by both applying the model to the test data set and the time dependent ROC curves. Our simulations also indicated better predictive performance of the proposed method than other penalization methods (results not shown due to page limitation).

As we observed in our analysis of the DLBCL data set, if there is a group of variables or genes among which the pairwise correlations are very high, the procedure with $\tau = 1$ tends to select only one variable from the group and does not care which one is selected. Such procedure, which is very closely related to Lasso, can give a good predictive performance. However, for genes sharing the same biological pathways, the correlations among them can be high. If the goal is to select important and relevant genes, one may want to include all these highly correlated genes, if one of them is selected. We observed that the TGD procedure with $\tau = 0.8$ precisely achieved this goal. In addition, we may expect more robust prediction since the gene expression levels of highly correlated genes are used in the model. Therefore, in practice, one should use cross validation to select the optimal threshold value $\tau$.

Some possible extensions of the proposed methods are worth mentioning. One is to study the treatment effect adjusting for genetic or genomic profiles. This can be done by including a treatment indicator variable in the Cox regression model. However, regularization by threshold gradient is only applied to the variables related the high-dimensional genomic profiles. A profile-type penalization can then be developed. Another interesting problem is to identify certain genetic profiles which may interact with the treatment in determining the risk of certain clinical event. This can be done by including all the treatment by gene interactions in the model and using the threshold gradient descent method to select the relevant genes and their interactions. Both approaches deserve further investigation.

In summary, the proposed threshold gradient descent method for the Cox model can be very useful in pharmacogenomics in building a parsimonious predictive model of risk based on the genetic or genomic profiles. The procedure is robust and numerically stable and can also be applied to select important genes that are related to patients' survival outcome.

## References

1. Golub, T.R., Slonim, D.K., Tamayo, P., *et al.* 1999 *Science* 286,531-537.
2. Rosenwald, A., Wright, G., Chan, W., *et al.* 2000 *The New England Journal of Medicine* 346,1937-1947.
3. Cox, D.R. 1972. *Journal of the Royal Statistical Society. Series B* 34,187-220.
4. Li, H. and Gui, J. 2004. *Bioinformatics* 20, i208-215.
5. Li, H. and Luan, Y. 2003. *Pacific Symposium on Biocomputing* 8,65-76.
6. Tibshirani, R. 1995. *Journal of Royal Statistical Society B* 58,267-288.
7. Efron B., Johnston I., Hastie T. and Tibshirani R. 2004. *Annals of Statistics.*
8. Gui J. and Li H. 2004. *Technical report, UC Davis.*
9. Friedman, J.H. and Popescu B.E. 2004. *Technical Report, Statistics Department, Stanford University.*
10. Heagerty, P.J., Lumley, T. and Pepe, M. 2000. *Biometrics* 56,337-344.

# PARAMETERIZATION OF A NONLINEAR GENOTYPE TO PHENOTYPE MAP USING MOLECULAR NETWORKS

JEAN PECCOUD, KENT VANDER VELDEN

*Pioneer Hi-Bred Int'l, Inc. 7250 NW 62$^{nd}$ Ave, Johnston, IA 50131-0552, USA*

*E-mail: jean.peccoud@pioneer.com , kent.vandervelden@pioneer.com*

Mathematical models of networks of molecular interactions controlling the expression of traits could theoretically be used as genotype to phenotype (GP) maps. Such maps are nonlinear functions of the environment and the genotype. It is possible to use nonlinear least square minimization methods to fit a model to a set of phenotypic data but the convergence of these methods is not automatic and may lead to a multiplicity of solutions. Both factors raise a number of questions with respect to using molecular networks as nonlinear maps. A method to fit a molecular network representing a bistable switch to various types of phenotypic data is introduced. This method relies on the identification of the model stable steady states and the estimation of the proportion of cells in each of them. By using environmental perturbations, it is possible to collect time-series of phenotypic data resulting in a smooth objective function leading to a good estimate of the parameters used to generate the simulated phenotypes.

## 1 Introduction

Pharmacogenomics' ambition is to relate a phenotype, the effect of a drug, to the genotype of patients exposed to environmental conditions partly defined by the drugs they receive [1]. For a geneticist this project requires building a genotype to phenotype map (GP map) of drug effects. Mathematically, a GP map is a function f such as $phenotype = f(genotype, environment)$. It maps into a phenotypic space, the product of a genetic space generated by the genetic diversity in a population by the space of environmental conditions to which individuals of this population can be exposed [2]. The simplest GP map is the one upon which relies Mendelian genetics. The function is Boolean indicating the presence or absence of a character. The environment is ignored and genes are considered independent of each other. Since most traits are quantitative and not binary, the genetics of quantitative traits relies on a more refined family of GP maps representing the phenotype as linear statistical models. In general multiple loci are assumed to contribute additively to the phenotype. In some cases terms representing digenic interactions are introduced. The effect of the environment on the phenotype is generally decomposed into an additive term and a genotype by environment term [3].

Just like complex interactions between multiple genetic loci generate a diversity of phenotypes for pathologies that were considered monogenic [4], responses to drugs are generally considered multigenic traits [1, 5]. Many of the genetic determinants controlling the response to drugs have been identified by a candidate-gene approach relying on the understanding of the molecular mechanisms of the drug action and metabolism. Integrating into a mathematical model the network of molecular interactions affecting the response to a drug is therefore an attractive avenue to build the GP map.

Using different approaches, a number of authors have recently demonstrated that it is possible to build mathematical models to predict the phenotype controlled by small artificial gene networks [6-10], larger natural networks [11, 12], or even genome-wide metabolic pathways [13, 14]. In order to use a mathematical model as a GP map it is necessary to bridge the molecular and population-levels views of the genotype-phenotype relationship. When using mass-action models of molecular interactions, it has proved possible to analyze the genetic properties of a molecular network by associating genetic polymorphism with discrete kinetic values of the parameter of each interaction [15]. The possibility of determining the kinetic parameters of each interaction is key to using molecular networks as GP maps.

One way to estimate the GP map parameters is to find a set of parameters minimizing the difference between the phenotype predicted by the model and the observed phenotype. Since the phenotype is a nonlinear function of the parameters, this problem can be addressed by using a nonlinear least-square approach [16, 17]. Nonlinear minimization methods are iterative algorithms that require a set of starting parameter values to converge to a local solution. Different starting values can result in different solutions with different quality of fit. This limitation has the potential to prevent a unique determination of the map parameters. The topology of the molecular network model and the experimental design both contribute to shape the objective function being minimized. The number and geometry of its local minima determines the possibility to find and identify solutions corresponding to the actual parameters' values that generated the set of observed phenotypes. Since for many real molecular networks, it is not possible to explore the entire parameter space, it is possible that no starting parameter values will converge toward the actual parameter set. It is also possible that many starting values will result in many solutions with similar fits making it impossible to distinguish the solutions closest to the actual parameter set. Few authors used nonlinear least-square minimization to estimate GP map parameters [12, 18] and it is likely that a number of people attempted this without success and never published these negative results.

This paper introduces an algorithm to estimate the parameters of a molecular network from time-series of molecular phenotypes collected after an environmental perturbation. The objective function used takes into consideration the possibility that phenotypic data collected at the cell population level result from a random distribution of the cells among multiple stable-steady states. The presence of a positive feed-back loop [19] creates the possibility of multistationarity. Multiple steady states have been observed in artificial gene networks [8-10, 20, 21] but also in natural regulatory networks [11], for which this possibility had not been considered even recently [12].

The algorithm considered in this article is automatic and can be applied to virtually any mass action model of molecular networks without requiring any manual mathematical derivation.

## 2 Methods

### 2.1 Model

The model used in this article is a mass action equivalent of a model of a bi-stable switch [9, 20, 21]. In the list of reactions below, $G_i$ and $GX_i$ refer to the active and inactive forms of the $i^{th}$ gene coding for the protein $P_i$, respectively, while $L_i$ represents the $i^{th}$ ligand and $PX_i$ the $i^{th}$ protein complexed with its ligand.

$$
\left.
\begin{aligned}
R_1 &: G_1 \xrightarrow{k_1} G_1 + P_1 \\
R_2 &: G_2 \xrightarrow{k_2} G_2 + P_2
\end{aligned}
\right\} \text{Gene expression}
$$

$$
\left.
\begin{aligned}
R_3 &: P_1 \xrightarrow{k_3} \varnothing \\
R_4 &: P_2 \xrightarrow{k_4} \varnothing
\end{aligned}
\right\} \text{Protein degradation}
$$

$$
\left.
\begin{aligned}
R_5, R_6 &: G_2 + 2P_1 \xrightleftharpoons[k_6]{k_5} GX_2 \\
R_7, R_8 &: G_1 + 2P_2 \xrightleftharpoons[k_8]{k_7} GX_1
\end{aligned}
\right\} \text{Repression}
$$

$$
\left.
\begin{aligned}
R_9, R_{10} &: P_1 + L_1 \xrightleftharpoons[k_{10}]{k_9} PX_1 \\
R_{11}, R_{12} &: P_2 + L_2 \xrightleftharpoons[k_{12}]{k_{11}} PX_2
\end{aligned}
\right\} \text{Repressor-ligand interaction}
$$

$$(1)$$

The time-evolution of the model is represented by mass-action differential equations. The set of coupled differential equations can automatically be derived from the chemical equations Eq. (1) [15].

Mass conservation relationships can be used to eliminate some variables from the model. Assuming that there is only one copy of each of the two genes in the system, the first mass-conservation relation makes it possible to eliminate the repressed forms of the genes. We also assume that the interaction between the small molecules representing the environment and the repressors are much faster than the other reactions. Using a quasi-steady state approximation, we eliminate $R_9$ to $R_{12}$ from the model. This results in the list of reaction rates below where $\mathbf{X}$ is the vector representing the state of the system and $r_i$ the rate of the reaction $R_i$:

$$\text{with } \mathbf{X} = \begin{pmatrix} G_1 \\ P_1 \\ P_2 \\ G_2 \end{pmatrix}, \left\{ \begin{array}{ll} r_1(\mathbf{X}) = k_1 G_1 & r_5(\mathbf{X}) = k_5 G_2 \left[\dfrac{1}{1+L_1} P_1\right]^2 \\ r_2(\mathbf{X}) = k_2 G_2 & r_6(\mathbf{X}) = k_6 (1-G_2) \\ r_3(\mathbf{X}) = k_3 P_1 & r_7(\mathbf{X}) = k_7 G_1 \left[\dfrac{1}{1+L_2} P_2\right]^2 \\ r_4(\mathbf{X}) = k_4 P_2 & r_8(\mathbf{X}) = k_8 (1-G_1) \end{array} \right. \tag{2}$$

The differential equation representing the time evolution of the system is derived from this list of reaction rates.

### 2.2 Numerical Identification of the Steady States

The most generic way of finding steady states is to find the solutions of Eq. (3) below. The notation below indicates that the reaction rates depend on the parameterization of the model, $\mathbf{K} = (k_1,...,k_8)$, and the environment, $\mathbf{E} = (L_1, L_2)$:

$$\frac{d\mathbf{X}}{dt} = \mathbf{F}(\mathbf{X},\mathbf{E},\mathbf{K}) = \begin{pmatrix} r_8(\mathbf{X}) - r_7(\mathbf{X}) \\ r_1(\mathbf{X}) - r_3(\mathbf{X}) + 2(r_6(\mathbf{X}) - r_5(\mathbf{X})) \\ r_2(\mathbf{X}) - r_4(\mathbf{X}) + 2(r_8(\mathbf{X}) - r_7(\mathbf{X})) \\ r_6(\mathbf{X}) - r_5(\mathbf{X}) \end{pmatrix} = 0 \tag{3}$$

Roots can be determined by minimizing $\|\mathbf{F}(\mathbf{X})\|$ starting from any point in the model state space. Since Eq. (3) is nonlinear, it is not possible to analytically find its solutions. In order to alleviate this limitation, a grid of starting points is created in a region of the state space expected to include all the biologically relevant steady states of the model.

Variables corresponding to conserved molecules are bounded by the initial conditions. Assuming that each gene in the model has a single copy, then $0 \le G_i \le 1$ with $i = 1,2$. The asymptotic values of the non-conserved molecules, i.e. proteins in this case, is somewhere between 0 and $k_{production}/k_{degradation}$, the asymptotic value corresponding to the maximum expression of the gene. Therefore, in the case of the model considered here, all the steady states are expected to be within $V = [0,1] \times [0, k_1/k_3] \times [0, k_2/k_4] \times [0,1]$.

It is therefore possible to regularly sample $V$ with a user-specified resolution. By starting the minimization algorithm from each point in this grid, a numerical solution to Eq. (3) will generally be found for each starting point. Numerical errors and differences of convergence toward the same limits will result in minor numerical differences between solutions reached from different starting conditions. If the distance between a solution and another previously found solution is less than some specified value, it is assumed that they are identical.

After the scan of $V$ is complete, the stability of the steady states is analyzed by computing, at the steady state, the eigenvalues of the Jacobian matrix associated to Eq. (3). If the real parts of all eigenvalues are negative, then the steady state is stable.

## 2.3    Fitting to Asymptotic Phenotypes

In the context of this article, "asymptotic phenotypes" refers to phenotypic data collected in the stationary regime [12, 22] in different environments $E_j$ with $j = 1,...,v$. Since in general, all variables of the model cannot be observed, the number of data points collected in each environment $\mu$ is less than $M$ the total number of state-variables of the model. It is convenient to represent asymptotic phenotypes as a $\mu X v$ matrix $\mathbf{P}$. Now that the experimental data set is structured, it is necessary to generate a predicted phenotype $\mathbf{Q}(\mathbf{K})$ corresponding to a given set of parameters $\mathbf{K}$. Assuming that it is possible to compute $\mathbf{Q}(\mathbf{K})$, then the least-square distance that needs to be minimized to fit the model to the phenotypes, $d(\mathbf{K},\mathbf{P})$, is:

$$d(\mathbf{K},\mathbf{P}) = \sum_{i=1}^{\mu}\sum_{j=1}^{v}\left[Q_i\left(K,E_j\right) - P_i\left(E_j\right)\right]^2 \qquad (4)$$

Computing the predicted phenotype for a specified environment and set of parameters is immediate if they result in a single stable steady state $\mathbf{S}$. In this case: $\mathbf{S}_i\left(\mathbf{K},\mathbf{E}_j\right) = \mathbf{Q}_i\left(\mathbf{K},\mathbf{E}_j\right)$ $i = 1,...,\mu$ $j = 1,...,v$.

In conditions where the model has two stable steady states $\mathbf{S}$ and $\mathbf{T}$, then the observed phenotype $\mathbf{P}$ is likely to result from a distribution of cells in the two steady states. So, instead of having a direct correspondence between the predicted phenotype and the observed phenotype, the predicted phenotype is a weighted average of the two stable steady states. What is not known, though, is the proportion of cells in each of the steady states. This proportion needs to be estimated by solving a linear constrained least-square problem:

$$\mathbf{Q}(\mathbf{K},\mathbf{E}) = \min_{\alpha \in [0,1]}\left(\alpha\mathbf{S}(\mathbf{K},\mathbf{E}) + (1-\alpha)\mathbf{T}(\mathbf{K},\mathbf{E}) - \mathbf{P}(\mathbf{E})\right) \qquad (5).$$

This approach can be generalized to more than two stable steady states.

## 2.4    Fitting to a Time Series of Phenotypes

Observing the model state variables at different points in time is a natural way of collecting data characterizing the model dynamics [9, 20]. Many experimental designs can lead to this type of data. Only a single simple experiment is considered in this paper but it demonstrates that system multi-stationarity needs to be considered to properly analyze the data.

A cell population is placed in a first environment $E_1$ until it reaches a stationary regime indicated by the stabilization of the phenotype. An instantaneous perturbation is applied to the environment creating a new environmental condition $E_2$. Phenotypic data are recorded at different time points while the population stabilizes toward a new stationary regime. For instance, cells can be grown in absence of ligands. One of the ligands is added to the growth medium creating a new environment. Samples of cell culture are taken and phenotyped at different points in time after the ligand has been added. This design can be generalized to multiple environmental perturbations. $E_{1,j}$ and $E_{2,j}$ refer to first and second environments of the $j^{th}$ perturbation. The first phenotype of each time series is collected in the stationary regime before the perturbation is applied. All other phenotypes are collected in the second environment and are indexed by the instant of observation. Similarly, it is necessary to compute a series of predicted phenotypes corresponding to the series of experimental data. The distance between the predicted and the observed phenotypes is computed by summing the distance over all time-points:

$$d(\mathbf{K},\mathbf{P}) = \sum_{k=1}^{\tau}\sum_{i=1}^{\mu}\sum_{j=1}^{\nu}\left[\mathbf{Q}_i\left(\mathbf{K},\mathbf{E}_{2,j},t_k\right) - \mathbf{P}_i\left(\mathbf{E}_{2,j},t_k\right)\right]^2 \tag{6}$$

Let $G(\mathbf{X}_0,\mathbf{K},\mathbf{E},t)$ be the solution of Eq. (3) starting from $\mathbf{X}_0$. Computing the predicted phenotype for a specified environmental perturbation and set of parameters is immediate if the initial environment and parameter set result in a single stable steady state $\mathbf{S}(\mathbf{K},\mathbf{E}_1)$. In this case the predicted phenotypes are extracted from the solution of Eq. (3) starting at $\mathbf{S}(\mathbf{K},\mathbf{E}_1)$: $G(\mathbf{S}(\mathbf{K},\mathbf{E}_1),\mathbf{K},\mathbf{E}_2,t_k) = \mathbf{Q}_i(\mathbf{K},\mathbf{E}_{2,j},t_k)$. If the parameter set leads to two steady states in the initial environment $\mathbf{S}(\mathbf{K},\mathbf{E}_1)$ and $\mathbf{T}(\mathbf{K},\mathbf{E}_1)$, then it is possible to estimate the proportion $\hat{\alpha}$ just as in Eq. (5). The predicted phenotype would then be a weighted average of trajectories starting from the two initial conditions $\mathbf{S}$ and $\mathbf{T}$.

*2.5 Application*

The number of variables observed in the phenotype and the number of environments where the phenotypes are observed are likely to have a significant impact on the possibility to match the model with phenotypic data. So, phenotypic data were simulated in different numbers of environments and by recording different numbers of observed variables.

Twelve series of phenotypic data were generated using the same set of parameters. The first 6 phenotypes were asymptotic phenotypes. The second group of 6 phenotypes were time series.

In both cases (asymptotic and time series), three of the phenotypes consisted in the observation of one protein, $P_1$. In the remaining three phenotypes the values of both proteins were recorded in the phenotype.

The asymptotic phenotypes were simulated in three different numbers of environments (3, 5, and 9 environments). Environments are represented by the concentrations of the two ligands, $(L_1, L_2)$. The first three environments were: (0,0), $(10^1, 0)$, and $(0, 10^1)$. In the 5 environments experiments, (1, 0) and (0, 1) were added to the first 3 environments. In the 9 environments experiments $(10^{-1}, 0)$, $(10^{-2}, 0)$, $(0, 10^{-1})$, and $(0, 10^{-2})$ were added to the five previous environments.

The times series phenotypes are transitions between two environments. In the first experiment, the transition from (10, 0) to (0, 10) was simulated. In the 2-transition experiment, the transition from (5, 0) to (0, 5) was added. In the 3-transition experiment, the transition from (1, 0) to (0, 1) was added to the two previous transitions.

The same set of 25 initial parameter values was used to fit the model to the asymptotic and time-series phenotypes resulting in a series of 300 optimizations.

## 3  Results

### 3.1  Numerical Identification of Steady States

The method to find the steady states of the model works well on this model. By using only the 8 "corners" of $V$, it seems that all the steady states of the system were found. Increasing the resolution of the grid did not result in a larger list of steady states. Depending on the environment and parameter values, two types of regimes were found: a single stable steady state or two stable steady states and one unstable steady state.

In the least-square minimization procedures, the specificity of this network made it possible to use only two initial conditions $\left(0.5, k_1/k_3, 0, 0.5\right)$ and $\left(0.5, 0, k_2/k_4, 0.5\right)$ to find the stable steady states of the system. This simplification speeds up the optimization process that often requires hundreds or even thousands of steady state determinations. These two initial conditions do not allow the identification of the unstable steady states of the system and this approach may not be applicable to other models.

A bifurcation diagram was generated by computing the steady states (stable and unstable) of the model over a range of $L_1$ concentrations in order to verify the steady state identification procedure while the concentration of the second ligand was kept at 0. The system is bi-stable for low concentrations of $L_1$ and beyond a critical concentration, the system becomes mono-stable. This result is consistent with the bifurcation diagram of a similar model [20] and also with our own bifurcation analysis run in XPP/AUT [23]. The positions of the stable steady states are not very much affected by the concentration of $L_1$, except in the vicinity of the critical concentration. This indicates the robustness of the phenotype to environmental perturbation.

## 3.2 Fitting to Asymptotic Phenotypes

An exploration of the neighborhood of the original set of parameters used to generate the phenotypes indicated that initial conditions very close to the original parameter set could not lead to a good fit (data not shown). This indicated that the objective function was rough and may be difficult to minimize. It turned out that convergence was much easier to achieve than initially anticipated. When the phenotype included the two protein concentrations a good fit was achieved for 1/3 of the initial conditions.

This can be explained by observing that an infinite number of parameterizations have the same steady states. Solutions of Eq. (3) verify:

$$r_8(\mathbf{X}) - r_7(\mathbf{X}) = k_8(1 - G_1) - k_7 G_1 \left[\frac{1}{1 + L_2} P_2\right]^2 = 0$$

$$r_1(\mathbf{X}) - r_3(\mathbf{X}) = k_1 G_1 - k_3 P_1 = 0$$

$$r_2(\mathbf{X}) - r_4(\mathbf{X}) = k_2 G_2 - k_4 P = 0 \tag{7}$$

$$r_6(\mathbf{X}) - r_5(\mathbf{X}) = k_6(1 - G_2) - k_5 G_2 \left[\frac{1}{1 + L_1} P_1\right]^2 = 0$$

In other words, the minimization problem defined by asymptotic phenotypes is unidentifiable. It is not possible to estimate the 8 kinetic parameters but only the 4 equilibrium constants.

## 3.3 Fitting to Time Series of Phenotypes

The convergence criteria used in this case was a root mean square of residuals less than $10^{-1}$. Using this criterion, 14 convergences were observed (9% of the 150 optimizations using time series phenotypes) that can be broken down into 13% of convergence when only one protein is observed and 5% when both proteins are recorded. These rates of convergence need to be confirmed by analyzing a larger number of initial conditions using a faster implementation of this algorithm. However, they are surprisingly high and indicative of a relatively smooth performance function.

All optimization solutions were indexed (not shown) for further analysis. In some cases very similar solutions were found. For instance solution 13 is very close to solution 14 and solution 11 is very close to solution 12. It is worth observing that if solutions 11, 13, and 14 all originated from the same initial condition, solution 12 was found using a different initial condition. Also solutions 11 and 12 are not very far in the parameter space from solutions 13 and 14. Solution 6 is also located in the same area. Interestingly, these 5 solutions are all very close to the original set of parameters used to generate the phenotype. The solutions were verified by plotting the time course of the two protein concentrations and the profiles are consistent with the objective function used to generate the solutions. Protein concentrations corresponding to solution 11 were plotted over a wide range of initial conditions. Visually they are indistinguishable from the plots generated by the original set of parameters (Figure 1).

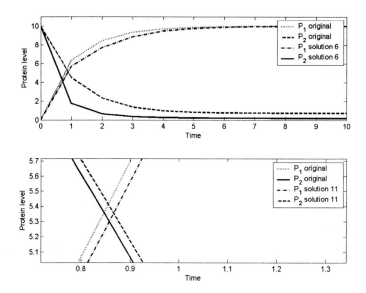

Figure 1: In order to visually assess the quality of the fit, the ODE was integrated using two solutions of the time-series optimization experiment and the original set of parameters used to generate the simulated phenotypes. The initial condition for the integration was set to (1, 0, 10, 0) and the environment to (0, 10). Solution 6 (top) was found when only one protein level was used in the phenotype. It is interesting to see that the fit for $P_1$ is better than the fit for $P_2$. The RMS computed using the two protein concentrations at the 11 time points is 0.83. Solution 11 (bottom) gives a very good fit of both of the protein expression profiles leading to a RMS of 0.06. It is necessary to zoom in on specific region of the plot to be able to visually distinguish the trajectories generated by the original parameter set and the trajectories generated by the parameters of Solution 11.

# 4 Discussion

## 4.1 Results

Even though this work focuses on a single molecular network model, results presented here are likely to be relevant to other models.

- The specific structure of molecular networks makes it possible to search for steady states in a limited volume of the model state space.
- The possibility of multi-stability should always be considered. In a population of cells observed in a stationary regime, cells can be randomly distributed between multiple steady states. Therefore, the measurement of a gene activity at the cell population level is a weighted average of the molecule concentrations corresponding to the different stable steady states of the model. For a given set of parameter values, different repartitions of the cells in the different steady states leads to different qualities of fit between the model parameterization and the observed phenotypes. In the context of this paper, a linear minimization step was introduced to find the repartition minimizing the distance between the model and the experimental data.
- Asymptotic phenotypic data can only lead to the determination of the equilibrium constants but not the kinetic constants.
- Environmental perturbations can be used to collect time-series of phenotypic data. The relaxation profile observed is a weighted average of trajectories originating from the different stable steady states in the first environment.

## 4.2 Necessary improvements of the algorithm

In order for this method to be used for routine analysis it will be necessary to address a few issues.

- The steady state finding algorithm needs to be systematically validated. In some cases very stiff parameter sets hampered the convergence of the steady state identification procedure. The reasons for this behavior need to be understood. Since the steady state identification algorithm is the bottleneck of the whole optimization process it is worth trying to improve it.
- Determining the stability of the steady states is also an important step of the algorithm. Numerical errors prevent an accurate determination of the stability in the vicinity of critical points. It is not clear what is the impact of this issue on the outcome of the minimization process. Limit cycles are not considered in this algorithm.
- The local optimization method described in this paper needs to be coupled to a global search strategy to explore the parameter space more systematically.
- In cases where the time of sampling cannot be controlled, it could be necessary to take the actual sampling time into consideration when fitting the model to the data.

- A random term representing the measurement error needs to be added to the phenotypic data. The effect of this term on the convergence of the least-square minimization should be characterized. The addition of an error term would transform the least-square minimization problem into a nonlinear regression problem that could lead to computing confidence intervals for the parameter estimates.

### 4.3 Research directions

We are working on a generalization of this algorithm to handle phenotypic data collected on a multiplicity of genotypes just like several environmental conditions have been considered in this paper. Along the same line, the current model assumes only one copy of each gene. Introducing a diploid genome with two homologous copies of each gene would require predicting the phenotype of heterozygous individuals, which requires developing a model of dominance at the parameter level. If only homozygous individuals are considered or a total dominance is assumed, the model would remain unchanged.

Geneticists have been building models of the genotype to phenotype relationship for traits of other organisms for more than a century. By deciphering networks of molecular interactions, they hope to be able to build nonlinear GP maps inspired by the mechanisms controlling the expression of complex traits. It is expected that these maps would capture epistatic interactions between the genetic determinants contributing to these traits. Such a map would help a plant breeder define more effective breeding strategies using molecular markers to manipulate alleles of genes contributing to trait variations or using transgene to introduce new sources of genetic variation, a human geneticist better understand how multiple genes can contribute to the development of a pathology, and pharmacogeneticists to customize a medication to the genotype of their patients. Mathematical methods, such as those described here, are needed to analyze molecular data. The next challenge may be to find ways of associating macroscopic phenotypes such as a patient response to a treatment, with the molecular data we collect and analyze.

### Acknowledgements

We are grateful to Mark Cooper, Chris Winkler, Dean Podlich, David Bickel, Bard Ermentrout and four anonymous reviewers for valuable comments and suggestions. This work would not have been possible without the support of Lane Arthur and Bob Merrill.

### References

1. W. E. Evans, M. V. Relling, *Nature* **429**, 464 (2004).

2. M. Cooper, S. C. Chapman, D. W. Podlich, G. L. Hammer, *In Silico Biol.* **2**, 151 (2002).

3. D. S. Falconer, T. F. C. MacKay, *Quantitative Genetics* (Longman Group Ltd., Harlow (U.K.), ed. 4, 1996).

4. D. J. Weatherall, *Nat. Rev. Genet.* **2**, 245 (2001).

5. W. E. Evans, H. L. McLeod, *N Engl J Med* **348**, 538 (2003).

6. J. Hasty, D. McMillen, J. J. Collins, *Nature* **420**, 224 (2002).

7. M. KAErn, W. J. Blake, J. J. Collins, *Annu. Rev. Biomed. Eng.* **5**, 179 (2003).

8. M. R. Atkinson, M. A. Savageau, J. T. Myers, A. J. Ninfa, *Cell* **113**, 597 (2003).

9. B. P. Kramer *et al.*, *Nat. Biotechnol.* (2004).

10. F. J. Isaacs, J. Hasty, C. R. Cantor, J. J. Collins, *Proc. Natl. Acad. Sci. U. S. A* **100**, 7714 (2003).

11. E. M. Ozbudak, M. Thattai, H. N. Lim, B. I. Shraiman, A. van Oudenaarden, *Nature* **427**, 737 (2004).

12. Y. Setty, A. E. Mayo, M. G. Surette, U. Alon, *Proc. Natl. Acad. Sci. U. S. A* **100**, 7702 (2003).

13. M. W. Covert, E. M. Knight, J. L. Reed, M. J. Herrgard, B. O. Palsson, *Nature* **429**, 92 (2004).

14. K. R. Patil, M. Akesson, J. Nielsen, *Curr. Opin. Biotechnol.* **15**, 64 (2004).

15. J. Peccoud *et al.*, *Genetics* **166**, 1715 (2004).

16. D. M. Bates, D. G. Watts, *Nonlinear regression analysis and its applications* (John Wiley & Sons, New-York, 1988).

17. J. E. Dennis, R. B. Schnabel, *Numerical methods for unconstrained optimization and nonlinear equations* (Prentice Hall, Inc., Englewood Cliffs, NJ, 1983).

18. M. Ronen, R. Rosenberg, B. I. Shraiman, U. Alon, *Proc. Natl. Acad. Sci. U. S. A* **99**, 10555 (2002).

19. D. Thieffry, R. Thomas, *Pac. Symp. Biocomput.* 77 (1998).

20. T. S. Gardner, C. R. Cantor, J. J. Collins, *Nature* **403**, 339 (2000).

21. R. N. Tchuraev, I. V. Stupak, T. S. Tropynina, E. E. Stupak, *FEBS Lett.* **486**, 200 (2000).

22. C. C. Guet, M. B. Elowitz, W. Hsing, S. Leibler, *Science* **296**, 1466 (2002).

23. B. G. Ermentrout, *Simulating, Analyzing, and Animating Dynamical Systems: A Guide to Xppaut for Researchers and Students* (SIAM, ed. 1, 2002).

# EXPLORATORY VISUAL ANALYSIS OF PHARMACOGENOMIC RESULTS

DAVID M. REIF[1,2], SCOTT M. DUDEK[2], CHRISTIAN M. SHAFFER[2], JANEY WANG[2], JASON H. MOORE[1,†]

[1]*Computational Genetics Laboratory, Department of Genetics,*
*Dartmouth Medical School, Lebanon, NH 03756*
[2]*Center for Human Genetics Research,*
*Vanderbilt University Medical School, Nashville, TN 37232*
*{reif, dudek, shaffer, wang}@chgr.mc.vanderbilt.edu, jason.h.moore@dartmouth.edu*

Comprehensive analysis of expansive pharmacogenomic datasets is a daunting challenge. A thorough exploration of experimental results requires both statistical and annotative information. Therefore, appropriate analysis tools must bring a readily-accessible, flexible combination of statistics and biological annotation to the user's desktop. We present the Exploratory Visual Analysis (EVA) software and database as such a tool and demonstrate its utility in replicating the findings of an earlier pharmacogenomic study as well as elucidating novel biologically plausible hypotheses. EVA brings all of the often disparate pieces of analysis together in an infinitely flexible visual display that is amenable to any type of statistical result and biological question. Here, we describe the motivations for developing EVA, detail the database and custom graphical user interface (GUI), provide an example of its application to a publicly available pharmacogenomic dataset, and discuss the broad utility of the EVA tool for the pharmacogenomics community.

## 1. Introduction

### 1.1. *Visualizing results*

Recent years have seen an explosion in the sheer volume of data generated by modern experimental methods. Analysis of pharmacological results can be a maze of complex spreadsheets and arbitrary statistical significance thresholds. Visualization is a proven solution to this challenge of scale. In his work on visualizing quantitative information, Tufte states that "the most effective way to describe, explore, and summarize a set of numbers—even a very large set—is to look at pictures of those numbers. Furthermore, of all methods for analyzing

---

† Address for correspondence:
706 Rubin Bldg, HB 7937
One Medical Center Dr.
Dartmouth Hitchcock Medical Center
Lebanon, NH 03756, USA
jason.h.moore@dartmouth.edu.

and communicating statistical information, well-designed data graphics are usually the simplest and at the same time the most powerful" [1]. In what would otherwise be a sea of numbers, analysis tools must organize and display results so that readily distinguishable patterns emerge.

## 1.2. *Other analysis tools currently available*

Other analysis tools have been developed in recent years, such as DAVID [2], FatiGO [3], GoMiner [4], GoSurfer [5], GOTree Machine [6], and Onto-Express [7]. Each of these tools fills a specific niche in the community of genomic analysis methods. However, there are significant limitations that must be addressed.

While the tree-like representations common to many of the aforementioned tools are suited to the hierarchical structure of the Gene Ontology, such a display cannot present the volume of simultaneous information necessary to see otherwise hidden patterns emerge. Many current applications are designed for a single type of input (*e.g.* only GenBank Accession IDs), rather than compatibility across multiple platforms. The sluggishness of many of the available web-based applications can make exploratory analysis frustratingly slow. Many of the current applications take raw data as input, which leaves the user dependent on the statistical tests built into the package, rather than having the flexibility to perform any sort of analysis. This is fundamentally different from a results-based tool that gives the user freedom to implement any type of analysis. Another major concern with exploratory tools is the lack of mechanisms to replicate findings. These gaps in the functionality of other tools currently available leave a critical void in genomic analysis.

## 1.3. *The motivation for developing EVA*

The Exploratory Visual Analysis software (EVA) was developed to address the limitations of other approaches to analysis of genomic results. Combining flexibility, speed, and visualization of both statistical and annotative information into a single package, EVA fulfills a crucial role in comprehensive pharmacogenomic analysis.

From its inception, EVA was intended to be flexible across a wide range of research goals—allowing a truly exploratory analysis. The software can take as input any kind of statistical result(s) for any number of experiments. The user is thus free to use any statistic of choice or to define a custom statistic, rather than be limited by those implemented in the software. EVA's graphical results display can be organized into nested groupings for any combination of six biological categories: Gene Ontology (GO) [8], Biopath [9], Domain [10], Map Location [11], Chromosome [11], and Phenotype [12]. The statistical significance of particular subsets of categories or particular genes within a category can be assessed through permutation testing. To complement the statistical analysis, EVA links to multiple annotation sources via Locus Link [11]. This aspect affords immediate evaluation of the pharmacological relevancy of candidate genes or groups of genes. To ensure that the user can replicate findings, EVA incorporates a printable command log feature.

The speed of the EVA tool enhances its interactive flexibility. Because all of the data and links are loaded into memory upon opening the software, the user can switch seamlessly between annotative groupings, statistics, significance levels, and display modes. The permutation testing and reporting features provide an immediate complement to visual exploration.

Visualization is the aspect that binds the various components of EVA into a single coherent exploratory tool. Color choices, sliding significance scales, category display thresholds, and other display parameters are modifiable in real-time according to user preferences and/or analysis goals. EVA can be adjusted to display highly significant genes, marginally significant genes, or both, with cut-offs decided by the user. Most importantly, the graphical nature of EVA facilitates interpretation of multitudes of information simultaneously. This is because the human eye is acutely trained to identify pictorial patterns, while digesting tables of lifeless numbers is a taxing exercise. EVA translates numbers into pictures and *vice-versa*. Thus, graphical discoveries can be verified statistically, and statistical significance can be verified graphically. Famous examples such as Anscombe's Quartet [13] validate the notion that both types of information are essential for confident analysis. The ability of graphics to rationally condense vast amounts of information is essential for evaluating biological systems in which the concerted action of numerous contributing factors is the final determinant of phenotype.

Taken together, the diverse abilities of EVA allow the kind of comprehensive analysis necessary to answer complex pharmacological questions. Because pharmacological phenotypes are the product of myriad interacting factors, the annotative groupings and immediate expert-knowledge links provided by EVA are essential to understanding biological questions of a systemic nature [14]. The synergistic pieces of EVA coalesce into an analysis tool wherein the visual, numerical, and annotative components are mutually complementary. EVA takes analysis beyond spreadsheets of flat statistical results and into the realm of integrated analysis.

### 1.4. *Application of EVA to a pharmacogenomic dataset*

To demonstrate the power of EVA, we apply the software to the leukemia drug response dataset published by Cheok *et. al* [15]. The study measured the expression of 9,600 genes in leukemia cells using oligonucleotide microarrays before and after *in vivo* treatment with methotrexate and mercaptopurine given alone or in combination. The gene expression changes induced by these two widely-used antimetabolites are important for understanding the pharmacological action—including side-effects—of common chemotherapy agents, as well as for identifying potential treatment targets. The original authors' conclusions were based upon a number of statistical approaches.

Using only the results of three simple statistical tests as input, we demonstrate the ability of EVA to not only replicate but also extend the findings of the original authors, and to draw new biologically plausible conclusions. By grounding the entire exploratory analysis in biological relevancy, EVA renders moot the usual analytical step of constructing a pharmacologically relevant explanation from a collection of statistical significance values after the fact. Additionally, because EVA displays information graphically, we can see unifying patterns in the results, rather than a list of disjointed significance levels. This lends greater confidence to our conclusions, as correlated patterns are more robust than thousands of measurements taken individually [16]. The exploratory capabilities of EVA allow pharmacological discovery with or without *a priori* hypotheses—giving the user increased power to elucidate novel mechanisms or target pathways. Details of our analysis are given in Section 2.3, and the results are described in Section 3.

## 2. Methods

### 2.1. *Details of the EVA database*

The EVA database was designed with versatility, speed, and user-friendliness in mind. The user downloads the client, which provides a portal to the EVA web server via the custom GUI. The web server accesses the EVA database (stored on an Oracle server). This architecture provides ease of security, distributability, and expandability. Changes made by the database administrator pass seamlessly to the user through the EVA web service. Thus, once the user has downloaded the client, updates or expansions of the EVA modules are transparent from the user's perspective. For speed considerations, everything is stored in memory upon loading a particular experiment so that performance is not limited by demand on the web server. This aspect negates the query lag time typical of other analysis packages.

The EVA database schema is available upon request. Briefly, the gene identifier chosen by the user (Affymetrix ID, GenBank ID, or user/custom-defined gene number) links particular genes to the various EVA modules. At present, these modules are Locus Link, Wormbase (*C. Elegans*), and Linkage.

## 2.2. *Details of the EVA GUI*

The GUI is the portal through which the user manipulates the components of the EVA package (see Figure 1). Presently, the interface is written in Visual Basic, with a platform-independent Java version in production. A brief overview of features is provided here, and a comprehensive, illustrated help menu is included with the software. Upon opening the software, the user supplies a username and secure password. Login grants the user access to interfaces for various administrative tasks, including creating, updating, deleting, or loading experiments and results. Defining a new experiment involves deciding upon a descriptive name, choosing the type of gene identifier (Affymetrix, Locus Link, Wormbase, or Linkage), selecting the statistical tests used, and uploading the text file of results. Once a new experiment has been defined or an existing experiment selected, all of the results and links for that experiment are loaded for viewing on the user's desktop.

EVA displays genes organized according to the selected biological category (Chromosome, Biopath, Domain, Map Location, GO, or Phenotype). Each time a relevant parameter is changed, EVA dynamically resizes all boxes displayed to provide the most efficient use of screen space possible. Additionally, the categories can be re-organized by a second biological group, which will re-partition the results into new group boxes based upon the intersection of the two categories. The differential color display is a function of the significance threshold selected for a particular statistical test. The threshold range is modified via a slider. In order to customize the display, the user can alter both the color palette and number of colors into which the chosen range is divided.

The volume of information displayed on the screen can be modified in a number of ways. The display threshold sets the minimum number of results a group box must contain for it to appear in the display panel. Only groups exceeding this threshold are shown. Additionally, the 'Filter' tool will restrict the categories displayed to those containing the search text.

The 'Find' tool will search the currently selected biological group for the text in question. Once found, the title bar of that results box will be highlighted in yellow, and the display panel will scroll to that box. Clicking 'Find Next' will zoom to subsequent boxes containing the search text.

Hovering over any individual gene square with the mouse brings up a text box containing summary annotation for that gene. Right-clicking accesses a wealth of biological annotation for the selected gene through Locus Link.

EVA includes command log and reporting features. The log can be cleared or saved to file at any time. The reporting feature generates reports listing all results from a single group, from all groups currently on display, or from all groups in the current biological category. The 'Preferences' menu includes tabs for 'Log,' 'Reports,' and 'Web Service' options. The log options govern which types of information are stored in the log. The reports options determine the organization of the printable analysis reports and the type of file to which they are saved. The web service tab stores the EVA web service URL. Taken together, these features allow findings and/or the accompanying EVA parameters to be written to external files for further study. They also afford a mechanism for replicating findings, which is a vital quality in an exploratory analysis tool.

EVA uses a permutation testing strategy to assess the significance of the statistical results for a biological group. This feature complements visual inspection and provides statistical validation for the relative enrichment of particular groups. A permutation testing significance range can be selected by one of three methods: clicking on a particular gene in a particular group, selecting a range of significance colors for a particular group, or running a batch permutation on a selected significance range for all visible groups. The procedure is as follows:

1. Count the actual number of results within the selected significance range for a particular category.
2. Determine the total number of genes in that category.
3. Fill a box of the same size as that category with a randomly selected set of results.
4. Increment a total counter every time the number of randomly assigned results within the selected significance range is greater than or equal to the actual count from Step 1 above.
5. Perform 1,000 iterations of Steps 3 and 4.
6. Calculate the final p-value by dividing the total counter from Step 4 by 1,000.

This p-value represents the probability of obtaining the observed number of genes within the selected significance range in a given category by chance alone. In the current implementation, it is left to the user to account for multiple testing and the fact that certain annotation categories (*e.g.* GO) may have the same genes appear within several groups. The permutation testing result(s) can be saved in the log and written to an external file for resorting.

## 2.3. *Details of the application to a pharmacogenomic dataset*

The pharmacogenomic dataset to which we applied EVA is described fully in [15]. Affymetrix oligonucleotide arrays measured the expression levels of

302

9,600 human genes in acute lymphoblastic leukemia cells isolated from patients both before and after treatment with common chemotherapy agents. Patients

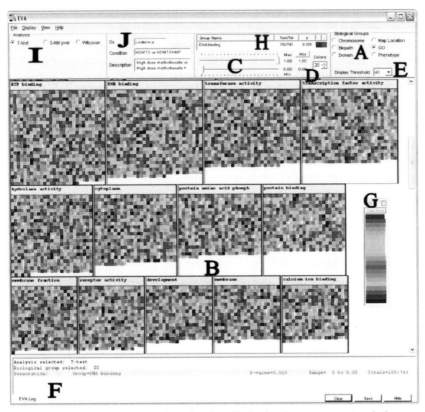

**Figure 1**. Details of the EVA graphical user interface. Each colored square represents a single gene. Visible features include: (A) – Biological group categories, (B) – Group boxes, (C) – Significance threshold color sliders, (D) – Number of color intervals displayed, (E) – Display threshold, (F) – Log, (G) – Significance range selector for permutation testing, (H) – Permutation testing results (also shown in log), (I) – Analysis types, (J) – Description of current experiment. In addition to options under the File, Display, View, and Help menus, right-clicking accesses a number of features not illustrated here.

were treated with either methotrexate ($n = 22$), mercaptopurine ($n = 12$), or a combination of both drugs ($n = 10$). This dataset provides the rare opportunity to measure the *in vivo* gene expression response of human patients to treatment with pharmacological agents.

The original authors' analysis involved a battery of statistical methods, including principal components analysis, hierarchical clustering, linear discriminant analysis, analysis of variance, support vector machines, and Fisher's exact test (for enrichment of selected GO groups). This approach, while grounded in statistical rigor, leaves the researcher to bridge the usual gap between statistical results and biological interpretation. This is exactly the role EVA was designed to fulfill.

We chose to demonstrate the power of EVA by feeding it the results of three basic statistical tests on the aforementioned dataset: the Student's t-test, the Wilcoxon Rank-Sum test, and the modified t-test implemented in the Significance Analysis of Microarrays (SAM) procedure. It would also have been possible to give EVA results from more complex statistical tests, such as those used in the original authors' analysis. Instead, we wanted to demonstrate the ability of EVA to translate widely-understood statistical results—of the type that do not require a thorough mastery of complex statistics—into plausible biological conclusions. For each of the three statistical tests, EVA was given the list of resultant p-values for all genes on the array. The p-values indicate the probability of obtaining the observed value of the test statistic for the given gene by chance alone. The statistical analyses were programmed in R version 1.8.1 [17]. The modified SAM t-test was implemented using the "siggenes" package, freely available from Bioconductor (www.bioconductor.org). Figure 2 depicts our statistical analyses and the treatment groups compared using EVA, which were: 1. methotrexate alone versus methotrexate plus mercaptopurine, 2. mercaptopurine alone versus methotrexate plus mercaptopurine, and 3. methotrexate alone versus mercaptopurine alone.

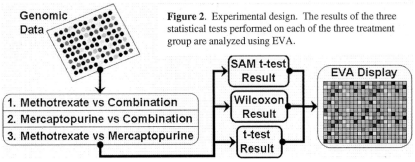

**Figure 2**. Experimental design. The results of the three statistical tests performed on each of the three treatment group are analyzed using EVA.

## 3. Results

### 3.1. *Replication of original authors' findings using EVA*

Using the results of the three basic statistical tests outlined above as input for EVA, we were able to replicate the major findings of the original authors. The original authors' main conclusions were that changes in gene expression are treatment-specific and that gene expression can illuminate differences in cellular response to drug combinations versus single agents. Through EVA's visualization of statistical results, these conclusions were immediately apparent.

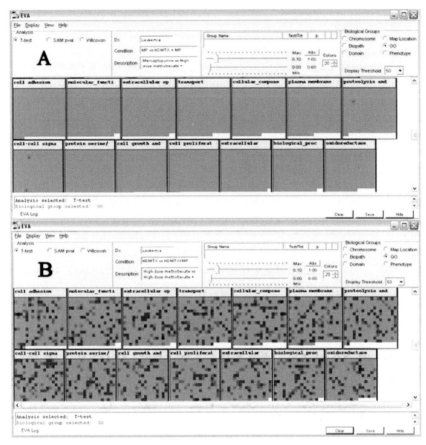

**Figure 3.** Visual inspection of statistical results for the treatment comparisons (A) – mercaptopurine versus combination and (B) – methotrexate versus combination. For each comparison, colored squares indicate p-values at or below the 0.05 (red) or 0.10 (orange) significance levels. The relative paucity of colored squares in (A) contrasted with the abundance in (B) indicates that there are fewer differentially expressed genes when statistically comparing mercaptopurine chemotherapy to combination chemotherapy.

The global patterns of gene expression were readily discernible between the three treatment groups, and the methotrexate plus mercaptopurine combination treatment exhibited an expression profile distinct from that of either agent given alone.

Demonstrating EVA's flexibility to incorporate any type of result into the analysis, we also took the list of genes highlighted by the original authors as input for the methotrexate alone versus combination chemotherapy comparison. Because EVA organized these genes into their annotation categories, we could instantly see where these individual genes fit into the broader biological picture.

EVA's reporting feature and links to the public annotation databases afforded one-stop evaluation of the biological relevancy of the genes in our list.

### 3.2. Novel findings using EVA

Drawing on exploratory visual, statistical, and annotative abilities of EVA enabled us to draw new biologically credible conclusions. Representative findings reached by starting at each of these three exploratory avenues are outlined below.

Upon visual inspection in EVA, it was immediately apparent that the gene expression pattern of the methotrexate alone treatment differs markedly from the combination chemotherapy, whereas the mercaptopurine alone treatment shows a gene expression pattern that was relatively closer to the combination chemotherapy (Figure 3). Corroborating visual evidence was provided by comparing the two treatments alone, where the global gene expression pattern resembled that of the methotrexate versus combination chemotherapy comparison. The next step called upon EVA's ability to back this conclusion statistically, and we found no genes significant at the 0.05 level and only a sparse few significant at 0.10 by any of the three statistical tests for the mercaptopurine versus combination comparison. This suggests that changes in gene expression induced by the combination chemotherapy are dominated by the action of mercaptopurine.

Starting with EVA's statistical capabilities, permutation testing for enrichment of biological categories with respect to the differences in gene expression patterns comparing methotrexate versus the combination treatment or mercaptopurine alone revealed a number of relevant biological findings. For example, there was a marked difference in expression of genes involved in cytoskeletal function. Statistically significant categories included the GO groups "epidermal differentiation" and "structural molecule activity," the Domain groups "kinesin like protein," "spectrin," "tubulin," and "gamma tubulin," and the Map Location 12q13, on which many of the genes in these categories are found. Keratins, a major component of hair follicles, are found throughout these annotative groups, which makes pharmacological sense because hair loss is one of the characteristic symptoms of methotrexate chemotherapy, though not mercaptopurine.

Incorporating expert-knowledge into the analysis, EVA's straightforward links to annotative information shed light on more interesting connections. For instance, because chemotherapy drugs affect the cell cycle, it is logical to look into that Biopath group. Sure enough, permutation testing showed the "cell cycle" group to be significantly enriched with respect to the number of genes differentially expressed at the 0.05 level of significance (Figure 4). Many of these genes also appear in the Domain groups "DNA topoisomerase," "G-protein beta subunit," and "NERF transcription factor," as well as the GO groups "DNA binding," "DNA topological change," "GTP binding," "GTPase activator activity," "kinase activity," and "nucleotide excision repair," all of which are significant or near-significant by permutation testing. Importantly, these relevant biological groups near the significance borderline would have been missed by a purely statistical analysis.

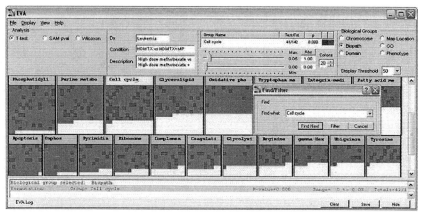

**Figure 4**. Permutation testing for significant enrichment of the Biopath group 'Cell cycle'. The Find tool was used to highlight the 'Cell cycle' group box. The permutation testing results are shown in the permutation testing results display and the log (see Figure 2). The probability of observing the given number of differentially expressed genes significant at the 0.05 level by chance alone is less than 0.001 for this group box.

## 4. Discussion

### 4.1. *Utility of EVA for pharmacogenomics*

As demonstrated by our results, EVA is adept at integrating multiple types of information to build cohesive biological conclusions supported by a variety of sources. This is vital in a field such as pharmacogenomics, where the cost of following false leads is prohibitively high. With EVA, the exploration of results can start down any of three avenues—visual, statistical, or annotative—to reflect the expertise or prior notions of the user. The various aspects of EVA are mutually complementary, and the flexibility, speed, and user-friendliness of the EVA interface allow users to move effortlessly between these three avenues.

EVA bridges the gap between raw statistical output and biological discovery, empowering the researcher to biologically validate statistical findings and to statistically test biological findings. The researcher can then plan the next experimental step by linking results to bodies of literature for particular genes.

## 4.2. *Future directions*

The development of EVA is an ongoing process. Future studies will incorporate results from machine learning and multivariate statistical methods. There are plans to integrate the tool with other publicly available data sources, including those for model organisms. Additionally, while EVA was developed for genomic applications, it can be naturally extended to genetic or proteomic analyses. A command line interface will allow programmable analyses and provide a mechanism to combine EVA with output from other analysis tools, such as sequence homology engines or alternative permutation testing strategies that address the issues of multiple testing and non-independence across tests. The new platform-independent Java version of the EVA GUI will be available at no cost to academic users. Contact the authors for distribution information.

## Acknowledgments

This work was supported by National Institutes of Health grants LM07613, HL68744, and AI59694. This work was also supported by generous funds from the Robert J. Kleberg, Jr. and Helen C. Kleberg Foundation.

## References

1. E. R. Tufte, *The Visual Display of Quantitative Information*, 2nd ed, (Graphics Press, Cheshire, CT, 2001).
2. G. Dennis, Jr. et al., *Genome Biol.* **4**, 3 (2003).
3. F. Al Shahrour, R. Diaz-Uriarte, and J. Dopazo, *Bioinf.* **20**, 578-580 (2004).
4. B. R. Zeeberg et al., *Genome Biol.* **4**, R28 (2003).
5. S. Zhong et al., *Applied Bioinf.* (*In press*), (2004).
6. B. Zhang, D. Schmoyer, S. Kirov, J. Snoddy, *BMC Bioinf.* **5**, 16 (2004).
7. S. Draghici et al., *Nucl. Acids Res.* **31**, 3775-3781 (2003).
8. M. A. Harris et al., *Nucl. Acids Res.* **32**, D258-D261 (2004).
9. M. Kanehisa, S. Goto, S. Kawashima, Y. Okuno, and M. Hattori, *Nucl. Acids Res.* **32**, D277-D280 (2004).
10. A. Marchler-Bauer et al., *Nucl. Acids Res.* **31**, 383-387 (2003).
11. K. D. Pruitt and D. R. Maglott, *Nucl. Acids Res.* **29**, 137-140 (2001).
12. McKusick-Nathans Institute for Genetic Medicine, *Online Mendelian Inheritance in Man*, National Center for Biotechnology Information, National Library of Medicine (2004).
13. F. J. Anscombe, *American Statistician* **27**, 17-21 (1973).
14. L. Hood, *Mech. Ageing Devel.* **124**, 9-16 (2003).
15. M. H. Cheok et al., *Nat. Genet.* **34**, 85-90 (2003).
16. P. O. Brown and D. Botstein, *Nat. Genet.* **21**, 33-37 (1999).
17. R. Ihaka and R. Gentleman, *J. Comp. Graph. Stat.* **5**, 3 (1996).

# STUDY OF EFFECT OF DRUG LEXICONS ON MEDICATION EXTRACTION FROM ELECTRONIC MEDICAL RECORDS

E. SIROHI, P. PEISSIG

*Marshfield Clinic Research Foundation, 1000 N. Oak Ave,*
*Marshfield, WI 54449, USA*
*sirohi.ekta@marshfieldclinic.org*

Extraction of relevant information from free-text clinical notes is becoming increasingly important in healthcare to provide personalized care to patients. The purpose of this dictionary-based NLP study was to determine the effects of using varying drug lexicons to automatically extract medication information from electronic medical records. A convenience training sample of 52 documents, each containing at least one medication, and a randomized test sample of 100 documents were used in this study. The training and test set documents contained a total of 681 and 641 medications respectively. Three sets of drug lexicons were used as sources for medication extraction: first, containing drug name and generic name; second with drug, generic and short names; third with drug, generic and short names followed by filtering techniques. Extraction with the first drug lexicon resulted in 83.7% sensitivity and 96.2% specificity for the training set and 85.2% sensitivity and 96.9% specificity for the test set. Adding the list of short names used for drugs resulted in increasing sensitivity to 95.0%, but decreased the specificity to 79.2% for the training set. Similar results of increased sensitivity of 96.4% and 80.1% specificity were obtained for the test set. Combination of a set of filtering techniques with data from the second lexicon increased the specificity to 98.5% and 98.8% for the training and test sets respectively while slightly decreasing the sensitivity to 94.1% (training) and 95.8% (test). Overall, the lexicon with filtering resulted in the highest precision, i.e., extracted the highest number of medications while keeping the number of extracted non-medications low.

## 1    Introduction

With the widespread use of computers in the healthcare domain, a large array of data – coded as well as free-text – is being stored digitally. Coded data can be easily interpreted by computer applications but free-text data poses a number of challenges[1]. Manual information extraction can be rather tedious and differences in style among providers means that document styles can vary widely. Added to that fact is the shear volume of clinical data that must be processed.

Natural Language Processing (NLP) has been showing promising results in solving this problem by extracting and structuring text-based biomedical or clinical information. To discover knowledge from free-text, researchers have been exploring NLP systems to facilitate Information Extraction (IE) and text mining. IE techniques allow users to automatically extract pre-defined information from free-text documents. Most NLP systems perform identification of terms in free text with entries from a lexicon[1]. Such dictionary-based entity name recognition studies extract information by searching the most similar (or identical) term in the dictionary to the target term. These extraction strategies have been used in other biological domains for extracting protein and gene names from biomedical literature[2-4]. The goal is to extract from the document salient facts about pre-specified types of events, entities or relationships. These facts are then usually used to populate clinical databases, which may then be used to analyze the data for trends. IE projects are currently being designed worldwide to summarize medical patient records by extracting diagnoses, symptoms, physical findings, test results and therapeutic treatments. Such systems can be used to assist health care providers with quality assurance studies or to support provider needs or to simply provide improved quality of service to patients.

## 2    Background and Related Work

It is a well established fact that different patients respond in different ways to the same medication. Previous studies have shown that genetics can account for 20 to 95 percent of variability in drug disposition and effects. While various non-genetic factors like age, organ function, drug interactions can  affect the response to medications, there are numerous cases in which differences in drug response have been attributed to genetic variations in genes encoding drug-metabolizing enzymes, drug transporters or drug targets[5]. Clinical observations of inherited differences in drug effects gave rise to the field of pharmacogenetics, which is the study of the hereditary basis for differences in a population's response to a drug[6]. While pharmacogenetics primarily focuses on the sequence variations in candidate genes suspected of affecting drug response, pharmacogenomics focuses on evaluation of the entire genome and the two terms are often used interchangeably[7]. Two main technologies are used to study the effect of genetic variations as basis for differences in drug response: genotyping and phenotyping. Genotyping is the study of the genetic variations while phenotyping is the study of observable physiological or biochemical measures.

The Marshfield Clinic Personalized Medicine Research Project (PMRP) (http://www.mfldclin.edu/pmrp) is an initiative to facilitate research in pharmacogenomics, epidemiology and population genetics. The primary goal of the project is to help researchers learn more about how genetic alterations cause diseases, how to use an individual's genetic information to predict which diseases he or she is likely to develop and which medications work best for a particular person. One of the objectives of PMRP is to develop a framework that has the ability to link genotype data with identified phenotypes.

The Marshfield Clinic is a fully integrated health care system that provides primary, secondary and tertiary care to patients living in Central and Northern Wisconsin. The Clinic has an integrated computerized system for automating the financial, practice management, clinical and real-time decision support processes and supports an electronic medical record that routinely captures clinical data like laboratory results, diagnosis, procedures, immunizations, vitals, etc. Medication information is currently stored as text in electronic clinical documents dating back to 1991. Although the medication information found in clinical notes is useful for patient care, it is not coded and has limited utility for research and computerized phenotyping. The amount of manual record abstraction needed to conduct pharmacogenomics studies is reduced significantly if coded medication information is available. Medication inventory and prescription systems are being deployed to capture coded medication information during a clinical visit. The current study is part of the initiative that aims to capture or extract medication information from historical clinical documents and convert it into a coded format that can be for research.

Automatic extraction of medications from free-text documents requires use of NLP systems. However, this application has various challenges because 1) new drugs are continually being created or older ones are renamed, 2) drug names are synonymous with other drug names, 3) drug names or its synonyms often have the same name as an English word, such as the drugs *Because* (a contraceptive) and *Duration* (nasal spray), and 4) terms in free text may be ambiguous and resolve to multiple senses, depending on the context in which they are used. Some of these challenges can be at least partially overcome by using good drug lexicons. A lexicon is a list of all the words used in a particular language or subject. In the current context, a drug lexicon contains a list of all the medications that we would like to extract from the text documents.

The lexicon-based approach is similar to previous dictionary-based studies that have extracted gene and protein names from literature[2-5,8], drug names and relationships from cancer literature[9] and studied indexing of entire documents by using special lexicons like UMLS[10]. However, our study differs from these works in that we are focusing on the automatic extraction of medication items from clinical documents for phenotypic development and the issues affecting performance of extraction, like the quality of the drug lexicon. We could not find any previous reports of medication extraction from clinical data.

FreePharma® (Language & Computing; http://www.landc.be) is a software that can automatically capture and structure medication information expressed in free-text natural language and link this information to existing drug databases. FreePharma® generates a structured XML representation of medication information derived from free-text documents, which can then be stored in databases for integration with host applications[11]. This product needs as input a drug lexicon containing a list of all medication items that must be extracted. This input lexicon is the largest factor in determining the extent and accuracy of extraction.

Marshfield Clinic uses the industry's most widely used source of up-to-date drug information, First DataBank's National Drug Data File (NDDF) Plus[TM], which delivers descriptive, pricing and clinical information on drugs, encompassing every drug approved by the Food and Drug Administration (FDA), over-the-counter drugs, plus information on herbals, nutraceuticals and dietary supplements[12]. We used data from this database to create the drug lexicons.

The purpose of the current study was to determine the best drug lexicon to use with FreePharma® to maximize extraction of prescription items, while reducing the extraction of non-medication items.

## 3  Methods

A sample of 52 documents was selected for this study by varying patients, service dates and providers to maximize the differences arising from varying styles of dictation and transcription. This convenience sample contained only clinic office visit notes and had been used in a pilot project prior to undertaking this study. This was the training set because the lexicon filtering techniques were refined based on results from this dataset. Each of the documents in this set contained at least one medication and this was verified by manual review. The manual review also yielded a list of 681 items (285 unique items) from the documents that were considered to be "true" medications. The documents were independently reviewed by a second reviewer to ensure that none of the valid medications were missed. These medications were collected from all sections of the documents and not just the discharge summary or medical history sections. The total number of terms in the documents was counted at 28496, including the medication items.

Based on FirstDataBank's NDDF data source, Marshfield Clinic's *drugs* database provides a *drug_name, short_name* and *generic_name* for each drug, in addition to information like American Hospital Formulary Service (AHFS) classification. We decided to use these columns for the purposes of creating a drug lexicon. Three sets of drug lexicons were prepared:

1. **Lexicon A**: unique terms from *drug_name* and *generic_name* columns of drugs database. Term count = 25907
2. **Lexicon B**: unique terms from *drug_name, generic_name* and *short_name* columns of drugs database. Term count = 29333
3. **Lexicon C**: unique terms from *drug_name, generic_name* and *short_name* columns of drug database followed by removal of items that met the filtering criteria in Table 1. Term count = 22345

**Table 1.** Filtering criteria for restricting non-medication terms in drug lexicon

| |
|---|
| Terms where AHFS classification is 'Devices', 'Dental Supplies' |
| Terms where generic_name is 'Organ Concentrates' or 'Homeopathic drugs' |
| Terms which contain only numerical values (such as '1', '3', etc.) |
| Terms that were ambiguous with general English words |

Lexicon C was created by applying the filtering criteria (Table 1) to lexicon B in two steps. First, we applied just the first three criteria to lexicon B to produce the interim version of lexicon C. These criteria were developed after a careful manual examination of results and terms from the use of the first lexicon. Second, to identify drug names that were ambiguous with general English language, we used a list of English words obtained from the SCOWL collection at SourceForge's wordlist website (http://wordlist.sourceforge.net/). This collection contained various wordlists and we analyzed the effectiveness of using them for our purposes. A wordlist was considered relevant for use if it contained the most frequently used words in general English and did not contain a high number of medication items. Based on this analysis, we combined the size 10, 20 and 35 "small" English lists to create a final list of 41,769 words. This wordlist was then compared with the interim version of lexicon C to find the set of terms common to both the lexicon and the wordlist. The set of 1170 common terms were then manually reviewed by a pharmacist to determine if the terms should actually be removed from the drug lexicon. The pharmacist identified 21 items in the set of common terms that should remain in the drug lexicon and the remaining 1149 were approved for removal from the lexicon C.

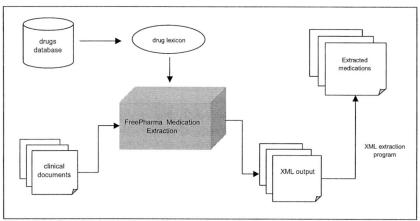

**Figure 1**. The process of medication extraction from clinical documents

We used 3 independent runs with FreePharma®, using a different drug lexicon in each, to extract medications from the documents. The XML outputs from all runs were analyzed by an automated program to extract medication names, which were then compared with the original medications from the 52 documents. The document extraction process is outlined in Figure 1.

Extraction results were summarized using the measures *sensitivity*, *specificity* and *precision*. Sensitivity (also known as *recall* in IE studies) for this study is the likelihood of retrieving medication items or the likelihood that the item extracted was truly a medication. Specificity is the true negative rate and determines how many of the non-medication items were truly not extracted. Another measure that is used frequently in IE studies is *precision*. Precision measures the proportion of true medications extracted out of all terms that were extracted. This is a fairly important measure for our study since our goal is to maximize precision, i.e., extract the maximum number of medications while reducing the number of non-medication terms that are extracted.

We are currently in the production phase of medication extraction at the Clinic and based on results from an internal study, we process only those document types that have a high likelihood of containing medications. Also, we perform weekly quality checks to eliminate or add terms to the drug lexicon. To assess the impact of lexicon filtering on an independent document set, we decided to process a randomized test set of 100 documents, selected from over 150,000 documents of 103 different document types that were processed with FreePharma® over a period of two weeks. Therefore, unlike the training set, this test contained a larger variety of document types. Of these 100 documents, 21 contained no medications and the remaining 79 contained a total of 641 medication terms (266 unique items). The number of non-medication terms in the entire test set was 41751. The medication items in this test set were independently reviewed and verified. We ran FreePharma® with each of the three lexicons separately to extract medications from the test set. The extracted results were analyzed as in the training set.

## 4    Results

The total number of documents used for the training set was 52, each of which contained at least one medication. There were a total of 28496 terms in these documents, of which 681 were medications. Tables 2a, 2b and 2c contain results for the extraction from the training set documents.

The data in Table 2a reveal that using only *drug_name* and *generic_name* yielded low sensitivity even though the specificity was fairly high. Providers often use short names for drugs (for example, *Nitro* for *Nitroglcerin*, *Metoprolol* for *Metoprolol Tartrate*) in patient notes for commonly used drugs. However, the first drug lexicon contained only brand names and generic names of drugs and therefore, missed all references to short names of drugs. This was a major factor in yielding low sensitivity.

Adding short names to lexicon A (Table 2b) increased the sensitivity of extraction but lowered the specificity and precision significantly. Inclusion of drug short names extracted 77 more medications than lexicon A. However, the short names list also added many terms that were ambiguous with English. This ambiguity resulted in extraction of a large number of terms that were not intended as medications in the patient documents, leading to low specificity and an even lower precision.

**Table 2a.** Training set results for medication extraction with Lexicon A (*drug_name + generic_name*). Sensitivity = 83.7% (570/681), Specificity = 96.2% (26782/27815), Precision = 35.6% (570/1603)

|  | Medication Items | Non-medication Items | Total count |
|---|---|---|---|
| Extracted by IE process | 570 | 1033 | 1603 |
| Not extracted by IE process | 111 | 26782 | 26893 |
| Total count | 681 | 27815 | 28496 |

**Table 2b.** Training set results for medication extraction with Lexicon B (*drug_name + generic_name + short_name*). Sensitivity = 95.0% (647/681), Specificity = 79.2% (22033/27815), Precision = 10.1% (647/6429)

|  | Medication Items | Non-medication Items | Total count |
|---|---|---|---|
| Extracted by IE process | 647 | 5782 | 6429 |
| Not extracted by IE process | 34 | 22033 | 22067 |
| Total count | 681 | 27815 | 28496 |

**Table 2c.** Training set results for medication extraction with Lexicon C (*drug_name + generic_name + short_name + filtering*). ). Sensitivity = 94.1% (641/681), Specificity = 98.5% (27400/27815), Precision = 60.7% (641/1056)

|  | Medication Items | Non-medication Items | Total count |
|---|---|---|---|
| Extracted by IE process | 641 | 415 | 1056 |
| Not extracted by IE process | 40 | 27400 | 27440 |
| Total count | 681 | 27815 | 28496 |

**Table 3a.** Test set results for medication extraction with Lexicon A (*drug_name + generic_name*). ). Sensitivity = 85.2% (546/641), Specificity = 96.9% (40444/41751), Precision = 29.5% (546/1853)

|  | Medication Items | Non-medication Items | Total count |
|---|---|---|---|
| Extracted by IE process | 546 | 1307 | 1853 |
| Not extracted by IE process | 95 | 40444 | 40539 |
| Total count | 641 | 41751 | 42392 |

**Table 3b.** Test set results for medication extraction with Lexicon B (*drug_name + generic_name + short_name*). ). Sensitivity = 96.4%% (618/641), Specificity = 80.1% (33431/41751), Precision = 6.9% (618/8938)

|  | Medication Items | Non-medication Items | Total count |
|---|---|---|---|
| Extracted by IE process | 618 | 8320 | 8938 |
| Not extracted by IE process | 23 | 33431 | 33454 |
| Total count | 641 | 41751 | 42392 |

**Table 3c.** Test set results for medication extraction with Lexicon C (*drug_name + generic_name + short_name + filtering*). ). Sensitivity = 95.8% (614/641), Specificity = 98.8% (41240/41751), Precision = 54.6% (614/1125)

|  | Medication Items | Non-medication Items | Total count |
|---|---|---|---|
| Extracted by IE process | 614 | 511 | 1125 |
| Not extracted by IE process | 27 | 41240 | 41267 |
| Total count | 641 | 41751 | 42392 |

Filtering the lexicon to reduce ambiguous non-medication items yielded high sensitivity and specificity (Table 2c), thus also increasing the corresponding precision value. Elimination of ambiguity with English and non-medication agents (filtering criteria in Table 1) created a lexicon with fewer non-medication items. The filtering reduced the number of non-medication items that were extracted from the documents leading to a much higher precision for extraction using lexicon C.

The total number of documents used for the test set was 100, of which only 79 contained at least one medication. There were a total of 42392 terms in these documents, of which 641 were medications. Results for the test set and the corresponding precision, specificity and sensitivity values are in Tables 3a, 3b and 3c. The test set results follow a pattern similar to that seen in the training set results. Lexicon A yielded a low sensitivity due to the absence of short names (Table 3a). Inclusion of short names (lexicon B) resulted in a higher sensitivity but decreased the specificity and precision (Table 3b). The filtered lexicon yielded both high specificity and precision (Table 3c)

## 5    Discussion

This study reveals some of the challenges in using drug lexicons to automatically extract medications from electronic medical records. Using existing drug sources without any attempts to remove non-specific or ambiguous terms will most likely extract many terms that are non-medication items. Such anomalies can only be revealed by a manual review of results to ascertain the quality of extraction. However, in the current study we have shown that identifying and filtering out non-medication items from the drug lexicon significantly enhances the sensitivity and precision of the results.

**Table 4.**    Extraction results summarizing the IE measures sensitivity, specificity and recall for each of the three lexicons. Lexicon A = drug_name + generic_name; Lexicon B = drug_name + generic_name + short_name. Lexicon C = Lexicon B + filtering techniques. The highest value for the measure among lexicons is marked in bold for both training and test sets.

|  | Training set | | | Test set | | |
|---|---|---|---|---|---|---|
|  | Lexicon A | Lexicon B | Lexicon C | Lexicon A | Lexicon B | Lexicon C |
| Sensitivity | 83.7% | **95.0%** | 94.1% | 85.2% | **96.4%** | 95.8% |
| Specificity | 96.2% | 79.2% | **98.5%** | 96.9% | 80.1% | **98.8%** |
| Precision | 35.6% | 10.1% | **60.7%** | 29.5% | 6.9% | **54.6%** |

Lack of drug short names in lexicon A resulted in a low sensitivity (recall) of only 83.7% in the training set and 85.2% in the test set (Table 4). When short names were added for lexicon B, the results showed an increase in recall to 95.0% (training set) and 96.4% (test set). This indicates that use of short names is fairly important for defining a drug lexicon. However, short name inclusion in lexicon B also resulted in lowering the specificity of the results. The results indicate that in the training set, 5782 non-medication items were also extracted when short names were added to the lexicon. This reduced the precision to only 10.1%, .i.e., only 10.1% of all extracted terms were real medications. Similarly, for the test set, 8320 non-medication items were extracted with the short name containing lexicon (Table 3b), resulting in a lower precision of only 6.9%.

An analysis of these non-medication items revealed that several were ambiguous with words in English (Table 5). Efforts to eliminate these ambiguous terms resulted in definition of the filtering criteria in Table 1. Use of these criteria yielded a smaller drug lexicon, lexicon C, containing only 22345 terms (compared to 29333 after inclusion of short names in lexicon B), which is a 23.8% reduction in lexicon size. Extraction from the training set with this lexicon yielded a high recall of 94.1%, a much higher specificity of 98.5% and the highest precision rate so far of 60.7%. Similarly, test set results with this lexicon yielded a high recall of 95.8%, a much higher specificity of 98.8% and the highest precision rate among lexicons so far of 54.6%. However, results with lexicon C (Tables 2c, 3c) showed a slightly lower recall than the results with lexicon B (Tables 2b, 3b). This is due to the fact that terms like *iron, influenza* and *tetanus* were removed during filtering from lexicon C, thus not extracting them as medications from expressions "liquid iron", "tetanus shots" and "influenza vaccine". These occurrences account for the difference in recall values between results from lexicons B and C.

Table 5. Examples of terms in drug lexicon B that contribute to ambiguity with English language. The last four columns indicate the data as present in FirstDataBank's drug database.

| Term in drug lexicon | Drug_name | Drug_short_name | Generic_name | AHFS_Category_desc |
|---|---|---|---|---|
| *The* | The Eliminator | The | Lecith/Pyridox HCL/I2/Cider | Miscellaneous Therapeutic Agents |
| *Benefit* | Benefit | Benefit | Nutritional Supplement | Electrolytic, Caloric and Water Balance |
| *Control* | Control Pads | Control | Incontinence Pad, Liner, Disp | Devices |
| *Pain* | Pain Reliever | Pain | Acetaminophen | Central Nervous System Agents |
| *Sleep* | Sleep Aid Formula | Sleep | Diphenhydramine HCL | Antihistamine Drugs |

Another source of missed extraction is the absence of medication in the source data from which the lexicons were constructed. In our analysis of results, we found many medications in the original documents which were not present in the FirstDataBank database and therefore not incorporated into any of the lexicons and consequently not extracted. Some of these were mis-spelt versions of existing medications, for example, "cyclosporin" is a mis-spelt variant of *cyclosporine*, "losartin" is *losartan* mis-spelt; some others like ASA (common short name for Acetyl Salicylic Acid) were simply missing from the database. To improve the drug lexicon further, we would need to add or remove these terms to maximize the values of our measures. Since we are already in production phase of the project, this is an essential step in our quality check procedure and involves active participation by pharmacists for evaluation of such terms for inclusion or removal from the lexicon.

One of the major goals of any IE study or application is to maximize precision, i.e., maximize the true positives while minimizing the false positives. For our study, this translates to applying additional filters to the drug lexicons or defining ways of filtering the extracted results based on other criteria like section headers. For example, our filtered lexicon contains the chemical term *potassium*. But *potassium* can mean a laboratory test item in "His potassium became elevated" or a drug item in "Potassium 10 mEq. three tablets q.a.m."[8]. In the current set of results, all occurrences of *potassium* are extracted, while only some of them are considered actual medications. To distinguish between the two occurrences of *potassium*, we can extract information about the sections in the document where they occur. Once we have that information, we can define the context in which they were used. In the example above then, *potassium* from the laboratory test item would then not be extracted because it occurred in the context of laboratory test and not a medication. This context filtering would lead to increasing the precision values.

One of the major limitations of this study is the fact that only one of the commercially available drug sources was used to evaluate the lexicon impact on medication extraction. However, we believe that the techniques used in this study can be used with other sources to achieve similar results. Other limitations include, but are not limited to the use of a convenience sample as a training set and the use of only a fixed set of filtering criteria to evaluate extraction precision.

## 6    Conclusions

Medication extraction is crucial for development of phenotypes and other pharmacogenomics studies. Drug dictionaries or lexicons are invaluable resources for extracting medication names from free-text documents. We have shown that drug lexicons can be used for medication extraction from clinical documents and that the precision and recall values for such studies can be considerably enhanced by defining filtering criteria to refine the drug lexicons. Future enhancements to the drug lexicon, such as specific additions and removals and section filtering would further increase the accuracy of results obtained from clinical documents..

## Acknowledgements

This project was funded in part by a grant from the Marshfield Clinic.  The authors acknowledge the contribution of Julie Stangl for data analysis, Peter Welch and Cynthia Motszko for data management and Dr. Russell Wilke and Dr. Gary Plank for assistance with validating removal of wordlist items from drug lexicon.

**References**

1. H. Liu, S.B. Johnson, C. Friedman, "Automatic Resolution of Ambiguous Terms Based on Machine Learning and Conceptual Relations in the UMLS" *J. Am. Med. Inform. Assoc.* 9(6):621-636 (2002)
2. Y. Tsuruoka, J. Tsujii, "Boosting Precision and Recall of Dictionary-based Protein Name Recognition", *Proceedings of the ACL 2003 Workshop on Natural Language Processing in BioMedicine.* 41-48 (2003)
3. A. Koike, T. Takagi, "Gene/Protein/Family Name Recognition in Biomedical Literature", *HLT-NAACL 2004 Workshop: BioLink 2004 – Linking Biological Literature, Ontologies and Databases.* 9-16 (2004)
4. L. Tanabe, W.J. Wilbur, "Tagging Gene and Protein Names in Biomedical Text", *Bioinformatics.* 18(8): 1124-1132 (2002)
5. W.E. Evans, H.L. McLeod, "Pharmacogenomics – Drug Disposition, Drug Targets and Side Effects", *N. Eng. J. Med.* 348(6):538-549 (2003)
6. D. Cooper. "What is Pharmacogenetics?" *Pharmacogenetics in Patient Care conference by American Association Clinical Chemistry*, Nov. 6, 1998.
7. L. Mancinelli, M. Cronic, W. Sadee. "Pharmacogenomics : the promise of personalized medicine." *AAPS PharmSci.* 2:30-37 (2000)
8. O. Tuason, L. Chen, H. Liu, J.A. Blake, C. Friedman, "Biological Nomenclatures: A Source of lexical knowledge and ambiguity", *Pacific Symposium on Biocomputing* 9:238-249 (2004)
9. T.C. Rindflesch, L. Tanabe, J.N. Weinstein, L. Hunter. "EDGAR: Extraction of Drugs, Genes and Relations from the BioMedical Literature", *Pacific Symposium on Biocomputing* 5:514-525 (2000)
10. S.L. Achour, M. Dojat, C Rieux, P. Bieling, E. Lepage, "A UMLS-based Knowledge Acquisition Tool for Rule-based Clinical Decision Support System Development", *J. Am. Med. Inform. Assoc.* 8(4):351-360 (2001)
11. "TeSSI® for Healthcare and Life Sciences", Language and Computing (http://www.landc.be) (2002)
12. "NDDF PLUS™ Documentation Feb04"(FirstDataBank, 2004)

# INTRODUCTION TO INFORMATICS APPROACHES IN STRUCTURAL GENOMICS: MODELING AND REPRESENTATION OF FUNCTION FROM MACROMOLECULAR STRUCTURE

PATRICIA C. BABBITT
*Departments of Biopharmaceutical Sciences and Pharmaceutical Chemistry,
University of California San Francisco
San Francisco, CA 94143 USA*


PHILIP E BOURNE
*Department of Pharmacology
The University of California San Diego
San Diego, CA 92093-0505 USA*


SEAN D. MOONEY
*Department of Medical and Molecular Genetics
Center for Computational Biology and Bioinformatics
Indiana University School of Medicine
Indianapolis, IN 46202 USA*

Despite the advantages provided by the enormous recent increases in the availability of structural information, functional assignment for the large number of proteins represented in the sequence and structural genomics projects remains a pressing problem for genomic era biology. This section describes work relevant to this problem from several perspectives, including new approaches that take advantage of combined structure and sequence-based classification. Leveraging of genomic context and evolutionary information to improve classification and predictive power is a second prominent theme in the papers represented here. Finally, issues in building a database for linking sequence, structural, and functional information are explored.

## 1.1. *Issues in Functional Inference for Structural Genomics*

As the structural genomics projects move beyond the initial implementation phase, functional assignment for *Initiative* targets and their homologs is an increasingly more important and frequent problem. For one Center for Structural Genomics, it has been estimated that approximately one-third of the solved structures are hypothetical proteins for which no functional information is available in sequence databases [1]. Currently, there are 1,227 structures in the Protein Databank [2] that are submitted from structural genomics projects, of which approximately 50% have unassigned functional classification (http://targetdb.rcsb.org and http://pdbbeta.rcsb.org). Moreover, because the core *Initiative* currently does not include functional characterization, there is no large-scale infrastructure to address the critical need for functional annotation of functionally uncharacterized protein structures. Thus, there is both a compelling need and an important opportunity for the biocomputing community to address the issue of modeling and representation of function from macromolecular structure.

Until recently, predicting the functions of gene products and identification of functionally important regions has largely been performed using sequence information alone. More sophisticated approaches that integrate sequence and structural data for functional characterization are being increasingly employed, however. Given the rise of structural genomics projects and related data, approaches that utilize three-dimensional structural information for inference of functional characteristics are an especially important area for research and represent a major theme in the work described here. This session is represented by five papers that address various issues in functional prediction and that directly or indirectly use three-dimensional structural information. In addition, the extended capabilities provided by incorporating evolutionary or other types of genomic context are explored in this session, along with issues in representation of structure-function linkage for database development.

### 1.2. A Short Description of Papers in This Section

Structural and functional similarity search methods enable functional annotation from a structural perspective. Chen, *et al.* describe a method for the comparison of structural motifs, or functional sites, in protein structures called the "Match Augmentation Algorithm." The authors find that statistically significant matches between two structures have functional significance, and this method significantly improves performance over geometric hashing methods.

An important problem in functional annotation is subfamily annotation. Brown, et al. show that subfamily hidden Markov models can classify the members of a family better than a single family HMM, and report a low expected error rate. They then apply the method to the serotonin receptors as a proof of concept.

Structural analysis of short protein segments can improve structure annotation and prediction methods. Tang, et al. provide an analysis of clusters of short protein segments. They find that their model improves both tertiary and secondary structure prediction methods.

Support vector machines continue to show improvement over other classification methods in bioinformatics problems. Nguyen and Rajapakse apply a multi class support vector machine to the problem of secondary structure prediction, showing improved results on several datasets.

In the last paper in this section, Pegg *et al.*, describe a new "Structure-Function Linkage Database," designed to aid in functional prediction from sequence and structure. Here, the authors compile sequence, structure, and functional information available for related proteins in large superfamilies of enzymes in order to associate and the sequence/structure variation in families within each superfamily with the similarities and differences among them. The result is a hierarchical system that distinguishes aspects of function common across all members of a superfamily from those common only to subgroups or families within the set. Computational issues in representing such structure-function linkage are also explored.

**References**

1. Laskowski, R. A., J. D. Watson, et al. *J. Struct. Funct. Genomics* **4**, 167-(2003).
2. H.M. Berman, J. Westbrook, Z. Feng, G. Gilliland, T.N. Bhat, H. Weissig, I.N. Shindyalov, P.E. Bourne. Nucleic Acids Research, 28, 235-242 (2000).

# SUBFAMILY HMMS IN FUNCTIONAL GENOMICS

DUNCAN BROWN, NANDINI KRISHNAMURTHY, JOSEPH M. DALE, WAYNE
CHRISTOPHER AND KIMMEN SJÖLANDER

*Department of Bioengineering, University of California*
*Berkeley, CA 94720*

The limitations of homology-based methods for prediction of protein molecular function
are well known; differences in domain structure, gene duplication events and errors in
existing database annotations complicate this process. In this paper we present a method
to detect and model protein subfamilies, which can be used in high-throughput, genome-
scale phylogenomic inference of protein function. We demonstrate the method on a set of
nine PFAM families, and show that subfamily HMMs provide greater separation of
homologs and non-homologs than is possible with a single HMM for each family. We
also show that subfamily HMMs can be used for functional classification with a very low
expected error rate. The BETE method for identifying functional subfamilies is illustrated
on a set of serotonin receptors.

## 1 Introduction

The vast majority of proteins have no experimentally determined function, and
prediction of molecular function by homology with functionally characterized
proteins has become *status quo*. Such predictions are used to obtain a
preliminary functional annotation, and thereby to guide wetbench experiments.
However, all homology-based methods of function prediction are known to be
prone to systematic errors of various types[1-4]. For a variety of reasons, including
domain fusion, gene duplication, and the undeniable presence of existing
database errors, inferring molecular function based on the annotated function of
the top hit in database search is fraught with potential hazards. Profile and
hidden Markov model (HMM) methods perform admirably in detecting
homologous proteins[5], but these generally afford only a very high level of
functional classification.

Phylogenomic analysis of a protein in the context of its entire family has
been demonstrated to improve both the accuracy and specificity of functional
annotation[2, 3, 6], but is time-consuming and not easily automated, and therefore is
generally applied to single families rather than at the genomic level.

We present here a method for obtaining a classification of sequences to
functional subfamilies that was used at Celera Genomics in the functional
classification of the human genome[7]. Subfamily HMMs model the functional
and structural variants of a protein family, so that regions of structural diversity
across the family are described by subfamily-specific amino acid preferences.

The ability to assign sequences to subfamilies automatically enables the
high-throughput application of phylogenomic inference of protein molecular
function which might otherwise be infeasible. If subfamily HMMs are

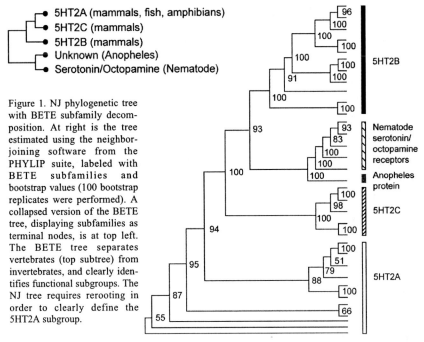

Figure 1. NJ phylogenetic tree with BETE subfamily decomposition. At right is the tree estimated using the neighbor-joining software from the PHYLIP suite, labeled with BETE subfamilies and bootstrap values (100 bootstrap replicates were performed). A collapsed version of the BETE tree, displaying subfamilies as terminal nodes, is at top left. The BETE tree separates vertebrates (top subtree) from invertebrates, and clearly identifies functional subgroups. The NJ tree requires rerooting in order to clearly define the 5HT2A subgroup.

constructed for the family, scores of sequences against subfamily HMMs can indicate a preliminary phylogenetic classification of a sequence, together with a more precise prediction of function.

The remainder of the paper is organized as follows. An illustration of the BETE subfamily decomposition is presented in section 2. Section 3 describes our method of constructing subfamily HMMs, and section 4 shows experimental results comparing the use of subfamily HMMs with single, family-level HMMs on several tasks: training sequence detection, remote homolog detection and classification accuracy. Discussion and future work are described in section 5.

## 2  BETE Subfamily Decomposition

In these experiments, we obtain a subfamily decomposition using Bayesian Evolutionary Tree Estimation (BETE)[8]. BETE estimates a phylogenetic tree using agglomerative clustering; subtrees are represented by profiles constructed using Dirichlet mixture densities[9] and symmetrized relative entropy is used as a distance metric between subtrees. Subfamilies are determined by a minimum-description-length cut of the tree into subtrees[8].

As presented elsewhere on *Src homology 2* (SH2) domains[8] and in the functional characterization of the proteins encoded in the human genome[7], the BETE subfamily decomposition corresponds closely to experimental data on

324

protein function and structure. We present in this section an illustration of the subfamily classification enabled by BETE, in application to the serotonin-receptor-related family of G-protein-coupled receptors. G-protein-coupled receptors are of enormous biomedical importance and include many pharmaceutical targets. Subfamily classification is particularly valuable in the context of this group due to the number of orphan receptors with unknown ligand specificity[10].

For this example, sequence homologs to serotonin receptor type 2B from human (SwissProt accession P41595) were gathered from the NR database using the FlowerPower program (in preparation). The homologs were aligned using MUSCLE[11], and trees were constructed using BETE and neighbor-joining (NJ, from the PHYLIP suite[12]). See Figure 1.

## 3    Subfamily HMM Construction Method

The method requires as input a multiple sequence alignment of a set of related proteins, with a specified decomposition into subfamilies. We use the same HMM architecture for each subfamily HMM (SHMM); a general HMM (GHMM) is constructed for the family as a whole and SHMMs are created by replacing the GHMM match state amino acid distribution at each position with a subfamily-specific distribution.

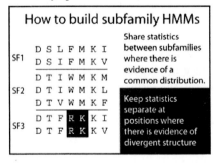

Figure 2. How to build subfamily HMMs. Amino acid distributions for positions defining the family as a whole are estimated once, and fixed within each subfamily. For non-globally conserved positions, examine the amino acids aligned by each of the other subfamilies. If a subfamily aligns similar amino acids, share statistics. Otherwise, keep statistics separate. In this toy example, the first two subfamilies will share statistics throughout the alignment. The last subfamily will share statistics with the first two (and vice-versa) at the black-on-white positions, but not at the white-on-black positions.

### 3.1 Estimating Subfamily Amino Acid Distributions

For each subfamily $s$, and at each column $c$ in the alignment, we compute a distribution over amino acids. We first discuss two special cases.

The first special case involves family-defining positions. Amino acid distributions at positions conserving the same amino acid across all sequences (allowing gaps) are fixed for all subfamilies. This enables subfamilies containing very few sequences to share in the knowledge of the critical residues (which might otherwise become generalized to allow substitutions). We first estimate the number of independent observations in the family as a whole; this number is used to weight the observed amino acids in deriving the posterior estimate of the amino acid distribution using Dirichlet mixture densities. The second case involves handling gapped positions. In these cases, the amino acid distribution is copied from the general HMM for that position.

*General case*: To allow us to recognize related family members, but still maintain specificity for each individual subfamily, we combine the amino acids in subfamily *s* at column *c* with amino acids from subfamilies aligning similar amino acids to those in subfamily *s* at position *c*. This is illustrated in Figure 2.

*Sequence weighting*: In common usage, sequence weighting is often restricted to deriving relative weights for a set of sequences, to down-weight sequences in highly populated subgroups and up-weight subgroups with few sequences. However, in using Dirichlet mixture densities to estimate amino acid distributions, the magnitude of the counts is also critical. In deriving subfamily HMM match state distributions, our approach involves estimating the number of independent observations in the alignment. We compute for every position in the alignment the frequency of the most frequent amino acid (ignoring gap characters) to derive the positional conservation propensity, and then compute the average conservation propensity over all columns ($P_{cons}$). The number of independent counts (NIC) can then be defined as $\text{NIC} = N^{1-P_{cons}}$, where $N$ is the number of sequences in the alignment. This has the effect of producing an NIC of 1 when the sequences in the alignment are 100% identical, and having NIC approach $N$ as the diversity in the alignment increases. The relative weights can then be derived independently. In the following equations, the notation $\bar{n} = (n_1, n_2, ..., n_{20})$ refers to the *weighted* counts of the amino acids seen at column *c* in subfamily *s*, and $n_i$ represents the weighted count of amino acid *i* at column *c* in subfamily *s*. The amino acid distribution at that position for subfamily *s* is estimated as follows:

*Step 1:    Obtain a Dirichlet mixture density posterior.*

We obtain a full Dirichlet mixture posterior density $\Theta^{Post}$ by combining the Dirichlet mixture prior $\Theta^{Prior}$ with the observed (weighted) amino acids seen in the column[9]. The mixture coefficients $q_j$, denoting the prior probability of each component *j* and the component density parameters $\bar{\alpha}_j$ of the Dirichlet mixture $\Theta^{Post}$ are set as follows:

$$q_j = \Pr(\bar{\alpha}_j \mid \bar{n}, \Theta^{Prior})$$
$$\alpha_{ji} = \alpha_{ji} + n_i \tag{1}$$

*Step 2:    Compute the family contribution from subfamilies $s' \neq s$.*

When we compute the contribution from other subfamilies to the profile for subfamily *s* at a fixed position, we add in amino acids from each subfamily proportional to the probability of the amino acids aligned by each subfamily at that position. Letting $\bar{n}_{s'}$ be the amino acids aligned at that column by subfamily $s'$, the "family contribution" is summed over all the subfamilies $s' \neq s$, creating a vector of amino acids $\bar{f}$, as follows:

$$\bar{f} = \sum_{s' \neq s} \Pr(\bar{n}_{s'} \mid \Theta^{Post}) \bar{n}_{s'} \qquad (2)$$

In this equation, $\Pr(\bar{n}_{s'} \mid \Theta^{Post})$ represents the posterior probability of $\bar{n}_{s'}$ given the posterior Dirichlet mixture density for subfamily $s$ at that position. In practice, we need to prevent the other subfamilies' contributions from swamping the amino acids observed in subfamily $s$; this is accomplished by capping the total $|\bar{f}|$ to a user-specified maximum

*Step 3: Combine the family contribution with the counts in subfamily $s$, to obtain the total counts $\bar{t} = (t_1, t_2, \dots t_{20})$, where $t_i = n_{si} + f_i$.*

*Step 4: Estimate the posterior amino acid distribution using Dirichlet mixture priors.* We modify the normal method for estimating the probability of amino acid $i$ at a position by substituting $t_i$ for $n_i$ and $|\bar{t}|$ for $|\bar{n}|$:

$$\hat{p}_i \propto \sum_j \Pr(\bar{\alpha}_j \mid \bar{n}, \Theta^{Prior}) \frac{t_i + \alpha_{ji}}{|\bar{t}| + |\bar{\alpha}_j|} \qquad (3)$$

Thus, we estimate the posterior probability of each component of the Dirichlet mixture prior using both the observed subfamily counts, and counts from all other subfamilies.

## 4  Experimental Validation

### 4.1 Data Chosen for Experiments

Subfamily HMMs are expected to contribute the most towards improved homolog detection when constructed for large and diverse protein families, and we selected a limited set of protein families with these characteristics at the outset. We chose entries from the list of PFAM[13] families beginning with letters A-C, based on the following criteria: (1) Each family had to have at least one member whose structure had been solved, and the alignment had to be at least 80 residues in length. This ensured that the selected family corresponded to a structural domain and was not simply a short repeat. (2) In order to provide informative comparisons between remote-homolog detection methods, the family-level HMM had to detect at least 10 homologs in the Astral PDB90 dataset of protein structural domains, and each family had to belong to a different SCOP superfamily[14]. (3) Finally, to ensure enough diversity in the family, the PFAM full alignment had to contain at least 600 sequences and have < 30% average pairwise identity (alignments with more than 3000 sequences after being made non-redundant at 95% identity were excluded). Details of PFAM families used in these experiments are provided in Table 1.

Table 1. PFAM family data. *SCOP* is the SCOP superfamily ID for the PFAM family; *# sequences, Gaps* and *Average %ID* refer to the number of sequences, fraction of gaps and average percent identity, respectively, within the UG95 MSA. *# subfamilies* is the number of subfamilies found by BETE for that family. *Subfam BL62* is the average per-position BLOSUM62 score in each subfamily MSA. *Full BL62* is the average per-position BLOSUM62 score in the UG95 MSA. *Full BL62 ≥ 1* is the fraction of columns in the UG95 MSA with average BLOSUM62 scores ≥ 1. *TS GHMM* and *TS SHMM* give the fraction of training sequences detected by the GHMM and SHMM within an E-value cutoff of 1e-10. *PDB GHMM* and *PDB SHMM* give the fraction of PDB90 homologs detected by the GHMM and SHMM within an E-value cutoff of 1e-10. *Class. Acc.* is the fraction of sequences correctly classified in the leave-one-out experiments.

| PFAM Family | SCOP | # sequences | Gaps | Average %ID | # subfamilies | Subfam BL62 | Full BL62 | Full BL62 ≥ 1 | TS GHMM | TS SHMM | PDB GHMM | PDB SHMM | Class. Acc. |
|---|---|---|---|---|---|---|---|---|---|---|---|---|---|
| AAA | c.37.1 | 1573 | 0.07 | 22 | 238 | 2.65 | 0.79 | 0.32 | 0.55 | 0.90 | 0.04 | 0.06 | 0.99 |
| Cadherin | b.1.6 | 2002 | 0.05 | 23 | 704 | 2.77 | 0.85 | 0.36 | 0.69 | 0.82 | 0.60 | 0.90 | 0.99 |
| Cytochrome C | a.3.1 | 725 | 0.14 | 17 | 245 | 2.57 | 0.58 | 0.18 | 0.12 | 0.68 | 0.10 | 0.54 | 0.93 |
| Alpha_amylase | c.1.8 | 874 | 0.13 | 17 | 184 | 2.48 | 0.52 | 0.21 | 0.98 | 0.99 | 0.34 | 0.34 | 0.98 |
| Aminotran_1_2 | c.67.1 | 1250 | 0.06 | 16 | 316 | 1.96 | 0.23 | 0.13 | 0.91 | 0.95 | 0.39 | 0.39 | 0.96 |
| C2 | b.7.1 | 865 | 0.06 | 21 | 263 | 2.69 | 0.77 | 0.33 | 0.38 | 0.88 | 0.69 | 0.77 | 1.00 |
| Aldo_ket_red | c.1.7 | 755 | 0.12 | 22 | 236 | 2.64 | 0.94 | 0.35 | 0.94 | 0.96 | 1.00 | 1.00 | 0.97 |
| Abhydrolase_1 | c.69.1 | 1209 | 0.06 | 14 | 687 | 2.57 | 0.02 | 0.13 | 0.41 | 0.84 | 0.18 | 0.27 | 0.99 |
| Amidohydro_1 | c.1.9 | 626 | 0.21 | 11 | 221 | 2.60 | 0.05 | 0.10 | 0.72 | 0.82 | 0.50 | 0.65 | 0.96 |

Figure 3 shows average per-position BLOSUM62 scores plotted for selected families, displaying both the high family diversity and within-subfamily conservation.

## 4.2 Technical Details

These experiments used the UCSC SAM software[15] for scoring and aligning proteins and to construct the general HMM. Several preprocessing steps were performed on the input sequences. The PFAM full alignment for each family was made non-redundant at 95% identity to create the NR95 multiple sequence alignment (MSA). We removed all columns having > 70% gaps and then all sequences matching fewer than 70% of the PFAM HMM match states, to give final un-gapped alignments (UG95). The UG95 MSA was used to derive a general HMM using the SAM w0.5 tool. The BETE algorithm[8] was used to identify subfamilies and SHMMs were constructed as described in Section 3.

SAM reverse scores were obtained using local-local scoring throughout. Sequences were assigned a GHMM score and a SHMM score (the best of the scores against all the SHMMs for the family). Subfamily HMM E-values were computed with respect to an extreme-value distribution fitted to SHMM scores against random sequences, as in the HMMER package. To ensure that E-values were comparable across families, an assumed database size of 100,000 (the

328

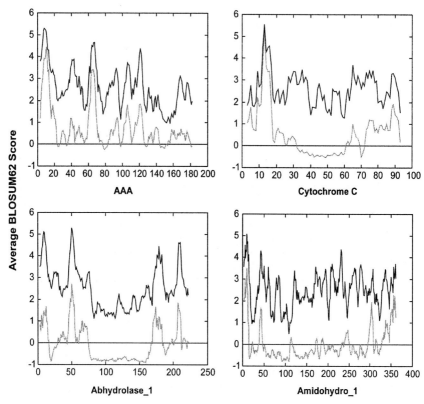

Figure 3. Average per-column pairwise BLOSUM62 scores for selected families. The X-axis shows the alignment column, and the Y-axis shows the average pairwise BLOSUM62 score for amino acids in that column. Within-subfamily scores are in black; scores across the whole family are in grey. Alignment columns with low average BLOSUM62 produce noisy HMM match-state distributions. Alignment quality of novel sequences to HMMs constructed from alignments with these characteristics can be correspondingly poor in regions of high structural variability across the family as a whole.

approximate size of SWISSPROT release 40) was used for training sequence detection experiments. For remote homolog detection experiments, we scored against Astral PDB90 release 1.65, and we used the true database size (8888).

### 4.3 Training Sequence Detection

Unaligned training sequences were extracted from the NR95 MSA and scored against the GHMM and SHMMs for that family. Figure 4 shows sequence coverage versus E-value summed over all nine PFAM families for both methods (results using an E-value cutoff of 1e-10 are summarized in Table 1). P-values were computed using the Wilcoxon signed-rank test to determine the significance of the differences between methods at different E-value cutoffs.

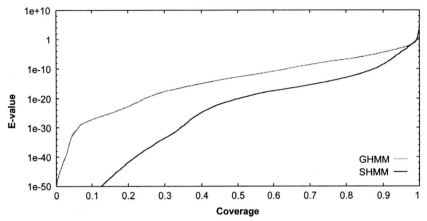

Figure 4. Training sequence detection using General and Subfamily HMMs. Shown here are results scoring training sequences against the general and subfamily HMMs using local-local scoring. The Y-axis gives the E-values for sequences against the different HMMs. The X-axis shows the fraction of sequences across all the families that scored at that level or better.

The SHMM method detects significantly more sequences than the GHMM at E-values below 1e-10, with a P-value of 0.002.

### 4.4 Remote Homolog Detection Tests

In these experiments, we compared subfamily and general HMMs on the ability to discriminate between homologs and non-homologs, using the Astral PDB90 database of structural domains[16], as classified by the Structural Classification of Proteins (SCOP) database[17]. The Astral PDB90 dataset is a subset of protein domains chosen so that no two are more than 90% identical when aligned. The Astral datasets have been widely used by the computational biology community to assess homology detection methods[5, 18, 19].

For computational efficiency, we first scored PDB90 with the general HMM for the family, and retrieved all sequences with E-values less than 100. These sequences were then scored against the subfamily HMMs, and the results were combined with the remaining GHMM scores. Preliminary results comparing this method with all-*vs*-all scoring of sequences against SHMMs indicated that there was little difference between the two. A more complete comparison is in preparation.

Each of the matches was marked as either True (classified to the same SCOP superfamily), False (classified to different SCOP folds) or Indeterminate (in the same SCOP fold but different SCOP superfamilies). We calculated normalized coverage and errors per query (EPQ) as described[20]. Results for each method were combined, sorted by e-value and assessed beginning with the most significant score. True positives were weighted such that each superfamily

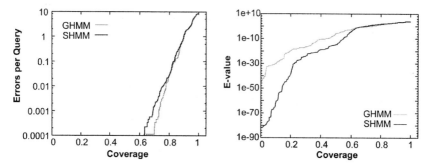

Figure 5. PDB90 discrimination experiments, comparing subfamily and general HMMs. For both methods, false positives appear at E-values of 0.1 and above, and both methods obtain similar coverage at this cutoff, indicating that SHMMs do not identify more homologs, but do afford a better separation between identified true positives and rejected sequences. SHMMs give stronger scores to clear homologs and reject non-homologs with large E-values.

contributed equally to the total coverage. The EPQ was calculated as the cumulative number of false positives divided by the total number of sequences in the database (8888 sequences for Astral PDB90 release 1.65).

In Figure 5, we show two plots to describe the combined results over all nine PFAM families: a standard Coverage *vs.* Error plot, and Coverage *vs.* E-value. Consistent with the results on training sequence detection, subfamily HMMs provide stronger scores to clearly homologous sequences, but general HMMs provide a small improvement over subfamily HMMs at detecting distant homologs. This improvement occurs at weak E-values of 0.1 and higher.

### 4.5 Classification Experiments

Homology-based functional annotation is at its heart a classification problem. Given a protein, even one for which the general family is known, determination of the functional subgroup to which it belongs is not a trivial task. As described in Section 2, the BETE subfamily decomposition correlates highly with subtypes already identified by biologists, and it is natural to use this breakdown to give a more precise prediction of protein molecular function. With this application in mind, we examined subfamily HMM performance in classifying previously unseen sequences. Note that although we used the BETE subfamily decomposition, the SHMM construction algorithm is independent of any particular sequence grouping; subfamilies identified by other methods (for example, manual annotation based on experimentally determined function) may also be used as input.

For each family, we took the BETE subfamily decomposition based on the UG95 MSA and removed one sequence at random. We required only that the subfamily to which the sequence belonged contained at least two sequences. Since the algorithm for constructing subfamily HMMs shares information across all subfamilies, we then rebuilt *all* of the subfamily HMMs using the modified alignment as input, but keeping the subfamily decomposition unchanged. We assumed that the *family* identification of a test sequence was given; withheld sequences were scored against the SHMMs from their family, and the top-scoring SHMM was identified. A sequence whose top-scoring subfamily was the one from which it had been extracted was counted as a success; any other result was a failure. We tested 10% of the sequences from each family in this way, for a total of 1035 sequences tested.

Results are shown in Table 1. Clearly, subfamily HMMs are proficient in recognizing sequences from their subgroup. The average success rate across the nine families was 97.4%, and the average for all sequences was 97.9% (22 sequences were incorrectly classified). In previous experiments (data not shown) classification errors typically come from one of several sources: alignment errors, subfamilies with few sequences losing sequences to larger subfamilies, fragments being misclassified and input multiple sequence alignments containing many gaps.

## 5. Discussion

Functional classification using homology-based methods is known to be prone to systematic errors of various types. Phylogenomic inference of protein molecular function has been shown by numerous investigators to improve the accuracy of functional classification, but is difficult to automate for high-throughput application.

This paper has described two tools to help automate phylogenomic inference of protein function, and demonstrated the use of these tools in predicting protein molecular function.

Bayesian Evolutionary Tree Estimation (BETE) identifies functional subfamilies given a multiple sequence alignment. BETE uses Dirichlet mixture densities and information theory to construct a phylogenetic tree and cut the tree into subtrees to obtain a subfamily decomposition. BETE has been demonstrated on a set of serotonin (and related) receptors; the subfamilies produced by BETE have been shown to correspond to known ligand receptor subtypes.

A novel method for constructing hidden Markov models for functional subfamilies has been described, given a subfamily decomposition and a multiple sequence alignment. Results have been provided on nine large and divergent PFAM families to demonstrate the use of subfamily HMMs for homolog detection, database discrimination and classification of novel sequences.

Subfamily HMMs have been demonstrated to provide better separation between homologs and non-homologs in database search than is possible using a single HMM for the family alone, recognizing more homologs with stronger scores and definitively rejecting non-homologs with large E-values. However, detection of true remote homologs is somewhat superior using HMMs constructed for the family as a whole, albeit with weak E-values and with a higher number of false positives. The advantages of using subfamily HMMs in detecting homologs are greater when the multiple sequence alignment used as the input for HMM construction contains a large number of variable sequences than when the sequences in the input alignment are more closely related.

Classification accuracy is high for subfamily HMMs. Given the biological validity of BETE-identified subfamilies, high classification accuracy makes subfamily HMMs a powerful tool for high-throughput functional genomics.

These results make sense in the light of a simple metric which can be used to compare the information content of BETE subfamily and whole-family alignments: the average pairwise BLOSUM62 score of the alignment columns, either within subfamilies or across the full alignment. As shown in Figure 3 and summarized in Table 1, within-subfamily scores are consistently higher than whole-family scores. This within-family similarity explains the high specificity of subfamily HMMs for closely-related sequences (such as training sequences, homologs, and sequences being classified). Conversely, the whole-family diversity explains the ability of general HMMs to improve upon SHMMs for remote homolog detection.

Our homolog detection results comparing subfamily and general HMMs illustrate the synergy between the two approaches for modeling protein families. General HMMs can be used as a first pass to detect related family members; subfamily HMMs then confer a more specific classification. Such a combined approach also minimizes the additional computational time required by our method, as only a very small fraction of sequences will be scored against the full set of subfamily HMMs.

Our results suggest several directions for future work. In particular, we chose the families in the current dataset based on our beliefs about what type of families would benefit from subfamily decomposition. Investigation of SHMM performance on a more representative set of families is an immediate priority. This will allow us to better ascertain how family characteristics such as size and diversity contribute to increased SHMM performance over GHMMs.

SHMMs require as input a subfamily decomposition; we have used the BETE method to derive this cut in these experiments. Other methods of obtaining this cut, including the use of standard phylogenetic tree construction algorithms, are also under investigation.

With regard to our SHMM construction algorithm, several issues should be investigated. Here we used a single method of sequence weighting; alternate algorithms may provide increased performance. Also, the homolog detection and classification performance of our SHMMs should be assessed against the 'naïve' SHMM method that does not share information between subfamilies. Such a method simply builds an HMM directly from the sequences in the subfamily.

## Acknowledgements

This work was supported by National Science Foundation Grant No. 0238311, and by the National Institutes of Health under Grant R01 HG002769-01. The authors wish to thank the reviewers for their insightful comments.

## References

1.  M. Y. Galperin, E. V. Koonin, *In Silico Biol* **1**, 55 (1998).
2.  J. A. Eisen, *Genome Res* **8**, 163 (Mar, 1998).
3.  K. Sjölander, *Bioinformatics* (2004).
4.  S. E. Brenner, *Trends Genet* **15**, 132 (Apr, 1999).
5.  J. Park *et al.*, *J Mol Biol* **284**, 1201 (Dec 11, 1998).
6.  J. A. Eisen, C. M. Fraser, *Science* **300**, 1706 (Jun 13, 2003).
7.  J. C. Venter *et al.*, *Science* **291**, 1304 (Feb 16, 2001).
8.  K. Sjölander, *Proc Int Conf Intell Syst Mol Biol* **6**, 165 (1998).
9.  K. Sjölander *et al.*, *Comput Appl Biosci* **12**, 327 (Aug, 1996).
10. A. Gaulton, T. K. Attwood, *Curr Opin Pharmacol* **3**, 114 (Apr, 2003).
11. R. C. Edgar, *Nucleic Acids Res* **32**, 1792 (2004).
12. J. Felsenstein. (2003).
13. A. Bateman *et al.*, *Nucleic Acids Res* **30**, 276 (Jan 1, 2002).
14. L. Lo Conte *et al.*, *Nucleic Acids Res* **28**, 257 (Jan 1, 2000).
15. R. Hughey *et al.*, "Sequence Alignment and Modeling Software System" *Tech. Report No. UCSC-CRL-99-11* (2000).
16. S. E. Brenner, P. Koehl, M. Levitt, *Nucleic Acids Res* **28**, 254 (Jan 1, 2000).
17. T. J. Hubbard, B. Ailey, S. E. Brenner, A. G. Murzin, C. Chothia, *Nucleic Acids Res* **27**, 254 (Jan 1, 1999).
18. J. Gough, C. Chothia, *Nucleic Acids Res* **30**, 268 (Jan 1, 2002).
19. L. Lo Conte, S. E. Brenner, T. J. Hubbard, C. Chothia, A. G. Murzin, *Nucleic Acids Res* **30**, 264 (Jan 1, 2002).
20. R. E. Green, S. E. Brenner, *Proc. IEEE.* **9**, 1834 (2002).

# ALGORITHMS FOR STRUCTURAL COMPARISON AND STATISTICAL ANALYSIS OF 3D PROTEIN MOTIFS

BRIAN Y. CHEN,[1] VIACHESLAV Y. FOFANOV,[2] DAVID M. KRISTENSEN,[3]
MAREK KIMMEL,[2] OLIVIER LICHTARGE,[3,4] LYDIA E. KAVRAKI[1,3]

*Rice University Departments of Computer Science[1] and Statistics[2],*
*Houston, TX 77005, USA*
*Baylor College of Medicine Struct. and Comp. Biol. and Mol. Biophys. Prog.[3] and*
*Department of Molecular and Human Genetics[4], Houston, TX 77030, USA*

The comparison of structural subsites in proteins is increasingly relevant to the
prediction of their biological function. To address this problem, we present the
Match Augmentation algorithm (MA). Given a structural *motif* of interest, such
as a functional site, MA searches a *target* protein structure for a *match*: the set
of atoms with the greatest geometric and chemical similarity. MA is extremely
efficient because it exploits the fact that the amino acids in a structural motif
are not equally important to function. Using motif residues ranked on functional
significance via the Evolutionary Trace (ET), MA prioritizes its search by initially
forming matches with functionally significant residues, then, guided by ET, it
augments this partial match stepwise until the whole motif is found. With this
hierarchical strategy, MA runs considerably faster than other methods, and almost
always identifies matches in homologs known to have cognate functional sites.
Second, in order to interpret matches, we further introduce a statistical method
using nonparametric density estimation of the frequency distribution of structural
matches. Our results show that the hierarchy of functional importance within
structural motifs speeds up the search within targets, and points to a new method
to score their statistical significance.

## 1. Introduction

Determining the function of proteins remains a primary goal of biology[1].
Tools such as PSI-BLAST[2], EMATRIX[3], and PROSITE[4] help predict function using sequence similarity. But with the increasing availability[5] of protein structures, other techniques have been developed to predict function
via geometric comparison of functional subsites, such as JESS[6], PINTS[7],
webFEATURE[8], and Geometric Hashing[9]. Starting from the basic observation that biological motifs are hierarchical in nature, this paper contributes a hybrid technique combining evolutionary information[10,11] with
protein geometry and chemical labels to efficiently identify proteins with
local structural similarity to a motif of interest.

**Contributions and Outline** We wish to annotate protein structures with
structural motifs that are functionally relevant. This problem naturally decomposes into the problem of motif design, and the problem of motif search.
The Evolutionary Trace[12,11,13] (ET) was developed to identify functionally
relevant motifs, and the topic of this paper is two novel algorithms for mo-

tif search and statistical interpretation of matches. Match Augmentation (MA) takes ET-based motifs, composed of subsets of three-dimensional (3D) protein structures ranked by evolutionary significance via ET and labeled by amino acid identity. MA hierarchically searches target structures of whole proteins, labeled with amino acids, for a match. Section 2.4 demonstrates that heuristic prioritization based on evolutionary rankings allows MA to identify cognate active sites in homologous proteins, with a 60 fold speed-up.

Much like sequence comparisons, MA may identify similar structures by chance alone. Hence, we ask if matches of cognate active sites have significantly greater structural similarity than what is expected by chance. In the second part of this paper, we apply Nonparametric Density Estimation (NDE) to attach statistical significance to a match, much like the BLAST $p$-value[2]. In Section 3.2, while current motifs are not yet optimal, and the test set is not complex, our results nevertheless verify that statistically significant matches are correlated with identifying cognate active sites.

## 2. Match Augmentation (MA)

Several methods exist for comparing protein structure, such as SSAP[14], DALI[15], tools for graph theoretical comparison[16], and Geometric Hashing[9], which was adapted to alignment by atom position[17], by backbone C-alpha[18], multiple structural alignment[19], and alignment of hinge-bending and flexible protein models[18]. More specifically, the problem of searching for motifs within protein structures, has also been approached using Geometric Hashing to find catalytic triads[20], using JESS[6], PINTS[7], and webFEATURE[8]. All structural comparison techniques share a dependence on heuristics, because the complexity of the structural pattern matching problem is at least NP-hard[21]. Heuristics are essential because biological input is too large for exhaustive approaches.

Our algorithm, MA, is unique because it uses the evolutionary significance of an amino acid in combination with structural and chemical data, stored as amino acid labels, to produce added performance with a novel combination of hashing and backtracking.

### 2.1. *Definition of Hierarchical Motifs*

Structure comparison is fundamentally hard[21], but heuristics using evolutionary data may improve performance. One source of such data is the Evolutionary Trace, which identifies functionally significant residues via Multiple Sequence Alignment (MSA) of homologous proteins[22,11]. For each residue, ET produces evolutionary significance *ranks*, which quantify the relative importance of individual residues to the function of the protein, and a list of functionally-compatible amino acid *alternates*, where mutation of a residue to its alternates is tolerated during evolution. MA can also accept rank information from other sources.

Ranks facilitate a prioritized attitude towards geometric and chemical comparison. Between high ranked motif points and corresponding target points, geometric and chemical similarity intuitively suggests greater functional similarity than correspondences involving low ranked motif points. Therefore, we seek correspondences for high ranking motif points before low ranking motif points, in a prioritized manner. Prioritization is the center point of our algorithmic design.

Our motifs $S = \{s_1, \ldots, s_m\}$ are sets of $m$ points in space whose coordinates are taken from backbone and sidechain atoms of high ranking residues around a ligand binding site, or other functional structure. Each point $s_i$ in the motif, or *motif point*, has an associated rank $p(s_i)$ and a set of alternate amino acid *labels* $l(s_i) = \{a_1, a_2, \ldots\}$ taken respectively from the significance rank and alternate sidechains generated by ET. Typically, our motifs are between 4 and 9 motif points.

## 2.2. The Problem

We seek a correspondence between $S$ and the target $T$, often hundreds of atoms encoded as $n$ *target points*: $T = \{t_1, \ldots t_n\}$, where each $t_i$ is taken from atom coordinates, and labeled $l(t_i)$ for the amino acid $t_i$ belongs to.

The correspondence is a *match* $M$, a bijection between $\{s_{M_i} \ldots s_{M_i}\} \in S$ and $\{t_{M_i} \ldots t_{M_i}\} \in T$ of the form $M = \{(s_{M_1}, t_{M_1}) \ldots (s_{M_m}, t_{M_m})\}$, with Euclidean distance between points $a$ and $b$ defined as $\|a - b\|$ and:

> **Criterion 1** $\forall i$, $s_{M_i}$ and $t_{M_i}$ are biologically *compatible*: $l(t_{M_i}) \in l(s_{M_i})$.
> **Criterion 2** LRMSD alignment, via rigid transformation A of $S$, causes $\forall i, \|A(s_{M_i}) - t_{M_i}\| < \epsilon$, our threshold for geometric similarity.

MA identifies the match with smallest LRMSD among all matches that have paired all $s_{M_i}$ to distinct $t_{M_i}$. Matches of subsets of $S$ to $T$ are rejected.

## 2.3. Description of the Algorithm

Following our prioritized data, we designed MA in a prioritized fashion, where correspondences with higher ranked points are identified first. MA is composed of two parts: *Seed Matching* and *Augmentation*. The purpose of Seed Matching is to identify a match for the *seed* $S' = \{s_1, s_2, s_3\}$, the three highest ranked motif points. The $k$ lowest LRMSD *seed matches* are passed to Augmentation to be iteratively expanded into matches for the remaining motif points, in descending rank order. Augmentation outputs the match with smallest LRMSD. We use $k = 30$ and $\epsilon = 3.0$Å.

**Seed Matching** We must find $k$ sets of 3 target points $T' = \{t_A, t_B, t_C\}$ which are compatible with $S' = \{s_1, s_2, s_3\}$, respectively, with similar inter-point distances as $S'$. Interpret $T$ as a geometric graph, where target points are vertices. Suppose $t_i, t_j$ are compatible with $s_1, s_2$. Then if $-2\epsilon \leq \|t_i - t_j\| - \|s_1 - s_2\| \leq 2\epsilon$, we define a green edge between $t_i$ and $t_j$. Similarly, red and blue edges are defined between target points compatible with $s_1, s_3$ and $s_2, s_3$ respectively, where again inter-point distances are within $2\epsilon$.

Edges are found by range search on a geometric data structure. When we find an edge, if it forms a triangle of all three colors, we have a seed match with compatible labels and similar inter-point distances. When we identify a triangle, LRMSD with $S'$ is calculated and if all points are aligned within $\epsilon$, the new seed match is stored. The $k$ lowest LRMSD seed matches are passed to Augmentation. Targets of size $n$ have at most $\binom{n}{3} = O(n^3)$ matching triangles, but this worst case would be a geometrically regular set of identical triangles, which never occurs in natural proteins. Performance on biological data is commonly $O(n^2)$.

**Augmentation** Augmentation is an application of depth first search. Given a seed match, we must find correspondences for unmatched motif points within the target. Considering the LRMSD alignment of the seed matches, we plot the position of the highest ranked unmatched motif point $s_i$ as if it were rigidly aligned with the rest of the seed. In the spherical vicinity $V$ of this position, we identify all $t_i \in T$ compatible with $s_i$. For each $t_i$, we calculate the LRMSD alignment $A$ of the seed with $(s_i, t_i)$. If $\|A(s_i) - t_i\| < \epsilon$, the seed match, with $(s_i, t_i)$, becomes a *partial match*.

$V$ often contains several $t_i$ compatible with $s_i$. We test all $t_i$, storing accepted partial matches on a stack. After all $t_i$ are tested, we pop the first partial match off the stack, and begin testing with the next unmatched motif point $s_{i+1}$. This is essentially depth first search (DFS), implemented with a stack. When no unmatched $s_i$ remain, or no compatible $t_i$ within $V$ can be aligned to satisfy $\|A(s_i) - t_i\| < \epsilon$, LRMSD is calculated for the entire match, and the match is stored. Final output is the match of all $s_i$ to distinct $t_i$ with lowest LRMSD.

Performance is dependent on the number of motif points $m$, and $c_r$, the number of compatible $t_i$ found in $V$, giving runtime $O(m^2(c_r^{m-3}))$. $c_r$ is bounded because repulsive Van der Waals forces limit the number of atoms found in $V$. The quadratic factor is the aggregate cost of LRMSD calculations, and the exponential is the cost of DFS with $c_r$ possibilities per iteration. With $m$ usually 4-9 points, MA is extremely efficient.

### 2.4. *Experimental Results*

To demonstrate the accuracy of MA, we searched for motifs within structures of evolutionarily related proteins. We use targets identified by sequence similarity because each residue in the motif has a cognate residue in the target: we know what match to expect beforehand. Using functional analogs may seem more relevant for functional annotation, but successfully matching analogs would only demonstrate how well our motifs represent function. Our focus is on methodology, and because analogs lack easily identifiable cognate residues, their use would sacrifice precise verifiability.

**Data Set** Our primary data (Figure 1) is 12 families of enzymes with known active sites. Each family is composed of a set of homologous sequences identified by BLAST, some of which have known structures in the Protein Data Bank[23] (PDB). Of the structures found, each family is assigned a *major* structure; the rest are *minor*. ET is applied on each family of sequences, and the significance ranks and labels generated are mapped onto the major structure for each family. Between 4 and 9 of the most functionally significant residues surrounding the active site on the major protein are selected, and their alpha carbons become the points in the motif. Specifics on amino acid selection and functional sites used for each motif can be found at http://www.cs.rice.edu/~brianyc/papers/PSB2005/.

Alpha carbons $(C_\alpha)$ were used in our motifs as preliminary data. Rather than debate the adequacy of $C_\alpha$ atoms to represent function, we seek only to document the correctness of our techniques. Future publications will comparatively document issues of motif design on a larger scale.

{**16pk**, 1vpe, 1php} {**1bqk**, 8paz, 1aaj, 1aan, 1ag6, 1b3i, 1baw, 1bxa, 1bxv, 1paz, 1pza, 1pzb, 1pzc, 1zia, 1zib, 2plt, 2rac, 3paz, 1aac} {**1amk**, 1tpe} {**1aky**, 5ukd, 1qf9, 1uke, 1zin, 1zio, 1zip, 2ak2, 2ukd, 3ukd, 4ukd, 1ak2} {**1a6m**, 1ymc, 1dwr, 1dws, 1dwt, 1m6c, 1mbs, 1mno, 1mwd, 1myg, 1pmb, 1wla, 1ymb, 1azi} {**1a3k**, 1slt, 1sla, 1slc, 1qmj} {**1finA**, 1hcl, 1hck, 1b38} {**1ukrA**, 1xyn, 1xnb, 1yna} {**3lzt**, 2ihl, 2lz2, 1jhlA, 1ghlA, 1fbiX, 1lz3, 1hhl, 1jug, 2eql, 1gd6A, 1f6rA, 1hfx} {**7a3hA**, 1g01A,1egzA} {**1juk**, 1j5tA, 1i4nA} {**1f8eA**, 1nn2, 1nsbA}

Figure 1. Families (bracketed) used in experimentation. Bolded proteins are major.

**Experimental Protocol** We search for each motif in the minor structures of the same family. These are homologous proteins (HPs). ET uses MSAs, so a functional residue in one sequence correlates with cognate residues of related function, at the same position, in all sequences of the family. Thus we can verify MA: if we find a *cognate match* where the target points are cognate to the motif points, we have a correct match, residue by residue. For comparison, we also searched for each motif in the minor proteins of the other families. These proteins are not homologous (NHPs).

**Results** In 69 out of the 73 motif-HP pairs (95.4%), MA matches 100% of the source motif with cognate residues in the target. Of the remaining four cases, two of the target structures (1m6c and 1mno) were experimental structures that had a point mutation which changed the label of residue 68 (in both cases) from a valine to an asparagine in order to over-stabilize oxygen binding in myoglobin (1a6m). As a result, the labels of the points corresponding to residue 68 in both 1m6c and 1mno were incompatible, and, correctly, the points were not matched. While this was not intended, it demonstrates the ability of our algorithm to eliminate potential matches with incorrect labels. In the other two cases, a match existed with lower LRMSD than the cognate match. These occurred between major protein 1amk with target 1tpe, and 1f8eA with 1nsbA. In each case the cognate match had a higher LRMSD (approx. .5Å) than the match MA identified. This is no fault of MA. Instead, it suggests that 1amk and 1f8eA are sub-optimal motifs, which bear accidental similarity to functionally unrelated structures: Ideally, motifs should have structural similarity only with proteins with functional similarity. True failures of MA would be the opposite: We would return a match with LRMSD higher than the cognate match, showing that the cognate match was overlooked. This never occurs. From our experiments, we found that MA is accurate and efficient on biological data, identifying cognate residue correspondences, except when the motif bears incidental structural similarity to unrelated residues.

Matches between motifs and HPs tended to have lower LRMSDs than between the same motif and NHPs. This is apparent in Figure 2, which plots LRMSD for all matches found. 9 out of 12 motifs considered had matches of HPs (Blue, Fig. 2) with LRMSD lower than most matches of NHPs (Red, Fig. 2). Two of the motifs breaking this trend were 1amk and 1f8eA, motifs which had incidental similarity with functionally unrelated residues, suggesting again that these motifs are not specific representatives of function. The remaining motif, 1finA, was defined on a flexible active site, so cognate active sites, flexible themselves, had less geometric similarity.

**Performance** We compared performance to our implementation of Geometric Hashing (GH), as described by Rosen[24], because the source code is not available. All published heuristics compatible with our data were implemented. GH has been applied many times[17,18,20,19], but cannot be prioritized as is the case with MA. GH identified identical HP matches and

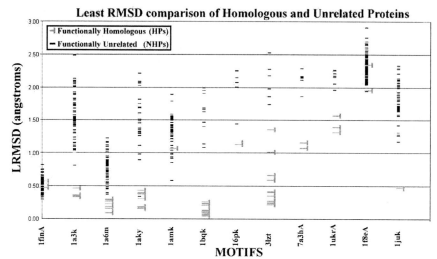

Figure 2.   Experimental Results: 12 motifs and 73 targets plotted by LRMSD

similar NHP matches, but on our motifs of 4 to 9 motif points, and targets with 123 to 398 target points, MA was about 60 times faster. Average execution time was 6.195 seconds for GH, and only 0.103 seconds for MA using identical thresholds. Without loss of accuracy, Seed Matching narrows the search to matches of the highest ranking motif points, whereas GH considers all points equally. Evolutionary prioritization seems to strongly improve performance. Experiments were run on Athlon 1900+ CPUs. GH and MA memory footprints varied between 5 and 20 megabytes, depending on input.

## 3. Statistical Analysis of Geometric Similarity

Structural similarity is important to functional annotation only if a strong correlation exists between identifiably significant structural similarity and functional similarity. However, as seen in Figure 2, algorithms like MA and GH can identify matches in NHPs with unrelated functions, so the existence of a match alone does not guarantee functional similarity. LRMSD can be a differentiating factor. If matches of HPs represent statistically significant structural similarity over what is expected by random chance, we could differentiate on LRMSD, as long as we can evaluate the statistical significance of the LRMSD of a match.

BLAST[2] first calculated the statistical significance of sequence matches with a combinatorial model of the space of similar sequences. Determining the statistical significance of structural matches has also been attempted. Modeling was applied for the PINTS database[7] to estimate the probability of a structural match given a particular LRMSD. An artificial distribution was parameterized by motif size and amino acid composition in order to fit a given data set, and the p-value is calculated relative to that distribution. Another approach was taken in the algorithm JESS[6], using comparative analysis to generate a significance score relative to a specific population of known motifs. Both methods have some disadvantages. The artificial models of PINTS are not parameterized by the geometry of motifs, and, all else equal, produce identical distributions for motifs of different geometry. JESS, on the other hand, is dependent on a set of known motifs; should this set change, all significance scores would have to be revised.

### 3.1. A Method for Characterizing Geometric Similarity

We begin by defining the statistical significance of geometric similarity. A match of motif $S$ to target $T$ with LRMSD $r$ is statistically significant if the *p-value*, the probability of finding a target $T'$ within a space of proteins $Z$, where the best match identified has LRMSD $r' < r$, is very low.

Determining the p-value is difficult because it requires the frequency distribution $D_S$ of match LRMSD between a given motif $S$ and all possible targets: all protein structures. But we have little knowledge of the space of protein structures; many proteins defy current techniques for structure determination. Rather than hypothesizing about unknown proteins, we use a set of known structures $Z$, and accept that our p-value reflects only this concrete set. Specifically, any $Z$ can have biases in the structural similarity of its members to some motif. This doesn't make the selection of $Z$ poor, because the primary purpose of our technique is to reflect that bias in the p-value calculated. We use the Protein Data Bank[23] as $Z$: the set $\{PDB\}$.

$D_S$ is essentially a histogram of how many proteins match $S$ at any LRMSD. Once $D_S$ is determined it can be interpreted as a distribution density function, which can be integrated to find the probability $P(r)$ of finding a match within $\{PDB\}$ with LRMSD less than a given $r$:

$$P(r) = \int_0^r D_S \tag{1}$$

One basic assumption of PINTS[7] and JESS[6] was that explicit calculation of $D_S$ is computationally infeasible; that running algorithms like MA with a given motif and every target would take too long. We tried this approach first, finding a match between $S$ and each member of $\{PDB\}$. This brute force approach generated $D_S$ in 3 hours for some motifs, or up to 631 for others. While this is acceptable for some applications, others, such as motif design, require frequent updates, for which this is too long. Scanning is embarrassingly parallel, but we provide a simpler solution first.

We used random sampling of the $\{PDB\}$ to avoid considering every target. $D_S$ was estimated using Nonparametric Density Estimation[25] (NDE). The distribution $D_S$ estimated from the sample needs to be smoothed, to neutralize spikes caused by the practice of submitting numerous similar structures to the PDB, and to interpolate between our sample points. Kernel Density Smoothing[25] was applied with a gaussian kernel to smooth the data. To avoid undersmoothing or oversmoothing, optimal bin-width determined by S-J estimation[26] was deemed best[27].

### 3.2. Experimental Results

**Nonparametric Density Estimation** We begin by demonstrating the effectiveness of sampling. We use a snapshot of the PDB from 8.17.2003. PDB files with multiple chains were divided into individual files, generating 55,305 structures. A handful of unparseable files were removed, and certain degeneracies were fixed, such as negatively indexed residues.

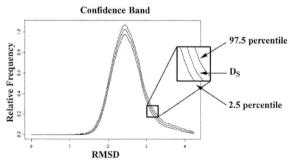

Figure 3. Quality of sampling fit

$\{PDB\}$ was scanned using MA, with each motif $S_i$ from Section 2.4. Brute force generated a reference distribution $D_{S_i}$. To verify sampling stability, each $D_{S_i}$ was sampled at 5%, 5,000 times. For all $S_i$, 95% of sampled curves fell within confidence bands tight around $D_{S_i}$. The confidence band, Figure 3, graphing frequency to LRMSD for $D_{3lzt}$, is typical of how tightly $D_{S_i}$ is approximated. Kolmogorov-Smirnoff[28] tests confirmed a lack of statistically significant differences between sampled distributions and $D_{S_i}$. Nonredundant PDB subsets produced no significant differences from $\{PDB\}$.

Random sampling directly improves performance. Brute force computation time was 12:48 (hrs:mins) on average, while sampling took 0:38 on average. The best case fell from 2:40 to 0:08 and the worst case from 631:41 to 31:30. Sampling cuts runtime by almost exactly 95%. Sampling does efficiently estimate $D_{S_i}$ without statistically significant loss of accuracy.

**Revisiting Earlier Results** After generating $D_S$ for all motifs from the previous section, in Figure 4, we calculated $p$-values for each LRMSD from Figure 2. The majority of $p$-values generated for HPs were between 1% and 0.01%. In contrast, most $p$-values generated for NHPs are above 10%. Notable exceptions are the $p$-values for matches of motifs 1amk and 1f8eA, which had accidental similarity to functionally unrelated structures. These had $p$-values above 10%. This verifies on a PDB-scale that 1amk and 1f8eA poorly represent functional sites: they have geometric and chemical similarity to 10% of all PDB proteins. The motif defined on 1finA, which had a flexible active site, also lacks statistical significance in its matches, because the geometry of functional residues may change relative to the motif. Matches of HPs represent identifiably significant structural similarity, except where the motif itself poorly represents protein function.

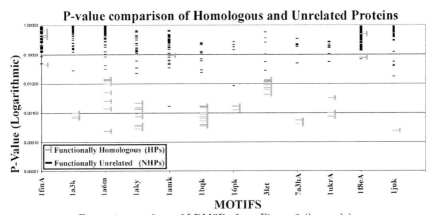

Figure 4.   $p$-values of LRMSDs from Figure 2 (log scale)

**Discussion** NDE avoids inflexibility, characteristic of parametric approaches[7], because it is not limited to a parametric model. While generally considered less powerful in other applications, NDE is appropriate here given breadth and complexity of protein structure space. Sampling, combined with NDE, greatly accelerates the process of calculating $p$-values.

On our data set of evolutionarily related proteins, our results show a correlation of statistically significant structural similarity to evolutionary relatedness between proteins, as long as the motifs properly represent function. This correlation indicates that statistically significant geometric and chemical similarity can be markers of cognate active sites.

## 4. Summary and Future Work

MA efficiently identifies homologs via rigid structural comparison. On our data set, 95.4% of active sites cognate to a given motif were correctly identified, and the remainder were not found because of the difficulty of representing some active sites with motifs. By optimizing on evolutionary data, MA is about 60 times faster than standard GH.

NDE via sampling calculates the statistical significance of matches identified. Testing indicates that we can drastically cut the number of calculations necessary to estimate $D_S$ without significant loss of accuracy. Furthermore, our results mesh with previous observations: matches between motifs and HPs were statistically significant, except for motifs which poorly represent protein function. Statistically significant LRMSD is correlated with the detection of cognate active sites.

This paper presents a fast method for identifying matches which permits an efficient statistical analysis of our data. Our future studies will develop methods for motif design, and test the sensitivity and specificity of these methods for functional annotation.

**Acknowledgements** This work is supported by the National Science Foundation NSF DBI-0318415. Additional support is gratefully acknowledged from training fellowships the Gulf Coast Consortia (NLM Grant No. 5T15LM07093) to B.C. and D.K.; from March of Dimes Grant FY03-93 to D.K.; from a Whitaker Biomedical Engineering Grant and a Sloan Fellowship to L.K; and from a VIGRE Training in Bioinformatics Grant from NSF DMS 0240058 to V.F. Experiments were run on equipment funded by AMD and EIA-0216467.

## References

1. Jones S. et. al. Searching for functional sites in protein structures. *Curr. Opin. Chem. Biol.*, 8(1):3–7, 2004.
2. Altschul S.F. et. al. Gapped blast and psi-blast: a new generation of protein database search programs. *Nucl. Acids. Res.*, 25(17):3389–3402, Sept 1997.
3. Wu T.D, Nevill-Manning C.G, and Brutlag D.L. Fast probabilistic analysis of sequence function using scoring matrices. *Bioinf.*, 16(3):233–44, 2000.
4. Hulo N. et. al. Recent improvements to the PROSITE database. *Nucl. Acids. Res.*, 32:D134–D137, 2004.
5. Blundell T.L., Jhoti H., and Abell C. High-throughput crystallography for lead discovery in drug design. *Nat. Rev. Drug Disc.*, 1:45–54, 2002.

6.  Barker J.A. and Thornton J.M. An algorithm for constraint-based structural template matching. *Bioinf.*, 19(13):1644–1649, 2003.

7.  Stark A., Sunyaev S., and Russell R.B. A model for statistical significance of local similarities in structure. *J. Mol. Biol.*, 326:1307–1316, 2003.

8.  Laing M.P. et.al. Webfeature: an interactive web tool for identifying and visualizing functional sites on macromolecular structures. *Nucl. Acids Res.*, 31(13):3324–7, 2003.

9.  Wolfson H.J. and Rigoutsos I. Geometric hashing: An overview. *IEEE Comp. Sci. Eng.*, 4(4):10–21, Oct 1997.

10. Lichtarge O. and Sowa M.E. Evolutionary predictions of binding surfaces and interactions. *Curr. Opin. Struct. Biol.*, 12(1):21–27, 2002.

11. Mihalek et. al. A family of evolution-entropy hybrid methods for ranking protein residues by importance. *J. Mol. Biol.*, 336:1265–82, 2004.

12. Lichtarge O., Bourne H.R., and Cohen F.E. An evolutionary trace method defines binding surfaces common to protein families. *J. Mol. Biol.*, 257(2): 342–358, 1996.

13. Yao H. et. al. An accurate, sensitive, and scalable method to identify functional sites in protein structures. *J. Mol. Biol.*, 326:255–261, 2003.

14. Taylor W,Orengo A. Protein structure alignment. *J.Mol.Biol*, 208:1–22, 1989.

15. Holm L. and Sander C. Protein structure comparison by alignment of distance matrices. *J. Mol. Biol.*, 233:123–138, 1990.

16. Artymuik P.J. et. al. A graph-theoretic approach to the identification of 3D patterns of amino acid side chains. *J.Mol.Biol*, 243:327–44, 1994.

17. Bachar O. et. al. A computer vision based technique for 3D sequence independent structural comparison of proteins. *Prot. Eng.*, 6(3):279–288, 1993.

18. Verbitsky G., Nussinov R., and Wolfson H. Structural comparison allowing hinge bending. *Prot: Struct. Funct. Genet.*, 34(2):232–254, 1999.

19. Liebowitz N et al. Automated multiple structure alignment and detection of a common substructural motif. *Prot: Struct. Func. Genet*, 43:235–245, 2001.

20. Wallace A.C., Laskowski R.A., and Thornton J.M. Derivation of 3D coordinate templates for searching structural databases. *Prot.Sci*, 5:1001–13, 1996.

21. Akutsu T. On determining the congruity of point sets in higher dimensions. In *Proc. ISAAC: 5th Symp. Alg. Comp.*, 1994.

22. Sowa M.E. et. al. Prediction and confirmation of a site critical for effector regulation of RGS domain activity. *Nat. Struct. Biol.*, 8:234–237, 2001.

23. H.M. Berman et al. The protein data bank. *Nucl.Acids Res*, 28:235–42, 2000.

24. Rosen M. et. al. Molecular shape comparisons in searches for active sites and functional similarity. *Prot. Eng.*, 11(4):263–277, 1998.

25. Silverman B.W. *Density Estimation for Statistics and Data Analysis*. Chapman and Hall: London, 1986.

26. Jones M.C., Marron J.S., and Sheather S.J. A brief survey of bandwidth selection for density estimation. *J. Amer. Stat. Assoc.*, 91:401–407, Mar 1996.

27. Sheather S.J. and Jones M.C. A reliable data-based bandwidth selections method for kernel density estimation. *J. Roy. Stat. Soc.*, 53(3):683–690, 1991.

28. Birnbaum Z.W. and Tingey F.H. One-sided confidence contours for probability distribution functions. *Ann. Math. Stat.*, 22(4):592–596, Dec 2003.

# TWO-STAGE MULTI-CLASS SUPPORT VECTOR MACHINES TO PROTEIN SECONDARY STRUCTURE PREDICTION

## M. N. NGUYEN AND J. C. RAJAPAKSE

*BioInformatics Research Centre,*
*School of Computer Engineering,*
*Nanyang Technological University, Singapore 639798*
*E-mail: asjagath@ntu.edu.sg*

Bioinformatics techniques to protein secondary structure (PSS) prediction are mostly single-stage approaches in the sense that they predict secondary structures of proteins by taking into account only the contextual information in amino acid sequences. In this paper, we propose two-stage Multi-class Support Vector Machine (MSVM) approach where a MSVM predictor is introduced to the output of the first stage MSVM to capture the sequential relationship among secondary structure elements for the prediction. By using position specific scoring matrices, generated by PSI-BLAST, the two-stage MSVM approach achieves $Q_3$ accuracies of 78.0% and 76.3% on the RS126 dataset of 126 nonhomologous globular proteins and the CB396 dataset of 396 nonhomologous proteins, respectively, which are better than the highest scores published on both datasets to date.

## 1 Introduction

One of the major goals of bioinformatics is to predict the three-dimensional (3-D) structure of a protein from its amino acid sequence. Unfortunately, the protein structure prediction problem is a combinatorial optimization problem, which so far has an eluded solution, because of the exponential number of potential solutions. One of the current approaches is to predict the protein secondary structure (PSS), which is linear representation of the full knowledge of the 3-D structure, and, thereafter, predict the 3-D structure [1,2]. The usual goal of secondary structure prediction is to classify a pattern of residues in amino acid sequences to a pattern of protein secondary structure elements: an $\alpha$-helix (H), $\beta$-strand (E) or coil (C, the remaining type).

Many computational techniques have been proposed in the literature to solve the PSS prediction problem, which broadly fall into three categories: (1) statistical methods, (2) neural network approaches, and (3) nearest neighbor methods. The statistical methods are mostly based on likelihood techniques [3,4,5]. Neural networks use residues in a local neighborhood or a window, as inputs, to predict the secondary structure at a particular location of an amino acid sequence by finding an appropriate non-linear mapping [6,7,8,9]. The nearest neighbor approach often uses the k-nearest neighbor techniques [10,11]. The consensus approaches that combine different classifiers, parallely, into a single superior predictor have been proposed for PSS prediction [12,13]. Support Vector Machines (SVMs) have been earlier applied to PSS prediction [14,15]; one of the drawbacks in these approaches is that the methods do not take into account the sequential relationship among the protein secondary structure elements. Additionally, SVM methods only construct a multi-class classifier by combining several binary classifiers.

Most existing secondary structure techniques are single-stage approaches, which are unable to find complex relations (correlations) among structural elements in the sequence. This could be improved by incorporating the interactions or contextual information among the elements of the sequences of secondary structures. We argue that it is feasible in enhancing the present single-stage MSVM approach farther by augmenting with another prediction scheme at their outputs and propose to use MSVM as the second-stage. By using the position specific scoring matrices generated by PSI-BLAST, the two-stage MSVM approach significantly achieves $Q_3$ accuracies of 78.0% and 76.3% on the RS126 and CB396 datasets, based on a seven-fold cross validation.

## 2  Two-Stage MSVM Approach

In the two-stage MSVM approach, we use two MSVMs in cascade to predict secondary structures of residues in amino acid sequences.

Let us denote the given amino acid sequence by $\mathbf{r} = (r_1, r_2, \ldots, r_n)$ where $r_i \in \Sigma_R$ and $\Sigma_R$ is the set of 20 amino acid residues, and $\mathbf{t} = (t_1, t_2, \ldots t_n)$ denote the corresponding secondary structure sequence where $t_i \in \Sigma_T$ and $\Sigma_T = \{H, E, C\}$; $n$ is the length of the sequence. The prediction of the PSS sequence, $\mathbf{t}$, from an amino acid sequence, $\mathbf{r}$, is the problem of finding the optimal mapping from the space of $\Sigma_R^n$ to the space of $\Sigma_T^n$.

Let $\mathbf{v}_i$ be the vector representing 21-dimensional coding of the residue $r_i$ where 20 units are the values from raw matrices of PSI-BLAST profiles ranging from $[0, 1]$ and the other is used for the padding space to indicate the overlapping end of the sequence [9]. Let the input pattern to the MSVM approach at site $i$ be $\mathbf{r}_i = (\mathbf{v}_{i-h_1^1}, \mathbf{v}_{i-h_1^1+1}, \ldots, \mathbf{v}_i, \ldots, \mathbf{v}_{i+h_2^1})$ where $\mathbf{v}_i$ denote the center element, $h_1^1$ and $h_2^1$ denote the width of window on the two sides; $w_1 = h_1^1 + h_2^1 + 1$ is the neighborhood size around the element $i$.

### 2.1  First Stage

A MSVM scheme has been proposed by Crammer and Singer [16]. For PSS prediction, this method constructs three discriminant functions but all are

obtained by solving one single optimization problem, which can be formulated as follows:
Minimize

$$\frac{1}{2} \sum_{k \in \Sigma_T} (\mathbf{w}_1^k)^T \mathbf{w}_1^k + \gamma^1 \sum_{j=1}^{N} \xi_j^1$$

subject to the constraints

$$\mathbf{w}_1^{t_j} \phi^1(\mathbf{r}_j) - \mathbf{w}_1^k \phi^1(\mathbf{r}_j) \geq c_j^k - \xi_j^1 \tag{1}$$

where $t_j$ is the secondary structural type of residue $r_j$ corresponding to the the training vector $\mathbf{r}_j$, $j = 1, 2, \ldots, N$, and $c_j^k = \begin{cases} 0 \text{ if } t_j = k \\ 1 \text{ if } t_j \neq k \end{cases}$

We find the minimization of the above formulation by solving the following quadratic programming problem [16]:

$$\max_{\alpha_j^k} -\frac{1}{2} \sum_{j=1}^{N} \sum_{i=1}^{N} \mathcal{K}^1(\mathbf{r}_j, \mathbf{r}_i) \sum_{k \in \Sigma_T} \alpha_j^k \alpha_i^k - \sum_{j=1}^{N} \sum_{k \in \Sigma_T} \alpha_j^k c_j^k \tag{2}$$

$$\text{such that } \sum_{k \in \Sigma_T} \alpha_j^k = 0 \text{ and } \alpha_j^k \leq \begin{cases} 0 \text{ if } t_j \neq k \\ \gamma^1 \text{ if } t_j = k \end{cases} \tag{3}$$

where $\mathcal{K}^1(\mathbf{r}_i, \mathbf{r}_j) = \phi^1(\mathbf{r}_i)\phi^1(\mathbf{r}_j)$ denotes the kernel function and $\mathbf{w}_1^k = \sum_{j=1}^{N} \alpha_j^k \phi^1(\mathbf{r}_j)$.

Once the parameters $\alpha_j^k$ are obtained from the optimization, the resulting discriminant function $f_1^k$ of a test input vector $\mathbf{r}_i$ is given by

$$f_1^k(\mathbf{r}_i) = \sum_{j=1}^{N} \alpha_j^k \mathcal{K}^1(\mathbf{r}_i, \mathbf{r}_j) = \mathbf{w}_1^k \phi^1(\mathbf{r}_i) \tag{4}$$

Let $f_1(\mathbf{r}_i) = \arg\max_{k \in \Sigma_T} f_1^k(\mathbf{r}_i)$. In the single-stage MSVM method, the secondary structural type $t_i$ corresponding to the residue at site $i$, $r_i$, is determined by

$$t_i = f_1(\mathbf{r}_i) \tag{5}$$

The function, $f_1$, discriminates the type of PSS, based on the features or interactions among the residues in the input pattern. With optimal parameters, the MSVM attempts to minimize the generalization error in the prediction. If the training and testing patterns are drawn independently and identically according to a probability distribution $\mathcal{P}$, then the generalization error, $\mathrm{err}_{\mathcal{P}}(f_1)$, is given by

$$\mathrm{err}_{\mathcal{P}}(f_1) = \mathcal{P}\{(\mathbf{r}, t) : f_1(\mathbf{r}) \neq t; (\mathbf{r}, t) \in \Gamma^1 \times \{H, E, C\}\}$$

where $\Gamma^1$ denotes the set of input patterns seen by the MSVM during both the training and testing phases, and $t$ denotes the desired ouput for input pattern $\mathbf{r}$.

## 2.2 Second Stage

We extend the single-stage MSVM technique by cascading another MSVM at the output of the present single-stage approach to improve the accuracy of prediction. This is because the secondary structure at a particular position of the sequence depends on the structures of the rest of the sequence, i.e., it accounts for the fact that the strands span over at least three adjacent residues and helices consist of at least four consecutive residues [6]. This intrinsic relation cannot be captured by using only single-stage approaches alone. Therefore, another layer of classifiers that minimize the generalization error of the output of single-stage methods by incorporating the sequential relationship among the protein structure elements improves the prediction accuracy.

Consider a window of $w_2$ size at a site of the output sequence of the first stage; the vector at position $i$, $\mathbf{d}_i = (d^k_{i-h_1^2}, d^k_{i-h_1^2+1}, \ldots, d^k_i, \ldots, d^k_{i+h_2^2})$ where $w_2 = 3(h_1^2 + h_2^2 + 1)$, $d^k_i = 1/(1 + e^{-f_1^k(\mathbf{r}_i)})$, and $f_1^k$ denotes the discriminant function of the first stage. The application of the logistic sigmoid function to the outputs of the first stage has the advantage of constraining the input units of the second stage to the $(0,1)$ interval that is similar to the range of the input units of the first stage. The purpose of this choice is to easier determine parameters for optimal performance. The MSVM converts the input patterns, usually linearly inseparable, to a higher dimensional space by using the mapping $\phi^2$ with a kernel function $\mathcal{K}^2(\mathbf{d}_i, \mathbf{d}_j) = \phi^2(\mathbf{d}_i)\phi^2(\mathbf{d}_j)$.

As in the first stage, the hidden outputs in the higher dimensional space are linearly combined by a weight vector, $\mathbf{w}_2$, to obtain the prediction output. Let the training set of exemplars for the second stage MSVM be $\Gamma^2_{\mathrm{train}} = \{\mathbf{d}_j : j = 1, \ldots, N\}$. The vector $\mathbf{w}_2$ is obtained by solving the following convex quadratic programming problem, over all the patterns seen in the training phase [16].

Let $f_2(\mathbf{d}_i) = \arg\max_{k \in \Sigma_T} f_2^k(\mathbf{d}_i)$. The secondary structural type $t_i$ corresponding to the residue $r_i$ is determined by

$$t_i = f_2(\mathbf{d}_i) \tag{6}$$

If the set of input patterns for the second stage MSVM in both training and testing phases is denoted by $\Gamma^2$, the generalization error of the two-stage MSVM approach, $\text{err}_{\mathcal{P}}(f_2)$, is given by

$$\text{err}_{\mathcal{P}}(f_2) = \mathcal{P}\{(\mathbf{d}, t) : f_2(\mathbf{d}) \neq t; (\mathbf{d}, t) \in \Gamma^2 \times \{H, E, C\}\}$$

If the input pattern $\mathbf{d}$ corresponds to a site $i$, then $\mathbf{d} = \mathbf{d}_i = ((1 + e^{-f_1^k(\mathbf{r}_{i-h_1^2})})^{-1}, (1 + e^{-f_1^k(\mathbf{r}_{i-h_1^2+1})})^{-1}, \ldots, (1 + e^{-f_1^k(\mathbf{r}_i)})^{-1}, \ldots, (1 + e^{-f_1^k(\mathbf{r}_{i+h_2^2})})^{-1})$. That is, the second stage takes into account the influences of the PSS values of residues in the neighborhood into the prediction. It could be easily shown that there exists a function $f_2$ such that $\text{err}_{\mathcal{P}}(f_2) = \text{err}_{\mathcal{P}}(f_1)$ when $h_1^2 = h_2^2 = 0$.

## 3 Minimal Generalization Error

In this section, we find a function $f_2$ that minimizes the generalization error $\text{err}_{\mathcal{P}}(f_2)$ when connecting another MSVM predictor at the output of the existing predictor. The optimal function $f_2$ providing the smallest $\text{err}_{\mathcal{P}}(f_2)$ ensures $\text{err}_{\mathcal{P}}(f_2) \leq \text{err}_{\mathcal{P}}(f_1)$. However, finding the global minimum of generalization error $\text{err}_{\mathcal{P}}(f_2)$ is not a trivial problem because the form of the probability distribution $\mathcal{P}$ is unknown. We can instead consider the probably approximately correct (pac) bound, $\epsilon(N, \delta)$, of the generalization error satisfying

$$\mathcal{P}\{\Gamma_{\text{train}}^2 : \exists f_2 \text{ such that } \text{err}_{\mathcal{P}}(f_2) > \epsilon(N, \delta)\} < \delta$$

This is equivalent to asserting that with probability greater than $1 - \delta$ over the training set $\Gamma_{\text{train}}^2$, the generalization error of $f_2$ is bounded by

$$\text{err}_{\mathcal{P}}(f_2) \leq \epsilon(N, \delta)$$

In the following proofs, we assume that both the training set $\Gamma_{\text{train}}^2 \subset \Gamma^2$ and the testing set $\Gamma_{\text{test}}^2 \subset \Gamma^2$ for the second stage contained $N$ patterns. For the MSVM technique at the second stage, let $\mathbf{w}_2^{k/l}$ be the weight vector $\mathbf{w}_2^k - \mathbf{w}_2^l$. Therefore, the secondary structure of a residue $r$ is not $l$ if $\mathbf{w}_2^{k/l}\phi^2(\mathbf{d}) > 0$ or not $k$ otherwise.

**Theorem 3.1.** [17] Let $\mathcal{F} = \{f_2^{k/l} : \mathbf{d} \to \mathbf{w}_2^{k/l}\phi^2(\mathbf{d}); \|\mathbf{w}_2^{k/l}\| \leq 1; \mathbf{d} \in \Gamma^2; k, l \in \Sigma_T\}$ be restricted to points in a ball of $m$ dimensions of radius $R$ about the origin, that is $\phi^2(\mathbf{d}) \in \mathbb{R}^m$ and $\|\phi^2(\mathbf{d})\| \leq R$. Then the fat-shattering dimension is bounded by

$$\text{fat}_{\mathcal{F}}(\eta_2^{k/l}) \leq \left(\frac{R}{\eta_2^{k/l}}\right)^2$$

**Theorem 3.2.** [18] Let $G$ be a decision directed acyclic graph on 3 classes $H$, $E$, and $C$, with 3 decision nodes, $H/E$, $E/C$, and $C/H$, with margins $\eta_2^{k/l}$ and discriminant functions $f_2^{k/l} \in \mathcal{F}$ at decision nodes $k/l$, where $\eta_2^{k/l} = \min_{\mathbf{d} \in \Gamma_{\text{train}}^2} \frac{|\mathbf{w}_2^{k/l} \phi^2(\mathbf{d})|}{\|\mathbf{w}_2^{k/l}\|}$, $k$ and $l \in \Sigma_T$. Then, the following probability is bounded by

$$\mathcal{P}\{\Gamma_{\text{train}}^2, \Gamma_{\text{test}}^2 : \exists\, G \text{ such that } \text{err}_{\Gamma_{\text{train}}^2}(G) = 0;\ \text{err}_{\Gamma_{\text{test}}^2}(G) > \epsilon(N, \delta)\} < \delta$$

where $\epsilon(N, \delta) = \frac{1}{N}\left(\sum_{k,l \in \Sigma_T} a^{k/l} \log \frac{4eN}{a^{k/l}} \log(4N) + \log \frac{2^3}{\delta}\right)$, $a^{k/l} = \text{fat}_{\mathcal{F}}\left(\frac{\eta_2^{k/l}}{8}\right)$, $\text{err}_{\Gamma_{\text{train}}^2}(G)$ and $\text{err}_{\Gamma_{\text{test}}^2}(G)$ are a fraction of points misclassified of $G$ on the training set $\Gamma_{\text{train}}^2$ and a random testing set $\Gamma_{\text{test}}^2$, respectively.

**Theorem 3.3.** Let $G$ be a decision directed acyclic graph with discriminant functions $f_2^{k/l} \in \mathcal{F}$ at nodes $k/l$, $k$ and $l \in \Sigma_T$. Then, the generalization error of $f_2$ where $f_2(\mathbf{d}) = \arg\max_{k \in \Sigma_T} f_2^k(\mathbf{d})$ in the probability distribution $\mathcal{P}$ is

$$\text{err}_{\mathcal{P}}(f_2) = \text{err}_{\mathcal{P}}(G)$$

**Proof.** This can be easily proved for an arbitrary example $\mathbf{d} \in \Gamma^2$, $f_2(\mathbf{d})$ equals to the secondary structural type of $\mathbf{d}$ predicted by the decision directed acyclic graph $G$. ∎

**Theorem 3.4.** [19] Let $\text{err}_{\mathcal{P}}(G)$ be the generalization error of $G$ at the output of the first stage. Then

$$\mathcal{P}\left\{\Gamma_{\text{train}}^2 : \exists G \text{ such that } \text{err}_{\Gamma_{\text{train}}^2}(G) = 0 \text{ and } \text{err}_{\mathcal{P}}(G) > 2\epsilon(N, \delta)\right\} \leq$$

$$2\mathcal{P}\left\{\Gamma_{\text{train}}^2, \Gamma_{\text{test}}^2 : \exists G \text{ such that } \text{err}_{\Gamma_{\text{train}}^2}(G) = 0 \text{ and } \text{err}_{\Gamma_{\text{test}}^2}(G) > \epsilon(N, \delta)\right\}$$

**Theorem 3.5.** Suppose we classify a random $N$ examples in the training set $\Gamma_{\text{train}}^2$ using the MSVM method at second stage with optimal values of weight vectors $\mathbf{w}_2^k$, $k \in \Sigma_T$. Then, the generalization error $\text{err}_{\mathcal{P}}(f_2)$ with probability greater than $1 - \delta$ is bound to be less than

$$\epsilon(N, \delta) = \frac{1}{N}\left(390R^2 \sum_{k \in \Sigma_T} \|\mathbf{w}_2^k\|^2 \log(4eN)\log(4N) + 2\log\frac{2(2N)^3}{\delta}\right)$$

**Proof.** Since the margin $\eta_2^{k/l}$ is the minimum value of the distances from the instances labeled $k$ or $l$ to the hyperplane $\mathbf{w}_2^{k/l}\phi^2(\mathbf{d}) = 0$ at the second stage, we have, $\eta_2^{k/l} = \min_{\mathbf{d}\in\Gamma_{\text{train}}^2} \frac{|\mathbf{w}_2^{k/l}\phi^2(\mathbf{d})|}{\|\mathbf{w}_2^{k/l}\|}$
$= \min_{\mathbf{d}\in\Gamma_{\text{train}}^2} \frac{|(\mathbf{w}_2^k - \mathbf{w}_2^l)\phi^2(\mathbf{d})|}{\|\mathbf{w}_2^k - \mathbf{w}_2^l\|} \geq \frac{1}{\|\mathbf{w}_2^k - \mathbf{w}_2^l\|}$. Therefore, the quantity
$M = \sum_{k,l}\frac{1}{(\eta_2^{k/l})^2} \leq \sum_{k,l}\|\mathbf{w}_2^k - \mathbf{w}_2^l\|^2 \leq 3\sum_k\|\mathbf{w}_2^k\|^2$. Solving the optimization problems at second stage results in the minimization of the quantity $M$ which is directly related to the margin of the classifier. Plugging the binary classifiers induced by $\mathbf{w}_2^{k/l}$ results a stepwise method for calculating the maximum among $\{f_2^k(\mathbf{d}) = \mathbf{w}_2^k\phi^2(\mathbf{d})\}$ that is similar to the process of finding the secondary structure in the decision directed acyclic graph $G$. Let us apply the result of Theorem (3.2) for $G$ with specified margin $\eta_2^{k/l}$ at each node to bound the generalization error $\text{err}_{\mathcal{P}}(G)$. Since the number of decision nodes is 3 and the largest allowed value of $a^{k/l}$ is $N$, the number of all possible patterns of $a^{k/l}$'s over the decision nodes is bounded by $N^3$. We let $\delta_i = \delta/N^3$ so that the sum $\sum_{i=1}^{N^3}\delta_i = \delta$. By choosing $\epsilon(N, \frac{\delta_i}{2})$

$$= \frac{1}{N}\left(195R^2 \sum_{k \in \Sigma_T} \|\mathbf{w}_2^k\|^2 \log(4eN)\log(4N) + \log\frac{2(2N)^3}{\delta}\right)$$

$$> \frac{1}{N}\left(\sum_{k,l \in \Sigma_T} \frac{R^2}{(\eta_2^{k/l}/8)^2} \log\frac{4eN}{a^{k/l}}\log(4N) + \log\frac{2^3}{\delta_i/2}\right)$$

from Theorem (3.1)

$$> \frac{1}{N}\left(\sum_{k,l \in \Sigma_T} a^{k/l}\log\frac{4eN}{a^{k/l}}\log(4N) + \log\frac{2^3}{\delta_i/2}\right)$$

Theorem (3.2) ensures that the probability of any of the statements failing to hold is less than $\delta/2$. By using the result of the Theorem (3.4), the probability $\mathcal{P}\{\Gamma_{\text{train}}^2 : \exists G; \text{err}_{\Gamma_{\text{train}}^2}(G) = 0; \text{err}_{\mathcal{P}}(G) > 2\epsilon(N, \delta_i/2)\}$ is bound to be less than $\delta$. From Theorem (3.3), the generalization error $\text{err}_{\mathcal{P}}(f_2)$ with probability greater than $1 - \delta$ is bound to be less than $2\epsilon(N, \delta_i/2)$. ∎

Minimizing the quantity $\sum_{k \in \Sigma_T} \left\| \mathbf{w}_2^k \right\|^2$, that is, maximizing the value of margin $\eta_2^{k/l}$ results in the minimization of the generalization error of the single stage MSVM method. Minimization of $\sum_{k \in \Sigma_T} \left\| \mathbf{w}_2^k \right\|^2$ is done by solving the convex quadratic programming problem of MSVM. As shown in the result of Theorem (3.5), two-stage MSVMs are sufficient for PSS prediction because they minimize both the generalization error $\mathrm{err}_{\mathcal{P}}(f_1)$ based on interactions among amino acids and the generalization error $\mathrm{err}_{\mathcal{P}}(f_2)$ of the output of the single-stage MSVM by capturing the contextual information of secondary structure.

## 4  Experiments and Results

### 4.1  Dataset 1 (RS126)

The set 126 nonhomologous globular protein chains, used in the experiment of Rost and Sander [6] and referred to as the RS126 set, was used to evaluate the accuracy of the classifiers. The RS126 set is available at http://www.compbio.dundee.ac.uk/~www-jpred/data/pred_res/126_set.html. The single-stage and two-stage MSVM approaches were implemented, with the position specific scoring matrices generated by PSI-BLAST, and tested on the dataset, using a seven-fold cross validation to estimate the prediction accuracy.

### 4.2  Dataset 2 (CB396)

The second dataset generated by Cuff and Barton [12] at the European Bioinformatics Institute (EBI) consisted of 396 nonhomologous protein chains and was referred to as the CB396 set. Cuff and Barton used a rigorous method consisting on the computation of the similarity score to derive their nonredundant dataset. The CB396 set is available at http://www.compbio.dundee.ac.uk/~www-jpred/data/. The single-stage and two-stage MSVM approaches have been used to predict PSS based on the position specific scoring matrices generated by PSI-BLAST.

### 4.3  Protein secondary structure definition

The secondary structure states for each structure in the training and testing sets were assigned from DSSP [20] that is the most widely used secondary structure definition. The eight states, H($\alpha$-helix), G($3_{10}$-helix, I($\pi$-helix), E($\beta$-strand), B(isolated $\beta$-bridge), T(turn), S(bend), and -(rest), were reduced to three classes, $\alpha$-helix (H), $\beta$-strand (E) and coil (C), by using the following method: H and G to H; E and B to E; all others states to C.

## 4.4 Prediction accuracy assessment

We have used several measures to evaluate the prediction accuracy. The $Q_3$ accuracy indicates the percentage of correctly predicted residues of three states of secondary structure [12]. The $Q_H, Q_E, Q_C$ accuracies represent the percentage of correctly predicted residues of each type of secondary structure [12]. Segment overlap measure (Sov) gives accuracy by counting predicted and observed segments, and measuring their overlap [21].

## 4.5 Results

For MSVM classifier at the first stage, a window size of 15 amino acid residues ($h_1^1 = h_2^1 = 7$) was used as input for optimal result in the [7, 21] range. At the second stage, the window size of width 21 ($h_1^2 = 2$ and $h_2^2 = 4$) in the [9, 24] range gave the optimal accuracy. The kernel selected here was the radial basis function $\mathcal{K}(\mathbf{x}, \mathbf{y}) = e^{-\sigma \|\mathbf{x} - \mathbf{y}\|^2}$ with the parameters: $\sigma = 0.05$, $\gamma^1 = 0.5$ for MSVM at the first stage, and $\sigma = 0.01$, $\gamma^2 = 0.5$ for two-stage MSVMs, determined empirically for optimal performance in the [0.01, 0.5] and [0.1, 2] ranges, respectively. We used BSVM library [22], which leads to faster convergence for large optimization problem, to implement the multi-class technique.

In tables 1 and 2, the results of Zpred, NNSSP, PREDATOR, DSC and Jpred methods on the RS126 and CB396 datasets were obtained from Cuff and Barton [12]. The results of the refined neural network proposed by Riis and Krogh, SVM method of Hua and Sun, dual-layer SVM of Guo, BRNN, and PHD methods were obtained from their papers [6,7,8,14,15].

Table 1 shows the performance of the different secondary structure predictors and two-stage MSVM approach on the RS126 set. The best algorithm was found to be the cascade of two MSVMs with the PSI-BLAST profiles, which achieved 78.0% of $Q_3$ accuracy. Comparing two-stage MSVMs to two multi-layer perceptron networks of PHD method proposed by Rost and Sander [6], a substantial gain of 7.2% of $Q_3$ accuracy was observed. Compared to SVM method of Hua and Sun [14], the two-stage MSVM method obtained 6.8% higher $Q_3$ score.

Table 2 shows the performance of two-stage MSVMs with the CB396 dataset based on multiple sequence alignments and PSI-BLAST profiles. Two-stage MSVMs with PSI-BLAST profiles achieved 76.3% of $Q_3$ accuracy that is

Table 1. Comparison of performances of single-stage and two-stage MSVM approaches in PSS prediction on the RS126 dataset. The notation - indicates that the result cannot be obtained from the papers.

| Method | $Q_3$ | $Q_H$ | $Q_E$ | $Q_C$ | Sov |
|---|---|---|---|---|---|
| Zvelebil *et al.* (Zpred) [23] | 66.7 | - | - | - | - |
| Rost and Sander (PHD) [6] | 70.8 | 72.0 | 66.0 | 72.0 | - |
| Salamov *et al.* (NNSSP) [10] | 72.7 | - | - | - | - |
| Frishman (PREDATOR) [24] | 70.3 | - | - | - | - |
| King and Sternberg (DSC) [25] | 71.1 | - | - | - | - |
| Riis and Krogh[7] | 71.3 | 68.9 | 57.0 | 79.2 | - |
| Baldi *et al.* (BRNN) [8] | 72.0 | - | - | - | - |
| Cuff and Barton (Jpred) [12] | 74.8 | - | - | - | - |
| Hua and Sun (SVM) [14] | 71.2 | 73.0 | 58.0 | 75.0 | - |
| Single-Stage MSVM | 76.2 | 69.6 | 63.5 | 83.1 | 68.8 |
| Two-Stage MSVMs | 78.0 | 73.1 | 65.7 | 83.8 | 72.6 |

Table 2. Comparison of performances of single-stage and two-stage MSVM approaches in PSS prediction on the CB396 dataset with PSI-BLAST profiles.

| Method | $Q_3$ | $Q_H$ | $Q_E$ | $Q_C$ | Sov |
|---|---|---|---|---|---|
| Zvelebil *et al.*(Zpred)[23] | 64.8 | - | - | - | - |
| Salamov *et al.* (NNSSP)[10] | 71.4 | - | - | - | - |
| Frishman·(PREDATOR)[24] | 68.6 | - | - | - | - |
| King and Sternberg (DSC)[25] | 68.4 | - | - | - | - |
| Guo *et al.*(Dual-Layer SVM)[15] | 74.0 | 79.3 | 69.3 | 72.0 | - |
| Single-Stage MSVM | 74.5 | 68.5 | 62.0 | 82.4 | 69.5 |
| Two-Stage MSVMs | 76.3 | 70.6 | 63.4 | 83.4 | 73.2 |

the highest scores on the CB396 set to date. Compared to the newest method of Guo *et al.* using dual-layer SVM [15], the two-stage MSVM method significantly obtained 2.3% higher $Q_3$ score. As shown, the prediction accuracy of two-stage MSVMs outperformed the result of single-stage MSVM method for PSS prediction.

In order to avoid to gross overestimates of accuracy, we performed another test on CB396 dataset: we selected best parameters within each cross-validation step by dividing the training data into one for SVM learning and another for selection of window size, $\sigma$ and $\gamma$ parameters. The accuracies of the new evaluation approach were not significantly different from those shown on Table 2 (76.5% of $Q_3$ and 72.9% of Sov). These results confirmed that the selected window size, sigma and gamma parameters in both learning stages were not biased by the test data chosen.

## 5 Discussion and Conclusion

We have introduced a two-stage MSVM approach to PSS prediction. With two-stage approaches, the accuracy of prediction is improved because secondary structure at a particular position of a sequence depends not only on the amino acid residue at a particular location but also on the structural formations of the rest of the sequence. Two-stage approach was first introduced in PHD approach which uses two MLPs in cascade for PSS prediction. MLPs are not optimal for this because the cannot generalize the prediction for unseen patterns. The outputs of single stages have been combined in parallel into a single superior predictor to improve upon the individual predictions [12,13]. However, these methods are depended on performances of individual single models and also do not overcome the limitation of single-stage methods. As shown, the MSVM method was an optimal classifier for the second stage because it minimizes not only the empirical risk of known sequences but also the actual risk of unknown sequences. Additionally, two stages were proven to be sufficient to find an optimal classifier for PSS prediction as the MSVM minimized the generalization error of the output of single-stage by solving the optimization problems at second stage.

Furthermore, we have compared two-stage SVM techniques for PSS problem: one method based on binary classifications of Guo [15] and the other approach for multi-class problem by solving one single optimization problem. We found that the two-stage MSVMs is more suitable for protein secondary structure prediction because its capacity to lead faster convergence for large and complex training sets of PSS problem and solve the optimization problem in one step.

As proved analytically, two-stage MSVMs have the best generalization ability for PSS prediction, by minimizing the generalization error made in the first stage MSVM. However, since this scenario could not be compared with the other techniques as they stick to seven-fold cross-validation for evaluation, which does not test true generalization capabilities. Further, our comparisons with the other techniques were not complete due to the inaccessibility of previously used data and programs. Also, the kernels and SVM parameters were empirically determined as there do not exist any simple methods to find them otherwise. Investigation on two-stage MSVM parameters could further enhance accuracies.

# References

1. P. Clote and R. Backofen, *Computational Molecular Biology*, Wiley and Sons, Ltd., Chichester, 2000.
2. D.W. Mount, *Bioinformatics: Sequence and Genome Analysis*, (Cold Spring Harbor Laboratory Press, 2001).
3. J. Garnier *et al*, Journal of Molecular Biology **120**, 97 (1978).
4. J.F. Gibrat *et al*, Journal of Molecular Biology **198**, 425 (1987).
5. J. Garnier *et al*, Methods Enzymol **266**, 541 (1996).
6. B. Rost and C. Sander, Journal of Molecular Biology **232**, 584 (1993).
7. S.K. Riis and A. Krogh, Journal of Computational Biology **3**, 163 (1996).
8. P. Baldi *et al*, Bioinformatics **15** 937 (1999).
9. D.T. Jones, Journal of Molecular Biology **292**, 195 (1999).
10. A.A. Salamov and V.V. Solovyev, Journal of Molecular Biology **247**, 11 (1995).
11. A.A. Salamov and V.V. Solovyev, Journal of Molecular Biology **268**, 31 (1997).
12. J. A. Cuff and G.J. Barton, Proteins **4**, 508 (1999).
13. M. Ouali and R. King, Protein Science **9**, 1162 (1999).
14. S. Hua and Z. Sun, Journal of Molecular Biology **308**, 397 (2001).
15. J. Guo *et al*, Proteins **54**, 738 (2004).
16. K. Crammer and Y. Singer, Computational Learing Theory , 35 (2000).
17. N. Cristianini and J. Shawe-Taylor, *An Introduction to Support Vector Machines*, (Cambridge University Press, 2000).
18. J.C. Platt *et al*, Proc. Advances in Neural Information Processing Systems 12. Cambridge, MA:MIT Press **12**, 547 (2000).
19. V. Vapnik, *Estimation of Dependences Based on Empirical Data*, (Springer-Verlag, New York, 1982).
20. W. Kabsch and C. Sander, Biopolymers **22**, 2577 (1983).
21. A. Zemla *et al*, Proteins: Structure, Function, and Genetics **34** ,220 (1999).
22. C.W. Hsu and C.J. Lin, IEEE Transactions on Neural Networks **13**, 415 (2002).
23. M.J.J.M Zvelebil *et al*, Journal of Molecular Biology **195**, 957 (1987).
24. D. Frishman and P. Argos, Proteins: Structure, Function, and Genetics **23**, 566 (1995).
25. R.D. King and M.J.E Sternberg*et al*, Protein Science **5** 2298 (1996).

# REPRESENTING STRUCTURE-FUNCTION RELATIONSHIPS IN MECHANISTICALLY DIVERSE ENZYME SUPERFAMILIES

SCOTT C.-H. PEGG[†], SHOSHANA BROWN[†], SUNIL OJHA[†], CONRAD C. HUANG[*], THOMAS E. FERRIN[*†], PATRICIA C. BABBITT[†*]

*†Dept. of Biopharmaceutical Sciences and *Dept. of Pharmaceutical Chemistry*
*University of California, San Francisco, 94143*

The prediction of protein function from structure or sequence data remains a problem best addressed by leveraging information available from previously determined structure-function relationships. In the case of enzymes, the study of mechanistically diverse superfamilies can provide a rich source of structure-function information useful in functional determination and enzyme engineering. To access these relationships using a computational resource, several issues must be addressed regarding the representation of enzyme function, the organization of structure-function relationships in the superfamily context, the handling of misannotations, and reliability of classifications and evidence. We discuss here our approaches to solving these problems in the development of a Structure-Function Linkage Database (SFLD) (online at http://sfld.rbvi.ucsf.edu).

## 1. Introduction

The solution of a protein's three-dimensional structure often does not immediately lead to the determination of its function.[1] Typically, we take the natural step of leveraging the information gained from experimental determinations of function by asking the question, "Is this structure and active site one that I've seen before, and if so, what does it do and how?" As the number of both solved protein structures and experimental determinations of function increase (with the former growing much faster than the latter), there is a growing need for computational methods for storing and searching representations of protein function in a way that correlates specific aspects of function with sequence and structural features. Ideally, such representations of function should go beyond simple identification of conserved and/or functionally validated residues in sequence alignments.

In the case of enzymes, the study of mechanistically diverse superfamilies—sets of homologous enzymes which, while often sharing very little sequence similarity to each other and often catalyzing different overall reactions with a variety of substrates and products, share the same fold and conserve a specific partial reaction (or some other aspect of mechanism) enabled by a conserved set of residues[2, 3]—allows us to leverage structure-function information at multiple levels. At the highest level, we can infer only a partial mechanistic step and the associated functionally important residues that are common across all members of the superfamily. At the lowest, most detailed level, we can determine the specific function of a single enzyme, including its mechanism as performed by specific active-site residues and co-factors. Often, however, because of the sophisticated level of chemical intuition required to identify the partial reaction(s) associated with conserved structural characteristics in such diverse proteins, only researchers who are intimately familiar with a given enzyme superfamily can take advantage of the structure-function information it contains. A resource that allows other investigators to utilize this information represents a valuable tool.

The terrain of mechanistically diverse enzyme superfamilies presents a number of obstacles that must be surmounted, some common to any database linking structural and functional information, others unique to mechanistically diverse enzyme superfamilies. The former include handling multi-functional enzymes, representing function in a computationally accessible format, and dealing with potential inaccuracies in annotation. Unique to analysis of mechanistically diverse enzyme superfamilies is the need to capture chemical function both in terms of the overall chemical reactions performed, but also at the level of the common partial reaction (or common chemical capability) associated with all of the different overall reactions represented in a superfamily. Partial reactions are captured in the Structure-Function Linkage Database (SFLD) by way of a partial reaction table in the relational database schema (see Fig. 4 below). Providing this information is especially important for the SFLD because it is only these partial reactions that correlate with active site similarities across all diverse members of a given superfamily. Identifying these partial reactions and linking them to structure provides a powerful tool for difficult problems in functional inference and protein engineering.[3, 4]

In this paper we discuss several major issues involved in the development of a computational resource for the storage and leverage of structure-function data in mechanistically diverse enzyme superfamilies and our specific approaches to handling these issues in the SFLD. Currently, five different superfamilies, representing over 3,800 sequences, are available in the database, with substantial expansion planned over the next year. Several of these superfamilies are used as examples here. Access to the SFLD is provided by a world-wide web based graphical user interface which accommodates many search and browse capabilities linking sequences, structures, and representations of the associated chemical transformations. A more detailed description of the content, uses, and the scientific principles motivating development of the SFLD will be presented elsewhere (manuscript in prepraration).

## 2. Representing Structure-Function Relationships in Mechanistically Diverse Enzyme Superfamilies

### 2.1. *Organization of Enzyme Superfamilies*

Within a mechanistically diverse enzyme superfamily, the elements of sequence and structure that deliver catalytic function are conserved to varying degrees. While all members of a given superfamily will possess the sequence and structural elements relating to the ability to perform the conserved mechanistic step (e.g., partial reaction) which helps define the superfamily, other subsets of the superfamily will possess a superset of conserved elements relating to other aspects of catalytic function. To clarify the distinction between these conserved elements, we have organized the enzyme superfamilies of the SFLD into a hierarchy of groupings. Figure 1 illustrates this hierarchy using the enolase superfamily[5] as an example.

At the top level of the hierarchy, enzymes are classified into the same superfamily if they appear to be evolutionarily related (based on sequence and structural information) and to share a common chemical capability (in the case of the enolase superfamily example, abstraction of a proton alpha to a carboxylic acid). The subgroup classification at the middle level of the hierarchy is superfamily-specific, and is defined by SFLD curators. In the enolase superfamily, enzymes are divided into subgroups based on active site residue motifs. At the next level of the hierarchy are families of enzymes each of whose members catalyze the same overall reaction. At the bottom of the hierarchy is a single enzyme, referred to as an enzyme functional domain (EFD). (See section 2.3 below for an example of how EFDs are defined.) According to the SFLD schema, an EFD need not be classified into a family to be classified into a subgroup or superfamily. Thus, if the full catalytic function of an EFD cannot be reliably determined, it may still be placed into a higher-level category in the hierarchy.

Because the hierarchical organization of EFDs into superfamilies, subgroups and families is based on functional as well as evolutionary criteria, functional classification of new sequences and structures is facilitated. For example, if an uncharacterized enzyme can be placed within a superfamily, the reaction catalyzed by the enzyme can be expected to utilize the chemical capability common to the superfamily. The overall reaction catalyzed by the enzyme may then be inferred based on additional information, such as operon context, or by further classifying the enzyme into a subgroup or family.

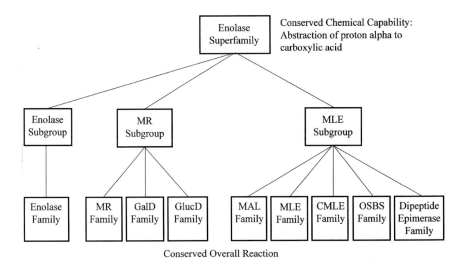

Figure 1. Hierarchical classification of EFDs in the enolase superfamily, based on sequence, structure and function. At the top level of the hierarchy, EFDs are classified into the same superfamily if they appear to be evolutionarily related based on sequence and structural information and to share a common chemical capability. The subgroup classification at the middle level of the hierarchy is superfamily-specific, and is defined by SFLD curators. At the bottom level of the hierarchy, enzymes in the same family are thought to catalyze the same overall reaction. Abbreviations used in this example: MR: mandelate racemase, GalD: galactonate dehydratase, GlucD, glucarate dehydratase, MAL: b-methylaspartate ammonia-lyase, MLE: muconate cycloisomerase, CMLE: chloromuconate cycloisomerase, OSBS: o-succinylbenzoate synthase.

## 2.2. Representation of Catalyzed Reactions

In order for the reactions catalyzed by the enzymes in mechanistically diverse superfamilies to be rapidly searched and compared to each other, they must be stored in a computationally accessible format that allows for not just comparisons of overall and partial reactions, but comparisons of substrate and product substructure. This issue has been addressed in the field of small molecule synthetic chemistry, where the standard has become the use of SMILES/SMARTS strings.[6] SMILES/SMARTS provides the type of functionality required to link enzyme chemistry to the sequence and structure information provided in the SFLD, as these strings of ASCII characters represent the chemical structures of participants in a reaction, including chirality. We have adopted this format for the SFLD, allowing users to search the overall and partial reactions using both reactions and substructures as queries. Figure 2 gives an example of some SMILES/SMARTS representations and queries.

| SMARTS Query | Meaning | Schematic |
|---|---|---|
| C([OD1]))>>C(=O) | Reaction contains the conversion of an alcohol to a ketone | 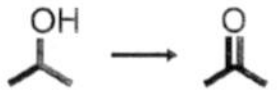 |
| >>c1ccccc1 | Product of the reaction contains a benzene ring | 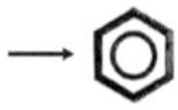 |
| C(=O)([OD1])C=CC=CC(=O)[OD1] | Either the substrate or the product of the reaction contains mandelate | 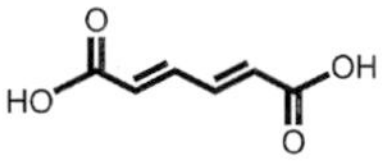 |

Figure 2. Examples of possible SMARTS queries and their chemical meanings.

While very flexible, and a good solution in the field of synthetic organic chemistry, SMILES/SMARTS does not provide a comprehensive solution for the study of enzyme chemistry, however. Extension of these representations, currently underway in our laboratory, will also be required. These include representation of the chemical contributions of active site residues as well as metals and complex cofactors.

## 2.3. Enzymes with Multiple Functions

A major hurdle in representing structure-function information is the issue of multi-functional enzymes—single protein sequences that catalyze multiple chemical reactions. Multi-functional enzymes can be viewed as one of three types. The first type consists of enzymes with multiple fused, but fundamentally independent domains, each with a separate active site. For example, phosphoribosylanthranilate isomerase (PRAI) and indoleglycerolphosphate synthase (IGPS), which catalyze two consecutive reactions in the tryptophan biosynthetic pathway, occur as a single protein chain in E. coli. Although the E. coli protein catalyzes both reactions, the n-terminal domain is responsible for the IGPS reaction, while the c-terminal domain is responsible for the PRAI reaction.[7-10] Furthermore, PRAI and IGPS occur as two physically separate protein chains in other organisms, such as T. maritima.

The SFLD accommodates this type of multifunctional protein by storing enzyme information at the level of the enzyme functional domain (EFD). An EFD is an enzyme, or part of an enzyme, that is capable of catalyzing a chemical reaction on its own. Thus, the E. coli IGPS-PRAI protein would be divided into two separate EFDs, one corresponding to the n-terminal domain of the protein, and one corresponding to the c-terminal domain of the protein.

The second type of multifunctional enzyme represents proteins that are capable of catalyzing an adventitious secondary reaction in the same active site responsible for catalyzing its primary reaction. One example, illustrated in Fig. 3, is the enolase superfamily enzyme, o-succinylbenzoate synthase (OSBS), from *Amycolaptosis sp*. In addition to the biologically relevant OSBS reaction, this enzyme also catalyzes the industrially important n-acylamino acid racemase (NAAAR) reaction.[11] The enzyme utilizes the same active site to catalyze both of these very different overall reactions. Catalytic promiscuity has been noted in other enzymes,[12, 13] but because it is difficult to determine possible secondary reactions based on the primary reaction of an enzyme, the overall incidence of this type of promiscuity is unknown. Many enzymes have also been noted to exhibit a third, and much more common form of catalytic promiscuity, e.g. turning over a variety of substrates related to their primary substrate.[12, 14, 15]

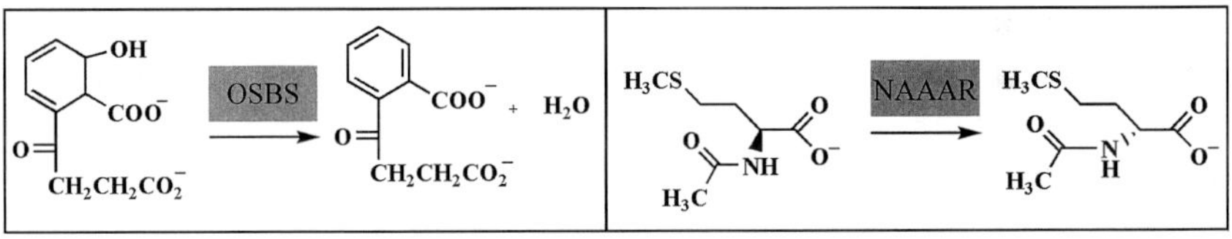

Figure 3. The O-succinylbenzoate synthase (OSBS) enzyme in *Amycolaptosis* sp. catalyzes both the OSBS reaction and the N-Acylamino Acid Racemase (NAAAR) reaction using the same active site.

The SFLD accommodates these latter two types of multifunctional enzymes by providing a many-to-many relationship between the EFD and Reaction tables—a single EFD can have an arbitrary number of Reaction entries, and can turn over multiple substrates. When known, the SFLD schema marks a single canonical reaction for each EFD to indicate which of the multiple reactions catalyzed has the greatest biological relevance. Figure 4 shows a simplified version of the SFLD schema. Note the inclusion of a partial reaction table, which, as discussed above, is key to the ability to correlate conserved structural elements/residues to the chemical functions conserved at the superfamily level.

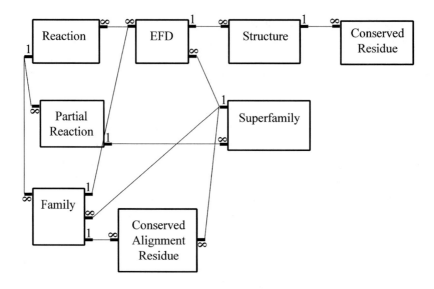

Figure 4. A simplified version of the SFLD schema.

## 2.4. *Functional Annotation and Misannotation*

The sheer size of many of the most commonly used sequence databases and sequencing projects requires the use of automated methods for assigning protein function to newly sequenced open reading frames (ORFs), leading to significant levels of misaannotation.[16-18] These typically sequence based methods face an especially difficult challenge when dealing with mechanistically diverse enzyme superfamilies.[2] This is because a superfamily member of unknown function can show a statistically significant level of sequence similarity to other members of the superfamily which, although they share a common mechanistic step, do not perform the same overall reaction. If, in the set of statistically significant matches, the closest characterized sequence represents an enzyme whose function is from a different family than the unknown sequence of interest, that sequence will often mistakenly be assigned the function of the characterized sequence.

The use of explicitly linked structure-function data helps us address the problems of annotation and misannotation. The SFLD allows users to place sequences of unknown function which appear to belong to an enzyme superfamily into a multiple sequence alignment of the superfamily, the subgroup, or family, thus providing an easily accessible basis for functional assignment. The superfamily level alignment includes information about the positions and residues of the superfamily that deliver catalytic function, allowing users to quickly evaluate whether the new sequence possesses the catalytic machinery required to perform partial reactions common to all members of the superfamily. Placement of these sequences in multiple alignments at the subgroup or family level aids in determining the likelihood that a given ORF has been accurately annotated with regard to the identity of its substrate and overall chemical reaction. This simple analysis, while not foolproof, has proven useful in evaluating the accuracy of annotations within a superfamily. Our lab was able to determine, for example, that 8 of the 30 sequences annotated in Genbank[19] as muconate cycloisomerases, while certainly members of the enolase superfamily, lack catalytic residues required to perform the specific annotated function.[20] To further enrich these capabilities, work is underway to link these alignments with the Chimera visualization software,[21] allowing users to view relevant three-dimensional structures simultaneously with related multiple sequence alignments.

Of course, multiple sequence alignments do not always provide enough information for a biologist to determine the precise function, e.g., the family level classification, of a new member of an enzyme superfamily. Often, only a subset of the functional and conserved residues of a family or superfamily can be identified in the new sequence. This only allows us to predict accurately that this enzyme will include the mechanistic step conserved throughout the superfamily, but not the substrate or the product. Alternatively, new sequences may perform new reactions for which the functionally important residues have not been identified (or cannot be inferred by alignment to known families). The SFLD classifies such sequences as within the proper superfamily, or in some cases within a subgroup labeled "unknown function", but not within a family. Information about the metabolic pathway, and when available, operon context, can also aid in the determination of an enzyme's primary catalytic role.[22-24] We are in the process of integrating these types of information into the SFLD.

### 2.5. Guidance for Protein Engineering

Protein engineering, whether to generate new functions or to improve on old ones, requires choosing a template protein for use as a scaffold. The kind of information captured in the SFLD can be used to guide this choice. Because mechanistically diverse enzyme superfamilies have been used by nature to evolve many different enzymatic reactions, it follows that superfamily members could be useful templates in the lab to re-engineer new and different enzymes as well. For example, it has been shown that two members of the enolase superfamily can be reengineered, via a single point mutation, to perform the very different reaction of a third member.[4] The key to success in this experiment was the recognition, provided by the superfamily context, of the common partial reaction all three members share. In effect, the active site of each of the superfamily members is already pre-organized to perform the proton abstraction step required for any other member, simplifying the re-engineering problem. The SFLD exploits these principles by allowing users to identify superfamily scaffold proteins potentially capable of performing a fundamental partial reaction required to generate entirely new chemical reactions.

## 2.6. *Grading the Reliability of Functional Information*

As mentioned above, the SFLD can facilitate the functional classification of an uncharacterized protein by placing that protein into the appropriate superfamily, subgroup, or family. The reliability of such a classification depends greatly on the quality of the classifications within the SFLD itself. For example, if an uncharacterized protein X closely resembles proteins in the adenosine deaminase family within the amidohydrolase superfamily, one might want to determine whether the closest relatives of protein X have been experimentally determined to perform the adenosine deaminase reaction or whether their family classification was made based merely on sequence similarity to other experimentally characterized members of the family. If protein X is closely related to experimentally characterized family members, that strengthens the argument for assigning protein X the adenosine deaminase function.

The SFLD uses evidence codes to indicate the type and reliability of functional information. SFLD evidence codes are based on those developed by the Gene Ontology consortium,[25] but have been modified to fit the requirements of representing structure-function information. Where applicable, evidence codes are paired with the literature references upon which they are based. The assignment of a particular EFD to a family, for example, comes with an evidence code and literature references deemed relevant by the curator to this assignment. This is similar to approaches adopted by other resources such as SwissProt.[26] Some examples of evidence codes that might be used for family assignments, ordered roughly in order of reliability, are:

> *IES (Inferred from Experiment and Sequence)*: Used when family membership is assigned based on an experimental assay that shows that the EFD in question catalyzes the canonical family reaction, and there is clear sequence and/or structural similarity to existing family members.

> *ISS (Inferred from Sequence or Structural similarity)*: Used when family assignment is based on overall sequence or structural similarity, reviewed for accuracy by a human curator, to existing family members.

*IEA (Inferred from Electronic Annotation)*: Used when family assignment is based on overall sequence or structural similarity to existing family members but has not been reviewed for accuracy by a human curator.

The evidence codes used in the SFLD and their definitions can be found at http://sfld.rbvi.ucsf.edu/ecodes.html.

## 2.7. Metadata

Whenever we try to place the boundaries of classification upon biological systems, we inevitably are confronted with cases that appear to stretch the rules. An enzyme, for example, may have as its biologically relevant function the catalysis of multiple reactions. The humulene synthase enzyme from *Abies grandis*, for instance, is known to catalyze reactions leading to at least 52 distinct products, only a fraction of which are of known biological importance.[27] The presence of such information, while often useful to users, is difficult to predict prior to curation. To handle these cases, nearly all of the SFLD tables contain a "metadata" field in which the curators of a family or superfamily can enter textual information.

## 2.8. Methods of Searching the SFLD

The main purpose of the SFLD is to facilitate the leveraging of structure-function data. Thus, it is of the highest importance that users be able to access the data via methods most informative from their own scientific perspectives. For example, a protein engineer looking to design an enzyme to perform a particular reaction might want to search the SFLD for enzymes catalyzing similar reactions or underlying partial reactions. Such searches can be performed by entering a SMARTS query, or by sketching chemical structures using a Java applet on the SFLD search page. Alternatively, users can also search the reactions by Enzyme Commission number[28] or simply browse a list of all the reactions in the database.

Those interested primarily in the determining the function of an uncharacterized protein can query the SFLD using its sequence. This sequence is matched to pre-generated hidden Markov models[29] representing the superfamilies, subgroups, and families in the SFLD. The resulting matches (with their scores and expectation values) are displayed, along with a hyperlink for each match leading to a dynamically generated alignment of the query sequence to the multiple sequence alignment used to construct the hidden Markov model. The alignments produced highlight the conserved residues that participate in enzymatic catalysis, and provide links to literature references of the experiments through which the structure-function relationship was determined. Users can also view a query sequence in the context of the superfamily/subgroup/family in the form of a dendrogram generated using ClustalW's neighbor joining algorithm[30] and can view relevant structures associated with the multiple alignments.

Users can also browse the SFLD in multiple ways. Lists of all superfamilies and EFDs within any hierarchical level of a superfamily are available, as well as lists of all reactions, and structures. Users can easily navigate the SFLD hierarchy of superfamily, subgroup, family, and enzyme functional domain levels.

When a three-dimensional structure is available, a link is provided allowing users to open and view the structure in Chimera[21] with a single mouse-click. Methods of searching the database using three-dimensional coordinates and new representations of enzyme function are currently being developed.

## 3. Conclusion

Developing a resource to represent structure-function relationships that can be leveraged for biological discovery in the context of mechanistically diverse enzyme superfamilies has required us to address many of the issues involved in making any biological database, including dealing with multi-functional enzymes and grading the reliability of data. It has also presented some unique, domain-specific challenges in terms of data organization and representation, such as the implementation of a structure-function knowledge hierarchy that reflects the patterns of conservation in enzyme superfamilies, and the representation of enzyme function itself. The SFLD is a first attempt at addressing some of these challenges, and provides a computational resource for those investigating enzyme structure-function relationships for applications that range from determination of the function of a new protein to providing guidance for engineering a new function into an existing enzyme scaffold.

## Acknowledgments

We thank John Gerlt for expert advice on the enolase and crotonase superfamilies, and Kinkead Reiling for his advice on the terpene cyclase superfamily. This work was supported by NIH R01-GM60595 and NSF DBI-0234768 to PCB, and NIH P41-RR01081 to TEF.

**References**

1.  Watson JD, et al, *IUBMB Life* 55, 249 (2003).
2.  Gerlt JA, Babbitt PC, *Genome Biol* 1, REVIEWS0005 (2000).
3.  Gerlt JA, Babbitt PC, *Annu Rev Biochem* 70, 209 (2001).
4.  Schmidt DM, et al, *Biochemistry* 42, 8387 (2003).
5.  Babbitt PC, et al, *Biochemistry* 35, 16489 (1996).
6.  Weininger D, *J. Chem. Inf. Comp. Sci.* 28, 31 (1988).
7.  Wilmanns M, Priestle JP, Niermann T, Jansonius JN, *J Mol Biol* 223, 477 (1992).
8.  Yanofsky C, Horn V, Bonner M, Stasiowski S, *Genetics* 69, 409 (1971).
9.  Kirschner K, Szadkowski H, Henschen A, Lottspeich F, *J Mol Biol* 143, 395 (1980).
10. Cohn W, Kirschner K, Paul C, *Biochemistry* 18, 5953 (1979).
11. Palmer DR, et al, *Biochemistry* 38, 4252 (1999).
12. Copley SD, *Curr Opin Chem Biol* 7, 265 (2003).
13. O'Brien PJ, Herschlag D, *Chem Biol* 6, R91 (1999).
14. Jensen RA, *Annu Rev Microbiol* 30, 409 (1976).
15. Miller BG, Raines RT, *Biochemistry* 43, 6387 (2004).
16. Bork P, Koonin EV, *Nat Genet* 18, 313 (1998).
17. Brenner SE, *Trends Genet* 15, 132 (1999).
18. Abascal F, Valencia A, *Proteins* 53, 683 (2003).
19. Benson DA, et al, *Nucleic Acids Res* 32 Database issue, D23 (2004).
20. Dodevski I, Pegg, S. C.-H., Babbitt, P. C. (unpublished)
21. Pettersen EF, et al, *J Comput Chem* 25, 1605 (2004).
22. Green ML, Karp PD, *BMC Bioinformatics* 5, 76 (2004).
23. Osterman A, Overbeek R, *Curr Opin Chem Biol* 7, 238 (2003).
24. Reed JL, Vo TD, Schilling CH, Palsson BO, *Genome Biol* 4, R54 (2003).
25. Ashburner M, et al, *Nat Genet* 25, 25 (2000).
26. Boeckmann B, et al, *Nucleic Acids Res* 31, 365 (2003).
27. Steele CL, Crock J, Bohlmann J, Croteau R, *J Biol Chem* 273, 2078 (1998).
28. Webb EC, NC-IUBMB. *Enzyme Nomenclature: Recommendations of the Nomenclature Committee of the International Union of Biochemistry and Molecular Biology on the Nomenclature and Classification of Enzymes.* New York, NY: Academic Press (1992).
29. Eddy SR, *Bioinformatics* 14, 755 (1998).
30. Thompson JD, Higgins DG, Gibson TJ, *Nucleic Acids Res* 22, 4673 (1994).

# DISCOVERING SEQUENCE-STRUCTURE MOTIFS FROM PROTEIN SEGMENTS AND TWO APPLICATIONS

THOMAS TANG, JINBO XU, MING LI

*School of Computer Science, University of Waterloo,*
*200 University Ave. W., Waterloo, Ont., N2L 3G1, Canada*
*{tcktang,j3xu,mli}@cs.uwaterloo.ca*

We present a novel method for clustering short protein segments having strong sequence-structure correlations, and demonstrate that these clusters contain useful structural information via two applications. When applied to local tertiary structure prediction, we achieve ~60% accuracy with a novel dynamic programming algorithm. When applied to secondary structure prediction based on Support Vector Machines, we obtain a ~2% gain in $Q_3$ performance by incorporating cluster-derived data into training and classification. These encouraging results illustrate the great potential of using conserved local motifs to tackle protein structure predictions and possibly other important problems in biology.

## 1 Introduction

A major obstacle for protein tertiary structure prediction lies in the complexity of modeling protein 3D conformations due to the large degree of structural freedom and complicated interactions among residues. Previous models of computation include a number of lattice as well as off-lattice models [1]. A recently emerging model treats a protein as a composition of small local structural motifs, a concept inspired by the conjecture that a newly created polypeptide forms local folds in parts before settling to its final fold [2]. This model manages to reduce the size of protein conformational space to a point where many search-based prediction strategies finally become feasible. As a result, extraction of local motifs through classification of protein segments has always been a subject of intense study.

We initially created RAPTOR [3], an innovative protein tertiary structure predictor based on optimal threading by linear programming. This development has stimulated our interest in *ab initio* structural prediction and led us to investigate local fold information through clustering. The current results to be presented include a novel method for clustering protein segments with strong sequence-structure correlations, and two applications of the resultant clusters to structural predictions for demonstrating their usefulness.

## 2 Clustering of Short Protein Segments

Methods for clustering short protein segments are generally divided into two groups: a) those with clustering based on structure alone [4, 5], and b) those with clustering based on both sequence and structure [6, 7]. Methods in the former omit sequence information, thus using the clusters they produce in *ab initio* structural prediction requires external guidance such as a global energy function. Since this study depends on sequence information to do prediction, we need a clustering method in the second group instead. Existing methods in this group perform clustering in two stages. Some of them first classify segments into clusters solely by sequence similarity and then sub-classify members in each cluster by structural similarity [6], while others did the reverse [7]. A problem associated with the two-stage approach is that segments with similar sequence patterns and folds might not as clearly reveal such a relationship when one looks at sequence and structure as separate entities. Those segments are likely to get misclassified in either or both stages. In this paper, we present a one-stage method, which will eliminate the deficiency by considering both sequence and structure together throughout the whole clustering process.

## 2.1 Segment Distance

For each residue $i$, its tertiary structure is represented by its *phi* ($\varphi_i$) and *psi* ($\psi_i$) angles in degrees, and its sequence information by frequency profiles comprising $f_{ij}$ for amino acid $j$. Given segments $x$ and $y$ of length $L$, their distance $D(x, y)$ is:

$$D(x, y) = \begin{cases} \sqrt{\sum_{i=0}^{L-1}\left(\left(\frac{\Delta\varphi_i}{360}\right)^2 + \left(\frac{\Delta\psi_i}{360}\right)^2 + \sum_{j=0}^{19}\Delta f_{ij}^{\ 2}\right)} & \text{if } \max(\Delta\varphi_i, \Delta\psi_i) \le \theta \ \forall i \\ \infty & \text{otherwise} \end{cases} \quad (1)$$

Symbol $\Delta$ denotes the absolute difference in the associated quantity. Value $\theta$ restricts the largest dihedral angle difference permitted, and is $L$-dependent so as to allow higher leniency for longer segments. Eq. (1) has two ideal properties. First, it encompasses differences in both sequence patterns and structures, hence allowing one-stage clustering. Second, it is the Euclidean distance between two points so it satisfies the triangular inequality, a qualifying condition for use in clustering [4]. Note that the validity of Eq. (2) justifies the assumption that contributions from differences in structure and in sequence have equal weights.

$$0 \le \left(\frac{\Delta\varphi_i}{360}\right)^2 + \left(\frac{\Delta\psi_i}{360}\right)^2 \le 2 \text{ and } 0 \le \sum_{j=0}^{19}\Delta f_{ij}^{\ 2} \le 2 \quad \forall i \in [0, L-1] \quad (2)$$

## 2.2 Cluster Radius

Besides a distance function, we need a threshold, called *cluster radius*, to tell if two segments are sufficiently close to be grouped together. The choice of cluster radius is crucial: being too small yields a handful of clusters capturing only the most conserved motifs, while being too large yields coarse clusters contaminated with irrelevant segments. A systematic way exists to determine a suitable radius for a given segment length. First, segments of that length are extracted from a large database of non-redundant proteins whose structures are known. An ideal database is PDB Select 25 [8]. The set of all segments are then divided in half, and distances between segments in different halves are computed. The resultant figures form a normal distribution with mean $\mu$ and standard deviation $\sigma$. The radius is set to $\mu - 3\sigma$, corresponding to a confidence interval of 99.73%. This choice of radius is found to consistently deliver clusters of reasonable quality.

## 2.3 Segment Preparation

Eq. (1) requires sequence profiles for both segments showing the frequency for each amino acid at each position. The profiles in this study are generated from multiple alignments in the HSSP database [9], and post-processed with the Voronoi Monte Carlo algorithm [10] to correct for unequal representations. Aside from profiles, secondary structure labels are also gathered, and for that the DSSP secondary structure labeling [11] is chosen due to its popularity.

## 2.4 Clustering Algorithm

The novel clustering algorithm is derived from the famous k-means algorithm [12], modified to allow a variable number of clusters [13]. It makes use of a special cluster called the *residue cluster* to hold segments failing to get classified due to their unique sequence patterns or shapes. The residue cluster is initially empty. Note that once a segment is placed into the residue cluster, it may no longer be used to start a new cluster. The algorithm is outlined below.

| **Protein Segment Clustering Algorithm** |
|---|
| *Input:* cluster radius $r$, minimum size $m$, segment set $S$, maximum trial count $t$ |
| 1. Create empty residue cluster $C_{res}$ |
| 2. Repeat until no changes or $t$ trials have been exhausted |
| 3.     For each segment $s \in S$ do |
| 4.         Find cluster closest to $s$, or set distance to $\infty$ if none exists yet |
| 5.         If distance $\leq r$ then move $s$ to new cluster and update old cluster |
| 6.         Otherwise, if $s \notin C_{res}$ then create new cluster with $s$ as centroid |
| 7.     Merge all nearby clusters (with distance $< 0.5r$) |
| 8. For each cluster smaller than $m$ do |
| 9.     Eliminate cluster and transfer all its segments to $C_{res}$ |
| 10. Return the final set of clusters |

## 2.5 Experiments and Results

This section presents results on clustering a set of 396 non-redundant protein peptides referred to as CB396 [14]. Segment length $L$ was set to 8, a value small enough to allow clusters of reasonable size but large enough to capture local residue interactions. In fact, it has been shown that segments of length 8 are very effective at preserving local sequence-dependent information [2]. The cluster radius was set to 1.2 based on the method described in Section 2.2. Both the minimum cluster size and maximum trial count were set to 5. Value of symbol $\theta$ in Eq. (1) was set to 120°, a reasonable limit for length-8 segments [6].

The output comprised 357 clusters, but the number of distinct structural motifs was much less since many clusters either had the same fold, or were overlapping images of the same motif. For instance, 89 clusters were helices, showing the motif's abundance and its variety in sequence patterns. In summary, all motifs in the I-sites library [6] had been discovered together with some new ones. Examples of new motifs are shown in Figure 1. The motif in Figure 1(a) is characterized by a strong preference for hydrophobic residues at position 3 followed by a strong preference against them at the next position, indicating a possible emergence from inside the protein to the surface. The motif in Figure 1(b) is characterized by a GLY at position 3, a conserved hydrophobic residue at position 4, and finally an ASN or ASP at position 5. Descriptions for the motif in Figure 1(c) and the rest are omitted due to space limitation.

## 3 *Ab Initio* Local Tertiary Structure Prediction

The first application of motif clusters is aimed at the *ab initio* prediction of local tertiary structures – the prediction of tertiary structures of short protein segments based solely on the sequence information contained in the segments. Success in resolving local structure prediction will be a major milestone in fold recognition, homology detection, and understanding of the protein folding process.

## 3.1 Assigning Clusters to Protein Segments

Scoring function $K_c(s)$, shown in Eq. (3), computes the likelihood of a length-$L$ segment $s$ belonging to cluster $c$ based on sequence composition. Symbols $s_{ij}$ and $c_{ij}$ denote the frequency of amino acid $j$ at position $i$ on $s$ and $c$'s centroid respectively. Symbol $b_j$ denotes the background frequency for amino acid $j$.

$$K_c(s) = \log_2 \left( \frac{\prod_{i=0}^{L-1} \sum_{j=0}^{19} s_{ij}\, c_{ij}}{\prod_{i=0}^{L-1} \sum_{j=0}^{19} s_{ij}\, b_j} \right) \tag{3}$$

a)  b)  c)

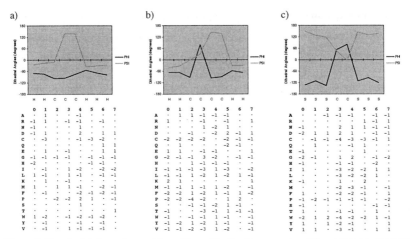

|   | 0 | 1 | 2 | 3 | 4 | 5 | 6 | 7 |
|---|---|---|---|---|---|---|---|---|
|   | H | H | C | C | C | M | H | H |
| A | · | 1 | · | · | -1 | · | · | · |
| R | -1 | 1 | · | -1 | -1 | · | -1 | · |
| N | -1 | · | · | 1 | · | · | · | · |
| D | -1 | 1 | · | · | 2 | · | 1 | 1 |
| C | · | -3 | · | · | -1 | -3 | -2 | · |
| Q | · | · | · | · | · | · | 1 | 1 |
| E | · | 1 | · | -1 | · | · | 1 | 1 |
| G | -1 | -1 | -1 | -1 | · | -1 | -1 | -1 |
| H | -2 | · | · | · | -1 | -1 | · | · |
| I | · | -1 | · | 1 | -2 | · | -2 | -2 |
| L | 1 | -1 | · | 1 | -1 | · | -1 | -1 |
| K | · | 1 | · | -1 | · | · | · | · |
| M | 1 | · | 1 | 1 | -1 | · | -2 | -1 |
| F | · | -1 | · | · | -2 | -1 | -2 | -1 |
| P | · | · | -2 | -2 | 2 | 1 | · | -1 |
| S | · | · | · | 1 | · | · | · | · |
| T | · | · | · | · | · | · | · | 1 |
| W | 1 | -2 | · | -1 | -2 | -1 | -2 | · |
| Y | · | -1 | · | · | -1 | · | -1 | · |
| V | · | -1 | · | 1 | -1 | -1 | -1 | · |

|   | 0 | 1 | 2 | 3 | 4 | 5 | 6 | 7 |
|---|---|---|---|---|---|---|---|---|
|   | H | H | C | C | C | C | H | H |
| A | · | 1 | 1 | -1 | -1 | -1 | · | · |
| R | 1 | · | · | -1 | · | -1 | · | 1 |
| N | · | · | · | 1 | -2 | 1 | · | · |
| D | · | · | -1 | · | -2 | 2 | -1 | 1 |
| C | -2 | -2 | -2 | -2 | · | -2 | -1 | -1 |
| Q | · | 1 | · | · | · | -2 | -1 | · |
| E | 1 | 1 | · | -1 | -1 | · | · | 1 |
| G | -2 | -1 | -1 | 3 | -2 | · | -1 | -1 |
| H | · | · | 1 | -1 | -1 | -1 | · | · |
| I | -1 | -1 | -1 | -3 | 1 | -3 | · | -2 |
| L | -1 | -1 | 1 | -2 | 1 | -2 | 1 | -1 |
| K | 2 | 1 | · | · | · | · | · | 1 |
| M | -1 | -1 | 1 | -1 | 1 | -2 | · | -1 |
| F | -2 | -2 | 1 | -2 | 1 | -1 | 1 | -2 |
| P | -2 | -2 | -4 | -2 | · | 1 | 2 | · |
| S | · | · | -1 | -1 | -2 | 1 | -1 | · |
| T | -1 | · | · | -3 | -1 | 1 | -1 | -1 |
| W | -1 | · | -1 | -1 | 1 | -1 | · | -1 |
| Y | -1 | -2 | 1 | -2 | 1 | -1 | · | -1 |
| V | -1 | -1 | -2 | -3 | 1 | -2 | · | -1 |

|   | 0 | 1 | 2 | 3 | 4 | 5 | 6 | 7 |
|---|---|---|---|---|---|---|---|---|
|   | S | S | S | C | C | S | S | S |
| A | · | · | -1 | -1 | -1 | · | -1 | -1 |
| R | · | · | · | · | 1 | 1 | -1 | 1 |
| N | -1 | · | · | 2 | 1 | 1 | -1 | -1 |
| D | -2 | 1 | 1 | 2 | 1 | · | -1 | -1 |
| C | · | -1 | -1 | -4 | -3 | -3 | -1 | 1 |
| Q | · | 1 | · | · | -1 | 1 | · | · |
| E | -1 | · | · | · | · | 1 | · | · |
| G | -2 | -1 | · | 1 | 2 | · | -1 | -2 |
| H | -1 | · | · | -1 | -1 | · | -2 | · |
| I | 1 | · | · | -3 | -2 | -2 | 1 | 1 |
| L | 1 | · | · | -3 | -2 | -2 | 1 | · |
| K | -1 | · | · | · | 1 | 2 | · | · |
| M | · | · | · | -2 | -3 | -1 | · | -1 |
| F | 1 | · | · | -2 | -2 | -1 | · | 1 |
| P | -1 | -2 | -1 | -1 | -1 | -1 | · | -2 |
| S | -1 | · | · | · | · | · | -1 | -1 |
| T | 1 | · | · | -1 | -1 | · | 1 | · |
| W | -2 | 1 | 2 | -4 | -2 | -2 | 1 | -1 |
| Y | 1 | · | 1 | -2 | -1 | · | · | 1 |
| V | 1 | 1 | · | -3 | -1 | · | 1 | 1 |

Figure 1. Structural information and log-odds profiles for three novel motifs not listed in the I-sites database. Dot (·) represents background frequency.

A *cluster assignment*, or *assignment*, refers to an instance when a cluster is assigned to a segment based on a score computed via Eq. (3). The assignment is said to "cover" the segment and its residues. Each assignment has three basic attributes: the cluster being assigned, the segment being covered, and the score associated with the pair. Finally, a *cluster assignment rank*, denoted by $R$, means that each segment is assigned the $R$ clusters yielding the $R$ highest scores. The highest scoring assignment is at rank 1, the second highest at rank 2, etc.

### 3.2 Evaluating Local Structure Prediction

The evaluation scheme for local tertiary structure prediction was invented by Lesk [15]. It takes two parameters, a window size $w$ and a RMSD threshold $t$. Given a true structure and its prediction, the scheme computes the percentage of residues found in length-$w$ segments whose predicted structures are within $t$ from the true structure after superposition. In this study, we use the same settings as Bystroff and Baker [6] to facilitate comparison (i.e. $w = 8$, $t = 1.4$ Å).

### 3.3 Eliminating Noise Clusters

In a large cluster set, some weak clusters capturing rare motifs possess similar sequence profiles as do the significant clusters capturing more common motifs. Those weak clusters tend to compete with the significant clusters for sequence similarity with target segments during cluster assignment, degrading prediction accuracy. Since they create "noise" that disturbs prediction, those weak clusters are called *noise clusters* and should be eliminated.

Clusters produced by the algorithm described in Section 2.4 are of minimum size $m$. If $m$ is set too small, many noise clusters result. If it is set too large, significant clusters are lost. To determine $m$ maximizing the predictive power for a set of clusters, the following empirical method is used:

| **Noise Cluster Elimination** |
| --- |
| *Input:* cluster set $C$, protein set $P$, minimum size bound $[m_l, m_h]$ |
| 1.  For each $m$ in range $[m_l, m_h]$ |
| 2.      Remove clusters of size less than $m$ from $C$ to get $C'$ |
| 3.      Get average prediction accuracy for $P$ using $C'$ as follows: |
| 4.          For each protein $p \in P$ do |
| 5.              Assign highest scoring cluster to each overlapping segment in $p$ |
| 6.              Sort all assignments by score |
| 7.              Assign structures to $p$ from highest scoring assignments |
| 8.              Evaluate prediction as described in Section 3.2 |
| 9.  Return $m$ and $C'$ resulting in highest average prediction accuracy |

Figure 2 shows the fluctuation in prediction accuracy as $m$ increased from 5 to 25 inclusive. While the accuracy remained rather constant in the middle stretch, it rose and fell sharply at both ends. Prediction was compromised by the presence of noise clusters for small $m$ ($< 8$) and the absence of significant clusters for large $m$ ($> 20$). The optimal minimum cluster size was $m = 16$, yielding a prediction accuracy of 54.66%.

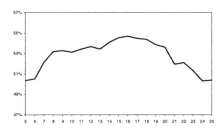

Figure 2. Fluctuation in prediction accuracy as minimum cluster size $m$ increases from 5 to 25.

### 3.4 Improving Cluster Likelihood Function

Eq. (3) has been extended with the addition of a new term, as shown below:

$$K_c(s) = \log_2 \left( \frac{\prod_{i=0}^{L-1} \sum_{j=0}^{19} s_{ij}\, c_{ij}}{\prod_{i=0}^{L-1} \sum_{j=0}^{19} s_{ij}\, b_j} \right) - base_c \tag{4}$$

The new term $base_c$ represents the basal cutoff score specific to cluster $c$. The old version (i.e. Eq. (3)) assumes a cutoff of 0 for all clusters, an intuitive choice for log-odds but nonetheless a crude assumption. The derivation procedure of cluster-specific cutoffs is based on a simple observation. Clusters capturing rare motifs are likely to introduce false positive ($F^+$) assignments, thus requiring a higher cutoff to avoid high a $F^+$ rate. On the other hand, clusters capturing common motifs are likely to introduce false negative ($F^-$) assignments and a lower cutoff is needed to suppress the $F^-$ rate. The actual procedure for deriving the cutoffs is relatively unimportant and hence omitted due to space limitation.

### 3.5 Predicting Local Structure using Dynamic Programming (DP)

The algorithm to be presented is inspired by two observations. First, the cluster assignment most appropriately capturing the shape of a segment might not be the optimal (i.e. highest-scoring) one but a sub-optimal one. Second, if overlapping assignments have serious structural conflicts, then they should not be adopted together. We propose here an objective function for measuring the quality of a set of assignments when used together to form a prediction, and a DP approach for maximizing it. The objective function for a target protein $p$ of length $n$ is:

$$F(X) = q \sum_{i=0}^{n-1} \text{score}(X,i) - \sum_{i=0}^{n-1} \text{conflict}(X,i) \tag{5}$$

Function $F(X)$ returns the objective score for a set of cluster assignments $X$. Symbol $q$ is a non-negative constant for balancing the two parts representing the total score and conflict induced by $X$. It is set to 70 in this study, a value found empirically to yield one of the best predictions. Function $\text{score}(X, i)$ is:

$$\text{score}(X,i) = \begin{cases} \text{score of highest scoring assignment in } X \text{ covering residue } i, \text{ or} \\ 0 \text{ if no assignment in } X \text{ covers residue } i \end{cases}$$

And function $\text{conflict}(X, i)$ is:

$$\text{conflict}(X,i) = \begin{cases} \overline{\Delta\varphi} + \overline{\Delta\psi} \text{ between all pairs of assignments in } X \\ \text{at positions covering residue } i, \text{ or} \\ 0 \text{ if at most 1 assignment in } X \text{ covers residue } i \end{cases}$$

Symbols $\overline{\Delta\varphi}$ and $\overline{\Delta\psi}$ denote the mean absolute difference in *phi* and *psi* angles respectively. Let $L$ be the segment length and $R$ be the assignment rank. To setup for the DP algorithm, the $R$ highest-scoring assignments are made to each overlapping length-$L$ segment along $p$. Let $a_{ir}$ denote the assignment at rank $r$ starting at position $i$, where $1 \le r \le R$ and $0 \le i \le n-L$. Define $A_i = \{a_{ir} \forall r\}$ and

$A = \{a_{ir}\}$. The algorithm is to find an assignment set $X^* \subseteq A$ such that $X^*$ covers all residues in $p$ and is optimal (i.e. maximizing objective function $F$).

Each assignment set is built progressively starting from the first residue by appending to the end one adjoining assignment at a time. Note that simply extending the current optimal set by adding to its tail the best available adjoining assignment does not guarantee optimality for the resultant set. The assignment just added may overlap with existing assignments in the set, introducing new conflicts that must be fixed by replacing those assignments, which in turn may cause more new conflicts with their prior overlapping assignments and necessitate further replacements. To avoid such propagation of conflicts, a more involved DP algorithm is needed.

When an assignment $\alpha \in A_i$ is appended to the end of assignment set $X$, it would come in contact with one or more trailing assignments in $X$. The arrangement of these trailing assignments and their ranks collectively form the *tail configuration* for $X$ with respect to $\alpha \in A_i$, denoted by $tail_i(X)$. We define $tail_i(X)$ to be an empty tail configuration if $X$ is too short to reach any assignment in $A_i$. For formulation purposes, we allow $tail_j(X)$ for $j > n-L$, in which case it is treated as if $A_j$ actually existed.

For each position $i$ starting from zero, the algorithm computes $V_i$, the collection of all optimal assignment sets $X$ each having a unique non-empty $tail_{i+1}(X)$. The following recurrence ensures the optimality for each $V_{(i+1)t'}$, and the uniqueness and non-emptiness of the associated tail configuration $t'$, so the inductive hypothesis holds for position $i+1$. Finally, we assign dihedral angles to residues in $p$ by back-tracking the creation of $X^*$.

---

**DP Recurrence for Local Tertiary Structure Prediction**

*Initial condition:*

$V_0 = \{\{\alpha\} \; \forall \; \alpha \in A_0\}$

*Inductive hypothesis for position $i$, $0 \leq i \leq n - L$:*

$V_i = \{$All optimal assignment sets $w$ each having a unique non-empty $tail_{i+1}(w)\}$

*Recurrence:*

Let $V_i' = \{w \cup \{\alpha\} \; \forall \; w \in V_i$ and $\alpha \in A_{i+1}\}$

For each unique non-empty tail configuration $t'$

$V_{(i+1)t'} = X \in \{w \in V_i \cup V_i' \mid tail_{i+2}(w) = t'\}$ s.t. $F(X)$ is maximized

$V_{i+1} = \{V_{(i+1)t'}\}$

*Final solution:*

$X^* = X \in \{w \in V_{n-L} \mid w$ has an assignment in $A_{n-L}\}$ s.t. $F(X)$ is maximized

---

Note that $|V_i|$ is bounded by $(R+1)^L - 1$, the number of all possible unique non-empty tail configurations. For each position $i$, the algorithm calculates the objective value for $|V_i|*R$ new assignment sets, where each calculation is $O(L^3)$ if done carefully. Hence, the total runtime is $O(n \ |V_i| \ R \ L^3) = O(n \ L^3 \ (R+1)^{L+1})$ for all $n$ positions. Fortunately, typical values for $R$ and $L$ are small enough to make the runtime acceptable (e.g. $R = 3$ and $L = 8$ in this study).

## 3.6 Experiments and Results

A jackknife test was performed on CB396 [14], a set of 396 peptides selected through a very stringent procedure to ensure non-redundancy between members. The entire test consisted of 10 iterations, each of which involved splitting CB396 into two disjoint subsets in 80/20 ratio by residue count. The larger subset was then used for training and the smaller one for testing.

Note that testing sets containing more helices tend to yield higher accuracies than those containing more coils. Consequently, for results to be consistent, all testing sets should contain similar proportions of each secondary structure (SS). To guarantee such condition, the background proportion of each SS was first estimated from the whole CB396. Each repetition of the jackknife test then produced 50 pairs of training and testing sets, and used the pair whose testing set exhibited SS proportions most closely resembling the background ones.

Results for rank $R = 3$ are listed in Table 1. Overall, an average of 58.21% of all residues was found in a length-8 segment within 1.4 Å of the true structure, measured in RMSD. This is significant considering that the prediction relied solely on sequence information, without taking into account global forces such as disulfide bridges, hydrophobic effects, inter-group charges, and so on. The result is also a great improvement over that published by Bystroff and Baker [6], which was 50%. A breakdown in overall prediction accuracy by SS states reveals the real strengths and weaknesses of prediction using clusters. Helices were by far the most accurately predicted because they were the most conserved and abundant local motifs. Strands, albeit well conserved, were a lot harder to predict as their formation involved long-range residue interactions, something not captured by local motif clusters. Coils were the most difficult to predict since most of them lacked virtually any kind of detectable conserved patterns.

Table 1. Prediction accuracy of the DP algorithm obtained from a ten-iteration jackknife test on CB396 and evaluated using the scheme described in Section 3.2.

|  | Helix (%) | Sheet (%) | Coil (%) | Total (%) |  | Helix (%) | Sheet (%) | Coil (%) | Total (%) |
|---|---|---|---|---|---|---|---|---|---|
| 1 | 86.22 | 44.91 | 40.58 | 58.51 | 6 | 88.11 | 42.92 | 43.05 | 59.69 |
| 2 | 84.54 | 39.61 | 40.52 | 56.19 | 7 | 84.83 | 44.35 | 40.99 | 57.76 |
| 3 | 84.27 | 44.71 | 41.07 | 58.03 | 8 | 84.94 | 43.98 | 42.90 | 58.83 |
| 4 | 84.65 | 43.22 | 40.19 | 57.45 | 9 | 86.09 | 43.22 | 42.78 | 58.86 |
| 5 | 86.12 | 43.63 | 42.87 | 59.15 | 10 | 83.68 | 45.30 | 40.61 | 57.62 |
|  |  |  |  |  | Mean | 85.35 | 43.59 | 41.56 | 58.21 |

## 4 Secondary Structure Prediction

The second application of clusters deals with enhancing secondary structure (SS) prediction. The target predictor [16] is the one based on Support Vector Machines (SVM) [17], so selected because it is one of the best available. As an overview, the procedure involves building a *Secondary Structure Confidence Profile* (SSCP) and using it as additional data for training and classification.

### 4.1 Secondary Structure Confidence Profile (SSCP)

The SSCP of a protein shows the confidence of each residue being in each of the three SS states, namely helix (H), strand (E), and coil (C). Given assignment rank $R$ and target protein $p$, the method for creating SSCP first makes the $R$ highest-scoring assignments to each length-$L$ overlapping segment in $p$. Then, for each residue $i$ and SS label $s \in \{H, E, C\}$, it computes $score_{is}$ by summing the scores of all assignments covering $i$ with label $s$ at the covering position. The value $score_{is}$ is then normalized to obtain $ssc_{is}$, the SS confidence for $i$ belonging to state $s$. That is, $ssc_{is} = score_{is} / (score_{iH} + score_{iE} + score_{iC})$. The set of all $ssc_{is}$ constitutes the SSCP for protein $p$.

### 4.2 Training of SVM Binary Classifiers

Fix a window half-width $h$ such that each residue is represented by the sequence profile spanning $(2h + 1)$ columns, with the said residue in the middle. Each column is coded using 21 entries, where the extra entry is set when the window is extended beyond the ends of a protein. Together, each residue is coded by a total of $(2h + 1) * 21$ entries. When SSCP is incorporated into training, each column is coded with four additional entries. Each of the first three holds the SSCP confidence value for a different SS state, and the last is again set for the case when the window is extended beyond the ends of a protein. Hence, each residue is now coded by a total of $(2h + 1) * 25$ entries.

### 4.3 SVM Predictor Construction

Han and Sun [16] have demonstrated that different arrangements of SVM binary classifiers contribute to varying performance for the resultant SS predictor. One of the most effective configurations found in their study is called SVM MAX, which comprises three SVM binary classifiers, namely H/~H, E/~E, and C/~C. Each target residue is fed in parallel to all three classifiers, and assigned the SS label corresponding to the one giving the largest value. For optimal prediction, the half-width $h$ for the three classifiers is set to 5, 4, and 3 respectively.

## 4.4 Experiments and Results

Three metrics were used to measure the quality of SS prediction. They were the $Q_3$, the Matthew's Correlation Coefficients (MCC) [18], and the Segment Overlap measure (SOV) [19]. The jackknife test and the target data set were as described in Section 3.6, except that each training set was also used to generate SSCP and train SVM in addition to creating motif clusters. Assignment rank $R$ was set to 6. Parameters for SVM binary classifiers were 1.5 for error trade-off and 0.1 for $\gamma$ in the radial basis function used as the kernel [16].

Table 2. Prediction accuracy of SVM MAX trained without SSCP (top values) and trained with SSCP (bottom values) in a ten-iteration jackknife test on CB396. A positive delta on the last row indicates an average improvement with SSCP (delta = mean bottom value – mean top value).

| | $Q_H$ (%) | $Q_E$ (%) | $Q_C$ (%) | $Q_3$ (%) | $C_H$ | $C_E$ | $C_C$ | SOV (%) |
|---|---|---|---|---|---|---|---|---|
| 1 | 74.30 | 55.44 | 78.49 | 72.15 | 0.63 | 0.51 | 0.53 | 68.57 |
| | 76.34 | 60.90 | 78.60 | 74.10 | 0.68 | 0.55 | 0.55 | 70.63 |
| 2 | 79.03 | 57.33 | 79.07 | 74.44 | 0.66 | 0.56 | 0.57 | 69.77 |
| | 78.85 | 63.39 | 79.86 | 75.97 | 0.69 | 0.59 | 0.58 | 71.58 |
| 3 | 80.36 | 50.26 | 76.87 | 71.94 | 0.64 | 0.51 | 0.53 | 69.50 |
| | 81.57 | 55.77 | 78.58 | 74.35 | 0.69 | 0.56 | 0.55 | 70.15 |
| 4 | 76.37 | 52.78 | 77.18 | 71.61 | 0.62 | 0.51 | 0.53 | 68.37 |
| | 76.92 | 58.07 | 77.99 | 73.29 | 0.65 | 0.54 | 0.54 | 70.52 |
| 5 | 77.21 | 52.96 | 78.32 | 72.36 | 0.65 | 0.51 | 0.54 | 70.23 |
| | 79.13 | 58.54 | 78.10 | 74.23 | 0.68 | 0.55 | 0.55 | 71.43 |
| 6 | 79.72 | 56.68 | 77.05 | 73.57 | 0.65 | 0.55 | 0.55 | 71.58 |
| | 81.89 | 61.96 | 77.79 | 75.84 | 0.71 | 0.59 | 0.57 | 73.01 |
| 7 | 78.23 | 52.07 | 77.62 | 72.09 | 0.64 | 0.51 | 0.53 | 70.11 |
| | 79.25 | 55.06 | 78.65 | 73.56 | 0.67 | 0.53 | 0.55 | 71.61 |
| 8 | 76.23 | 51.23 | 76.56 | 70.81 | 0.62 | 0.50 | 0.51 | 66.54 |
| | 77.50 | 56.07 | 77.80 | 72.86 | 0.66 | 0.53 | 0.53 | 68.57 |
| 9 | 76.25 | 56.36 | 79.52 | 73.13 | 0.64 | 0.55 | 0.55 | 70.78 |
| | 77.82 | 62.51 | 79.33 | 75.01 | 0.68 | 0.59 | 0.56 | 72.92 |
| 10 | 74.93 | 50.48 | 77.72 | 70.70 | 0.61 | 0.49 | 0.52 | 66.34 |
| | 77.24 | 56.39 | 77.72 | 72.86 | 0.65 | 0.53 | 0.54 | 67.97 |
| Mean | 77.26 | 53.56 | 77.84 | 72.28 | 0.64 | 0.52 | 0.53 | 69.18 |
| | 78.65 | 58.87 | 78.44 | 74.21 | 0.68 | 0.56 | 0.55 | 70.84 |
| Delta | 1.39 | 5.31 | 0.60 | 1.93 | 0.04 | 0.04 | 0.02 | 1.66 |

The results are listed in Table 2. By combining SSCP with sequence profile for training and classification, SVM MAX predictor showed improvements in all $Q_3$, MCC and SOV measures. Specifically, SSCP contributed to an average $Q_3$ improvement of 1.93% by boosting the accuracy for helixes and strands, the latter in particular. In other words, SSCP helped the predictor be more certain when determining if a residue was part of a helix or strand. Moreover, the use of SSCP also resulted in visible improvements in all aspects of MCC and SOV. Unfortunately, improvements to $Q_C$ and $C_C$ were only minimal. After all, motif clusters could only identify regions with strong sequence-structure correlations, a condition excluding most coils. Consequently, assignments made to segments in coil regions were mostly incorrect, producing unreliable SS confidence values.

## 5 Conclusion and Future Work

Motivated by the substantial improvement to secondary structure prediction [20], a repeat of this study using sequence profiles generated by PSI-BLAST [21] is currently underway. In the longer run, we are going to investigate another kind of sequence-structure motifs for capturing long-range inter-residue interactions. Our current sequence-structure motifs can only capture local interactions, so they are not very helpful for beta sheet prediction. Nevertheless, the partition of short protein segments into clusters of local sequence-structure motifs does have profound applications. These motif clusters achieve discretization of protein conformational space and provide an effective mapping between sequence and structure, all contributing to the success of their employment to both secondary and local tertiary structure prediction. The promising results obtained could mark the beginning of a wide range of potential applications for motif clusters, which include fold recognition, domain detection, and functional annotation.

## References

1. X. Yuan, Y. Shao, C. Bystroff. *Comp Funct Genom*, 4(4):397-401, 2003
2. C. Bystroff, et al. *Curr Opin Biotechnol*, 7:417-421, 1996
3. J. Xu, M. Li, D. Kim, Y. Xu. *JBCB*, 1(1):95-117, 2003
4. R. Kolodny, P. Koehl, L. Guibas, M. Levitt. *J Mol Biol*, 323:297-307, 2002
5. C. G. Hunter, S. Subramaniam. *Proteins*, 50:580-588, 2003
6. C. Bystroff, D. Baker. *J Mol Biol*, 281:565-577, 1998
7. A. G. de Brevern, C. Etchebest, S. Hazout. *Proteins*, 41:271-287, 2000
8. U. Hobohom, M. Scharf, R. Schneider, C. Sander. *Protein*, 1:409-417, 1992
9. C. Sander, R. Schneider. *Proteins*, 9:56-68, 1991
10. P. Sibbald, P. Argos. *J Mol Biol*, 216:813-818, 1990
11. W. Kabach, C. Sander. *Biopolymers*, 22:2577-2637, 1983
12. B. S. Everitt. *Cluster Analysis 3$^{rd}$ Edition*, Halsted Press, New York, 1993
13. M. Anderbert. *Cluster analysis for applications*, Academic Press, NY, 1973
14. J. A. Cuff, G. J. Barton. *Proteins*, 34:508-519, 1999
15. A. M. Lesk. *Proteins Suppl*, 1:151-166, 1997
16. S. Hua, Z. Sun. *J Mol Biol*, 308:397-407, 2001
17. C. Burges. *Data Mining and Knowledge Discovery 2*, 121-167, 1998
18. B. W. Matthews. *Biochim Biophys Acta*, 405:442-451, 1975
19. A. Zemla, C. Venclovas, K. Fidelis, B. Rost. *Proteins*, 34:220-223, 1999
20. D. T. Jones. *J Mol Biol*, 292:195-202, 1999
21. S. F. Altschul et al. *Nucleic Acids Res*, 25(17):3389-402, 1997

# INFERRING SNP FUNCTION USING EVOLUTIONARY, STRUCTURAL, AND COMPUTATIONAL METHODS

MATTHEW W. DIMMIC

*Dept. of Biological Statistics and Computational Biology, Cornell University*
*Ithaca, NY 14853, USA*

SHAMIL SUNYAEV

*Dept. of Medicine, Brigham & Women's Hospital, Harvard Medical School*
*Cambridge, MA 02115, USA*

CARLOS D. BUSTAMANTE

*Dept. of Biological Statistics and Computational Biology, Cornell University*
*Ithaca, NY 14853, USA*

Single nucleotide polymorphisms (SNPs) are the most prevalent form of genetic variation within populations. Recent technological advances have enabled the accumulation of massive amounts of data on SNPs—more than 15 million entries in dbSNP alone—from within a range of species (e.g., human, *Drosophila*, *Anopheles*, mouse, dog, *Arabidopsis*, maize, and *Plasmodium*). Efficient and accurate prediction of a mutation's effect promises to accelerate research in a myriad of fields, ranging from medicine and agriculture to basic genetics and evolutionary biology.

Functional SNPs affect the structure or function of DNA, RNA, or proteins. The effect on the molecular function is in some cases translated further into an effect on the organism phenotype. If the phenotypic effect impacts survival and reproduction, natural selection operates on the SNP alleles. Consequently, evidence of the functional importance of SNP variants can come from three different sources. Structural biology and biochemistry can detect the influence of amino acid or nucleotide substitutions at the molecular level; association (linkage) studies pursue the detection of correlation among SNP variants with specific phenotypes of interest; evolutionary and population genetics detects natural selection by means of statistical analysis. Bioinformatics makes possible a correlated study of the bulk of recent data on human SNPs through a variety of computational approaches, which include ideas from all of the above fields. The papers in this session represent a diversity of statistical and computational advances towards elucidating the evolutionary and biophysical functional significance of particular SNPs.

The frequency of a SNP is related to both the rate of mutation and the selective forces acting on that mutation; the relative contributions of each process can affect analysis of a SNP's functional impact. To tease apart mutation bias from natural selection, Yampolsky and Stolztfus employ a novel measure of amino acid exchangeability (EX), which is based on the results of thousands of mutagenesis experiments on protein activity. Applying this measure to both between-species and within-species variation, they find that the apparent contribution can change quite drastically with increasing evolutionary distance. Their analysis on hominid variation, for example, finds a sharp cutoff point in the relationship between fixation probability and amino acid exchangeability.

Presumably, amino acid exchangeabilities are governed by the interplay of physiochemical features of amino acids, location in protein structure, and molecular function. Methods which explicitly account for these features should aid in predicting which SNPs will have an effect on protein function. In their contribution to this session, Karchin and co-workers use mutual information to measure which changes in amino acid features correlate most strongly with *in vivo* functional effects. When used as inputs in a support vector machine (SVM) classifier, the most mutually informative measures were also better at predicting which SNPs were most likely to affect function in several different protein families.

The degree of selection acting on functional SNPs is difficult to calculate without an accurate estimate of the underlying mutation rate, which can vary widely across the genome. In their session paper, Rogozin and co-workers focus on the biochemical mechanisms of the nucleotide mutational process at mutational hotspots. Examining a variety of mechanisms, they find that a large proportion of mutations can be explained by oxidative damage on the nucleotide level. An analysis of mutations in human mitochondrial genes finds that the molecular mechanism of mutation differs significantly between hypervariable and coding regions, hypothesizing that this difference may be caused by dislocation mutagenesis.

The paper by Webb-Robertson *et al.* describes a Bayesian formulation to compare the power of molecular evolutionary models to predict the distribution of observed SNPs. Their analysis indicates that simple models which treat only mutational effects perform roughly as well as the current models of amino acid exchangeability, evidence that there is still a great deal of room for improvement on current methodologies.

These identification and modeling methods assume that an organism's genetic variation has already been described and that a set of SNPs is already available. One promising technique for assembling the map of an individual's genetic variation is optical mapping, where fragments of DNA are bound to a surface, cleaved, and visualized using light microscopy. Anantharaman and co-workers describe an algorithm for reassembling the ordering of these fragments based on the location of the cleavage sites. They demonstrate how this algorithm can be used to infer the parental haplotypes of a diploid organism, an advance which holds great promise for the genome-wide study of how variation is maintained in species over generational time.

## Acknowledgments

The session organizers would like to thank all those who submitted manuscripts to this session, and we are grateful to the anonymous referees for their careful reviews.

# FAST AND CHEAP GENOME WIDE HAPLOTYPE CONSTRUCTION VIA OPTICAL MAPPING*

T.S. ANANTHARAMAN*, V. MYSORE†, AND B. MISHRA†

*Wisconsin Biotech Center, Univ. Wisc., Madison WI, U.S.A

† Courant Institute of Mathematical Sciences, NYU, New York, NY, U.S.A.

E-mail:tsa@biostat.wisc.edu; {vm40,mishra}@nyu.edu

We describe an efficient algorithm to construct genome wide haplotype restriction maps of an individual by aligning single molecule DNA fragments collected with Optical Mapping technology. Using this algorithm and small amount of genomic material, we can construct the parental haplotypes for each diploid chromosome for any individual. Since such haplotype maps reveal the polymorphisms due to single nucleotide differences (SNPs) and small insertions and deletions (RFLPs), they are useful in association studies, studies involving genomic instabilities in cancer, and genetics, and yet incur relatively low cost and provide high throughput. If the underlying problem is formulated as a combinatorial optimization problem, it can be shown to be NP-complete (a special case of $K$-population problem). But by effectively exploiting the structure of the underlying error processes and using a novel analog of the Baum-Welch algorithm for HMM models, we devise a probabilistic algorithm with a time complexity that is linear in the number of markers for an $\epsilon$-approximate solution. The algorithms were tested by constructing the first genome wide haplotype restriction map of the microbe *T. pseudoana*, as well as constructing a haplotype restriction map of a 120 Mb region of Human chromosome 4. The frequency of false positives and false negatives was estimated using simulated data. The empirical results were found very promising.

## 1. Introduction

Diploid organisms, such as humans, carry two mostly similar copies of each chromosome, referred to as haplotypes. Variations in a large population of haplotypes at specific loci are called polymorphisms. The co-associations

---

*The work reported in this paper was supported by grants from NSF's Qubic program, NSF's ITR program, Defense Advanced Research Projects Agency (DARPA), Howard Hughes Medical Institute (HHMI) biomedical support research grant, the US Department of Energy (DOE), the US air force (AFRL), National Institutes of Health (NIH) and New York State Office of Science, Technology & Academic Research (NYSTAR).

of these variations across the loci indices are of intense interest in disease research.

The main limitation of most SNP based approaches is that each SNP is assayed separately without the related phasing information. Instead, the phase is inferred statistically from a large population of SNP data and employ certain simplifying assumptions such as: parsimony in the total number of different haplotypes in the population, the Hardy-Weinberg equilibrium, perfect phylogeny to combinatorially constrain the possible haplotypes. See the full paper [4] for a detailed survey of the literature.

For a genotyping method to be able to correctly determine the phasing between neighboring polymorphic markers in every individual haplotype map, it must ultimately be able to test single DNA fragments containing 2 or more heterozygous polymorphic markers in a single test. It is possible, of course, to assemble individual haplotype maps by sequencing the individual's entire genome using a modified sequence assembly algorithm [7,9] but the cost of doing this is prohibitive[a].

Here, we propose a direct and more cost-effective approach using the fairly well developed single molecule technology of Optical Mapping.

Each individual haplotype map of restriction sites will only detect a small fraction of all polymorphisms in the human genome, but using a commonly accepted linkage disequilibrium assumption (see[4]), approximately 8 individual haplotype restriction maps will contain more than the 300,000 SNPs required to infer all other known polymorphisms in the individual genome. Even with 50 fold data redundancy required, all date required for 8 individual haplotype restriction maps can be collected for under $1000.

## 2. Problem Formulation

Our problem can be formulated mathematically as follows: We assume that all individual single molecule DNA fragments are derived from a diploid genome (ignoring the case of sex chromosomes) with two copies of homologous chromosomes. Each DNA fragment is further mapped by cleavage with a restriction enzyme of choice and imaged by an imaging algorithm to produce an ordered sequence of "restriction fragment lengths" or equivalently, "restriction sites." The variations in these restriction fragment lengths are primarily due to RFLPs as well as SNPs at the restriction sites. Additionally, there are further variations introduced by the experimental

---

[a]This cost has been estimated to be over $10 million per individual.

process and could be assumed due to: sizing errors, partial digestion, short missing restriction fragments, false cuts, ambiguities in the orientation, optical chimerisms, etc. Thus, the genomes may be represented as two haplotype restriction maps, $H_1$ and $H_2$, for the same individual which differ only slightly from a genotype restriction map $H$ by a small number of short insertions, deletions and SNPs that coincide with restriction sites. All such maps, $H$, $H_1$ and $H_2$, are assumed to be representable as a sequence of restriction sites (e.g. $H_{2,i}$, with indices $0 \leq i \leq (N+1)$, where $H_{2,0}$ and $H_{2,N+1}$ represent the chromosome ends), but are unknown. However, short DNA fragments of around 500 Kb derived from such maps, and further corrupted by experimental noise processes can be readily generated at high throughput and very low cost using a technology like Optical Mapping (see the full paper [4], and additional references therein). These short DNA fragments will be written as $D_k$, with indices $1 \leq k \leq M$, where $M$ is the number of data fragments and each data fragment is in turn represented as a sequence of restriction sites (e.g. $D_{k,j}$, $0 \leq j \leq m_k + 1$ ) and can be aligned globally to create an estimate of genotype map $H$ using algorithms described previously [2].

The algorithmic problem, we wish to study, is to further separate $H$ into two maps $H_1$ and $H_2$ in such a manner that each data fragment $D_k$ is aligned well to one haplotype or other and that $H_1$ and $H_2$ differ from $H$ only by modifications consistent with SNPs or RFLPs polymorphisms.

Thus, ultimately, this problem corresponds to a problem of refining a multiple map alignment into two families, starting with one global alignment. A combinatorial generalization, where the number of such families is arbitrarily large ($k > 1$) and the cost of each alignment is arbitrarily unconstrained, has been shown to lead to computationally infeasible problems. See [8] for the proof of NP-completeness as well as a probabilistic analysis to show conditions under which the problem can be solved efficiently with a probability close to one. The key to an effective solution of these problems relies on careful experiment design (e.g., choice of coverage, restriction enzyme, experimental conditions, etc.) to ensure conditions under which a polynomial time probabilistic algorithm will work with high probability in conjunction with a Bayesian error model that encodes the error processes properly.

To construct individual haplotype maps from Optical Mapping data we use a mixture hypothesis of pairs of maps $H_1$ and $H_2$ for each chromosome, corresponding to the correct restriction map of the two parental chromosomes. We first assemble the data into a regular map of the entire genome and use this assembly to separate the data into distinct chromosome sets: all maps from the same chromosome belonging to a pair will be included in the same set. We then use a probabilistic model of the errors in the data to derive conditional probability density expressions $f(D_k|H_1)$ and $f(D_k|H_2)$, and apply Bayes rule to maximize a score for the best alignment with respect to proposed $H_1$ and $H_2$, Equation 1.:

$$f(H_1, H_2|D_1, \ldots, D_M) \propto f(H_1, H_2)f(D_1, \ldots, D_M|H_1, H_2) \qquad (1)$$

The first term on the right side is the prior probability of $H_1$ and $H_2$ and we just use a low prior probability for each polymorphism (difference in $H_1$ vs. $H_2$). For the conditional probability term, we can assume each map is a statistically independent sample from the genome and that the mapping errors are drawn from i.i.d. distributions and hence write:

$$f(D_1, \ldots, D_M|H_1, H_2) = \prod_{k=1}^{M} \frac{[f(D_k|H_1) + f(D_k|H_2)]}{2} \qquad (2)$$

The conditional terms of the form $f(D_k|H_i)$ above can be written as a summation over all possible (mutually exclusive) alignments between the particular $D_k$ and $H_i$, and for each alignment the probability density is based on an enumeration of the map errors in the alignment and multiplying together the probability associated with each error under some suitable error model. The exact form of the error models suitable for Optical Mapping is described in the next section, but for almost any error models used the sum of the probability for all alignments can be computed effectively using dynamic programming.

Other methods for assembling Optical Mapping data for relatively short clones into ordered restriction maps exist, as detailed in the full paper [4], but they only focus on genotyping, and with widely varying degrees of success. See the full paper [4] for a detailed survey of the literature. We believe that this paper describes the first published algorithm for assembling single molecule data into haplotype maps.

## 3. Algorithm

Following theorems form the basis for computing various conditional probabilities for a hypothesis.

**Theorem 3.1.** *Consider an arbitrary alignment between the data $D$ and the hypothesis $H$, $J^{\text{th}}$ restriction site of $D$ matching the $I^{\text{th}}$ restriction site of $H$. We will denote this aligned pair by $J \mapsto I$.*

Figure 1. To define the notation required we consider a single arbitrary alignment between a particular data $D$ and hypothesis $H$. Recall that $N$ is the number of restriction sites in $H$ and $m$ the number of restriction sites in $D$. Any arbitrary alignment between $D$ and $H$ can be described as a list of pairs of restriction sites from $H$ and $D$ that describes which restriction site from $H$ is aligned with which restriction site from $D$. As an example, Here the alignment consists of 4 aligned pairs $(4, 2)$, $(5, 2)$, $(I, J)$ and $(P, Q)$. Notice that not all restriction sites in $H$ or $D$ need be aligned. For example between aligned pairs $(I, J)$ and $(P, Q)$ there is one misaligned site on $H$ and $D$ each, corresponding to a missing site (false-negative) and extra-site (false-positive) in $D$. In this alignment a true small fragment between sites 4 and 5 in $H$ are missing from $D$, which is shown by aligning both sites 4 and 5 in $H$ with the same site 2 in $D$. Note that if two or more consecutive fragments in $H$ are all missing in $D$, this would be described by aligning all sites for the missing fragments in $H$ with the same site in $D$ (rather than showing only the outermost of this set of consecutive sites in $H$ aligned with $D$, for example). The expression for the conditional probability density of any alignment, such as the one here, can be written as the product of a number of probability terms corresponding to the regions of alignment between each pair of aligned sites, plus one probability term for each unaligned region at the two ends of the alignment.

*Let the probability density of the unaligned portion on the left and right end of such an alignment be denoted by $f_{ur}(I, J)$ on the right end if $J \mapsto I$ is the rightmost aligned pair, and $f_{ul}(I, J)$ on the left end if $J \mapsto I$ is the leftmost aligned pair.*

*In addition, the following probability density functions $f_m$ and $f_a$ denote the following:*

$$f_m(I, P) = Pr[H[I..P] \text{ is missing in the observed data } D].$$
$$f_a(I, J, P, Q) = Pr[H[I..P] \text{ is an aligned region but not a}$$
$$\text{missing fragment with respect to}$$
$$\text{the observed data region } D[J..Q]].$$

*We assume that $I < P$ and $J < Q$.*
*Then the following holds:*

$$f(D|H) = \sum_{I=1}^{N} \sum_{J=0}^{m+1} f_{ul}(I, J) f(D[J..m + 1] | H[I..N] \wedge J \mapsto I).$$

$$f(D[J..m+1]|H[I..N] \wedge J \mapsto I)$$
$$= f_{ur}(I,J) + f_m(I,I+1)f(D[J..m+1]|H[I+1..N] \wedge J \mapsto (I+1))$$
$$+ \sum_{P=I+1}^{N} \sum_{Q=J+1}^{m+1} f_a(I,J,P,Q)f([Q..m+1]|H[P..N] \wedge Q \mapsto P)$$

*In particular, if the intermediate values are kept in a DP table* $\mathsf{A}_{\mathrm{suf}}[I,J]$

$$\mathsf{A}_{\mathrm{suf}}[I,J] = f(D[J..m+1]|H[I..N] \wedge J \mapsto I)$$

*then it is easily seen that* $f(D|H)$ *can be computed exactly in* $O(m^2 N^2)$ *time and* $O(mN)$ *space, assuming that* $f_m$ *and* $f_a$ *are* $O(1)$ *time functions and* $f_{ul}$ *and* $f_{ur}$ *are* $O(N)$ *time functions.* $\square$

In a later section we will see how to reduce the complexity to linear time when we only require an $\epsilon$-approximate value $\tilde{f}$

$$f(D|H) - \epsilon < \tilde{f}(D|H) < f(D|H) - \epsilon,$$

for the probability density function arising in the context of optical mapping as follows:

$$f_m(I,I+1) = P_\nu^{H_{I+1}-H_I}$$
$$f_a(I,J,P,Q) = \lambda^{Q-J-1} P_d(1-P_d)^{P-I-1}$$
$$(1-P_\nu)^{H_P-H_I} \, G_{(H_P-H_I),\sigma^2(H_P-H_I)}(D_Q - D_J),$$

where $P_d$ = the digest rate, $\lambda$ = the false-positive site rate, $\sigma^2 h$ = the Gaussian sizing error variance for a fragment of size $h$, $P_\nu$ = the probability of missing a fragment of unit size, and $R_e$ = the breakage rate of DNA (the inverse of the expected fragment size). For a random variable $x$ following a Gaussian distribution $\mathbf{N}(\mu, \sigma^2)$, the probability density value at $d$ is $G_{\mu,\sigma^2}(d) = \exp[-(d-\mu)^2/2\sigma^2]/(\sqrt{2\pi}\sigma)$.

The exact form of the functions for $f_{ul}$ and $f_{ur}$ for Optical Mapping are complicated, but do not affect the complexity of the algorithm; thus a detailed discussion is omitted here, but can be seen in the full paper [4]. The key assumption required is that $f_{ul}$ and $f_{ur}$ permit $O(1)$ $\epsilon$-approximation.

As it has been shown elsewhere [1], a good approximate location of the best alignment between $D$ and $H$ can be determined in $O(1)$ expected time, if the conditional probability density has been previously evaluated for a similar $H$ or alternatively, through a geometric hashing algorithms. Only a $O(1)$-width band of the DP table needs to be evaluated to compute an $\epsilon$-approximation $\tilde{f}(D|H)$. In particular, the band width of the DP table used in practice is usually about $\Delta = 8$; more generally for Optical mapping $\Delta$ is bounded by

$$(1 - P_d)^{\Delta - 1} = \epsilon, \quad \text{or} \quad \Delta = 1 + \frac{\ln(\epsilon)}{\ln(1 - P_d)}.$$

With this approach we achieve a reduced time complexity of $O(\min(m, N))$ (more explicitly, $O(\min(m\Delta^3, N)))$.

Now we show how we may recompute conditional probabilities for a modification to hypothesis: How can one re-evaluate the conditional probability distribution function, $f(D|H' = p(H))$ when the new hypothesis, $H'$, has been obtained by locally changing $H$ in just one place (corresponding to a polymorphism). There are three cases to consider. We study one of the three cases here in detail and refer the reader to [4] for the remaining cases. The omitted cases are similar but tedious.

We may obtain $H'$ by

(1) Deleting one of the existing restriction sites in $H$, as the site may contain a heterozygous SNP;
(2) Adding a new restriction site at a specified location in $H$, symmetrical to the previous case;
(3) Increasing or decreasing a restriction fragment length in $H$, an RFLP;

Consequently, we may also need to compute the first and second derivative of $f(D|H)$ relative to the change in any fragment size in $H$.

**Theorem 3.2.** *Consider an arbitrary alignment between the data $D$ and the hypothesis $H$, $J^{\text{th}}$ restriction site of $D$ matching the $I^{\text{th}}$ restriction site of $H$. Using the notations of the previous discussion, we write:*

$$\mathsf{A}_{\text{suf}}[I, J] = f(D[J..m + 1]|H[I..N] \wedge J \mapsto I),$$

*and*

$$\mathsf{A}_{\text{pref}}[I, J] = f(D[0..J]|H[1..I] \wedge J \mapsto I).$$

*Then*

$$\mathsf{A}_{\text{suf}}[I, J] = f_{ur}(I, J) + f_m(I, I + 1)\mathsf{A}_{\text{suf}}[I + 1, J]$$
$$+ \sum_{P=I+1}^{N} \sum_{Q=J+1}^{m+1} f_a(I, J, P, Q)\mathsf{A}_{\text{suf}}[P, Q],$$

*and similarly,*

$$A_{\text{pref}}[I, J] = f_{ul}(I, J) + A_{\text{pref}}[I - 1, J]f_m(I - 1, I)$$
$$+ \sum_{P=1}^{I-1} \sum_{Q=0}^{J-1} A_{\text{pref}}[P, Q]f_a(I, J, P, Q),$$

*If $H \setminus \{H_K\}$ is obtained from $H$ by deleting the site $H_K$, then*

$$f(D|H \setminus \{H_K\})$$
$$= Pr[\text{Alignments with rightmost aligned } I < K]$$
$$+ Pr[\text{Alignments with leftmost aligned } J > K]$$
$$+ Pr[\text{Alignments with a fragment spanning } H[K - 1..K + 1]]$$
$$= \sum_{I=1}^{K-1} \sum_{J=0}^{m+1} A_{\text{pref}}[I, J]f_{ur}^{(-H_k)}(I, J) + \sum_{I=K+1}^{N} \sum_{J=0}^{m+1} f_{ul}^{(-H_k)}(I, J)A_{\text{suf}}[I, J]$$
$$+ \mathbb{I}_{K<N} \sum_{J=0}^{m+1} A_{\text{pref}}[K - 1, J]f_m(K - 1, K + 1)A_{\text{suf}}[K + 1, J]$$
$$+ \sum_{J=0}^{m+1} \sum_{I=1}^{K-1} \sum_{P=K+1}^{N} \sum_{Q=J+1}^{m+1} A_{\text{pref}}[I, J]\frac{f_a(I, J, P, Q)}{1 - P_d}A_{\text{suf}}[P, Q],$$

*where $f_{ul}^{(-H_k)}$ and $f_{ur}^{(-H_k)}$ are computed respectively from $f_{ul}$ and $f_{ur}$ by suitable simple modifications.*

*Then it is seen that $f(D|H \setminus \{H_K\})$, $\forall K$ $1 \le K \le N$, can be computed exactly in $O(m^2N^2)$ time and $O(mN)$ space, assuming that $f_m$ and $f_a$ are $O(1)$ time functions and $f_{ul}$ and $f_{ur}$ are $O(N)$ time functions.*

If we only wish to compute an $\epsilon$-approximation $\tilde{f}$, for some consecutive range of $m$ different K values, one can compute these $m$ probabilities $f(D|H \setminus \{H_K\})$ for each kind of modification in $O(min(m, N))$ time [4].

### 3.1. *Search Algorithm for Haplotypes*

The recurrence equations of the previous subsections and the dynamic programming algorithms based on those allow us to efficiently compute the posterior probability for a single possible pair of maps $H_1$ and $H_2$ and their modifications

$$\begin{bmatrix} H_1^{(0)} \\ H_2^{(0)} \end{bmatrix} \Rightarrow \begin{bmatrix} H_1^{(1)} \\ H_2^{(1)} \end{bmatrix} \Rightarrow \begin{bmatrix} H_1^{(2)} \\ H_2^{(2)} \end{bmatrix} \Rightarrow \cdots$$

The computationally expensive part of computing the haplotype map algorithm is the search over possible maps $H_1$ and $H_2$ in order to find the one with the highest posterior probability.

Initially, we assume that a single genotype map hypothesis $H$ has been computed and it has been determined that $H$ best matches all data. The algorithms to compute such maps have been developed [3,2] and have been in use for more than five years. The speed of the main algorithm, GenTig, has been improved through an important heuristic stage that relies on geometric hashing to quickly identify the maps that overlap, and can also be used in the context of haplotyping. The time complexity of this geometric-hashing-stage is super-linear and is given as

$$T_H = O(N + M_D^{4/3}), \quad \text{where } M_D = \sum_{j=1}^{M} m_j + 1,$$

i.e., $M_D$ is the total number of fragments in the optical mapping data. We will see that the actual time for this stage $T_H$ is dominated by the remaining computation involving search over possible haplotype pairs $H_1$ and $H_2$, unless the genome we are dealing with is much larger than the human genome; see next subsection.

If our initial hypothesis is $H$, then $H_1^{(0)} = H_2^{(0)} = H$, and at each stage $H_1^{(i)}$ and $H_2^{(i)}$ must then be refined by trying to add or delete restriction sites and by adjusting the distance between restriction sites by doing a gradient optimization of the probability density of all maps for each fragment size. The result is $H_1^{(i+1)}$ and $H_2^{(i+1)}$.

Note that at each hypothesis-recomputation step, trying each new restriction site polymorphism involves modifying $H_1$ or $H_2$ by adding or deleting a restriction site from $H_1$ (or $H_2$) only, while trying an RFLP involves modifying the same interval in both $H_1$ and $H_2$ by adding some $\delta h$ to $H_1$ and subtracting the same $\delta h$ from $H_2$. In each case both possible "phases" of each polymorphism is to be accounted for, reversing the use of $H_1$ and $H_2$ above. Since both phases must be tested and the better scoring one selected, except when adding the first polymorphism to $H_1$ and $H_2$, the search process can easily turn in to $2^{O(N)}$.

Note also that if the data cannot allow the phasing to be determined because there are no (or insufficient) data molecules spanning both polymorphisms, both phases (orientations) will score almost the same. This fact is also recorded since it marks a break in the phasing of polymorphisms.

Further note that RFLP polymorphisms are more expensive to score, since in addition to the phasing (whether $H_1$ or $H_2$ has the bigger fragment) it is necessary to determine the amount of the fragment size difference for $H_1$ and $H_2$ (the $\delta h$ value), which can be searched for in $O(1)$ expected time, and the constant is essentially logarithmic in the ratio of the expected fragment length to the resolution of optical mapping. More precisely, this step involves trying a number of different multiples of $\delta h$ values that is logarithmic in the number of total possible values using the well known unimodal function maximization algorithm based on the golden mean ratio. As an example, the total number of $\delta h$ values required for any fragment can be bounded by about 20 if the resolution of $\delta h$ is set at 0.1Kb and the largest restriction fragment length is 50Kb; usually, this number is extremely small: just 1 or 2 small $\delta h$ values are sufficient to verify that no polymorphism exists.

A purely greedy addition of polymorphisms to $H_1$ and $H_2$ is not sufficient to get the phases correct as the search can get stuck in local maxima when two or more polymorphisms are nearby. We avoid this problem by using a heuristic look ahead distance of $w$ restriction sites, and scoring all combinations of polymorphisms in this window, before committing the best scoring set of polymorphisms in $H_1$ and $H_2$. With a sufficiently large window size $w$, the fraction of the polymorphic sites the algorithm misses or phases incorrectly can be made negligible. Since this heuristic can increase the worst case complexity of the algorithm exponentially with the window size $w$ we heuristically determine the smallest possible window $w$ by using simulated data and search the space of possible polymorphisms within a window by adding/deleting just one or two polymorphisms at a time until no further improvement in the probability density occurs.

The overall algorithm must try every possible restriction site and fragment as a possible polymorphic SNP or RFLP respectively using a rolling window of size $w$ restriction sites. This process must be repeated a few times until no further polymorphisms are detected. Typically just two to three iterations of scanning all restriction sites suffice.

The overall complexity of the basic haplotype search algorithms described here, just using the basic DP algorithm from Theorem 3.1, is $O(M_D^2/C + N)$, where $C = $ coverage and $C = (1/N) \sum_{j=1}^{M} m_j$. Several simple tricks to speed up the evaluation of conditional probabilities, coupled with a judicious applications of dual DP tables ultimately improves the asymptotic time complexity to $O(M_D)$. Detailed analysis is in the full-paper [4].

## 4. Empirical Results

In this section, we summarize experiments constructing a haplotype map of *T. pseudoana* and of a 120 Mb region of the human chromosome to demonstrate the feasibility and relevance of our approach. Finally, we also summarize results based on simulation to provide insight into the accuracy of our results. See the full-paper [4] for a detailed description of these studies.

- The optical mapping data for *T. Pseudoana* (Diatom) was analyzed by our algorithm; for all except chromosome 19, it successfully phased all polymorphisms and generated two separate maps.
- Our algorithm found 233 restriction site polymorphisms and 12 fragment length polymorphisms in the human chromosome 4 data, and was able to phase all polymorphisms into 2 contiguous regions. The nature of the polymorphisms detected was somewhat surprising and is discussed in details in the full paper [4].
- The simulated data, with statistical characteristics derived from human chromosome 21, was assembled using different data redundancy of 6×, 12×, 16×, 24×, 50× and 100× (per haplotype). The results are summarized in Table 1. From the simulated data we can infer that 16× redundancy is required to eliminate most errors in SNPs and about 50× redundancy is required to eliminate most errors in indels (RFLPs)

| Redundancy | fp SNPs | fn SNPs | fp RFLPs | fn RFLPs | Phase err | Molecules |
|:---:|:---:|:---:|:---:|:---:|:---:|:---:|
| 6x | 5 | 5 | 1 | 18 | 7/26 | 30 |
| 12x | 4 | 2 | 4 | 16 | 2/55 | 60 |
| 16x | 2 | 1 | 0 | 12 | 2/71 | 80 |
| 24x | 2 | 1 | 1 | 11 | 3/111 | 120 |
| 50x | 0 | 1 | 1 | 5 | 4/228 | 250 |
| 100x | 0 | 0 | 2 | 1 | 2/441 | 500 |

Figure 2. Haplotyping algorithm performance for 16 SNPs and 24 RFLPs.

## 5. Discussions and Future Work

Single molecule mapping technologies, such as Optical Mapping, are ideal for detecting genetic markers with phasing information and without population-based assumptions. It elegantly circumvents many problems that have proven unsurmountable in all other population-based approaches (see discussion in [4]).

Furthermore, we estimate that our approach is currently the only approach that can produce a genome wide individual haplotype map for under $1000 (based on 8 restriction enzyme haplotype maps). The dominant SNP based approach requires testing of about 300,000 SNPs which costs at least ten times more per person. Our approach can be applied to other single molecule mapping technologies. When applied to single molecule technologies to map short 6–8bp LNA hybridization probes, it can be used to sequence the entire human genome: With 50× coverage the location of probes can be determined to within about 200bp. Hence well known error tolerant SBH (Sequencing by Hybridization) algorithms [6] can be used to determine the sequence within any 200bp window from maps of a universal set of about 2048 probes of 6bp, allowing a draft quality individual haplotype sequence to be assembled for about $20,000.

## References

1. T. ANANTHARAMAN, B. MISHRA, AND D.C. SCHWARTZ, "Genomics via Optical Mapping II: Ordered Restriction Maps," *J. of Comp. Bio.*, 4(2):91–118, 1997.

2. T. ANANTHARAMAN, B. MISHRA, AND D.C. SCHWARTZ, "Genomics via Optical Mapping III: Contiging Genomic DNA and Variations," ISMB '99 , 7:18–27, 1999

3. T. ANANTHARAMAN, AND B. MISHRA, "A Probabilistic Analysis of False Positives in Optical Map Alignment and Validation," WABI '01,**LNCS 2149**:27–40, 2001

4. T. ANANTHARAMAN, V. MYSORE, AND B. MISHRA, *Fast and Cheap Genome wide Haplotype Construction via Optical Mapping*, NYU Tech. Report# TR 2004-852, 2004.
   http://cs.nyu.edu/web/Research/technical_reports.html.

5. R. BRITTEN *et al.*, "Majority of Divergence between Closely Related DNA Samples is due to Indels," *PNAS*, 100(8):4461–4465, 2003.

6. E. HALPERIN *et al.*, "Handling Long Targets and Errors in Sequencing by Hybridization," *J. of Comp. Bio.*, 10:3–4, 2003.

7. G. LANCIA *et al.*, "Practical Algorithms and Fixed-Parameter Tractability for the Single Individual SNP Haplotyping Problem," *WABI '02*: 29–43.

8. B. MISHRA AND L. PARIDA, "Partitioning Single-Molecule Maps into Multiple Populations: Algorithms And Probabilistic Analysis," *Discrete Applied Mathematics*, 104(1-3):203-227, 2000.

9. M. WATERMAN *et al.*, "Haplotype Reconstruction from SNP Alignment," *RECOMB 03*: 207–216, 2003

# IMPROVING FUNCTIONAL ANNOTATION OF NON-SYNONOMOUS SNPs WITH INFORMATION THEORY

R. KARCHIN, L.KELLY, A. SALI

*Departments of Biopharmaceutical Sciences and Pharmaceutical Chemistry, and*
*California Institute for Quantitative Biomedical Research*
*Mission Bay Genentech Hall, 600 16th Street, Suite N472D*
*University of California, San Francisco*
*San Francisco, CA 94143-2240*

Automated functional annotation of nssNPs requires that amino-acid residue changes are represented by a set of descriptive features, such as evolutionary conservation, side-chain volume change, effect on ligand-binding, and residue structural rigidity. Identifying the most informative combinations of features is critical to the success of a computational prediction method. We rank 32 features according to their *mutual information* with functional effects of amino-acid substitutions, as measured by *in vivo* assays. In addition, we use a greedy algorithm to identify a subset of highly informative features [1]. The method is simple to implement and provides a quantitative measure for selecting the best predictive features given a set of features that a human expert believes to be informative. We demonstrate the usefulness of the selected highly informative features by cross-validated tests of a computational classifier, a support vector machine (SVM). The SVM's classification accuracy is highly correlated with the ranking of the input features by their mutual information. Two features describing the solvent accessibility of "wild-type" and "mutant" amino-acid residues and one evolutionary feature based on superfamily-level multiple alignments produce comparable overall accuracy and 6% fewer false positives than a 32-feature set that considers physiochemical properties of amino acids, protein electrostatics, amino-acid residue flexibility, and binding interactions.

## 1. Introduction

Over 70,000 coding non-synonymous SNPs (single nucleotide polymorphisms that produce a changed amino acid residue in a gene's protein product) have been identified in the human genome to date [2]. These changes can alter protein function and thus contribute to variation in disease susceptibility, drug efficacy, and drug toxicity [3,4].

In the past few years, a number of groups have developed computational methods to predict the functional effects of coding nssNPs [5-9]. The first studies in this area relied on sequence information only. Cargill *et al.* classified substitutions as conservative or non-conservative using BLOSUM62 scores [10]. Ng and Henikoff substantially increased prediction accuracy with their SIFT algorithm, by computing a combination of a position-specific substitution score and a conservation score from a multiple alignment [6]. Several studies have attempted to understand how nsSNPs affect protein stability and function by mapping them onto protein structures, and developing rules to predict functionally important mutations [8,9]. Most recently, a few groups have applied machine learning methods to this problem [11,12].

Progress in this field depends on selection of a feature set that best represents the effects of amino-acid substitution in a protein. Chasman *et al.* applied a standard statistical analysis to 16 structure- and phylogeny-based features describing nsSNPs [5]. They tested each feature for association with functional effects by comparing tolerated and deleterious substitutions (from studies of T4 lysozyme and lac repressor [13-16]) using ANOVA (analysis of variance) and $\chi^2$ statistics. They selected a measure of residue flexibility, an entropy-based evolutionary conservation measure, solvent accessibility, buried charge, and "unusual amino acid". Saunders *et al.* evaluated a set of 12 features, using a similar dataset [7]. Each feature was assessed according to its ability to discriminate between tolerated and deleterious mutations in a 20-fold cross-validation test, with thresholds set to minimize classification error using a held-out partition of the data. They identified two optimal features: SIFT score (a measure of evolutionary conservation) [6] and a solvent-accessibility measure, based on the density of $C_B$ atoms in the neighborhood of each residue.

Although no published studies have applied mutual information analysis to this problem, feature selection algorithms using mutual information have been used in many machine-learning applications, such as computer-aided medical diagnosis and text categorization [17,18]. These algorithms are well suited to the problem of predicting the functional consequences of nsSNPs. The mutual information metric is supported by rigorous mathematics and does not make assumptions that data fit a known family of distributions or that there are linear relationships between the features and classes to be predicted [1,19].

In this study, we rank a representative set of 32 candidate features according to their mutual information with categories of functional effects (*mutation classes*) observed in the T4 lysozyme and lac repressor assays (Section 2.2). We use the greedy MIFS algorithm to find an "optimized" set of features having large mutual information with the mutation classes and low redundancy with each other [1,20]. To validate that we are measuring the correct quantity, we evaluate the relationship between the performance of a state-of-the-art supervised learning method, a support vector machine (SVM), as well as the mutual information of its inputs (features) and desired outputs (mutation classes). We compare a variety of feature sets:

features selected by MIFS, features selected according to their mutual information rank, and a large set of all 32 features.

## 2. Methods

### 2.1. Dataset

Our evaluations were done on 6044 experimentally characterized point mutations in bacteriophage T4 lysozyme and *E. coli* lac repressor (data courtesy of Pauline Ng) [13-16]. 2015 mutations were from lysozyme and 4029 from lac repressor. Because the mutations were introduced in a systematic and unbiased fashion, this data has become a standard benchmark for methods that predict nsSNP functional effects. It has been used in several published studies, although direct comparison of results is difficult, as some groups have chosen to filter out mutations characterized as moderately deleterious or as temperature-sensitive [5-7,11].

### 2.2. Mutation Classes

The dataset is based on a plaque assay of bacteriophage T4 lysozyme mutants [13,14] and a colorimetric assay of *E. coli* lac repressor mutants [15,16]. Mutants were ranked according to reduced (or enhanced) function and assigned to a *mutation class*. In both experiments, mutants were assigned to four classes.

Because the mutation classes were based on visual inspection of plated cell cultures, these example labels are noisy. To deal with the problem, some groups reduce the four mutation classes to two by lumping all varieties of functional effect into a single class [5,6,11]. Others drop examples with moderate functional effect from the data set [7]. We chose the former definition (ANY-EFFECT, NO-EFFECT) because we are interested in predicting moderate as well as severe functional effects. Many disease susceptibility and drug response phenotypes are believed to result from interactions of SNPs in different genes that individually have moderate effects [21,22].

### 2.3. Candidate Features

We evaluated 32 features potentially useful for computational prediction of the functional effects of nsSNPs, supplementing those found in the literature with a few of our own design. For detailed descriptions of each feature, see Appendix A. All features can be calculated inexpensively by a computer program or database lookup and thus are suitable for large-scale projects.

## 2.4. Feature Evaluation

In a computational classification method, each nsSNP is represented by a set of categorical- or numerically-valued features. From an information theoretic perspective, the classifier is a system that reduces our initial uncertainty about the experimentally-characterized mutation class of the nsSNP by "consuming" the information in the features. If there is sufficient information and it is used efficiently by the classifier, classification errors will be minimized [1]. The information that a feature $X$ reveals about the mutation class $Y$ can be quantified as *mutual information* (in units of bits):

$$I(X,Y) = \Sigma_{X,Y} \, p(X,Y) \log_2 \left( p(X,Y) / p(X)p(Y) \right) \qquad (1)$$

In this setting, the sum is over the cross-product of 6044 observations of feature $X$ and mutation class $Y$ in our data set.

Because we do not know the feature probability density functions p(X) and p(Y), we perform a histogram analysis to assign continuous data to discrete categories (bins). We build contingency tables for each (X,Y) pair to obtain the empirical estimates $\hat{p}(X,Y)$, $\hat{p}(X)$, $\hat{p}(Y)$ and $\hat{I}(X,Y)$.

Given the limited size of our data set, we opted to use a small number of bins in our histograms, rather than a large number of sparsely populated bins. All continuous-valued features were partitioned into five equal-frequency bins. Some of the tested features were categorical, such as buried vs. solvent-exposed residue position. These features had between two and five categories.

Mutual information is overestimated when sample size is small [23]; the effect becomes more pronounced as the number of bins increases and a greater number bins are undersampled. A feature with five bins will get a more inflated mutual information estimate than a feature with two bins. To deal with this problem, we applied a correction described by Cline *et al.* to compute *excess mutual information* $I_E$ for each feature [24]:

$$\hat{I}_E(X,Y) = \hat{I}(X,Y) - E[\hat{I}_R(X,Y)] \qquad (2)$$

The correction term on the right-hand side of Eq. (2) is the expected value of *random mutual information* $E[I_R(X,Y)]$, where $E[I_R(X,Y)]$ is the mutual information between X and Y after the pairs are scrambled. (If our sample was sufficiently large, we should always have $I_R(X,Y) = 0$, because the scrambling destroys all associations between X and Y.) We get stable estimates of $E[\hat{I}_R(X,Y)]$ by using 5,000 scramblings.

Features that are individually most informative about the mutation classes may be redundant, so we looked at how to select features that are most informative as a group. Our approach is to identify a subset of the candidate features that maximizes the joint *excess* mutual information of features and mutation classes [25].

$$J_E = I_E(X_1,..., X_k; Y) \tag{3}$$

Given that we have 32 candidate features, the space of possible subsets is too large for exhaustive search, and we approximate maximization of Eq. (4) with the MIFS algorithm, a greedy algorithm that selects a subset of features $S$ having high $I_E$ with the mutation classes and low $I_E$ with each other [1]. The algorithm iteratively selects the feature that maximizes Eq. (4). The parameter ß, which is chosen empirically, controls the relative importance of the two selection criteria. We experimented with values of 0.25-1.0 and obtained best results with ß=0.5, so that each feature's mutual information with the mutation classes is given twice the importance of the penalty for feature redundancy (data not shown). Eq. (4) is a modification of the original MIFS objective function and produced superior results in our tests. It uses the excess mutual information correction (Eq. 2) and an improved feature redundancy measure suggested by Kwak et al. [20].

$$\hat{J}_E = \hat{I}_E(X,Y) - \beta \sum_{s \in S} \frac{\hat{I}_E(Y,s)}{H(s)} \hat{I}_E(X_i, s), \quad 0 \le \beta \le 1 \tag{4}$$

To select a desired number of features $m$, the algorithm initially sorts the features in the candidate set $F$ by $I_E(X,Y)$. The top ranked feature is moved from $F$ to the subset of selected features $S$. It proceeds for $m$ iterations; at each iteration, the feature $X_i$ in $F$ that maximizes Eq. (4) is selected and moved from $F$ to $S$, until finally $m$ features are selected.

### 2.5. Testing Protocol

We expect that a computational classifier will perform better when the most informative features are used as inputs. To test the predictive value of features selected by mutual information, we compared performance of: a large set of 32 features (Table 1); the top-ranked two features, the top-ranked three features, a set of five features selected by maximizing Eq. (4); and 28 sets of five features selected according to mutual information ranking. In the "ranked sets", the features were ordered according to their excess mutual information with the mutation classes (Table 1). Features ranked 1-5 were assigned to set number one; features ranked 2-6 were assigned to set number two, and so forth.

The features were used as inputs to a computational classifier known as a *support vector machine* (*SVM*) [26]. The SVM uses a *kernel function* to map the feature vectors into a high-dimensional space and find an optimal separation of examples from the different classes. We chose the SVM to reduce the possibility that classification errors were produced by inefficient classifier operation. SVMs are state-of-the-art classifiers and have been shown to be robust to noise and overfitting. In the results reported here, we used a radial basis kernel function and a 1-norm soft margin (with C=1) [27]. Although these choices produced our best results, they have not been carefully optimized.

To identify informative features relevant to both lac repressor and lysozyme, we used all of our mutation data in the feature ranking and selection process. We applied a stringent *heterogeneous* cross-validation protocol in which the SVM was trained on mutation data from one of the proteins and then tested on data from the other (and *vice versa*). Lac repressor and lysozyme are not structurally or functionally similar, so a classifier that does well on such a test is potentially suitable for predicting nsSNP functional effects in a wide range of globular proteins. *Homogeneous* cross-validation (training and testing on data from a single protein) achieves 15-20% higher classification accuracy, but these SVMs generalize poorly when tested on the other protein (data not shown).

Our experiments were done with in-house tools coded in Perl and Java. All software and alignments used in this analysis are available from the authors upon request.

**Results**

Table 1 shows the 32 evaluated features, ranked by mutual information with the mutation classes. If a particular mutation class $Y$ and feature $X$ always occurred together, such as all residues with functional effect having buried charges, the feature could be used to predict the correct mutation class to a certainty. In this case, $p(X,Y)=p(X)$ and the feature would have approximately 2 bits of information. This result can be derived by making the substitution in Eq (1), given the distributions of $\hat{p}(X)$ and $\hat{p}(Y)$. Here, the individual features are weakly informative – the best ones have only 0.1 bits. A select combination of several such features is required for accurate prediction of the mutation class.

**Table 1.** Thirty-two tested features, ordered by excess mutual information with the mutation classes (in bits). We performed a histogram analysis to assign continuous feature data to discrete categories (bins). WT=wild-type. MUT=mutant. HMM=hidden Markov model. For feature descriptions, see Appendix A. The top-ranked and bottom-ranked five-feature sets are shaded in gray.

| Features | Bins | Bits | Features | Bins | Bits |
|---|---|---|---|---|---|
| Fractional solvent accessibility WT | 5 | 0.104 | Change in solvent accessibility | 5 | 0.052 |
| HMM PHC score superfamily | 5 | 0.103 | Buried charge | 2 | 0.045 |
| Fractional solvent accessibility MUT | 5 | 0.101 | Standardized residue B-factor | 5 | 0.044 |
| Solvent accessibility WT | 5 | 0.096 | Change in residue hydrophobicity | 5 | 0.026 |
| Solvent accessibility MUT | 5 | 0.089 | Average residue B-factor | 5 | 0.014 |
| HMM entropy subfamily | 5 | 0.087 | Change in fractional solvent acc. | 5 | 0.014 |
| HMM relative entropy superfamily | 5 | 0.083 | EC/EU subfamily | 2 | 0.007 |
| Buried/exposed residue MUT | 2 | 0.079 | SIFT score | 2 | 0.005 |
| HMM PHC score subfamily | 5 | 0.079 | Grantham values | 5 | 0.004 |
| HMM entropy superfamily | 5 | 0.073 | Change from buried to exposed | 3 | 0.002 |
| HMM relative entropy subfamily | 5 | 0.073 | Unusual residue | 2 | 0.002 |
| Buried/exposed residue WT | 2 | 0.072 | Change in residue formal charge | 5 | 0.002 |
| HMM relative entropy family | 5 | 0.071 | Domain interface contact | 2 | 0.002 |
| HMM PHC score family | 5 | 0.071 | Change in residue volume | 5 | 0.001 |
| HMM entropy family | 5 | 0.071 | Turn breaker (P or G in turn) | 2 | -0.001 |
| DNA or small ligand contact | 2 | 0.070 | Helix breaker (P or G in helix) | 2 | -0.001 |

The top-three ranked features are fractional solvent accessibility (wild-type residue), HMM PHC score (superfamily alignments), and fractional solvent accessibility (mutant residue). Most of the information in the solvent-accessibility features comes from the fact that buried residue positions are most likely to be adversely effected by amino-acid substitutions, due to loss of structural stability [14,16,28,29]. Our structural modeling of amino-acid substitutions appears to capture some information about these energetic changes. This is reflected in the 2-3% accuracy increase between the top-two and top-three features shown in Table 2.

The superfamily-level PHC score, which considers several aspects of evolutionary conservation, is more informative than the simpler conservation measures we evaluated and more informative than family- or subfamily-based conservation measures (Appendix). The difference is not dramatic but suggests that superfamily alignments containing paralogous sequences can be useful in predicting untolerated amino-acid substitutions.

Table 2 shows the classification accuracy and false positive rate (fraction of mutations with effect that are incorrectly classified) of the SVM when tested on six feature sets. Accuracy is consistently better for lysozyme because a larger fraction of point mutations have no functional effect (0.68) than in lac repressor (0.56). If we "play the odds" (*e.g.,* randomly classifying effect and no-effect 32% and 68% of the time, respectively) for lysozyme, our classification accuracy will be 0.57 (vs. 0.51 for lac repressor). These accuracies are approximately what we get using the SVM with the five least informative features.

**Figure 1.** Support vector machine classification accuracy (fraction of correctly classified point mutations) is strongly correlated ($r$=-0.8) with the mutual-information rank of the five-feature sets described in Section 2.5. The line y = -0.006 x + 0.681 shows where the points would lie if correlation was at a maximum ($r$=-1.0). The outliers are four sets (ranks 12-15) that contain the feature "Ligand-binding-with-DNA". The feature is very informative for lac repressor, a DNA-binding protein, but not for lysozyme, which has no ligand or DNA contacts. The SVM cannot predict EFFECT or NO-EFFECT at the DNA-binding and ligand sites, given insufficient information. In this case, the default prediction value is assigned. In our implementation, the default prediction is EFFECT, producing the outliers.

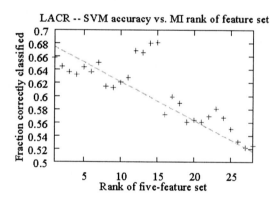

We get best results with a redundant set of features (the five "top-ranked") that includes four descriptions of solvent accessibility. The top-three ranked features achieve classification accuracy comparable to the top-five ranked features, the optimized subset of five features, and the large set of 32 features. The remaining 29 features, which include buried charge, residue flexibility, subfamily-level evolutionary conservation, and physiochemical properties, make an insignificant contribution to classifier accuracy. In addition, classification specificity improves when the top-three features are used instead of all 32. The false positive rate (the fraction of mutations with no effect that are incorrectly classified) is reduced by more than 6%. Since experimental validation of predicted functional effects is often expensive and time-consuming, a low false-positive rate is desirable.

**Table 2.** Classification accuracy of SVM (fraction of correctly classified mutations) and false positive rate (fraction of mutations with no functional effect that are incorrectly classified) for six feature sets in cross-validation experiments described in Section 2.5. The dataset contains 4029 functionally characterized point mutations for lac repressor (LACR) and 2015 for lysozyme (LYS). The "optimized subset of five features" selected by the MIFS algorithm is shown in Table 3.

|  | Classification accuracy | | | **False positive rate** | | |
|---|---|---|---|---|---|---|
|  | LACR | LYS | Both | LACR | LYS | **Both** |
| Top-five ranked features | 0.658 | 0.714 | 0.677 | 0.158 | 0.149 | 0.155 |
| Top-three ranked features | 0.668 | 0.702 | 0.679 | 0.172 | 0.153 | 0.165 |
| Top-two ranked features | 0.650 | 0.673 | 0.658 | 0.182 | 0.156 | 0.172 |
| "Optimized" subset of five features | 0.655 | 0.709 | 0.675 | 0.204 | 0.178 | 0.194 |
| Large set of 32 features | 0.654 | 0.748 | 0.685 | 0.272 | 0.157 | 0.229 |
| Bottom-five ranked features | 0.525 | 0.570 | 0.540 | 0.362 | 0.236 | 0.314 |

**Table 3.** Subset of five features selected by the MIFS algorithm, with $\hat{J}_E$ values (Eq. 4). Computations were done using all 6044 mutations in the lac repressor-lysozyme set.

| Feature subset selected by MIFS algorithm | $\hat{J}_E$ (bits) |
|---|---|
| Fractional solvent accessibility (wild-type) | 0.104 |
| HMM PHC score, superfamily alignments | 0.102 |
| HMM entropy, subfamily alignments | 0.081 |
| Fractional solvent accessibility (mutant) | 0.074 |
| Residue in contact with DNA or within 5Å of small ligand | 0.069 |

The "optimized" subset selected by MIFS does poorly compared to the five top-ranked features (the top-five set is equivalent to running MIFS with $\beta=0$). In this setting, reducing feature redundancy does not give an advantage, possibly because the information in individual features is weak and solvent accessibility is the most important predictor of deleterious amino-acid substitutions.

Figure 1 shows SVM classification accuracy (fraction of point mutations with correctly classified functional effects) vs. rank of the 28 five-feature sets for lac repressor (Section 2.5). Results are very similar for lysozyme. SVM classification accuracy is highly correlated with the excess mutual information ranking of the selected input features. Pearson's correlation coefficient $r$ is -0.8 for lac repressor and -0.87 for lysozyme.

## Summary

We have shown that mutual information is a useful tool in identifying biologically important features, given a set of functionally characterized point mutations. The strongest signals in the lac repressor/lysozyme set are solvent accessibility and superfamily-level evolutionary conservation. In cross-validated tests with a SVM classifier, using the top five ranked features gave us the lowest false positive rate.

We are currently working on applying this method to membrane proteins, using point mutation datasets generated by the Pharmacogenomics of Membrane Transporters project [30].

## Acknowledgments

This work was supported by NIH grant U01-GM-61390-04. Thanks to Drs. K. Giacomini, D. Kroetz and the PMT project, Dr. K. Karplus for contributing to the script used in mutual information evaluations, and Dr. P. Ng for mutation data.

## Appendix

*Sequence-based features* used in our study are listed in Table 2.

**Table 2.** Features based on amino-acid sequence only.

| | |
|---|---|
| *Grantham values* | physiochemical difference between sidechains [31] |
| *Net residue charge change* | formal charge change between wild-type and mutant |
| *Residue volume change* | change in van der Waals volume [32] |
| *Residue hydrophobicity change* | change in hydrophobicity values [33] |
| *Unusual residues* | proline/glycine |
| *Helix/Turn-breaker* | proline or glycine in a helix or turn as defined by DSSP [5] |

*Evolutionary-conservation features* were extracted from multiple alignments of sequences related at the superfamily level (common structure and function but low sequence similarity), the family level (paralogs and orthologs), and the subfamily level (orthologs only). Superfamily alignments and hidden Markov models (HMMs) were built with the SAM-T02 webserver [31]. Family and subfamily alignments and HMMs were constructed manually from superfamily alignments.

*EC/EU:* Defines an alignment column as either 100% conserved or unconserved.

*Shannon entropy.* A measure of conservation in a column of interest, where $\hat{p}(\bar{x})$ are the observed frequencies. Computed as $H(\bar{x}) = -\sum_i \hat{p}(x_i) \log \hat{p}(x_i)$ [34].

*Relative entropy.* Difference in entropy between $\hat{p}(\bar{x}$ and the background distribution of amino acids $\hat{p}(\bar{v}$ , estimated from a large sample of alignments. Computed as $R(\bar{x},\bar{v}) = \sum_i \hat{p}(x_i, v_i) \lg \hat{p}(x_i) / \hat{p}(x_i, v_i$ .

*PHC score.* A score that considers the difference in conservation between wild-type (*W*) and mutant (*M*) amino acids and the conservation of the most-probable (*consensus*) amino acid *C*.

$$\text{PHC} = \log(|\, p(W) - p(M)\,|) + \log(p(W)) + \log(P(C)) - \log(p(M)) \quad (5)$$

*SIFT score.* SIFT is an automated method that builds a multiple sequence alignment and computes the probability that a mutation is deleterious [6]. We obtained SIFT mutation scores from the SIFT site at http://blocks.fhcrc.org/sift.

**Structural features** are shown in Table 3. The features are based on PDB structure 1efa for *E. coli* lac repressor (2.6 Å resolution) [35] and 2lzm for bacteriophage T4 lysozyme (1.7 Å) [36]. For each point mutation, we used MODELLER (version 7.0) to perform sidechain replacement on the crystal structure [37].

**Table 3.** Features based on amino-acid residue solvent accessibility.

| | |
|---|---|
| *Solvent accessibility of wild-type/mutant* | calculated by DSSP [38] |
| *Change in solvent accessibility* | between wild-type and mutant |
| *Fractional solvent accessibility (FSA)* *of wild-type/mutant* | normalized by maximum solvent accessibility for each residue type, using values from Rost [39] |
| *Change in FSA* | between wild-type and mutant |
| *Buried or exposed wild-type/mutant* | buried defined with FSA < 16% |
| *Change in buried/exposed state* | between wild-type and mutant |

*Residue B-factor.* Average crystallographic temperature factor of residue backbone and sidechain atoms (proxy for residue rigidity) [5,7].

*Standardized residue B-factor.* Obtained by subtracting the mean and dividing by the standard deviation of residue B-factors in a protein of interest.

*Buried charge.* Charged residue in position with FSA<16%.

*Turn/Helix breaker.* Proline/glycine in a turn (or helix) as identified by DSSP [5].

**Interaction features**

*Ligand-binding.* We used the *LigBase* database to identify ligand-binding residues with atoms within 5Å of any HETATM listed in the PDB structure [40]. Lac repressor residues in contact with DNA were annotated using PDBSUM [41].

*Domain-interface.* We used the *PIBase* database to identify interface residues with atoms within 6Å of atoms in the residue of an oligomeric partner [40].

**References**
[1]     Battiti, R. (1994) IEEE Trans. Neural Networks 5, 537-550.
[2]     UCSC Genome Browser (build hg16) http://genome.ucsc.edu.
[3]     McKusick, V. (2000) OMIM http://ncbi.nlm.nih.gov/omim.
[4]     Klein, T.E. *et al.* (2001) Pharmacogenomics J 1, 167-70.
[5]     Chasman, D. and Adams, R.M. (2001) J.Mol.Biol. 307, 683-706.
[6]     Ng, P.C. and Henikoff, S. (2001) Genome Res. 11, 863-874.
[7]     Saunders, C.T. and Baker, D. (2002) J.Mol.Biol. 322, 891-901.
[8]     Sunyaev, S. *et al.* (2001) Hum.Mol.Genet. 10, 591-597.
[9]     Wang, Z. and Moult, J. (2001) Hum.Mutat. 17, 263-270.
[10]    Cargill, M. *et al.* (1999) Nat.Genet. 22, 231-238.
[11]    Krishnan, V.G. and Westhead, D.R. (2003) Bioinformatics 19, 2199-209.
[12]    Yue, P., Li, Z. and Moult, J. (2004).
[13]    Alber, T. *et al.* (1987) Biochemistry 26, 3754-8.
[14]    Rennell, D. *et al.* (1991) J Mol Biol 222, 67-88.
[15]    Suckow, J. *et al.* (1996) J Mol Biol 261, 509-23.
[16]    Markiewicz, P. *et al.* (1994) J Mol Biol 240, 421-33.
[17]    Tourassi, G.D.F. *et al.* (2001) Med. Phys. 28, 2394-2402.
[18]    Dumais, S.P. *et al.* (1998) in: 7th ACM International Conference on Information and Knowledge Management, pp. 148-155 ACM Press.
[19]    Li, W. (1990) J Stat Phys 60, 823-837.
[20]    Kwak, N. and Choi, C.H. (1999) in: IJCNN, Vol. 2, pp. 1313-1318.
[21]    Evans, W.E. and Relling, M.V. (1999) Science 286, 487-91.
[22]    Dean, M. (2003) Hum Mutat 22, 261-74.
[23]    Wolpert, D.H.W., D.R; (1995) Phys. Rev. E. 52, 6841-6854.
[24]    Cline, M.S. *et al.* (2002) Proteins 49, 7-14.
[25]    Cover, T.M. and Joy, T.A. (1991) John Wiley and Sons, New York.
[26]    Vapnik, V. (1995) Springer-Verlag.
[27]    Cristianini, N. *et al.* (2000) Cambridge University Press.
[28]    Sunyaev, S..*et al* (2000) Trends Genet 16, 198-200.
[29]    Bowie, J.U. *et al.* (1990) Science 247, 1306-10.
[30]    Leabman, M.K. et al. (2003) Proc Natl Acad Sci U S A 100, 5896-901.
[31]    Grantham, R. (1974) Science 185, 862-864.
[32]    Zamyatin, A.A. (1972) Prog. Biophys. Mol. Biol. 24, 107-123.
[33]    Engelman, D.M. *et al.*(1986) Annu Rev Biophys Chem 15, 321-53.
[34]    Shenkin, P.S. *et al.* (1991) Proteins 11, 297-313.
[35]    Bell, C.E. *et al.*(2000) Cell 101, 801-11.
[36]    Weaver, L.H. and Matthews, B.W. (1987) J Mol Biol 193, 189-99.
[37]    Fiser, A. and Sali, A. (2003) Methods Enzymol 374, 461-91.
[38]    Kabsch, W. and Sander, C. (1983) Biopolymers 22, 2577-637.
[39]    Rost, B. and Sander, C. (1994) Proteins 20, 216-26.
[40]    Pieper, U. *et al.* (2004) Nucleic Acids Res. 32, D217-D222.
[41]    Laskowski, R.A. *et al.* (1997) Trends Biochem Sci 22, 488-90.

# FROM CONTEXT-DEPENDENCE OF MUTATIONS TO
# MOLECULAR MECHANISMS OF MUTAGENESIS

IGOR B. ROGOZIN[†]

*National Center for Biotechnology Information NLM, National Institutes of Health,
Bethesda, MD 20894, USA; rogozin@ncbi.nlm.nih.gov*

BORIS A. MALYARCHUK

*Institute of Biological Problems of the North, Far-East Branch of the Russian Academy
of Sciences, Magadan 685000, Russia; malyar@ibpn.kolyma.ru*

YOURI I. PAVLOV

*University of Nebraska Medical Center, Omaha, NE 68198, USA; ypavlov@unmc.edu*

LUCIANO MILANESI

*Institute of Biomedical Technologies CNR, Milano 20090, Italy;
luciano.milanesi@itb.cnr.it*

Mutation frequencies vary significantly along nucleotide sequences such that mutations
often concentrate at certain positions called hotspots. Mutation hotspots in DNA reflect
intrinsic properties of the mutation process, such as sequence specificity, that manifests
itself at the level of interaction between mutagens, DNA, and the action of the repair and
replication machineries. The nucleotide sequence context of mutational hotspots is a
fingerprint of interactions between DNA and repair/replication/modification enzymes,
and the analysis of hotspot context provides evidence of such interactions. The hotspots
might also reflect structural and functional features of the respective DNA sequences and
provide information about natural selection. We discuss analysis of 8-oxoguanine-
induced mutations in pro- and eukaryotic genes, polymorphic positions in the human
mitochondrial DNA and mutations in the HIV-1 retrovirus. Comparative analysis of 8-
oxoguanine-induced mutations and spontaneous mutation spectra suggested that a
substantial fraction of spontaneous A•T→C•T mutations is caused by 8-oxoGTP in
nucleotide pools. In the case of human mitochondrial DNA, significant differences
between molecular mechanisms of mutations in hypervariable segments and coding part
of DNA were detected. Analysis of mutations in the HIV-1 retrovirus suggested a
complex interplay between molecular mechanisms of mutagenesis and natural selection.

## 1. Mutation spectra and mutation hotspots

Genetic variation is a necessary prerequisite of evolution. Genomes are
replicated at a level of fidelity that "determined" by deep evolutionary forces, by

---

[†] Adjunct research scientist at the Institute of Cytology and Genetics RAS,
Novosibirsk, Russia.

the life history it has adopted, and by accidents of its evolutionary history [1]. The mechanisms of spontaneous and induced mutagenesis are complex, and much research is devoted to understanding of these mechanisms and the factors that alter mutation rate. Mutation spectra (distributions of mutations along nucleotide sequences of a target gene) are frequently used for such studies (Figure 1) [2,3]. Mutations in target sequences are usually revealed by either phenotypic selection in experimental test systems or, in case of disease-causing genes in humans, by clinical studies in which certain genes are sequenced in groups of patients and in control groups. Both the experimental test systems and the clinical studies rely on detectable (mutable) positions, which are sites where DNA sequence changes cause phenotypic changes [2,3].

A standard representation of a mutation spectrum is a nucleotide sequence of a target gene with all changes detected put above this sequence. The base substitution mutation spectrum [4,5] in Figure 1 includes two principal elements: (i) the target sequence (Fig. 1; lower line of continuous DNA sequence) and (ii) the mutations in the target sequence (Fig. 1). Mutations in DNA/RNA molecules are classified as point mutations, deletions/insertions, duplications, inversions, and chromosomal rearrangements. Point mutations are further classified as base pair substitutions, including transitions (purine [R = A/G] mutates to R or pyrimidine [Y = C/T] mutates to Y) and transversions (R mutates to Y or Y mutates to R), and +1 and −1 frameshifts (insertions and deletions of a single base pair). Complex mutations include combinations of several point mutations and are relatively rare.

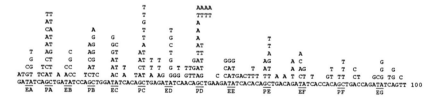

Figure 1. Somatic hypermutation spectrum in an artificially synthesized EPS sequence [4,5]. Potential hotspot positions within AGCT and TA mutable sequences are underlined. The AGCT and TA mutable sequences match well-known RGYW/WRCY and WA/TW mutable motifs [6,7].

Mutability varies significantly along nucleotide sequences: mutations, whether induced or spontaneous, occur at higher frequencies at certain positions of a nucleotide sequence (mutation "hotspots") [8]. Some mutation hotspots are thought to depend on the nucleotide sequence and the mechanism of mutagenesis *per se*; these hotspots are called intrinsic mutation hotspots. In contrast, some hotspots may be due to preferential expansion of mutants with high fitness, for example hotspots in the p53 gene might reflect both intrinsic mutability and selection during tumor progression [3]. Thus, study of mutation hotspots can help reveal mutagenic mechanisms, or can reveal information about the functional domains of a target protein [2,3,9].

## 2.  Nucleotide context of mutation hotspots

### 2.1. Local context

Intrinsic mutation hotspots are frequently caused by mutable motifs (hotspot motifs) (reviewed in [2,3,10]). One well-known example is CpG dinucleotides which are correlated with mutation hotspots in mammalian genomes [11]. The mutational mechanism for this effect is likely to involve deamination of 5-methyl cytosine, which is frequently found at CpG dinucleotides. Thus, C•G→T•A mutations occur at <u>C</u>G mutable motifs (hotspot bases are underlined) due to deamination of 5-methyl cytosine followed by replication of the resulting T/G mispair. Another well-known example of obvious mutational hotspots is hotspots of somatic mutations in immunoglobulin V genes [12]. In this case mutation hotspots are associated with R<u>G</u>YW/WR<u>C</u>Y and W<u>A</u>/<u>T</u>W mutable motifs (potential hotspot sites are underlined, W = A/T) [6,7]. Usually mutation hotspots emerged at a specific position of a mutable motif, for example, only G•C bases are mutation-prone within R<u>G</u>YW/WR<u>C</u>Y motifs (potential hotspot sites are underlined). Many other nucleotide sequence context effects on mutation rate have been studied and characterized, some examples are shown in the Table 1.

Table 1. Examples of mutable motifs.

| Spectrum/test system/mutagen | Mutable motif | Comments |
|---|---|---|
| Spontaneous G•C→A•T mutations in mammalian genomes | <u>C</u>G | May result from the spontaneous deamination of 5-methylcytosine [11] |
| Somatic mutations in immunoglobulin V genes | R<u>G</u>YW W<u>A</u> | A<u>G</u>YW is more mutable compared to G<u>G</u>YW T<u>A</u> is more mutable compared to A<u>A</u> [7] |
| Hotspots of error produced by cytidine deaminase APOBEC3G | <u>G</u>G | *in vivo* experiment |
| 8-Oxoguanine induced hotspots | A<u>A</u> | This motif was found to be mutable in pro- and eukaryotic genes |
| Target signal of retroposable elements | TTAAAA | LINEs and SINEs [17] |

Hotspot positions are underlined. R = A or G; Y = T or C; S = G or C; W = A or T; K = G or T; M = A or C; B = T, C or G; H = A, T or C; V = A, C or G; D = A, T or G.

412

Alternatively, repetitive sequences such as homonucleotide runs, direct and inverted repeats and microsatellite repeats are involved in specific types of high frequency mutational events (reviewed in [13]). For these mutations, the exact DNA sequence is not critical but only the fact that a sequence motif is repeated. The theoretical basis of these observations was suggested by Streisinger and co-workers [14]: it was proposed that short deletions and insertions within homonucleotide or homopolymeric tracts arise by misalignment of DNA strands during replication. This misalignment can lead to heterogeneity in the length of homopolymeric tracts; similar arguments apply to the more complex tandemly repeated structures of microsatellites (reviewed in [2,13]). Dislocation mutagenesis is similar to misalignment mutagenesis, but involves transient strand slippage in a monotonous run of nucleotides in the primer or template strand which is followed by incorporation of the next correct nucleotide (Figure 2) [15]. This mechanism was proposed based on studies of the *in vitro* mutation spectra of DNA polymerase β [15]. Dislocation mutagenesis may also play an important role *in vivo* generating base substitution hotspots in the control region of human mitochondrial DNA [16].

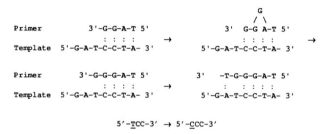

Figure 2. Dislocation mutagenesis. The primer strand dislocation, a three-nucleotide subsequence of the template strand is shown below a schematic representation of dislocation model.

There is strong evidence that short direct repeats mediate deletions and duplications in DNA [13]. Two possible mechanisms for these events are: 1) recombination between short homologous repeats or 2) DNA polymerase slippage between short repeated sequences [13]. In addition, if heteroduplexes form between imperfect direct repeats, repair of the mismatches could cause base substitutions and frameshift mutations in a concerted manner [18]. This mechanism has been suggested for some classes of spontaneous mutations in bacterial and eukaryotic genes [19] and somatic mutations in immunoglobulin genes [20].

Long inverted repeats (40-150 bases) are also particularly unstable in bacterial cells [21]. This instability is likely due to formation of hairpin structures in single-stranded DNA and/or DNA polymerase "jumps" [13]. Correction of a quasipalindrome to a perfect inverted repeat may occur by either inter- or intramolecular strand switch [18]. Many mutations of this type have been observed in bacteria, yeast and human cells [18].

## 2.2. Global factors

Mutable motifs alone are not enough for emergence of hotspots, this is illustrated by the distribution of somatic mutations across AGCT sites in an artificially synthesized EPS sequence inserted into immunoglobulin gene (Figure 1). The EPS sequence contains AGCT sequences matching well-known RGYW/WRCY mutable motifs [6] repeated six times, respectively (PA-PF monomeric units in Figure 1) [4,5]. The number of mutations at G:C bases within AGCT motifs varied from 4 (the PF monomer) to 21 mutations (the PD monomer). Two significantly different classes of AGCT motifs was revealed by the CLUSTERM program (www.itb.cnr.it/webmutation/) [22], the hotspot class includes PA, PB, PC, and PD sequences (Figure 1), while another class consists from PE and PF sequences which have significantly lower frequency of mutations. This result shows that a significant heterogeneity of the mutation rate exists even in monotonously repeated AGCT motifs. Notably, the frequency of mutations in AGCTs significantly dropped at the end of the EPS sequence. This illustrates that mutation hotspots are not equivalent to mutable motifs. Emergence of hotspots is a complex process depending on high-order structures which are hard to detect.

Many factors may influence mutation frequency in a particular nucleotide sequence. However, in most cases, only local nucleotide sequence context was studied. It is likely that other higher-level features of gene or chromatin structure also have significant influence on mutation frequency of a mutable motif at a specific site. An important factor could be the rate of DNA repair. DNA repair rates vary for transcribed and non-transcribed strands of the same gene and for more and less highly expressed genes [23]. Inherent asymmetry between the two DNA strands at the replication fork could also influence mutation frequency and specificity [24]. Other potential factors include asymmetric base composition or higher order chromatin structure (reviewed by Boulikas [25]). In general, the impact of mutation rate heterogeneity is not clear, and there are some contradictions about neutral mutation rate variation across genomes. It was suggested that mutation rates differ substantially among regions of mammalian genomes [26]. However, analysis of genomic alignments of human, chimpanzee, and baboon suggested that since the time of the human-chimpanzee ancestor, there has been little or no regional variation in mutation [27]. The controversy about mutation rate variation can be crucial for estimates of a fraction of human non-coding DNA which is under purifying selection. In general, the problem of mutation rate variation is important for understanding of fundamental problems of molecular biology and evolution. In this paper we will discuss three examples of mutation spectra analysis.

## 3. 8-Oxoguanine-induced mutations

Chemical agents, ionizing radiation and oxidative stress cause DNA oxidation [28]. 8-Oxoguanine (8-oxoG) is one of the most prominent base oxidation products and has been implicated in mutagenesis, carcinogenesis and aging [29]. It has been shown to cause G•C→T•A and A•T→C•G mutations *in vivo* and *in vitro*, depending whether guanine is oxidized in DNA or in the DNA precursor pools, respectively [30]. A spontaneous mutation spectrum in the *mutT* deficient *E.coli* strain (186 mutations; *lacI*$^d$ test system) is composed almost exclusively, of A•T→C•G transversions which is in general consistent with mutagenic properties of 8-oxoGTP [31]. Hotspot context analysis of these transversions [31] using the CLUSTERM program [22] and regression trees [32] revealed AA mutable sequence (the hotspot position is underlined) (Figure 3). Comparison of the *mutT* spectrum and A•T→C•T transversions in a spectrum of spontaneous mutations in the *lacI* gene (*lacI*$^d$ test system) [33] did not reveal significant differences between them (probability that these two spectra are different [34] $P(\chi^2)$ = 0.69). Furthermore, a highly significant positive correlation was found (Kendell's ι correlation coefficient [35] = 0.65, P < 0.01). This result suggested that a substantial fraction of spontaneous A•T→C•T mutations in *E.coli* is caused by 8-oxoGTP in nucleotide pools.

Table 2. Comparison of A•T→C•G transversion in *lacI* gene from *mutT* and wild-type strains of *E.coli* [31, 33].

| Position | | | | | | | | | | | | | | | | | | | |
|---|---|---|---|---|---|---|---|---|---|---|---|---|---|---|---|---|---|---|---|
| 41 | 81 | 72 | 64 | 87 | 168 | 79 | 189 | 192 | 195 | 167 | 83 | 117 | 96 | 128 | 177 | 77 | 141 | 105 | 54 |
| A•T→C•G mutations in *mutT* strain | | | | | | | | | | | | | | | | | | | |
| 4 | 10 | 5 | 2 | 4 | 9 | 7 | 23 | 18 | 10 | 37 | 5 | 1 | 5 | 4 | 7 | 20 | 5 | 2 | 8 |
| A•T→C•G spontaneous mutations | | | | | | | | | | | | | | | | | | | |
| 2 | 3 | 2 | 2 | 1 | 2 | 2 | 8 | 1 | 6 | 10 | 0 | 1 | 0 | 0 | 3 | 4 | 0 | 1 | 3 |

Positions of AA mutable motifs are underlined.

Reconstructed spontaneous mutations in human pseudogenes [36] (ftp.bionet.nsc.ru/pub/biology/dbms/PSEUDO.ZIP) were also analyzed, and the frequencies of nucleotides surrounding A•T→C•T transversions are shown in Table 3. Notably, AA and TT are the most frequent dinucleotide combinations. Such excess is statistically significant ($P(\chi^2) < 0.01$) as compared to dinucleotide frequencies in reconstructed ancestral sequences (Table 3). A substantially higher frequency of A in the position +1 was observed for A/C SNPs (A >> T ≈ C >>G) [37] which is consistent with the A̲A mutable motif.

Table 3. Frequencies of bases in position +1 and -1 in a set of spontaneous A•T → C•G transversions found in human pseudogenes [36].

| Substitution | Position −1 | | | | Position +1 | | | |
|---|---|---|---|---|---|---|---|---|
| | A | T | G | C | A | T | G | C |
| A→C | | | | | 0.35 | 0.24 | 0.17 | 0.24 |
| T→G | 0.25 | 0.32 | 0.21 | 0.22 | | | | |
| Expected | 0.25 | 0.22 | 0.22 | 0.31 | 0.26 | 0.23 | 0.29 | 0.22 |

Expected values (frequencies of A̲N and N̲T dinucleotides, N=A/T/G/C) were calculated in ancestral sequences used for reconstruction of spontaneous mutations [36].

These results suggested that mutagenesis due to 8-oxoG is significantly influenced by nearest neighboring bases and the context is quite evolutionarily stable. The revealed context properties could be fingerprints of interactions between DNA and repair/replication/modification enzymes. There is, at present, no data on evolutionary conservation of context specificity of such interactions between pro- and eukaryotes [3]. Thus, the alternative view that the revealed context properties reflect intrinsic properties of interactions between 8-oxoG and DNA might be the current model of choice.

## 4. Human mitochondrial DNA

Most of mitochondrial DNA (mtDNA) variability studies have been based on sequence variation of the fast-evolving major non-coding (or control) region, which spans 1122 bases between the tRNA genes for proline (tRNA$^{Pro}$) and phenylalanine (tRNA$^{Phe}$) [38]. The majority of mutations are concentrated in two hypervariable segments, HVS I (positions 16024-16365) and HVS II (positions 73-340) [39]. Our analysis of phylogenetically reconstructed mutation spectra of the mtDNA HVS I and II regions has suggested that the dislocation mutagenesis (Figure 2) plays an important role for generating base substitutions in these regions [16,40]. However, an impact of the dislocation mutagenesis on the remaining part of mtDNA remains unclear.

To study spontaneous base substitutions in regions of human mtDNA other than HVS I and II, we reconstructed mutation spectra of the mtDNA region containing ND3, tRNA$^{Arg}$, and ND4L genes (positions 10171-10659) using published data on polymorphisms in various human populations. We have analyzed different phylogenetic haplogroups of mtDNA revealed by means of median network analysis [39] (http://fluxus-engineering.com, the Network 3.1 program). We used only the published population data comprising mtDNA sequences with known phylogenetic status as described by Malyarchuk and co-workers [16,40]. The reconstructed mutation spectrum contained 93 mutations in 489 bases.

The dislocation mutagenesis model was analyzed for the reconstructed mutation spectrum using a Monte-Carlo procedure [40]. No statistically significant support for this model was found ($P(W \leq W_{random}) = 0.68$). This result suggested that the dislocation mutagenesis does not play an important role for generating substitutions in the coding regions of mtDNA. A higher rate of molecular evolution in HVS regions than in the remaining part of mtDNA can be explained by differences in either mutation or selection pressure [39]. The observed differences in dislocation mutagenesis suggested that a higher rate of mutations in HVS regions is caused by intrinsic properties of mutations. HVS regions are associated with initiation/termination of mtDNA replication and RNA/DNA transition, and one of these processes may be error-prone for the DNA strand dislocation mutagenesis.

## 5. Hypermutation in HIV-1

Genomic heterogeneity is a hallmark of retroviruses, especially of HIV, which helps virus to escape the host immune system. Hypermutability is linked to pathogenicity and it was generally attributed to relatively low fidelity of reverse transcriptase [41]. A high rate of mutagenesis in retroviruses, in addition to being a way to elude the immune system, may lead to their low viability. It was hypothesized that even a relatively small increase of the mutation rate in a retrovirus will lead to the accumulation of many deleterious mutations and virus extinction. Indeed, the treatment of HIV-infected human cells by mutagenic nucleoside analogs resulted in the loss of viral replicative potential [42]. Recent discoveries suggest that nature already exploited this mechanism for protection from retroviruses. The unique cellular gene CEM15 was found that conferred resistance to HIV. Its antiviral action could be overcome by the presence of virion infectivity factor (Vif), encoded by the viral genome [43]. CEM15 appeared to be identical to the cytidine deaminase APOBEC3G. It is already known that APOBEC3G is a strong mutator when expressed in *E. coli*, suggesting that it could deaminate cytosines in DNA [44]. The viruses without Vif experienced hypermutation and all these mutations were transitions that could be explained by deamination of a (-) DNA strand of the virus [45]. The current model for APOBEC3G antiviral action proposes that the deaminase is packaged into $Vif^-$ HIV virions and induces massive deamination in the viral (-) strand [45,46]. This deamination can lead to hypermutagenesis or the destruction of the viral genome during repair of uracil [46].

We analyzed the published spectra of APOBEC3G-induced mutations (434 mutations) in the GFP gene [47] (Fig. 3A). Analysis of nucleotide context of mutation hotspots using regression trees [32] suggested that almost all APOBEC3G hotspots are located in the GG motif (Fig. 3A). Not all GG motifs appeared to be hotspots for mutations. This can be explained by the GFP selection system (mutations in some GG sites cannot be detected by this system), however some unknown global context properties of GG hotspot sites might modulate mutability in some GG sites. We analyzed a correlation between this motif and mutations of HIV-1 DNA (30 mutations) in the absence of the Vif protein [45] (Fig. 3B) using the CONSEN program [6,7]. A low probability $P_{W<Wrandom}$ ($P_{W<Wrandom}$ < 0.001) indicated that there is a highly significant correlation between the APOBEC3G mutable motif GG and hypermutation of HIV-1. This result suggested that APOBEC3G caused the mutations in HIV-1 as was originally speculated by Lecossier and co-workers [45].

Specificity of APOBEC3G for GG sequences, which is frequently a part of TGG tryptophan codons, results in a frequent generation of TAG nonsense codons which leads to a premature termination of protein synthesis. This might be a genetic strategy to kill the virus, and the APOBEC3G hotspot specificity is determined by its biological role. The mutagenic activity of APOBEC3G might have perspectives for novel anti-HIV therapies that interfere with this virus-host interaction.

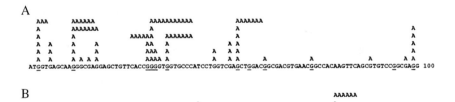

Figure 3. Example of mutation spectra generated by APOBEC3G expression in *E. coli* and mutation spectrum in HIV-1 retrovirus in the absence of the Vif protein. Potential hotspot positions within GG mutable motifs are underlined. (A) Spectrum of mutations in the first 100 bases of the *GFP* gene induced by the expression of APOBEC3G [47]. (B) Spectrum of mutations in HIV-1 DNA in the absence of the Vif protein [45].

## 6. Discussion

Many mutable motifs are short (Table 1), thus a lot of caution needed while analyzing correlation between these motifs and distribution of mutations (e.g., the CONSEN program) due to a high rate of false positives expected for short motifs. Therefore, such indirect comparisons should be validated by direct comparisons of mutational spectra produced in the same sequence using statistical methods to compare two or more spectra based on an exact or pseudo-probability test (a Monte Carlo modification of the exact test) [34] or correlation analyses [35,48]. This strategy was employed in studies of somatic hypermutation in immunoglobulin genes. Mutational spectra analysis of 15 immunoglobulin genes suggested that the WA mutable motif is a universal descriptor of somatic hypermutation at A•T pairs. Highly mutable sites, "hotspots", that matched WA were preferentially found in one DNA strand. Analysis of base-substitution hotspots in DNA polymerase error spectra showed that 33 of 36 hotspots in the human polymerase η spectrum conformed to the WA consensus. A correlation between the W<u>A</u> motif and the error specificity of human DNA polymerase η suggested that this polymerase contributes to the W<u>A</u> hotspots [7]. Additional analysis of this correlation using the same mouse immunoglobulin target sequence for *in vivo* and *in vitro* spectrum generation combined with studies of mutable motifs and frequencies of substitutions greatly improved the power of comparisons, allowing use of the pairwise and multiple regression analysis [48]. Obtained results supported the hypothesis that DNA polymerase η contributes to somatic hypermutation of immunoglobulin genes at A•T pairs via short patches of low fidelity DNA synthesis of both strands, but with a preference for the non-transcribed strand [48].

It should be emphasized that the context of hotspots may be very helpful in deep understanding of underlying molecular mechanisms of mutagenesis. However, the determinants of mutation frequency and specificity are complex and there are many analytical methods for their study. The most reliable results can be obtained if several methods are combined or used sequentially and if many different sources of information are considered. It is a challenging task to analyze mutation spectra, and in some cases, the effort will primarily be descriptive in nature. However, in several well-documented studies, the analysis of mutation spectra has contributed substantially to understanding molecular mechanisms of mutagenesis. As analytical methods continue to be developed, more theoretical and experimental studies will contribute insights into the complex process of mutagenesis.

## Acknowledgements

We thank Eugene Koonin and Galina Glazko for helpful comments on the manuscript. This work was partially supported by RFBR, by Far-East Branch of the Russian Academy of Sciences, MIUR "Functional Genomics" 449/97, "High through put Grid computational platforms for virtual scalable organizations" and "Bioinformatics" Italian FIRB projects.

## References

1. J.W. Drake, B. Charlesworth, D. Charlesworth and J.F. Crow, *Genetics* **148**, 1667 (1998).
2. J.H. Miller, *Annu. Rev. Genet.* **17**, 215 (1983).
3. I.B. Rogozin, V.N. Babenko, L. Milanesi and Y.I. Pavlov, *Brief. Bioinform.* **4**, 210 (2003).
4. U. Storb, E.L. Klotz, J. Hackett, K. Kage, G. Bozek and T.E. Martin, *J. Exp. Med.* **188**, 689 (1998).
5. N. Kim, G. Bozek, J.C. Lo and U. Storb, *J. Exp. Med.* **190**, 21 (1999).
6. I.B. Rogozin and N.A. Kolchanov, *Biochim. Biophys. Acta* **1171**, 11 (1992).
7. I.B. Rogozin, Y.I. Pavlov, K. Bebenek, T. Matsuda and T.A. Kunkel, *Nat. Immunol.* **2**, 530 (2001).
8. S. Benzer, *Proc. Natl. Acad. Sci. USA* **47**, 403 (1961).
9. A. Siepel and D. Haussler, *Mol. Biol. Evol.* **21**, 468 (2004).
10. M. Zavolan and T.B. Kepler, *Curr. Opin. Genet. Dev.* **11**, 612 (2001).
11. D.N. Cooper and H. Youssoufian, *Hum. Genet.* **78**, 151 (1988).
12. A.G. Betz, C. Rada, R. Pannell, C. Milstein and M.S. Neuberger, *Proc. Natl. Acad. Sci. USA* **90**, 2385 (1993).
13. D.A. Gordenin and M.A. Resnick, Mutat Res **400**, 45 (1998).
14. G. Streisinger, Y. Okada, J. Emrich, J. Newton, A. Tsugita, E. Terzaghi and M. Inouye, *Cold Spring Harb. Symp. Quant. Biol.* **31**, 77 (1966).
15. T.A. Kunkel, *J. Biol. Chem.* **260**, 5787 (1985).
16. B.A. Malyarchuk, I.B. Rogozin, V.B. Berikov and M.V. Derenko, *Hum. Genet.* **111**, 46 (2002).
17. J. Jurka and P. Klonowski, *J. Mol. Evol.* **43**, 685 (1996).
18. L.S. Ripley, *Proc. Natl. Acad. Sci. USA* **79**, 4128 (1982).
19. G.B. Golding and B.W. Glickman, *Proc. Natl. Acad. Sci. USA* **82**, 8577 (1985).
20. N.A. Kolchanov, V.V. Solovyev and I.B. Rogozin, *Dokl. Akad. Nauk* **281**, 994 (1985).
21. D.M. Lilley, *Nature* **292**, 380 (1981).
22 G.V. Glazko, L. Milanesi and I.B. Rogozin, *J. Theor. Biol.* **192**, 475 (1998).
23. P.C. Hanawalt, *Environ. Health Perspect.* **76**, 9 (1987).
24. X. Veaute and R.P. Fuchs, *Science* **261**, 598 (1993).

25. T. Boulikas, Evolutionary consequences of nonrandom damage and repair of chromatin domains, *J. Mol. Evol.* **35**, 156 (1992).

26. K.H. Wolfe, P.M. Sharp and W.H. Li, *Nature* **337**, 283 (1989).

27. M.T. Webster, N.G. Smith and H. Ellegren, *Mol. Biol. Evol.* **20**, 278 (2003).

28. R. Adelman, R.L. Saul and B.N. Ames, *Proc. Natl. Acad. Sci. USA* **85**, 2706 (1988).

29. B.N. Ames, *Mutat. Res.* **214**, 41 (1989).

30. M.L. Michaels and J.H. Miller, *J. Bacteriol.* **174**, 6321 (1992).

31. R.G. Fowler and R.M. Schaaper, *FEMS Microbiol. Rev.* **21**, 43 (1997).

32. V.B. Berikov and I.B. Rogozin, *Bioinformatics* **15**, 553 (1999).

33. A.R. Oller and R.M. Schaaper, *Genetics* **138**, 263 (1994).

34. N.F. Cariello, W.W. Piegorsch, W.T. Adams and T.R. Skopek, *Carcinogenesis* **15**, 2281 (1994).

35. V.N. Babenko and I.B. Rogozin, *Biofizika* **44**, 632 (1999).

36. M.A. Pozdniakov, I.B. Rogozin, V.N. Babenko and N.A. Kolchanov, *Dokl. Akad. Nauk.* **356**, 566 (1997).

37. Z. Zhao and E. Boerwinkle, *Genome Res.* **12**, 1679 (2002).

38. S. Anderson, A.T. Bankier, B.G. Barrell, M.H. de Bruijn, A.R. Coulson, J. Drouin, I.C. Eperon, D.P. Nierlich, B.A. Roe, F. Sanger, P.H. Schreier, A.J. Smith, R. Staden and I.G. Young, *Nature* **290**, 457 (1981).

39. L. Vigilant, M. Stoneking, H. Harpending, K. Hawkes and A.C. Wilson, *Science* **253**, 1503 (1991).

40. B.A. Malyarchuk and I.B. Rogozin, *Ann. Hum. Genet.* **68**, 324 (2004).

41. J.D. Roberts, K. Bebenek and T.A. Kunkel, *Science* **242**, 1171 (1988).

42. L.A. Loeb, J.M. Essigmann, F. Kazazi, J. Zhang, K.D. Rose and J.I. Mullins, *Proc. Natl. Acad. Sci. USA* **96**, 1492 (1999).

43. A.M. Sheehy, N.C. Gaddis, J.D. Choi and M.H. Malim, *Nature* **418**, 646 (2002).

44. R.S. Harris, S.K. Petersen-Mahrt and M.S. Neuberger, *Mol. Cell* **10**, 1247 (2002).

45. D. Lecossier, F. Bouchonnet, F. Clavel and A.J. Hance, *Science* **300**, 1112 (2003).

46. R.S. Harris, K.N. Bishop, A.M. Sheehy, H.M. Craig, S.K. Petersen-Mahrt, I.N. Watt, M.S. Neuberger and M.H. Malim, *Cell* **113**, 803 (2003).

47. R.S. Harris, A.M. Sheehy, H.M. Craig, M.H. Malim and M.S. Neuberger, *Nat. Immunol.* **4**, 641 (2003).

48. Y.I. Pavlov, I.B. Rogozin, A.P. Galkin, A.Y. Aksenova, F. Hanaoka, C. Rada and T.A. Kunkel, *Proc. Natl. Acad. Sci. USA* **99**, 9954 (2002).

# A BAYESIAN FRAMEWORK FOR SNP IDENTIFICATION

B.M. WEBB-ROBERTSON, S.L. HAVRE

*Computational Biology & Bioinformatics, Pacific Northwest National Laboratory*
*Richland, WA 99352, USA*


D.A. PAYNE

*Information Analytics, Pacific Northwest National Laboratory*
*Richland, WA 99352, USA*

As evolutionary models for single-nucleotide polymorphisms (SNPs) become available, methods for using them in the context of evolutionary information and expert prior information is a necessity. We formulate a probability model for SNPs as a Bayesian inference problem. Using this framework we compare the individual and combined predictive ability of four evolutionary models of varying levels of specificity on three SNP databases (two specifically targeted at functional SNPs) by calculating posterior probabilities and generating Receiver Operating Characteristic (ROC) curves. We discover that none of the models do exceptionally well, in some cases no better than a random-guess model. However, we demonstrate that several properties of the Bayesian formulation improve the predictability of SNPs in the three databases, specifically the ability to utilize mixtures of evolutionary models and a prior based on the genetic code.

## 1. Introduction

Interest in single-nucleotide polymorphisms (SNPs) has exploded in recent years. This interest is evident from the large number of SNP databases publicly available on the web: NCBI SNP database (http://www.ncbi.nlm.nih.gov/SNP), Human SNP database (http://www.broad.mit.edu/snp/human/), hemoglobin database (http://globin.cse.psu.edu), SNP Consortium Ltd. (http://snp.cshl.org/), and many others. The compilation of SNPs is vital to studying important biological problems such as the identification of biomarkers for disease and evolution at the molecular level. High-throughput technologies, such as proteomics via mass spectrometry (MS) hold promise to identify SNPs rapidly at a global scale, but identification by these approaches using brute force is computationally unattractive. Thus, accurate methods to assign probabilities to potentially polymorphic sites are necessary.

Evolutionary information is generally captured by estimating model parameters associated with a set of biosequences, DNA [1-5] or proteins [6-7], from one or multiple organisms. These models are then used in the context of some specific framework, for example phylogenetics [8-9] or sequence alignment [10-12]. Limited evolutionary models have been developed for

SNPs, capturing information at the amino acid [13] and codon [14] levels. No framework for assigning probabilities to individual polymorphic events exists and thus little comparison of both SNP specific and general evolutionary models has been performed.

This paper presents the general framework for SNP identification in the context of Bayesian inference through the use of evolutionary models to assign probabilities to all possible SNPs (confined by an application to MS). The Bayesian framework allows both individual and mixtures of evolutionary models to be used. Additionally, it allows for the injection of additional information in the form of a prior. We demonstrate the Bayesian framework by comparing four specific evolutionary models (one SNP specific [13], two nucleotide evolutionary rate matrices [1,4], and one inter-species amino acid model [7]) on three SNP databases. The first two are databases of disease causing or enhancing SNPs for the human proteins hemoglobin [15-17] and p53 [18-19]. The third is a set of genes that characterize SNPs in eight inbred mouse strains [20]). Lastly, we explore the benefits observed from the Bayesian formulation related to the use of mixtures of evolutionary models and the use of the genetic based prior in comparison to a neutral based prior.

## 2. Methods

Bayesian statistics is an attractive approach for making probabilistic inferences from biological data because it supports the injection of information related to the data, for example, expert opinion or evolution constraints based on sequence composition or length [21-22]. Due to the uncertainty associated with biological data and the frequent availability of expert opinion, many biological problems can be more easily modeled by Bayesian methods than by other approaches.

The Bayesian framework for SNPs attempts to quantify the belief that a nucleotide at a given position in a genetic sequence underwent a polymorphism. We model the problem at the codon level to observe both individual nucleotides and amino acids. In a Bayesian formulation both the observed and unobserved data are treated as random variables. The general formulation defines the annotated genomic data $(G)$ as the observed data. The unobserved data are the codons $(S)$ and two types of background information – an evolutionary model $(\Psi)$ and a mutational descriptor $(M)$.

### 2.1. Background Information

Evolutionary information is depicted by matrices at either the amino acid or nucleotide level describing the likelihood of one residue being substituted by another. In this study four evolutionary models ($\Psi$) are evaluated for comparative value. We first describe each of these four models; two amino acid and two nucleotide. Subsequently, we describe the SNP mutation variable ($M$).

**Amino Acid Matrices.** The first model is from the *BLOSUM* [7] series of scoring matrices commonly used in sequence alignment. This series is generated from a large set of sequences from multiple species at various levels of sequence identity and thus represents a complex ancient history over speciation. It is believed to be inappropriate for intra-species evaluation so a less divergent matrix, *BLOSUM80* (referred to as *BL80*), is included for comparative value. The second model is a newly developed substitution matrix by Majewski and Ott [13] (referred to as *M-O*). It is based on identified SNPs in the human genome and thus captures recent evolutionary changes.

**Nucleotide Matrices.** The last two models are based on continuous-time Markov chain models that describe the evolutionary rate of substitution between two nucleotides [9]. A fully parameterized model, a 4x4 rate matrix, requires the estimation of 12 parameters (16 possible substitutions minus the four changes to the same nucleotide). To reduce the parameterization several nested evolutionary models have been developed based on possible transitions between purines and pyrimidines at various levels [1-4]. The number of parameters estimated and their estimation values are dependent upon two factors – the data selected and the model. We use the parameter values estimated by Suchard et al. [9] for two models reflecting different levels of evolution. The first of these is the Tamura and Nei model (*TN93*) [4], parameterized into three rates based on data from the "Tree of Life", representing organisms across all living kingdoms. The last model is the Hasegawa et al. model [1], which calculates the rate matrix for each codon position using two parameters based on primate data, resulting in a less general model.

**Mutation Variable.** The background information ($M$) is a binary variable that describes a detectable SNP event; defined as a mass changing substitution. Undetectable SNPs include silent mutations (SNPs resulting in no change at the amino acid level) and mutations between leucine ($L$) and isoleucine ($I$) (whose mass is indistinguishable by MS). Additionally, we define mutations to and from *STOP* codons as invalid, assuming that such mutations are typically detrimental to the protein. Additionally, we assume that an observed amino acid change is the result of one SNP per codon and not from double mutations. Thus, given two codons, $c_i$ and $c_j$, where $a_{(i)}$ and $a_{(j)}$ define their respective amino acids, a valid SNP is expressed explicitly as,

$$M(c_i, c_j) = \begin{cases} 1 & \text{if } c_i \text{ and } c_j \text{ differ by only one nucleotide and} \\ & m(a_{(i)}) \neq m(a_{(j)}) \text{ and } a_{(i)} \neq a_{(j)} \neq STOP \\ 0 & \text{otherwise} \end{cases} , \quad (1)$$

where $m(a_{(i)})$ and $m(a_{(j)})$ are the masses associated with $a_{(i)}$ and $a_{(j)}$, respectively. There are several benefits of defining the mutation variable in this manner. For example, in lieu of a binary definition, probabilities could be defined for SNP events based on mass difference. Also, multiple types of peptide variants could be defined, for example $M_1$ for SNPs, $M_2$ for frameshifts, and $M_3$ for multiple nucleotide polymorphisms.

### 2.2. Bayesian Formulation

In our basic Bayesian formulation (1) the genomic data ($G$) consists of $I$ codons, (2) the codon substitution ($S$) represents a mutation to a codon $j$, $j=1,\ldots,64$ ($s_j$), (3) the evolutionary model ($\Psi$) describes the probability ratio or rate of mutation between two residues, and (4) the mutation variable ($M$) describes the probability of a substitution between two codons. The Bayesian formulation is described as the joint distribution of the observed and unobserved data: $P(G,S,M,\Psi)$. This is the product of the likelihood ($L$) and the prior ($P$):

$$P(G, S, M, \Psi) = L(S, M, \Psi; G)P(S, M, \Psi).$$

The likelihood is the probability of the observed data given the unobserved data:

$$P(G, S, M, \Psi) = P(G \mid S, M, \Psi)P(S, M, \Psi). \quad (2)$$

The prior $P(S,M,\Psi)$ can be decomposed into easily calculable probabilities. The evolutionary model $\Psi$ does not change based on the genetic code or the type of mutation being observed. Thus, we assume independence from $S$ and $M$: $P(S,M,\Psi)=P(S,M)P(\Psi)$. Lastly, returning to the genetic code, we observe that the probability of observing a given codon is dependent on $M$, and given that there is only one type of mutation event in this case: $P(S,M)P(\Psi)=P(S|M)P(\Psi)$. Hence the Bayesian formulation observed in Eq. 2 can be expressed as:

$$P(G, S, M, \Psi) = P(G \mid S, M, \Psi)P(S \mid M)P(\Psi). \quad (3)$$

The calculations for implementation occur at the individual codon level. Thus, the above general representation can be given in terms of the individual elements of the observed and unobserved data. The joint distribution of a specific mutation in the genome, observing codon $j$ at the $i^{th}$ position in the genome given a specified evolutionary model, $\psi_k$, is defined as:

$$P\big(g_i, s_j, M, \psi_k\big) = P\big(g_i \mid s_j, M, \psi_k\big) P\big(s_j \mid M\big) P\big(\psi_k\big). \tag{4}$$

The joint distribution in Eq. 4 allows easy calculation of posterior probabilities of interest; for example, specific SNPs describing the probability of observing a SNP in the form of codon $s_j$ at the $i^{th}$ position in the genome. Given a specific evolutionary model, $\psi_k$, this is formulated in terms of Bayes theorem:

$$P\big(s_j \mid g_i, \psi_k\big) = \frac{P\big(g_i \mid s_j, M, \psi_k\big) P\big(s_j \mid M\big) P\big(\psi_k\big)}{\sum_j P\big(g_i \mid s_j, M, \psi_k\big) P\big(s_j \mid M\big) P\big(\psi_k\big)}. \tag{5}$$

Additionally, the Bayesian formulation allows the probability of observing a specific SNP independent of the evolutionary model to be calculated:

$$P\big(s_j \mid g_i\big) = \frac{\sum_k P\big(g_i \mid s_j, M, \psi_k\big) P\big(s_j \mid M\big) P\big(\psi_k\big)}{\sum_k \sum_j P\big(g_i \mid s_j, M, \psi_k\big) P\big(s_j \mid M\big) P\big(\psi_k\big)}. \tag{6}$$

To obtain these values of interest, the likelihood and priors much each be calculated.

**The Likelihood.** There is no loss in information by transforming evolutionary models in the form of symmetric 4x4 or 20x20 matrices into 64x64 codon matrices. We perform this conversion for consistency to make the likelihood calculation straight forward. Accordingly, $\psi_k$ describes the probability ratio or rate of substitution between two codons $g_i$ and $s_j$. The likelihood also includes the mutation event variable, Eq. 1, which allows only valid SNPs to have non-zero probabilities. Thus, the likelihood can be described as:

$$P\big(g_i \mid s_j, M, \psi_k\big) = \psi_k\big(g_i, s_j\big) * M\big(g_i, s_j\big).$$

**The Prior.** There are two prior in Eqs 3-6, where $P(\psi_k)$ is the prior belief that the codon mutation model $\psi_k$ fits the data and $P(s_j \mid M)$ is the probability of observing a given codon $s_j$ (independent of the genomic data) given a mutation of type $M$. The prior on the evolutionary model can be either defined by the user or *a priori*. We assume *a priori* – all models are equally likely. By assuming that mutations at all positions in the genome are equally likely, the prior $P(s_j \mid M)$ can be defined directly from the genetic code. Because there are 64 codons, there are 576 possible SNPs between all codons, 385 of these are valid as described by $M$, Eq. 1. The prior probabilities are calculated for each codon, observing that a given codon, $P(s_j \mid M)$, can only result from a SNP to nine other possible codons. As illustrated in Figure 1 [14], of the nine possible

SNPs that could result in the codon *AGA*, two are products of silent mutations and one is a *STOP* codon. Thus, in this example, the probability of observing *AGA* given *M* (all possible valid SNPs) is $P(AGA|M)=^6/_{385}$. Alternatively Bayes theorem can be used to arrive at this answer in perhaps a more intuitive manner:

$$P(s_j \mid M) = \frac{P(M \mid s_j)P(s_j)}{P(M)} = \frac{P(M \mid s_j)P(s_j)}{\sum_j P(M \mid s_j)P(s_j)} .$$

The benefit of this approach is that it incorporates prior information on the probability of observing a given codon, $P(s_j)$. For instance, codon frequency information on a specific species could be incorporated.

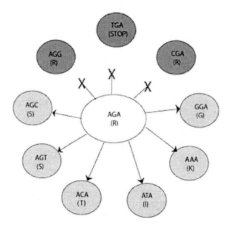

Figure 1. A schematic of the nine possible codons and corresponding amino acids from a SNP to codon *AGA*. In this case there are 6 nonsynonymous SNPs (resulting in an amino acid change), 2 synonymous SNPs, and 1 SNP from a *STOP* codon.

## 3. Results and Discussion

The human ability to treat disease has developed man beyond a simple 'survival of the fittest' species. Accordingly, due to the potential for identifying biomarkers to treat disease, the majority of SNP data is human focused. We focus on two of these compilations for the proteins hemoglobin [15-17] and p53 [18-19]. These SNPs are functionally important SNPs as they are disease causing or enhancing. Additionally, we observe a third database that characterizes observable SNPs in eight inbred mouse strains [20]. We first discuss the specifics of each SNP databases. Secondly, we evaluate the predictive ability of each individual evolutionary model using the Bayesian framework. Lastly, we assess the benefits of the Bayesian formulation, specifically the inclusion of the genetic code based prior and the posterior based on mixtures of evolutionary models.

### 3.1. *The SNP Databases*

Hemoglobin is a protein that transports oxygen from the lungs to the peripheral tissues to maintain the viability of cells. It has been largely studied because diseases such as sickle cell anemia have been linked to variants of the protein. The genomic sequence (including the $\alpha$ and $\beta$ chains) consists of 287 codons, resulting in 1871 valid SNPs. The *"Syllabus of Human Hemoglobin Variants"* [15] is a comprehensive listing of all known human hemoglobin variants. Approximately 541 (29%) of the possible 1871 SNPs are represented in this databases (http://globin.cse.psu.edu).

The p53 protein is the result of a tumor suppressor gene located on human chromosome 17. The p53 gene has been largely studied because mutations of this gene are often accompanied by cancer. This protein is 393 codons in length and has 2559 valid SNPs. From the database, 776, or approximately 30%, of the 2559 SNPs are represented (http://p53.curie.fr/).

The Mouse database contains SNPs identified in eight inbred strains of mice. Although its generation is quite different than that of freely mating populations, it is included for comparative value to determine if any of the defined evolutionary models hold predictive power despite the forced inbreeding. This database covers a much larger genomic space than the human protein-specific databases. The database contains 1307 sequences with 71,798 codons yielding 439,202 possible valid SNPs. Only 1822 SNPs, or 0.4%, of the valid SNPs are represented (http://www.broad.mit.edu/snp/moue).

### 3.2. *Predictive Ability of Individual Evolutionary Models*

Each database has one or more genes associated with it; from these genes all valid SNPs can be calculated. Each database consists of a list of observed SNPs (a subset of all valid SNPs), which are presumed to be true positives. All the remaining SNPs not represented in a database are assumed to be true negatives. The hemoglobin, p53, and mouse databases have 541, 776, and 1881 true positives and 1330, 1783, and 437,380 true negatives, respectively. To observe the predictive capability of each evolutionary model with the Bayesian framework we use it as a classifier. Given the probabilities assigned to each SNP from the Bayesian model, the true positives and negatives can be used to generate Receiver Operating Characteristic (ROC) curves [23] for each database.

The ROC curve gives a graphical representation of the trade-off between sensitivity and specificity. The plot displays the false positive rate (ratio of false positive to total negatives) versus the true positive rate (ratio of true positives to total positives) at all possible cut-off values. A completely random predictor would give a straight line at a 45° angle – *TP rate* equal to *FP rate*. This is the *Baseline* model. Figure 2 shows ROC curves generated from the individual SNP posterior probabilities (Eq. 5) obtained for each of the four described evolutionary models. Since there are many SNPs with the same posterior probability we generate points on the ROC curve by randomly shuffling the order of the SNPs within any given probability 100 times and display the average.

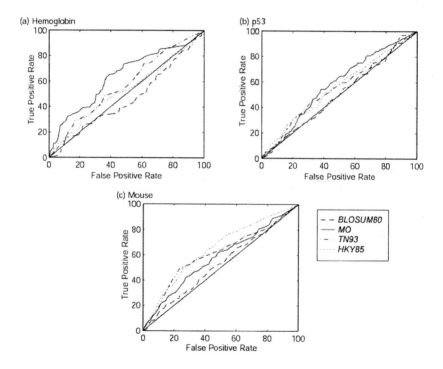

Figure 2. The ROC curves for the (a) hemoglobin, (b) p53, and (c) mouse databases to assess the evolutionary models (1) *BL80*, (2) *M-O*, (3) *TN93*, and (4) *HKY85*.

At first glance, Figure 2 appears to show that the *M-O* model performs the best for both of the disease databases and the *HKY85* model performs best for the mouse database. None of the evolutionary models are exceptionally good classifiers, however it is of interest to compare and contrast these models. The area under the ROC curve is the most common measure used to compare the discriminative effectiveness of classifiers. DeLong et al. [24] is a standard statistical test for comparing correlated ROC curves that have a large sample size. Tables 1-3 display the area under the ROC curve for each of the evolutionary models and the associated $p$-values. The largest of each database is highlighted in bold.

Table 1. The area under the ROC curves (in parentheses) and $p$-values
for primary model comparison for the hemoglobin SNP database.

| | BL80 (0.461) | M-O (0.639) | TN93 (0.571) | HKY85 (0.528) |
|---|---|---|---|---|
| Baseline (0.500) | 0.0111 | <0.0001 | <0.0001 | 0.0586 |
| BL80 (0.461) | | <0.0001 | <0.0001 | 0.0008 |
| M-O (0.639) | | | 0.0001 | <0.0001 |
| TN93 (0.571) | | | | 0.0074 |

Table 2. The area under the ROC curves (in parentheses) and $p$-values
for primary model comparison for the p53 SNP database.

| | BL80 (0.507) | M-O (0.580) | TN93 (0.550) | HKY85 (0.561) |
|---|---|---|---|---|
| Baseline (0.500) | 0.5563 | <0.0001 | <0.0001 | <0.0001 |
| BL80 (0.507) | | 0.0001 | 0.0045 | 0.0009 |
| M-O (0.580) | | | 0.0518 | 0.2177 |
| TN93 (0.550) | | | | 0.4230 |

Table 3. The area under the ROC curves (in parentheses) and $p$-values
for primary model comparison for the mouse SNP database.

| | BL80 (0.527) | M-O (0.577) | TN93 (0.609) | HKY85 (0.646) |
|---|---|---|---|---|
| Baseline (0.500) | 0.0005 | <0.0001 | <0.0001 | <0.0001 |
| BL80 (0.524) | | <0.0001 | <0.0001 | <0.0001 |
| M-O (0.577) | | | 0.0006 | <0.0001 |
| TN93 (0.609) | | | | <0.0001 |

Applying a standard threshold of 0.05, all models are significantly different than the *Baseline* model, except *HKY85* for hemoglobin and *BL80* for p53. Additionally, the low $p$-value observed for *BL80* on hemoglobin is a result of this model performing significantly worse than *Baseline* (Figure 2a). The most surprising result is that the human amino acid model does not outperform either of the nucleotide models on mouse despite many mouse genes being orthologous to human. In fact the global *TN93* model has a consistent performance on all three databases. It has the second largest area under the ROC curve for both hemoglobin and mouse. This is significant since in practice evolutionary models specific to the organism under study may not be available. The parameters for *TN93* were generated from species drawn from eukaryotes, eubacteria, halobacteria, and eocytes.

### 3.3. *Assessing the Benefits of the Bayesian Formulation*

The Bayesian formulation has several benefits, specifically the ability to obtain a posterior probability using multiple evolutionary background information parameters and the ability to define priors. It has been shown in sequence alignment that sensitivity can be improved by summing over multiple substitution matrices [12]. We assess the improvement in sensitivity observed from mixtures of evolutionary models, as well as when the codon defined prior is used in lieu of a neutral defined prior.

**Inclusion of Multiple Evolutionary Models.** ROC curves were constructed using posterior probabilities calculated for the eleven possible model combinations (Eq. 6). The areas under the ROC curves were compared to the best individual model using DeLong et al. [24]. Table 4 gives these results for p53 and mouse – no model combinations gave an area under the ROC curve greater than $M\text{-}O$ for hemoglobin. Not surprisingly, the predictability of the $M\text{-}O$ model, tuned specifically to human SNP data, is not improved for hemoglobin or p53 by including additional models. For mouse, two mixture models perform significantly better than the best individual model; (1) $M\text{-}O$ and $HKY85$, and (2) $BL80$, $M\text{-}O$, and $HKY85$. It appears that the decision to utilize multiple models may be subjective to the evolutionary distance between the organism being studied and the organisms used to generate the models.

Table 4. Comparison of the area under the ROC curve for each model combination that had a larger area than the best individual model and the corresponding $p$-values.

|  | p53 ($M\text{-}O = 0.580$) | | Mouse ($HKY85 = 0.646$) | |
|---|---|---|---|---|
|  | ROC area | $p$-value | ROC area | $p$-value |
| $M\text{-}O/TN93$ | 0.586 | 0.6241 | 0.634 |  |
| $M\text{-}O/HKY85$ | 0.588 | 0.5284 | **0.655** | **0.0001** |
| $BL80/M\text{-}O/HKY85$ | 0.586 | 0.6244 | **0.654** | **0.0008** |
| $M\text{-}O/TN93/HKY85$ | 0.594 | 0.2166 | 0.650 | 0.3222 |
| $All$ | 0.592 | 0.3139 | 0.647 | 0.7798 |

**The Prior.** The Bayesian SNP model includes a prior on the probability of a codon given a valid SNP, $P(s_j|M)$. This prior is based on the genetic code (Figure 1). Table 5 gives the area under the ROC curve for the genetic code defined prior and for a neutral prior (all codons are equally likely to undergo mutation), as well as the $p$-value comparing the ROC area difference. The results are startling. In all cases the area under the ROC curve for the neutral prior is less than or equal to the genetic code prior, in most cases returning significant $p$-values. Of the twelve generated $p$-values, nine have a $p$-value of less than 0.1. Thus, the inclusion of this prior is beneficial to the overall predictability of the model.

Table 5. The area under the ROC curve generated from the genetic code (C prior) and the *neutral* (N prior) and associated *p*-values.

| | Hemoglobin | | | p53 | | | Mouse | | |
|---|---|---|---|---|---|---|---|---|---|
| | C Prior | N Prior | *p*-value | C Prior | N Prior | *p*-value | C Prior | N Prior | *p*-value |
| BL80 | 0.461 | 0.437 | <0.01 | 0.507 | 0.486 | <0.01 | 0.524 | 0.525 | 0.72 |
| M-O | 0.639 | 0.627 | <0.01 | 0.580 | 0.563 | <0.01 | 0.577 | 0.577 | 0.34 |
| TN93 | 0.571 | 0.540 | <0.01 | 0.550 | 0.514 | <0.01 | 0.609 | 0.597 | <0.01 |
| HK85 | 0.528 | 0.525 | 0.21 | 0.561 | 0.544 | <0.01 | 0.646 | 0.645 | 0.40 |

## 4. Conclusions

We propose a Bayesian methodology for assigning posterior probabilities to individual SNPs. We evaluate the model using posterior probabilities associated with one or more evolutionary models on three databases of functional SNPs. These probabilities were used to classify the SNPs represented in each database and observe the sensitivity versus specificity (ROC curves). We observe that none of the models hold strong predictive power (Figure 2). Not surprisingly, the best single model for hemoglobin and p53 is *M-O*, which was generated from human SNP data (Tables 1 and 2). Surprisingly, both nucleotide models outperform the amino acid *M-O* model for mouse; the largest area under all ROC curves for the individual models was 0.646 on *HKY85* for mouse (Table 3).

We also demonstrate that two properties unique to the Bayesian framework improve SNP identification. First, the Bayesian formulation allows inference to be made over mixtures of evolutionary models. Given the specialty of *M-O* to hemoglobin and p53 no improvement in specificity was observed, but for the mouse database two mixture models found an area under the ROC curve that was significantly better than the best individual model *HKY85* (Table 4). Finally, we focus on the prior, a special feature of the Bayesian model. We define our prior as the probability of observing a specific codon given the SNP model from the genetic code. We compare the results of applying our prior to the results using the neutral prior. We observe that the area under the ROC curve for the neutral prior is always smaller than or equal to the genetic code based prior. Furthermore, 75% of the time the area under the curve is significantly smaller for the neutral prior at a *p*-value of 0.05 (Table 5). Although none of the evolutionary models were highly accurate predictors, the Bayesian formulation gives a framework under which prior knowledge or more advanced evolutionary models can be incorporated to assign probabilities to individual polymorphic sites.

## Acknowledgments

This work was supported by the U.S. Department of Energy (DOE) through the Computational Sciences and Engineering Initiative Laboratory Directed Research and Development program at Pacific Northwest National Laboratory (PNNL). PNNL is a multiprogram national laboratory operated by Battelle Memorial Institute for the U.S. DOE under contract DE-AC06-76RLO 1830.

## References

1. M. H. Hasegawa et al., *J. Mol. Evol.* **22**, 160-74 (1985).
2. M. Kimura, *J. Mol. Evol.* **16**, 111-20 (1980).
3. M. Kimura, *PNAS.* **78**, 454-58 (1981).
4. K. Tamura and M. Nei, *Mol. Biol. Evol.* **10**, 512-26 (1993).
5. Z. Yang and R. Nielsen, *Mol. Biol. Evol.* **17**, 32-43 (2000).
6. M. O. Dayhoff, R. M. Schwartz and B. C. Orcutt, *Atlas of Pro. Seq. Str.* 345-52 (1978).
7. S. Henikoff and J. G. Henikoff, *PNAS.* **11**, 725-36 (1994).
8. T. R. Buckley and C. W. Cunningham, *Mol. Biol. Evol.* **19**, 394-405 (2002).
9. M. A. Suchard et al., *Mol. Biol. Evol.* **18**, 1001-13 (2001).
10. S. F. Altschul et al., *J. Mol. Biol.* **215**, 403-10 (1990).
11. T. F. Smith and M. S. Waterman, *J. Mol. Biol.* **147**, 195-97 (1981).
12. B. M. Webb et al., *Nucleic Acids Res.* **30**, 1268-77 (2002).
13. J. Majewski and J. Ott, *Gene.* **305**, 167-73 (2003).
14. N. Goldman and Z. Yang, *Mol. Biol. Evol.* **11**, 725-36 (1994).
15. T. H. J. Huisman et al., *A Syllabus of Human Hemoglobin Variants* (1996).
16. R. Hardison et al., *Genomics* **47**, 429-37 (1998).
17. R. Hardison, et al., *Hemoglobin* **22**, 113-27 (1998).
18. C. Beroud, et al., *Hum. Mutat.* **15**, 86-94 (2000).
19. T. Soussi, et al., *Hum. Mutat.* **15**, 105-13 (2000).
20. K. Linblad-Toh et al., *Nat. Genet.* **24**, 381-86 (2000).
21. R. D. Knight et al., *Genome Biol.* **2**, (2001).
22. D. J. Lipman et al., *BMC Evol. Biol.* **2**, (2002)
23. J. P. Eagen, (1975). Signal Detection Theory and ROC Analysis (New York: Academic Press.
24. E. R. DeLong et al., *Biometrics* **44**, 387-45, (1988).

# UNTANGLING THE EFFECTS OF CODON MUTATION AND AMINO ACID EXCHANGEABILITY

L. Y. YAMPOLSKY

*Department of Biological Sciences,*
*East Tennessee State University, Johnson City, TN 37614-1710*

A. STOLTZFUS

*Center for Advanced Research in Biotechnology,*
*9600 Gudelsky Drive, Rockville, MD 20850*

Determining the relative contributions of mutation and selection to evolutionary change is a matter of great practical and theoretical significance. In this paper, we examine relative contributions of codon mutation rates and amino acid exchangeability on the frequencies of each type of amino acid difference in alignments of distantly related proteins, alignments of closely related proteins, and among human SNPs, using a model that incorporates prior estimates of mutation and exchangeability parameters. For the operational exchangeability of amino acids in proteins, we use EX, a measure of protein-level effects from a recent statistical meta-analysis of nearly 10,000 experimental amino acid exchanges. EX is both free of mutational effects and more powerful than commonly used "biochemical distance" measures (*1*). For distant protein relationships, mutational effects (genetic code, transition/transversion bias) and operational exchangeability (EX) account for roughly equal portions of variance in off-diagonal values, the complete model accounting for $R^2 = 0.35$ of the variance. For human/chimpanzee alignments representing closely related proteins relationships, mutational effects (including CpG bias) account for 0.52 of the variance; adding EX to the model increases this to 0.67. For natural variation in human proteins, the variance explained by mutational effects alone, and by mutational effects and operational exchangeability are, respectively, 0.66 and 0.70 for SNPs in HGVBase, and 0.56 and 0.60 for disease-causing missense variants in HGMD. Thus, exchangeability has a stronger relative effect for distant protein evolution than for the cases of closely related proteins or of population variation. A more detailed model for the hominid data suggests that 1) there is a threshold in EX below which substitutions are highly unlikely to be accepted, corresponding to roughly 30 % relative protein activity; 2) selection against missense mutants is a slightly convex function of protein activity, not changing much as long as protein activity is low; and 3) the probability of disease-causing effects decreases nearly linearly with EX.

## 1.  Introduction

The evolution of molecular sequences, including the change of one amino acid for another, appears to reflect a two-step process of mutation and fixation, in which first, a mutation introduces a new allele into a population, and second, the allele rises to fixation by some combination of drift and selection. In the case of population variation, such as missense SNPs, the fixation process has not gone to completion. In either case, the observation of an inter-specific difference, or an intra-specific variant, reflects these two primary factors, mutation and natural selection. For missense changes, or missense variants, the mutational factor is the rate of mutation from one DNA codon to another, while the primary selective factor is the operational exchangeability of the amino acids in their protein context. Other potentially relevant fitness consequences of such a change are differences in the metabolic costs of amino acids, or in translation efficiencies of codons, but here such effects are assumed to be secondary to the effect of an exchange on the operation of a protein.

The goal of this work is to make a preliminary estimate of the relative contributions of mutational and selective factors to observed amino acid changes in protein evolution, as well as to observed amino acid variation in the human population (i.e., missense SNPs). These effects can be confounded easily. Therefore, *to be effective, any effort to untangle the relative contributions of mutation and selection must rely on some predictive model that incorporates prior knowledge of parameter values.* Mutation parameters are available from the comparative analysis of pseudogene divergence (2).

Although measures of amino acid similarity or distance have existed for a long time (3), the concept of a measure of the exchangeability of amino acids in a protein context is problematic. In the past, analyses that call for such a measure (4-6) have relied on so-called "biochemical distances" (7, 8). However, these are not pure measures of amino acid exchangeability, but attempts to fit observed propensities of evolution using a small number of biochemical parameters, as is clear from the original work of Grantham (7). With respect to the present problem of untangling mutational and selective effects, one cannot use an exchangeability measure that is based on observed propensities of evolution, because this would confound precisely the effects that must be treated separately. Therefore, the analysis here uses EX, a measure of relative effect on protein activity (derived from a statistical meta-analysis of nearly 10,000 experimental amino acid exchanges) that is free of mutational effects and which, in tests of power, out-performs "biochemical distance" measures (1).

We first analyze amino acid differences in variation and evolution using a formal *ad hoc* statistical model in which mutational biases and amino acid exchangeability act as factors. This approach allows us to assess the relative contributions of mutational effects and fitness effects in accounting for patterns of amino acid sequence differences in data sets representing close and distant inter-specific divergence, and in data sets representing intra-specific variation in humans (i.e., missense SNPs), including disease-associated variants in HGMD, as well as the general population sample in HGVBase. Then, we analyze a more specific set of models that includes the CpG mutational effect, applied to the data on human intra-specific variation, as well as to data from human/chimp divergence.

## 2.  Methods

**2.1.  *Sources and treatment of data on divergence and variation.*** The BLOSUM series of matrices used to represent distant protein divergence were taken from the supplementary material to (9), which provides five digits of precision in log-of-odds ratios ($s_{ij}$). Close divergences are represented by the set of human-chimpanzee alignments of over 7000 coding regions from Clark, et al. (10), using mouse as the out-group to polarize substitutions, and including only those codons in which all three sequences are known, polarization is non-ambiguous, and the gene alignment includes the start codon. The resulting set of data comprises 821,180 codons, representing roughly 5% of hominid genome. These dataset will be referred to as HomoPan dataset. Human polymorphism data came from two databases, the "Proven" subset of SNPs from HGVBase (11) and the disease-associated missense variants from HGMD (12). Both HGVBase and HGMD frequencies were divided by frequencies of source amino acid in human coding regions inferred from a codon usage database (13). Combined samples sizes are 1628 and 15373 for HGVBase and HGMD, respectively.

**2.2.** *Prior estimates of parameters of the prediction model.* Estimates of mutation parameters for hominids are taken from (2). To represent the frequencies of CpG sites in hominid genes, including CpG sites that straddle adjacent codons, we computed the frequency distribution of pentamers consisting of a codon with the 5' and 3' flanking nucleotides, from the complete set of coding sequences in the human RefSeq standard (14), omitting any entries with non-canonical start or stop codons, or with nucleotide ambiguities. The exchangeability of amino acids is parameterized in terms of the EX measure of Yampolsky and Stoltzfus (1). Since this measure is not well known, we describe it briefly. EX is based on a statistical meta-analysis of published data on the effects of 9671 amino acid changes in experimental studies carried out on 12 different proteins. Data on mutant protein activity from a subset of the studies provides the basis for a model of the frequency distribution of effects on protein activity; this model is then used to assign scores on a common scale for all of the exchanges. Taken literally, an $EX_{ij}$ value of 0.42 means that, on average, a variant protein with a residue j replacing the wild-type residue i has 42 % of the activity of the original protein. The mean value of EX is 0.28. EX out-performs Grantham's distances and Miyata's distances in an unbiased test of the ability to predict effects of experimental exchanges, and in a test that incorporates a measure of amino acid distance into the mutation-acceptance model of (15) implemented in the PAML package (5).

**2.3.** *Statistical analysis.* The mutational effects that are considered are the effect of the genetic code in imposing a minimum number of mutational steps ("minimum mutational distance") of 1, 2, or 3 (16), which we refer to here as "singlet", "doublet" and "triplet" exchanges; the effect of a transition/transversion bias; and for hominid data, the effect of a CpG context. Transition/transversion and CpG biases are considered only within singlet exchanges. "Transition" factor is assigned to a level of 1 for any singlet exchange that can occur by a transition, and a level of 0 otherwise. The "CpG" factor was a continuous factor represented by the combined frequency of all CpG containing codons and cGNNn and nNNCg pentamers (codons with flanking neighbors) among the codons of a given source amino acid that can mutate in a single step to each codon of the destination amino acid. For pairs of amino acids that cannot mutate into each other by a mutation at a CpG site such frequency is 0.

Since the overwhelming majority of hominid data are singlet differences, we performed two types of analyses. First, we included the genetic code effect, treating lack of observations of doublet and triplet exchanges as zero frequencies. Second, we considered only singlet exchanges, considering only transition/transversion and CpG effects among such exchanges.

Thus, the statistical analysis includes two groups of factors, mutational effects and exchangeability (EX), and four sets of response variables, representing frequencies of amino acid differences in distantly related proteins (BLOSUM30 through 100), in closely related proteins (human-chimpanzee alignment data), among human missense SNPs (HGVBase data) and among disease-associated human missense SNPs (HGMD data).

GLM models including these factors and their interactions were evaluated using JMP statistical package (17). For each test, the first model includes only the genetic code effect, then we add transition/transversion bias and (for human data) CpG bias, and finally, EX and its interactions. $R^2$ values associated with each model are the measure of relative contribution of each factor to the variance in the response variable.

For the case of hominid data (human-chimp divergence, and human missense variation), we also consider a more sophisticated model that takes into account the relevant target size for each individual mutational path, considering enhanced rates of transition and transversion at CpG sites. The target size is simply the relative frequency of codons that participate in a particular mutational path from one amino acid to another, e.g., for the Val-to-Leu change there is some subset of GTN Val codons that are preceded by a C, and thus are subject to an enhanced rate of mutation from GTN to CTN, specifically the transversion rate at CpG sites.

For each of three types of human data a simple model predicting substitution frequencies has been constructed. For human-chimpanzee substitutions we assume that differences are proportional to rates of change, which are in turn described by an origin-fixation process with a rate equal to the rate of mutational origin multiplied by the probability of fixation (18). Then the occurrence of some type of difference is proportional to

$$P_{fix}\, \mu\, (T_{11} + T_{12}t + T_{21}c_v + T_{22}c_t), \qquad (1)$$

where $P_{fix}$ is the unscaled mean probability of a given type of mutant being fixed; $\mu$ is the mutation rate for non-CpG transversions; $t$ is transition/transversion bias; $c_t$ and $c_v$ are the biases in transitions and transversions, respectively, at CpG sites; and $T_{11}$, $T_{12}$, $T_{21}$ and $T_{22}$, respectively, are the target sizes representing the sums over the frequencies of codons that, when subjected to each kind of mutation (non-CpG transversions, non-CpG transitions, CpG transversions and CpG transitions) produce the amino acid exchange of interest. The values of mutational biases used were: $t = 2.4$; $c_t = 23.0$, and $c_v = 7.0$ (2). A logistic function was used to describe relationship between $P_{fix}$ and EX:

$$P_{fix} = k/(1 + exp(-a(EX-b))). \qquad (2)$$

This constrains the function to be between 0 and 1, and to increase, but allows it to take nearly any shape. The meaning of parameters $a$ and $b$ is steepness of the curve and location of the inflection point, respectively. The meaning of $k$ is simply the number of generations since the common ancestor (twice that number for both lineages combined). This analysis has been done for substitution frequencies in human and chimpanzee lineages separately and for both lineages combined.

An essentially identical model has been used to fit HGMD data, only instead of the probability of being accepted, the HGMD model has the probability of having severe effects, assumed to be related to EX through a logistic function, this time a decreasing one:

$$P_s = k - k/(1 + exp(-a(EX-b))). \qquad (3)$$

The parameter $k$ here is a scaling factor reflecting how well the human populations have been screened for deleterious variants. A slightly different model was utilized for HGVBase data. First, we assume that majority of known human SNPs are recessive deleterious mutations segregating at mutation-selection balance, i.e., their frequencies are at $\sqrt{(\mu/s)}$, where s is selection against the mutant variant (18). Second, we assume that the probability of finding a variant is proportional to its frequency. This representation of ascertainment bias is reasonable when most SNPs do not have any clinically important effects and are discovered in population genetics or genomic screens. This seems to be the case for SNPs in HGVBase (11). Then, the probability of observing a change is proportional to

$$k\sqrt{\frac{\mu}{s}}(T_{11} + T_{12}\sqrt{t} + T_{21}\sqrt{c_v} + T_{22}\sqrt{c_t}) \qquad (4)$$

It is reasonable to assume that selection against mutant variants with relative activity equal to that of the wild type is 0, so the function describing the relationship between s and EX must contain $(0,1)$ point. Thus, logistic function cannot be used. Instead, a power function has been used:

$$s = b(1 - EX^a). \qquad (5)$$

The quantity in Equation 1 was fitted to the observed frequencies of substitutions in human and chimpanzee lineages and to HGMD frequencies, substituting Equations 2 and 3 for $P_{fix}$ and $P_s$, respectively. Quantity (4), substituting (5) for $s$, was fitted to observed frequencies in HGVBase. For HGMD and HGVBase data, target size was recalculated for the entire genome, assuming that sampled 821180 codons in human/chimpanzee alignment represent 5% of the entire length of coding regions and 25% of Clark at al alignments. All fitting was done in the non-linear fit platform in JMP (17) using the least-squares model. Profile likelihood confidence intervals were calculated

438

iteratively when possible; when there was no convergence, approximate standard errors were calculated by the derivative cross-product inverse matrix method (*17*).

## 3. Results

Contributions of mutational biases, amino acid exchangeability and their interactions to the explained portion ($R^2$) of variance among amino acid substitution rates are shown in Figure 1. Adding EX to the model more than doubles $R^2$ for distant protein substitutions represented by BLOSUM log-of-odds values (fig. 1A), raising it from around 0.15 to about 0.35. Note that the variance of off-diagonal values explained by this model increases monotonically from BLOSUM30 to BLOSUM100 (an issue addressed further below). EX alone explains much smaller portion of the variance of substitution frequencies: 0.036, 0.002 and 0.076 for HGVBase, HGMD and HomoPan frequencies, respectively and 0.12-0.20 for BLOSUM. It is worth mentioning that EX is a lot more successful in predicting the ratio between HGMD and HGVBase frequencies ((*1*)), explaining nearly 50% of the variance. This is because the contributions of mutational biases (presumably identical in both datasets) cancel out, leaving the exchangeability effect intact. Table 1 provides ANCOVA results for BLOSUM62 (the matrix most commonly used for protein alignments).

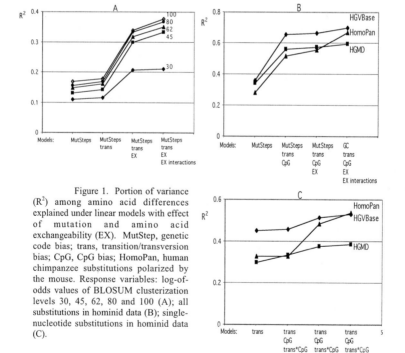

Figure 1. Portion of variance ($R^2$) among amino acid differences explained under linear models with effect of mutation and amino acid exchangeability (EX). MutStep, genetic code bias; trans, transition/transversion bias; CpG, CpG bias; HomoPan, human chimpanzee substitutions polarized by the mouse. Response variables: log-of-odds values of BLOSUM clusterization levels 30, 45, 62, 80 and 100 (A); all substitutions in hominid data (B); single-nucleotide substitutions in hominid data (C).

Table 1. ANCOVA of the effects of mutation biases and EX on log-of-odds score of BLOSUM62, with and without interactions. Abbreviations as on Fig. 1A. P values < 0.015 in bold.

| Response: BLOSUM62 | Without interactions | | | With interactions | | |
|---|---|---|---|---|---|---|
| Source | df | F | P | df | F | P |
| MutSteps | 2 | 19.1 | **1.0E-08** | 2 | 1.4 | 0.24 |
| trans | 1 | 4.2 | 0.04 | 1 | 2.1 | 0.15 |
| EX | 1 | 90.0 | **3.0E-19** | 1 | 57.6 | **3.0E-13** |
| EX*MutSteps | | | | 2 | 8.1 | **0.0004** |
| EX*trans | | | | 1 | 0.7 | 0.39 |
| Error | 369 | | | 366 | | |

Transition/transversion bias contributes strongly to the power of the model when applied to close (hominid) divergence data and human variation data (fig. 1B), unlike the case for distant divergence data. Most of the increases in $R^2$ values are due to the transition/transversion factor, while the CpG factor adds very little (data not reported). Adding EX and its interactions with mutational biases to the model improves the ability of the model to predict SNP frequencies surprisingly little, although the effect of EX is highly significant (Table 2). There is, however, a large increase in predicting power of the model attributable to EX when the response variable is frequency of human/chimpanzee substitutions, particularly when single-nucleotide substitutions alone are considered.

Although incorporating the interactions between mutational biases and EX to the models adds little to the explained variance, some of these interactions are significant and of interest (Tables 1 and 2). In particular, it is striking that well known genetic code component of BLOSUM scores statistically is made up entirely of the interaction with amino acid exchangeability. The nature of this interaction is the presence of a strong EX effect among single nucleotide substitutions, a weaker effect among double-nucleotide substitutions and lack of such effect among triple-nucleotide substitutions. Interaction between EX and transition/transversion bias is ubiquitous and highly significant in hominid data (Table 2). The nature of this interaction is that there is a covariance between observed frequency of substitutions and EX for substitutions that can occur through a transition, but is absent or much weaker in the group of substitutions that can only occur through a transversion.

Results of fitting non-linear models to hominid data are shown on Figure 2. The probability of fixation as a function of EX appears to have a critical range of mutant protein activity in which the drop of $P_{fix}$ occurs quickly (Fig. 2A). This inflection point is located around EX = 0.3. Fig. 2A shows the fit to combined substitution frequencies in both human and chimpanzee lineages; parameters of the model fitted to these lineages separately are not significantly different from each other or from the combined date fit. The curve for the chimpanzee is slightly less steep and inflection point is shifted slightly towards higher EX values, possibly indicating stronger stabilizing selection in chimpanzee lineage than in human one. The third parameter of the fitted model, K, has the meaning of the number of generations since the common ancestor. The best fit for this parameter is, assuming baseline mutation rate of $5*10^{-9}$, 250,000 for humans and 300,000 for chimpanzee. Assuming $5*10^6$ my since the common ancestor, this corresponds to generation times of 20 years for humans and 17 years for chimpanzees, a remarkably meaningful estimate.

Table 2. ANCOVAs of the effects of mutation biases and EX (continuous variable) on frequencies of single-nucleotide amino acid substitutions in hominid data, with and without interactions. Abbreviations as on Fig. 1C.

| Source | Without interactions | | | With interactions | | |
|---|---|---|---|---|---|---|
| Response: HGMD | df | F | P | df | F | P |
| trans | 1 | 46.1 | **2.7E-10** | 1 | 23.1 | **4.0E-06** |
| CpG | 1 | 2.2 | 0.14 | 1 | 2.0 | 0.16 |
| trans*CpG | 1 | 8.0 | **0.0053** | 1 | 0.023 | 0.88 |
| EX | 1 | 19.9 | **1.7E-05** | 1 | 3.6 | 0.058 |
| EX*trans | | | | 1 | 7.6 | **0.007** |
| EX*CpG | | | | 1 | 0.79 | 0.37 |
| EX*trans*CpG | | | | 1 | 0.44 | 0.50 |
| Error | 145 | | | 142 | | |
| Response: HGVBase | Df | F | P | df | F | P |
| trans | 1 | 74.6 | **9.5E-15** | 1 | 0.44 | 0.51 |
| CpG | 1 | 1.9 | 0.17 | 1 | 0.0034 | 0.95 |
| trans*CpG | 1 | 0.4 | 0.52 | 1 | 2.3 | 0.13 |
| EX | 1 | 18.5 | **3.1E-05** | 1 | 3.3 | 0.072 |
| EX*trans | | | | 1 | 4.7 | 0.032 |
| EX*CpG | | | | 1 | 0.2 | 0.64 |
| EX*trans*CpG | | | | 1 | 3.2 | 0.076 |
| Error | 145 | | | 142 | | |
| Response: HomoPan | df | F | P | df | F | P |
| trans | 1 | 66.1 | **1.7E-13** | 1 | 6.3 | **0.014** |
| CpG | 1 | 0.16 | 0.69 | 1 | 0.02 | 0.89 |
| trans*CpG | 1 | 0.004 | 0.95 | 1 | 1.6 | 0.20 |
| EX | 1 | 44.1 | **5.8E-10** | 1 | 8.1 | **.0052** |
| EX*trans | | | | 1 | 30.8 | **1.4E-07** |
| EX*CpG | | | | 1 | 0.03 | 0.87 |
| EX*trans*CpG | | | | 1 | 1.2 | 0.28 |
| Error | 145 | | | 142 | | |

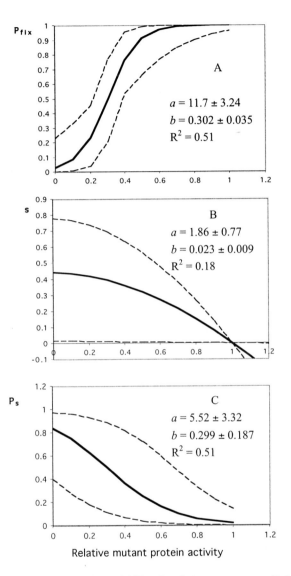

Figure 2. Probability of evolutionary acceptance ($P_{fix}$), mean selection coefficient (s) and probability of having clinically severe effects ($P_s$) as functions of relative mutant protein activity. Dashed lines, 95 % confidence intervals.

The shape of the relationship between selection coefficient and EX estimated from HGVBase frequencies under the assumption of mutation selection balance is slightly convex (fig. 2B). 95% confidence intervals are quite large, so this result needs to be taken with caution. The value of $b$ reported on fig. 2B is an arbitrary value calculated for the scaling factor $k = 0.1$. The relationship between $b$ and $k$ is such it is a possible to estimate $k$ from very broad limits on $b$. For $b = 1$ (all loss of function mutations lethal, clearly the upper limit of $b$) $k = 0.15$. For $b = 0.01$ (probably a gross underestimate of selection against loss of function mutations) $k = 0.015$. In other words, the set of "proven" SNPs in HGVBase constitutes somewhere between 1.5 and 15% of all non-synonymous SNPs in existence.

The results of fitting the model to HGMD frequencies are shown in Figure 2C. Again, as for HGVBase data, the confidence intervals are quite high, but one can conclude that $P_s$ decreases more or less linearly from over 0.5 for mutations with the lowest protein activities to about 0.2 for the ones with relative activity 0.5 of the wild type or more. Mean $P_s$ (averaged across all amino acid substitutions types) is 0.52 (standard dev. 0.14).

## 4. Discussion

Mutational biases and amino acid exchangeability have roughly equal effects on frequencies of amino acid substitutions among distantly related proteins, while on a smaller evolutionary scale, mutational biases add a relatively higher portion to the amount of variance explained by the model. The effect of amino acid exchangeability is seldom a pure effect, acting most often through interactions with mutational biases. Specifically, exchangeability matters for singlet exchanges more than for doublet and triplet exchanges, and for transitions more than for transversions. The first of these two interactions is easy to explain. While singlet differences may often reflect a single origin-fixation event, doublets and triplets probably only do so rarely, resulting instead from multiple changes such that the exchangeability of the original source and final destination amino acids is largely irrelevant.

This interaction is observed in all BLOSUM matrices except BLOSUM30 and is illustrated by Figure 3 showing regression coefficients of the effect of EX on off-diagonal BLOSUM values. Regression of the EX effect on singlet pairs of amino acids monotonically and significantly increases when the BLOSUM level is increased from 30 to 100; this increase is less significant for doublet exchanges and is reversed (though insignificantly) for triplet exchanges.

There is, therefore, also an interaction between the strength of the effect of EX on off-diagonal BLOSUM elements and the level of BLOSUM clustering, with the effect being strongest in the BLOSUM matrix that represents all degrees of relationship among aligned sequences (BLOSUM100), and the weakest in the matrix based only on sequences that are 30 % similar or less (BLOSUM30; (9)). Presumably, the reason for this is that a difference in closely related proteins (these being more strongly emphasized in BLOSUM100 than in BLOSUM30) is a difference in a nearly identical protein context, thus the pattern of occurrence of differences in closely related proteins is restricted to amino acids that are compatible with the same contexts, whereas for distantly related proteins, the context has been degraded so that it is no longer so well shared, but instead each protein is exploring a different region of the possibility-space for the family, so that there are essentially more degrees of freedom in the pattern of divergence.

It is much more difficult to explain why exchangeability has a greater effect on singlet differences that arise via transitions than on those that can only arise via transversions. If this result were found only in a single dataset, we would be inclined to treat it as an artifact of smaller sample size of transversions. But this effect is present in all three hominid datasets, including the very large human-chimpanzee alignments dataset. At present, we don't have a biological explanation for this observation.

More detailed models connecting observed frequencies of amino acid substitutions with mutational biases and amino acid exchangeability can yield information about the shape of relationship between mutant protein activity and how human medicine (HGMD data) or stabilizing selection (HGVBase and HomoPan data) perceive the strength of the deleterious effect such mutations. Of the three datasets analyzed, only the HomoPan dataset has enough data to yield both reasonably high $R^2$ of the fitted model and tangible narrow confidence intervals around the best fit. Fitting an evolutionary model to this dataset results in the conclusion that there is a relatively sharp transition from mutations with low probability of fixation to those with a high probability, and that the critical value of mutant protein activity is approximately 30% of the wild type activity. Note that the fitted logistic function was not forced to be close to 0 when EX is low or close to 1 when EX is high.

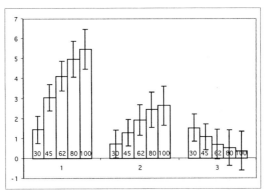

Figure 3. Regression coefficients (slopes) of the regression of BLOSUM $s_{ij}$ values on EX for singlet, doublet, and triplet exchanges. BLOSUM clusterization levels of 30, 45, 62, 80 and 100 % identity cut-off are used.

For the model fitted to HGVBase data, the main assumption is that all substitutions are recessive deleterious alleles at mutation-selection balance. Clearly, since some portion of known SNPs (unknown and probably over-represented) are entirely neutral (or epistatically deleterious) alleles segregating at much higher frequencies, and others almost certainly have some co-dominant deleterious effects and therefore segregate at much lower frequencies, the results of the fit based on mutation-selection balance is almost certainly biased and it is impossible to estimate whether these two violations of the assumptions tend to compensate for each other. Therefore, it is not surprising that, of the three models, the mutation-selection balance model fitted to HGVBase data has the lowest $R^2$ values and the broadest confidence intervals.

The model fitted to HGMD data is free from the assumptions about allele frequencies, since each substitution at a given site is reported at HGMD only once. Just as in the case of hominid model, the logistic function shown on fig. 2C was not forced to be close to 1 when EX is small or to be close to 0 when EX is large. Although it is hard to make conclusions about the exact shape of the function relating $P_s$ to EX, there is clearly a strong effect.

## 5.  Acknowledgements

This work was supported by the East Tennessee State University, and by the Center for Advanced Research in Biotechnology (a research institute jointly supported by the National Institute of Standards and Technology and the University of Maryland Biotechnology Institute). The identification of specific commercial software products in this paper is for the purpose of specifying a protocol, and does not imply a recommendation or endorsement by the National Institute of Standards and Technology.

## 6.  References

1.      L. Y. Yampolsky, A. Stoltzfus, (in review).
2.      I. Ebersberger, D. Metzler, C. Schwarz, S. Paabo, *Am J Hum Genet* **70**, 1490-7 (Jun, 2002).
3.      P. H. A. Sneath, *Journal of Theoretical Biology* **12**, 157 (1966).
4.      M. Krawczak, D. N. Cooper, *Hum Mutat* **8**, 23-31 (1996).
5.      Z. Yang, R. Nielsen, M. Hasegawa, *Mol Biol Evol* **15**, 1600-11. (1998).
6.      D. Vitkup, C. Sander, G. M. Church, *Genome Biol* **4**, R72 (2003).
7.      R. Grantham, *Science* **185**, 862-864 (1974).
8.      T. Miyata, S. Miyazawa, T. Yasunaga, *J Mol Evol* **12**, 219-36 (Mar 15, 1979).
9.      S. Henikoff, J. G. Henikoff, *Proc Natl Acad Sci U S A* **89**, 10915-9. (1992).
10.     A. G. Clark *et al.*, *Science* **302**, 1960-3 (Dec 12, 2003).
11.     D. Fredman *et al.*, *Nucleic Acids Res* **30**, 387-91 (Jan 1, 2002).
12.     P. D. Stenson *et al.*, *Hum Mutat* **21**, 577-81 (Jun, 2003).
13.     Y. Nakamura, T. Gojobori, T. Ikemura, *Nucleic Acids Res* **28**, 292 (Jan 1, 2000).
14.     K. D. Pruitt, D. R. Maglott, *Nucleic Acids Res* **29**, 137-40 (Jan 1, 2001).
15.     N. Goldman, Z. Yang, *Mol Biol Evol* **11**, 725-36 (1994).
16.     W. M. Fitch, *J. Mol. Biol.* **16**, 9-16 (1966).
17.     SAS_Institute, *JMP ® Statistics and Graphis Guide* (SAS Institute, Cary, NC, 2002).
18.     J. F. Crow and M. Kimura. An Introduction to Population Genetics Theory. Burgess Publ., Minneapolis, 1970.

# JOINT LEARNING FROM MULTIPLE TYPES OF GENOMIC DATA

ALEXANDER J. HARTEMINK
*Department of Computer Science and*
*Center for Bioinformatics and Computational Biology*
*Duke University, Box 90129*
*Durham, NC 27708-0129*
*amink@cs.duke.edu*

ERAN SEGAL
*Department of Computer Science*
*Stanford University*
*Stanford, CA 94305*
*eran@cs.stanford.edu*

Recent technological advances enable us to collect many different types of data at a genome-wide scale, including DNA sequences, gene and protein expression measurements, protein-protein interactions, information regarding protein structure and localization, and protein-DNA binding data. These data provide us with a means to begin elucidating the large-scale modular organization of the cell. Indeed, much recent work has been devoted to the analysis of these data for this purpose. However, most of this work has been devoted to the analysis of a single type of data at a time, using other types of data only for validation.

In contrast, results jointly learned from more than one type of data are likely to lead to new insights that might not be as readily available from analyzing one type of data in isolation. For instance, experimental genomic datasets often contain errors arising from imperfections in the applied technology. Thus, some of the findings of methods that analyze a single type of data may be erroneous. If we assume that technological errors across different genomic datasets are largely independent, then the probability of error in results that are supported by two different types of data is dramatically reduced.

The Joint Learning from Multiple Types of Genomic Data session first appeared at PSB 2004 to provide a forum for novel methods that use more than one type of data in their analysis, and do so jointly; this year's session represents a continuation of that conversation. Our goal in organizing these sessions at PSB is two-fold: first, we hope to encourage the computational biology community to develop methods that are capable of integrating the large number of different types of data that are becoming increasingly available; second, we hope to stimulate the discovery of new biological insights that would be difficult or impossible to identify in the analysis of only single types of data.

445

Based on the number of excellent papers submitted, the session has clearly tapped into a growing interest in such joint methods. We received 28 submissions of very high quality; we were able to accept 8 papers for publication, 6 of which were accepted for oral presentation. The papers represent a wide range of goals and approaches. Some examples include: combining DNA sequence data and gene expression data for improved detection of transcription factor binding sites, and at a higher level of detection, combining sequence data from multiple organisms for the task of identifying genomic regions that are bound by multiple transcription factors; predicting the function of uncharacterized genes by combining protein-protein interaction data, protein complex data and gene expression data; using gene expression data to better predict gene structure; and combining protein-DNA binding data, *cis*-regulatory binding data and gene expression for predicting regulatory networks. The methods employed for the joint learning were also very diverse, and included support vector machines, Bayesian networks and other Bayesian methods, non-negative matrix factorization, random forests, techniques from the data-mining community, and methods from combinatorial optimization. Taken together, these papers represent a fairly thorough cross-section of the most promising directions in this field. As more types of data become widely available, it is our belief that these kinds of unified approaches are likely to produce great insights into the complex biological systems that we are trying to better understand.

The session co-chairs are grateful to those who submitted papers to the session for their contributions in advancing the field of joint learning, and especially grateful to those who reviewed submissions for their contributions in selecting the most outstanding papers to present this year, which was a challenging task given the large number of excellent submissions.

# SPARSE FACTORIZATIONS OF GENE EXPRESSION DATA GUIDED BY BINDING DATA

LIVIU BADEA and DOINA TILIVEA

*AI Lab, National Institute for Research and Development in Informatics*
*8-10 Averescu Blvd., Bucharest, Romania, badea@ici.ro*

Existing clustering methods do not deal well with overlapping clusters, are unstable and do not take into account the robustness of biological systems, or more complex background knowledge such as regulator binding data. Here we describe a nonnegative sparse factorization algorithm dealing with the above problems: cluster overlaps are allowed by design, the nonnegativity constraints implicitly approximate the robustness of biological systems and regulator binding data is used to guide the factorization. Preliminary results show the feasibility of our approach.

## 1    Introduction and Motivation

The advent of microarray technology has allowed a revolutionary transition from the exploration of the expression of a handful of genes to that of entire genomes. However, despite its enormous potential, microarray data has proved difficult to analyze, partly due to the significant amount of noise, but also due to the large number of factors that influence gene expression (many of which are *not* at the mRNA/transcriptone level) and the complexity of their interactions.

One of the most successful microarray data analysis methods has proved to be *clustering* (of genes and/or samples), and a large variety of such methods have been proposed and applied to real-life biological data. This large body of work, impossible to extensively review here, has emphasized important limitations of existing clustering algorithms:

(1)    Most clustering methods produce *non-overlapping* clusters. However, since genes are typically involved in several biological processes, "non-overlapping" clustering methods, such as hierarchical clustering (HC) [2], self-organizing maps (SOM) [12], k-means clustering, etc., tend to be unstable, producing different gene clusters for only slightly different input samples (e.g. in the case of HC), or depending on the choice of initial conditions (as in the case of SOM [5], or k-means).

Algorithms allowing for overlapping clusters, such as fuzzy k-means [4] achieved significant improvements w.r.t. "non-overlapping" clustering, but they still have the problems discussed below.

(2)    Most algorithms perform clustering along a single dimension comparing e.g. genes w.r.t. *all* the available samples, whereas in reality genes have coordinated expression levels only for certain subsets of conditions. Algorithms dealing with this problem, such as biclustering [13], coupled-two way clustering (CTWC) [3], ISA (iterative signature) [1] have other problems mostly related to the control of overlap between biclusters.

447

(3)  Although genes are subject to both positive and negative influences from other genes, the *robustness* of biological systems requires that an observed change in the expression level of a given gene is the result of *either* a positive *or* a negative influence rather than a complex combination of positive and negative influences that partly cancel out each other (as in the case of Principal Component Analysis).

Nonnegative Matrix Factorization (NMF) [9] deals with this problem by searching for *nonnegative* decompositions of (nonnegative) data. The observed localized nature of the decompositions seems to be a biproduct of the nonnegativity constraints [9].

Recently, Brunet at al [5] applied NMF for clustering samples in a *non-overlapping* mode for three cancer datasets. While oligonucleotide arrays used in that work produce *positive* data, which lend themselves naturally to nonnegative decompositions, clustering genes in an analogous manner would ignore potential negative influences (i.e. genes downregulating other genes).

On the other hand, Kim and Tidor [6] used NMF to cluster genes in the context of a large dataset of yeast perturbation experiments (spotted arrays) [7]. Although NMF has the tendency of producing sparse representations, the factorizations obtained were subjected to thresholding and subsequent reoptimization to obtain sufficiently sparse clusters.

Unfortunately however, microarray data is noisy and it might be useful to be able to take into account any background knowledge that may be available. For example, Lee et al [8] have published binding (location analysis) data for a large number (106) of transcription factors in the yeast S. Cerevisiae.

Elsewhere [in preparation] we have observed that transcription factor (TF) expression levels are not always good predictors for the expression levels of their targets. (The presence of the transcription factor is of course required for the target to be expressed, but very frequently the TF is activated by a different signaling molecule, e.g. a kinase.) Therefore using the binding data directly as background knowledge may not be very helpful in practice.

In fact, it seems that although TFs are not well correlated with their targets, the targets themselves seem to be much better correlated among each other.

In the following, we show how TF binding data can be used as background knowledge in a novel nonnegative factorization algorithm $NNSC_B$ (Nonnegative Sparse Coding with Background Knowledge) designed to address the above-mentioned problems of existing clustering algorithms.

Our algorithm is an improvement of the nonnegative sparse coding (NNSC) algorithm of Hoyer [11]. It produces overlapping clusters that are much more stable than those generated with other algorithms, while also being able to take background knowledge into account.

## 2    The Data Sources

Since the most extensive background knowledge is available for the yeast S. Cerevisiae, in this paper we use the Rosetta Compendium [7], the largest publicly available gene expression dataset for yeast perturbations, as well as the binding (location analysis) data of Lee et al. [8].

The Rosetta Compendium contains expression profiles over virtually all yeast genes (6315 ORFs) corresponding to 300 diverse mutations and chemical treatments (276 deletion mutants, 11 tetracycline-regulatable alleles of essential genes and treatments with 13 well-characterized compounds) of S. Cerevisiae grown under a single (normal) condition. The data contains log-expression ratios $\log_{10} r = \log_{10} \frac{g(tf\Delta)}{g(WT)}$, where $g(tf\Delta)$ and $g(WT)$ are the mRNA concentrations of gene g in the $tf\Delta$ mutant and the wild type respectively.

The location analysis data of Lee et al. contains information about binding of 106 transcriptional regulators to upstream regions of target genes.

Since log-ratios can be negative, we cannot directly apply a nonnegative factorization algorithm to the log-ratio dataset. On the other hand, although the ratios $r$ (or maybe $r$-1) *are nonnegative,* applying nonnegative factorization on them would only uncover the positive influences, while in practice the low level of certain genes is due to them being *downregulated* by other genes.

To address problem (3) mentioned in the Introduction, we separate, as in [6], the up-regulated from the down-regulated part of each gene, i.e. obtain two entries $g^+$ and $g^-$ for each gene $g$ from the original gene expression matrix:

$$r(g^+) = \begin{cases} r(g) - 1 & r(g) \geq 1 \\ 0 & otherwise \end{cases} \qquad r(g^-) = \begin{cases} \dfrac{1}{r(g)} - 1 & r(g) < 1 \\ 0 & otherwise \end{cases}$$

Note that in this representation, a significantly downregulated gene will require a non-negligible contribution in the factorization. (We use the ratios rather than log-ratios as in [6], since linear combinations of log-ratios amount to products of powers of ratios rather than additive contributions.)

## 3    Nonnegative Sparse Coding

Hoyer's NNSC algorithm [11] factorizes a $n_s \times n_g$ matrix $X \approx A \cdot S$ as a product of an $n_s \times n_c$ matrix $A$ and a $n_c \times n_g$ matrix $S$ by optimizing (minimizing) the following objective function: [1]

---

[1] To achieve information compression, the number of internal dimensions $n_c$ must verify the constraint $n_c(n_s + n_g) < n_s n_g$.

$$C(A,S) = \frac{1}{2}\|X - AS\|_F^2 + \lambda \sum_{c,g} S_{cg} \qquad (1)$$

with respect to the *nonnegativity constraints*    $A_{sc} \geq 0,\ S_{cg} \geq 0$    (2)

The objective function combines a *fitness* term involving the Frobenius norm of the error and a *size* term penalizing the non-zero entries of S. (The Frobenius norm is given by $\|E\|_F^2 = \sum_{s,g} E_{sg}^2 = Tr(E^T E).$)

The *Nonnegative Matrix Factorization* (NMF) of Lee and Seung [10] is recovered by setting the size parameter $\lambda$ to zero, while a non-zero $\lambda$ would lead to sparser factorizations.

The objective function (1) above has an important problem, due to the invariance of the fitness term under diagonal scalings. More precisely we have the following result.

*Proposition.* The fitness term $\frac{1}{2}\|X - AS\|_F^2$ (i.e. the NMF objective function) is invariant under the following transformations:

$$A \leftarrow A \cdot D^{-1},\ S \leftarrow D \cdot S, \qquad (3)$$

where $D = \mathrm{diag}(d_1,\ldots,d_{nc})$ is a positive diagonal matrix ($d_c > 0$).

Note that such positive diagonal matrices are the most general positive matrices whose inverses are also positive (thereby preserving the nonnegativity of A and S under the above transformation).

The scaling invariance of the fitness term in (1) makes the size term ineffective, since the latter can be forced as small as needed simply by using a diagonal scaling D with small enough entries. Additional constraints are therefore needed to render the size term operational. Since a diagonal matrix D operates on the rows of S and on the columns of A, we could impose unit norms either for the rows of S, or for the columns of A.

Unfortunately, the objective function (1) used in [11] has an important flaw: it produces decompositions that depend on the scale of the original matrix X (i.e. the decompositions of X and $\alpha X$ are essentially different), regardless of the normalization scheme employed. For example, if we constrain the rows of S to unit norm, then we cannot have decompositions of the form $X \approx A \cdot S$ and $\alpha X \approx \alpha A \cdot S$, since at least one of these is in general non-optimal due to the dimensional inhomogeneity of the objective function w.r.t. A and X

$$C_{\alpha X}(\alpha A, S) = \alpha^2 \frac{1}{2}\|X - AS\|_F^2 + \lambda \sum_{c,g} S_{cg}.$$

On the other hand, if we constrain the columns of A to unit norm, the decompositions $X \approx A \cdot S$ and $\alpha X \approx A \cdot \alpha S$ cannot be both optimal, again due to the dimensional inhomogeneity of C, now w.r.t. S and X:

$$C_{\alpha X}(A, \alpha S) = \alpha^2 \frac{1}{2}\|X - AS\|_F^2 + \alpha\lambda \sum_{c,g} S_{cg}$$

Therefore, as long as the size term depends only on $S$, we are forced to constrain the columns of $A$ to unit norm, while employing an objective function that is *dimensionally homogeneous* in $S$ and $X$. One such dimensionally homogeneous objective function is:

$$C(A, S) = \frac{1}{2}\|X - AS\|_F^2 + \lambda\|S\|_F^2 \tag{1'}$$

which will be minimized w.r.t. the nonnegativity constraints (2) and the constraints on the norm of the columns of $A$:

$$\|A_c\| = 1 \quad (\text{i.e. } \sum_s A_{sc}^2 = 1) \tag{4}$$

It can be easily verified that this produces *scale independent decompositions*, i.e. if $X \approx A \cdot S$ is an optimal decomposition of $X$, then $\alpha X \approx A \cdot \alpha S$ is an optimal decomposition of $\alpha X$.

The constrained optimization problem could be solved with a gradient-based method. However, in the case of NMF, faster so-called "multiplicative update rules" exist [10,11], which we have generalized to the NNSC problem as follows. (These methods only produce local minima, but the solutions tend to be quite 'stable' – see also Section 5 below.)

**Modified NNSC algorithm**

Start with random initial matrices $A$ and $S$
**loop**

$$S_{cg} \leftarrow S_{cg} \frac{(A^T \cdot X)_{cg}}{(A^T \cdot A \cdot S + \lambda S)_{cg}}$$

$$A \leftarrow A + \mu(X - A \cdot S) \cdot S^T$$

normalize the columns of $A$ to unit norm: $A \leftarrow A \cdot D^{-1}$, $D = diag\left(\sqrt{\sum_s A_{sc}^2}\right)$

**until** convergence.

In the following, we assume that the gene expression data is given in an $n_s \times n_g$ matrix $X$, where $n_s$ and $n_g$ are the numbers of samples and genes respectively, so that $X_{sg}$ represents the expression level of gene $g$ in sample $s$.

A sparse factorization $X \approx A \cdot S$ will be interpreted as a generalization of *clustering the genes* (i.e. the columns $X_g$ of $X$) into overlapping clusters $c$ corresponding to the rows $S_{cg}$ of $S$. More precisely, a non-zero value of $S_{cg}$ (or at least an $S_{cg}$ larger than a given threshold) will be interpreted as the gene $g$ belonging to cluster $c$. Note that clusters can be overlapping, since the columns of $S$ may have several significant entries.

Although overlaps are allowed, NNSC will not produce highly overlapping clusters, due to the sparseness constraints. This is unlike many other clustering algorithms that allow clusters to overlap, which have to resort to several parameters to keep excessive cluster overlap under control.)

Also note that we factorize $X$ rather than $X^T$ since the sparsity constraint should affect the clusters of genes (i.e. $S$) rather than the clusters of samples $A$. (This is unlike NMF, for which the factorizations of $X$ and of $X^T$ are completely symmetrical.)

## 4    Nonnegative Sparse Coding using Background Knowledge

Transcription factor binding data can be represented by a $n_f \times n_g$ Boolean matrix $B$, such that $B_{fg}=1$ iff the transcription factor $f$ binds to the upstream region of gene $g$.

As already mentioned in the Introduction, although transcription factor expression levels are not always good predictors of the expression levels of their targets, the targets are frequently much better correlated among themselves. This suggests using the co-occurrence matrix $K=B^T B$ rather than $B$ as background knowledge. $K$ is an $n_g \times n_g$ square matrix, in which $K_{g'g''} \neq 0$ iff genes $g'$ and $g''$ are both targets of some common transcription factor $f$.

Our idea of exploiting background knowledge during clustering is quite simple. Normally, non-zero entries in the "gene cluster" matrix $S$ are penalized for size. However, if certain entries conform to the background knowledge, they will be exempted from size penalization. We thus need to modify the size term in $C(A,S)$ to take into account $B$.

Implementing this simple idea involves however certain subtleties. Assume that some gene cluster $c$ (i.e. set of genes $g$ for which $S_{cg} \neq 0$, or at least $S_{cg} > T$ for some given threshold $T$) contains many genes that are targets of several TFs (e.g. Figure 1b below). Although this cluster is preferable from the point of view of the background knowledge to the one from Figure 1a, it is worse than the one from Figure 1c, in which all the genes are the targets of a single TF.

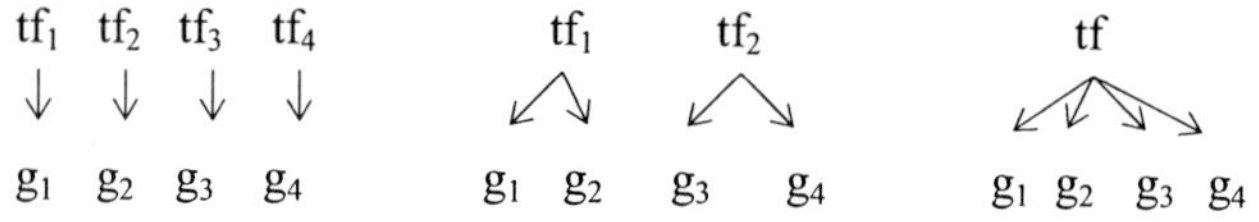

tf$_1$  tf$_2$  tf$_3$  tf$_4$          tf$_1$          tf$_2$                    tf

↓   ↓   ↓   ↓          ↗↖          ↗↖               ↙↗↘↘

g$_1$  g$_2$  g$_3$  g$_4$          g$_1$  g$_2$  g$_3$  g$_4$          g$_1$  g$_2$  g$_3$  g$_4$

(a)                            (b)                            (c)

**Figure 1.** Conformance of clusters to the binding data: (c) is preferable to (b), which is preferable to (a).

Thus, the size term cannot be simply a sum of overlaps of the clusters (i.e. rows of $S$) with groups of TF-targets (rows of $B$), since

$$\sum_c \sum_f \sum_g S_{cg} B_{fg} = \sum_g \left( \sum_c S_{cg} \right) \left( \sum_f B_{fg} \right)$$

does not depend on the way the genes are distributed in groups of TF-targets for different TFs.

Clusters like the one in Figure 1c can be highly evaluated if cross-terms between genes controlled by the same TF are added. We thus encourage genes controlled by the same TF in the binding data, while penalizing the size of $S$ using an objective function of the form:

$$C(A, S) = \frac{1}{2}\|X - AS\|_F^2 + \frac{\lambda}{2}\left(\|S\|_F^2 - Tr(SKS^T)\right) \qquad (5)$$

with $K = \gamma \cdot \left(\sigma(B^T B) - diag\left(\sigma(\sum_f B_{fg})\right)\right)$, where $\gamma$ is a factor such that

$\frac{1}{n_g}\sum_{g',g''} K_{g'g''} = 1$ and $\sigma(\cdot)$ is the Heaviside step function (applied element-wise).

Of course, optimization of (5) is attempted in the context of the constraints (2) and (4). The algorithm below solves the above optimization problem using combined multiplicative and additive update rules. (The final normalization of the rows of $S$ renders the resulting clusters comparable.)

**NNSC$_B$ algorithm**

start with random initial matrices $A$ and $S$

**loop**

$$S_{cg} \leftarrow S_{cg} \frac{(A^T \cdot X + \lambda S \cdot K)_{cg}}{(A^T \cdot A \cdot S + \lambda S)_{cg}}$$

$$A \leftarrow A + \mu(X - A \cdot S) \cdot S^T$$

normalize the columns of $A$ to unit norm: $A \leftarrow A \cdot D^{-1}$, $D = diag\left(\sqrt{\sum_s A_{sc}^2}\right)$

**until** convergence

$S \leftarrow D^{-1} \cdot S$ and $A \leftarrow A \cdot D$, where $D = diag\left(\sqrt{\sum_g S_{cg}^2}\right)$

To test our approach, we have applied the NNSC$_B$ algorithm on a synthetic dataset with several highly overlapping clusters. NNSC$_B$ has been able to consistently recover the clusters, or close approximations thereof even in the presence of noise. (See http://www.ai.ici.ro/psb05/synthetic.pdf for more details.)

## 5    Clustering the Rosetta Dataset w.r.t. the Binding Data of Lee *et al.*

Although the main goal of this paper is the presentation of a new clustering algorithm able to deal with background knowledge rather than obtaining new biological insights, we also briefly discuss our initial attempts at applying our algorithm to yeast microarray data.

The binding data of Lee et al. contains the targets of 106 transcription factors, roughly about half the total number of yeast transcription factors. In order not to introduce a bias towards the targets of these TFs due to the background knowledge, we have selected from the Rosetta dataset only these targets. (We also eliminated the genes that had unreliable measurements in the Rosetta dataset – dealing with missing values in our context is a matter of future work.) This left us with a set of 99 TFs and 2099 genes. The matrix $X$ to be factorized was constructed by duplicating genes as described in Section 2 ($X$ has therefore 4198 columns). Duplicating genes $g$ into their positive ($g^+$) and negative parts ($g^-$) may raise potential problems with possible conflicts between nonzero $S_{cg+}$ and $S_{cg-}$ entries, as a gene cannot be *both* up- and down regulated in a given cluster. The fact that our decompositions never have both $S_{cg+}$ and $S_{cg-}$ nonzero (significant) shows that the approach is biologically sensible.

An important parameter of the $\text{NNSC}_\text{B}$ factorization is its *internal dimensionality* (the number of clusters $n_c$). A useful estimate of the internal dimensionality of a dataset can be obtained from its singular value decomposition (SVD).

A more refined analysis [6] determines the number of dimensions around which the root mean square error (RMSE) *change* of the real data and that of a randomized dataset become equal. Kim and Tidor's analysis estimated the internal dimensionality of the Rosetta dataset to be around 50. We performed this analysis for our restricted dataset and obtained a similar dimensionality around 50 (see Figure 2).

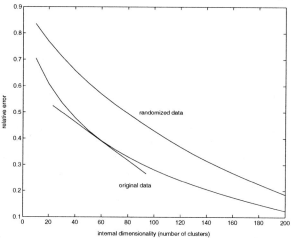

**Figure 2.** Determining the internal dimensionality of the Rosetta dataset using the method of Kim and Tidor

We also considered another approach: for a given set of clusters we determined the fraction of *intra-cluster* correlated pairs of genes (i.e. the number of correlated pairs of genes divided by the total number of correlated pairs of genes). This fraction was estimated for various correlation thresholds $r_0$. (More precisely, a gene pair $(g_1,g_2)$ is called correlated w.r.t. threshold $r_0$ iff $|r(g_1,g_2)| \geq r_0$.) Figure 3 depicts the $r_0$-dependence of the fraction of intra-cluster correlated pairs of genes for various dimensionalities. Notice that $n_c$=50 is a reasonable choice as the above mentioned fraction approaches 90% for a large range of $r_0$. (This means in other words that most of the correlated gene pairs are *within* clusters.)

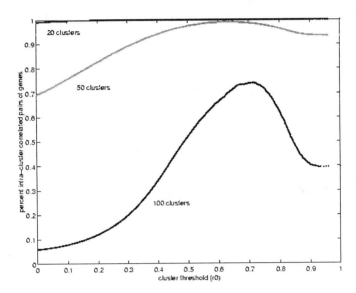

**Figure 3.** The correlation-threshold dependence of the fraction of intra-cluster correlated gene pairs.

Next we studied the *stability* of the algorithm w.r.t. the initial starting point. (The algorithm was run disregarding the background knowledge, i.e. taking $\lambda$=0, in order to avoid any possible influences of the background knowledge on the stability of some solution fragment.) The Table below lists the relative errors ($\|X\text{-}AS\|_F / \|X\|_F$) obtained in 7 runs for 7 different initializations.

| run no. | 1 | 2 | 3 | 4 | 5 | 6 | 7 |
|---|---|---|---|---|---|---|---|
| relative error | 0.1544 | 0.1677 | 0.1636 | 0.1718 | 0.1592 | 0.1606 | 0.1782 |

To asses the stability of the decomposition in the case of $k$ runs, we determined the best matches among the $k$ sets of clusters. The following Table shows the numbers of matching clusters for a progressively larger number of runs. Note the (relative) stabilization of the number of matching clusters w.r.t. the numbers of runs:

| Numbers of runs | 1 | 2 | 3 | 4 | 5 | 6 | 7 |
|---|---|---|---|---|---|---|---|
| Matching clusters (avg.) | 50 | 48.4 | 42.2 | 40.1 | 37.2 | 36 | 35 |

Next, we studied the influence of the background knowledge on the solution. In order to separate the variability of the results due to the initialization from that due to the background knowledge, we performed all the following tests using the same initialization. We ran NNSC$_B$ with several values of the parameter $\lambda$, ranging from 0 (background knowledge is not taken into account) to 0.75 (background knowledge has comparable weight to the data fit term) and tested the overlap of the resulting clusters with the background knowledge. Briefly, we observed a clear increase of this overlap with larger $\lambda$. More precisely (more details can be found in the Table below):

- the average fraction of cluster genes controlled by TFs increased from about 47% for $\lambda=0$ to 66% for $\lambda=0.75$. (Only the TFs controlling at least two genes from the cluster were counted.)
- the average number of TFs per cluster controlling at least two genes increased from around 8 for $\lambda=0$ to 14 for $\lambda=0.75$.

| $\lambda$ | 0 | 0.1 | 0.2 | 0.5 | 0.75 |
|---|---|---|---|---|---|
| avg. number of genes in a cluster | 27.8 | 32 | 36.4 | 43.6 | 78.7 |
| avg. fraction of cluster genes controlled by TFs | 0.47 | 0.46 | 0.51 | 0.60 | 0.66 |
| avg. number of TFs per cluster | 7.9 | 8.6 | 10.4 | 13.4 | 17.9 |
| avg. cluster overlap | 1 | 1.3 | 2.1 | 3.6 | 10 |
| avg. cluster overlap (when only overlapping pairs are averaged) | 2.7 | 3.3 | 4.3 | 6.2 | 14.7 |
| Relative error | 0.16 | 0.18 | 0.23 | 0.33 | 0.41 |

Of course, conformance to the data and/or to the background knowledge does not imply the biological relevance of the results. To estimate the latter, we searched for significant Gene Ontology (GO) annotations [14] of the genes in the clusters. More precisely, we employed the hypergeometric distribution to compute p-values representing likelihoods that specific GO annotations and a given cluster share a given number of genes by chance and retained only the annotations with a p-value less than $10^{-3}$. (This p-value threshold was chosen so that not more than 1 or 2 annotations are false positives, given that the genes in our dataset share 2422 GO annotations – some of which may of course be related by sub- or super-class relationships. We did not use a lower threshold in order to avoid many missing annotations.)

To demonstrate the biological relevance of the factorizations using background knowledge, we performed alternative factorizations for randomized background knowledge (more precisely, by randomly permuting the lines of $B$ independently of each other). This lead to a drop in the average number of significant GO annotations per cluster from 8.16 to 4.88 (for $\lambda=0.75$).

We also looked at a few clusters in more detail. For example, cluster 47 had 18 genes, involved in the STE12 control of pheromone response, among which, for instance, AGA1, FIG1, FUS1 are involved in cell fusion, while GPA1, FUS3 and PRR2 are involved in the pheromone signal transduction pathway, KAR4 is a regulatory protein required for pheromone induction of karyogamy genes and SST2 is involved in desensitization to alpha-factor pheromone. The mating a-factor genes MFA1 and MFA2 are also in the cluster. The entire cluster is presented in the

Annex, together with the associated significant GO annotations and the cluster coefficients from the $S$ matrix. (The threshold used for extracting clusters from the $S$ matrix was $1/\sqrt{n_g} = 0.0154$.)

## 5    Conclusions

Despite their wide-spread use in microarray data analysis, existing clustering algorithms have serious problems, the most important one being related to the fact that biological processes are overlapping rather than isolated. The impact of microarray technology is also limited by the noisy nature of measurements, which can only be compensated by additional background knowledge. Here we have shown how these important problems faced by microarray data analysis can be dealt with in the context of a sparse factorization algorithm capable of dealing with regulator binding data as background knowledge. A key ingredient of this algorithm is the nonnegativity constraint. Such an approach, for example using NMF, has been mostly advocated in connection with oligonucleotide array data, which are (at least theoretically) nonnegative. However, this viewpoint is only partially correct, since downregulated genes would not be explained in such a framework. Actually, we argue that nonnegative factorizations are appropriate due to the robustness of biological systems, in which an observed change in a gene's expression level is the result of *either* a positive *or* a negative influence rather than a complex combination of the two. Although our preliminary results are encouraging, a more detailed biological analysis should be the focus of subsequent work. (The clusters obtained by our algorithm for various parameter settings can be found online at http://www.ai.ici.ro/psb05/.)

## References

1.  Bergmann S., J. Ihmels, N. Barkai. Iterative signature algorithm for the analysis of large-scale gene expression data. Physical Review E 67, 031902 (2003).
2.  Eisen M.B., P.T. Spellman, P.O. Brown, D. Botstein. Cluster analysis and display of genome-wide expression patterns, PNAS Vol.95, 14863-8, Dec.1998.
3.  Getz G., Levine E., Domany E. Coupled two-way clustering analysis of gene microarray data, PNAS Vol. 97, 12079-84, 2000.
4.  Gasch A.P, Eisen M.B. Exploring the conditional coregulation of yeast gene expression through fuzzy k-means clustering. Genome Biol. 2002, Oct 10;3(11).
5.  Brunet J.P., Tamayo P., Golub T.R., Mesirov J.P. Metagenes and molecular pattern discovery using matrix factorization. PNAS 101(12):4164-9, 2004, Mar 23.
6.  Kim P.M., Tidor B. Subsystem identification through dimensionality reduction of large-scale gene expression data. Genome Res. 2003 Jul;13(7):1706-18.
7.  Hughes T.R. et al. Functional discovery via a compendium of expression profiles. Cell. 2000 Jul 7;102(1):109-26.

8. Lee T.I. et al. Transcriptional regulatory networks in Saccharomyces cerevisiae. Science. 2002 Oct 25; 298(5594):799-804.
9. Lee D.D., H.S. Seung. Learning the parts of objects by non-negative matrix factorization. Nature, vol. 401, no. 6755, pp. 788-791, 1999.
10. Lee D.D., H.S. Seung. Algorithms for non-negative matrix factorization. In Advances in Neural Information Processing 13 (Proc. NIPS*2000), MIT Press, 2001.
11. P. O. Hoyer. Non-negative sparse coding. Neural Networks for Signal Processing XII, 557-565 Martigny, Switzerland, 2002.
12. Tamayo P. et al. Interpreting patterns of gene expression with self-organizing maps: methods and application to hematopoietic differentiation, PNAS 96(6):2907-12 (1999).
13. Cheng Y, Church GM. Biclustering of expression data. Proc. ISMB 2000; 8:93-103.
14. Gene Ontology Consortium. Gene Ontology: tool for the unification of biology. Nature Genet. 25:25-29, 2000.

## Annex. The 18 genes of cluster 47 and the TFs controlling them

**Significant GO annotations (with p-values)**: conjugation(0),conjugation with cellular fusion(0),sexual reproduction(0),reproduction(2.30e-11),response to pheromone(4.96e-11),response to chemical substance(2.24e-9),response to abiotic stimulus(2.89e-9),development(1.07e-8),response to pheromone during conjugation with cellular fusion 1.13e-8),response to external stimulus(1.34e-8),cell communication(2.51e-8),signal transduction during conjugation with cellular fusion(1.57e-6),response to stimulus(2.11e-6),signal transduction(3.85e-6),G-protein coupled receptor protein signaling pathway(4.27e-6),cell surface receptor linked signal transduction(9.42e-6),shmoo tip(9.42e-6),signal transducer activity(4.42e-5),pheromone activity(1.97e-4),receptor binding(1.97e-4),cellular process(3.05e-4),receptor signaling protein serine/threonine kinase activity(3.92e-4),site of polarized growth(9.32e-4),site of polarized growth (sensu Fungi)(9.32e-4),site of polarized growth (sensu Saccharomyces)(9.32e-4),receptor signaling protein activity(9.70e-4)

STE12 targets: STE12, SST2, TEC1, FUS1, KAR4, GPA1, MFA2
MCM1 targets: MFA1, STE6, MFA2, AGA1
PHD1 targets: PRR2, GPA1

| | | | |
|---|---|---|---|
| +: 0.000000 | -: 0.970648 | PRR2 | strong similarity to putative protein kinase NPR1 |
| +: 0.000000 | -: 0.146131 | YDL133W | |
| +: 0.000000 | -: 0.132290 | FUS1 | cell fusion protein |
| +: 0.064781 | -: 0.000229 | PRY3 | |
| +: 0.000000 | -: 0.056725 | FIG1 | required for efficient mating |
| +: 0.000000 | -: 0.040888 | AGA1 | a-agglutinin anchor subunit |
| +: 0.029486 | -: 0.000000 | HSP12 | heat shock protein |
| +: 0.000000 | -: 0.027441 | MFA2 | mating pheromone a-factor 2 |
| +: 0.000000 | -: 0.023362 | GPA1 | GTP-binding protein alpha subunit of the pheromone pathway |
| +: 0.000000 | -: 0.022420 | questionable ORF | |
| +: 0.000062 | -: 0.022318 | STE6 | ATP-binding cassette transporter protein |
| +: 0.000000 | -: 0.020931 | STE12 | transcriptional activator |
| +: 0.000000 | -: 0.020907 | FUS3 | mitogen-activated protein kinase (MAP kinase) |
| +: 0.000000 | -: 0.020485 | MFA1 | mating pheromone a-factor 1 |
| +: 0.000000 | -: 0.020241 | SST2 | involved in desensitization to alpha-factor pheromone |
| +: 0.000000 | -: 0.020059 | YLR343W | |
| +: 0.000000 | -: 0.019342 | TEC1 | Ty transcription activator |
| +: 0.000103 | -: 0.018020 | KAR4 | regulatory protein required for pheromone induction of karyogamy genes |

Parameters: λ=0.1, μ=5·10⁻⁶, S-threshold=0.0154, 300 iterations.

# INFORMATIVE STRUCTURE PRIORS: JOINT LEARNING OF DYNAMIC REGULATORY NETWORKS FROM MULTIPLE TYPES OF DATA

ALLISTER BERNARD, ALEXANDER J. HARTEMINK

*Duke University, Dept. of Computer Science, Box 90129, Durham, NC 27708*

We present a method for jointly learning dynamic models of transcriptional regulatory networks from gene expression data and transcription factor binding location data. Models are automatically learned using dynamic Bayesian network inference algorithms; joint learning is accomplished by incorporating evidence from gene expression data through the likelihood, and from transcription factor binding location data through the prior. We propose a new informative structure prior with two advantages. First, the prior incorporates evidence from location data probabilistically, allowing it to be weighed against evidence from expression data. Second, the prior takes on a factorable form that is computationally efficient when learning dynamic regulatory networks. Results obtained from both simulated and experimental data from the yeast cell cycle demonstrate that this joint learning algorithm can recover dynamic regulatory networks from multiple types of data that are more accurate than those recovered from each type of data in isolation.

## 1  Introduction and motivation

Discovering networks of transcriptional regulation is an important problem in molecular biology, and progress toward this goal has been accelerated by the advent of new technologies for collecting high-throughput data. Data collected using different technologies offer different perspectives on a problem, but jointly analyzing such data in a single framework enables a consensus perspective to emerge. In addition, joint analysis is likely to produce more accurate results since noise characteristics and biases of the various technologies should be largely independent. Here, we present a framework for jointly learning dynamic models of transcriptional regulatory networks from both gene expression data and transcription factor binding location data.

Most early research on automatic learning of transcriptional regulatory networks used only gene expression data.[1,2] However, recent simulation studies suggest that regulatory networks learned from gene expression data alone can be considerably obscured by the recovery of spurious interactions when the number of observations is small,[3,4] although a few methods have been developed to address this problem.[4] Joint learning from multiple types of data can alleviate the problem further.

Joint learning in the context of transcriptional regulation has primarily developed around two somewhat related approaches. In one approach, various types of data are used to identify *sets of genes* that interact together in the cell (pathways), share common roles (processes), or are regulated in concert (modules). Recent noteworthy examples include Segal et al.[5] and Bar-Joseph et al.[6] In the other approach, various types of data are used to supplement gene expression data when learning regulatory network models directly. This latter approach was developed by Hartemink et al. and applied in the context of supplemental data describing transcription factor binding location.[7] More recently, Imoto et al. proposed a similar method that applies in the context of supplemental data describing interactions that are restricted to be binary and symmetric,[8] such as the presence of a protein-protein interaction.[9]

Hartemink et al.[7] had two significant limitations that we overcome here. First, because Bayesian networks must be directed acyclic graphs, the resultant models were previously incapable of representing feedback and other dynamic processes. In contrast, here we use a dynamic Bayesian network (DBN) framework to learn regulatory network models from time series data. DBNs are a class of Bayesian network models that permit cyclic structures like regulatory feedback loops; DBNs have been advocated for use in a biological context,[10] and have been used before to analyze time series data in contexts from transcription in *E. coli*[11] to brain electrophysiology in songbirds.[12]

Second, because computational efficiency is a serious concern, the location data were previously treated as infallible: models that failed to include an edge where the location data suggested one should be were eliminated from consideration *a priori*. In formal terms, the prior probability for such networks was forced to zero. While this enabled computationally efficient search, it is inconsistent with the notion that genomic data are generally quite noisy. In contrast, here we present a new informative prior over network structures that is capable of incorporating the location data using a smooth probabilistic model: location data provides evidence as to whether a regulatory relationship exists, and now the more significant the location data (lower the *p*-value), the more likely the edge is to be included. As a consequence, this prior is subtler and more robust; nevertheless, it is designed to be factorable in the context of a DBN, enabling computationally efficient local search. In fact, we can learn a DBN model using the new structure prior in the same amount of time as before. Moreover, unlike in Imoto et al.,[8] our prior is suited to handle the kinds of asymmetric interactions that exist between regulators and their targets.

## 2 Modeling framework

A Bayesian network encodes a probability distribution over a set of random variables $X = \{X_1, \ldots, X_N\}$. The encoding of this probability distribution consists of two components: a network structure $S$ and a set of parameters $\Theta$. The network structure $S$ describes the qualitative nature of the dependencies between the random variables in the form of a directed acyclic graph; the vertices of the graph represent the random variables in $X$ and the directed edges represent the dependencies between those variables. In particular, each variable $X_i$ is assumed to be independent of its non-descendants given its set of parents, denoted $\mathbf{Pa}(X_i)$. Under such a Markov assumption, the joint probability distribution can be written:

$$\mathcal{P}(x) = \prod_{i=1}^{N} \mathcal{P}(x_i \mid \mathbf{pa}(X_i)) \qquad (1)$$

where lowercase variables denote values of the corresponding uppercase random variables. The set of parameters $\Theta$ describes the quantitative nature of the dependencies between the random variables by characterizing the individual probability distributions in this product.

## 2.1 Dynamic Bayesian networks

A dynamic Bayesian network[13] extends the notion of a Bayesian network to model the stochastic evolution of a set of random variables over time; the structure of a DBN thus describes the qualitative nature of the dependencies that exist between variables in a temporal process. We use $X_i[t]$ to denote the random variable $X_i$ at time $t$ and the set $X[t]$ is defined analogously. The evolution of the temporal process is assumed to occur over discrete time points indexed by the variable $t \in \{1, \ldots, T\}$. Under such an assumption, we now have $T \times N$ interacting random variables where previously we had $N$. To simplify the situation, we make two further assumptions. First, we assume that each variable can only depend on variables that temporally precede it. This fairly innocuous assumption allows us to model natural phenomena like feedback loops, but still guarantee that the underlying graph will be acyclic. Second, we assume that the process is a stationary first-order Markov process, which means that $\mathcal{P}(X[t] \mid X[t-1], \ldots, X[1], t) = \mathcal{P}(X[t] \mid X[t-1])$. This assumption is somewhat less innocuous, and we discuss relaxations later. Given these two assumptions, the resultant joint probability distribution can be written:

$$\mathcal{P}(x[1], \ldots, x[T]) = \prod_{i=1}^{N} \left[ \mathcal{P}(x_i[1]) \prod_{t=2}^{T} \mathcal{P}(x_i[t] \mid \mathbf{pa}(X_i[t])) \right] \quad (2)$$

where we note that the first-order Markov assumption means the variables in the set $\mathbf{Pa}(X_i[t])$ are a subset of $X[t-1]$. The underlying acyclic graph with $T \times N$ vertices can now be compactly represented by a graph with $N$ vertices that is permitted to have cycles (see Figure 1 for an example).

## 2.2 Learning the structure of a dynamic Bayesian network

The goal when learning the structure of a dynamic Bayesian network is to identify the network structure $S$ that is most probable given some observed data $D$; in the context of a DBN, the data $D$ typically consists of the $T$ observations of the $N$ variables. The notion of the most probable network structure is made formal by the Bayesian scoring metric (BSM), which is simply the log posterior probability of $S$ given $D$:[14]

$$\mathrm{BSM}(S : D) = \log \mathcal{P}(S|D) = \log \mathcal{P}(D|S) + \log \mathcal{P}(S) + c \quad (3)$$

where the constant $c$ is the same for all structures and can be safely ignored. In a fully Bayesian treatment, the calculation of the log likelihood, $\log \mathcal{P}(D|S)$, involves marginalizing over the distribution of possible parameters $\Theta$, which is analytically tractable when the variables in the network are discrete,[14] as we assume here. Because the expression in (2) is factorable as a product over the variables, the resultant closed-form expression for the log marginal likelihood can be written as a sum of terms where each term corresponds to one variable. Thus, a local change to the network structure—adding or deleting a single edge—affects only one term in this sum.

An especially common choice for the log prior over structures, $\log \mathcal{P}(S)$, is to assume that it is uninformative: every structure is equally likely; in this case, the prior term can be safely ignored since it is the same for all structures. In the rare instance where an informative prior is chosen, it is typically hand-constructed by domain experts.[14] In the next section, we develop a new approach for constructing an informative prior over regulatory network structures automatically from location data.

## 3 Informative structure priors

Transcription factor binding location data provides (noisy) evidence as to the existence of a regulatory relationship between a transcription factor and genes in the genome. This evidence is reported as a $p$-value, and the probability of an edge being present in the true regulatory network is inversely related to this $p$-value: the smaller the $p$-value, the more likely the edge is to exist in the true structure. A more precise formulation of this relationship is provided below.

### 3.1 Probability of an edge being present

We first need to derive a function to map $p$-values to corresponding probabilities of edges being present in structure $S$. Let us define the $p$-value for the location data corresponding to edge $E_i$ in terms of a random variable $P_i$ defined on the interval $[0, 1]$. In this interval, $P_i$ has been previously assumed to be exponentially distributed[16] if the edge $E_i$ is present in $S$, and uniformly distributed if the edge $E_i$ is absent from $S$ (by the definition of a $p$-value). Formally, we have $\mathcal{P}_\lambda(P_i = p \mid E_i \in S) = \lambda e^{-\lambda p}/(1 - e^{-\lambda})$, where $\lambda$ is the parameter controlling the scale of the truncated exponential distribution, and $\mathcal{P}(P_i = p \mid E_i \notin S) = 1$.

Let us use $\beta$ to denote $\mathcal{P}(E_i \in S)$, the probability that edge $E_i$ is present before observing the corresponding $p$-value. Using Bayes rule, we can show that the probability that edge $E_i$ is present after observing the corresponding $p$-value is:

$$\mathcal{P}_\lambda(E_i \in S \mid P_i = p) = \frac{\lambda e^{-\lambda p} \beta}{\lambda e^{-\lambda p} \beta + (1 - e^{-\lambda})(1 - \beta)} \tag{4}$$

As the parameter $\lambda$ increases, the mass of this distribution becomes more concentrated at smaller values of $P_i$; conversely, as $\lambda$ decreases, the distribution spreads out and flattens. The role of the parameter $\lambda$ can be more clearly understood by considering the value $p^*$ obtained by solving the equation $\mathcal{P}_\lambda(E_i \in S \mid P_i = p^*) =$

$\mathcal{P}_\lambda(E_i \notin S \mid P_i = p^*)$, which yields:

$$p^* = \frac{-1}{\lambda} \log \left[ \frac{(1 - e^{-\lambda})(1 - \beta)}{\lambda \beta} \right] \qquad (5)$$

For any fixed value of $\lambda$, an edge $E_i$ is more likely to be present than absent if the corresponding $p$-value is below this critical value $p^*$ (and vice versa). As we increase the value of $\lambda$, the value of $p^*$ decreases and we become more stringent about how low a $p$-value must be before we consider it as prior evidence for edge presence. Conversely, as $\lambda$ decreases, $p^*$ increases and we become less stringent; indeed, in the limit as $\lambda \to 0$, we can show that $\mathcal{P}_\lambda(E_i \in S \mid P_i = p) \to \beta$ independent of $p$, revealing that if we have no confidence in the location data, the probability that edge $E_i$ is present is the same value $\beta$ both before and after seeing the corresponding $p$-value, as expected. Thus, $\lambda$ acts as a tunable parameter indicating the degree of confidence in the evidence provided by the location data; this allows us to model our belief about the noise level inherent in the location data and correspondingly, the amount of weight its evidence should be given.

### 3.2  Bayesian marginalization over parameter $\lambda$

One approach to suitably weighing the evidence of the location data would be to somehow select a single value for $\lambda$, either by guessing or by some other heuristic like finding the value of $\lambda$ that corresponds to a certain "magic" value for $p^*$ like 0.001. Instead, we adopt a more robust Bayesian approach that avoids the selection of a single value and instead marginalizes over $\lambda$. For convenience, we assume that $\lambda$ is uniformly distributed over the interval $[\lambda_L, \lambda_H]$ and then integrate $\lambda$ out of (4) to yield:

$$\mathcal{P}(E_i \in S \mid P_i = p) = \frac{1}{\lambda_H - \lambda_L} \int_{\lambda_L}^{\lambda_H} \frac{\lambda e^{-\lambda p} \beta}{\lambda e^{-\lambda p} \beta + (1 - e^{-\lambda})(1 - \beta)} \, d\lambda \qquad (6)$$

Although (6) cannot be solved analytically, it can be solved numerically for fixed $P_i = p$. Since we have a finite set of $p$-values for a given set of location data, we precompute this integral for each $p$-value and store the results in a table for later use. The computational overhead associated with marginalizing over $\lambda$ is thus constant. The net effect of marginalization is an edge probability distribution that is a smoother function of the reported $p$-values than without marginalization (the tail is much heavier; for a visual depiction, please see the figure in the supplemental material).

### 3.3  Prior probability of a structure

We express the complete log prior probability over structures using the following edge-wise decomposition:

$$\log \mathcal{P}(S) = \sum_{E_j \in S} \log \mathcal{P}(E_j \in S \mid P_j = p) + \sum_{E_k \notin S} \log \mathcal{P}(E_k \notin S \mid P_k = p)) \qquad (7)$$

464

Figure 1. Simplified schematic of a first-order Markov DBN model of the cell cycle. On the left, variables $X1$ through $X4$ are shown both at time $t$ and $t + 1$; variable $\phi$ represents the cell cycle phase; dashed edges are stipulated to be present whereas solid edges are recovered by the learning algorithm. On the right, a compact representation of the same DBN model in which the cycle between $X4$, $X3$, and $X2$ is apparent.

where the term corresponding to the normalizing constant has been dropped since it is the same for all structures. Analogous to the likelihood calculations, the calculations of the prior under this formulation are computationally efficient: a local change to the network structure affects only one term in this sum. As a result, we need not recompute the entire prior with each local change.

Note that in the absence of location data pertaining to a particular edge, we simply use the probability $\mathcal{P}(E_i \in S) \equiv \beta$ for that edge. Our informative prior is thus a natural generalization of traditional priors: in the absence of any location data whatsoever, the prior probability of a network structure is exponential in the number of edges in the graph, with edges favored if we choose $\beta > 0.5$ and edges penalized if we choose $\beta < 0.5$. In the special case where $\beta = 0.5$, the prior over structures is uniform.

## 4 Learning dynamic models of the cell cycle

We assume above that the stochastic process regulating the expression of genes throughout the cell cycle is stationary. This poses a bit of a problem since we may ha‵e a different underlying genetic regulatory network during each phase of the cell cycle. To overcome this problem, we employ an additional variable $\phi$ that can be used by the model to explain how each variable's regulators depend on the cell cycle phase. The phase variable $\phi$ is multinomial and has as many states as there are phases in the cell cycle, allowing us to model a different stationary process within each phase. If we can label each of the time points with the appropriate phase, the inference problem reduces to learning network structure with complete data. We prefer this option to the alternative of learning a hidden phase variable because in our context, the quantity of expression data that is available is quite limited; besides, the state of $\phi$ changes smoothly and predictably so labeling each time point with the appropriate phase is fairly straightforward. A simplified schematic of such a DBN model of the cell cycle is depicted in Figure 1.

Because space is quite limited here, we provide only a brief description of the basic structure of each of our experiments (for further details, please see the supplemental material). The experimental gene expression data are discretized into three states using interval discretization;[4] the simulated data are discretized into two states because the generating model is Boolean. The discretized data in each case are used to compute the log likelihood component of the BSM. The log prior component is computed from the location data $p$-values using (6) and (7). The parameter $\lambda$ is marginalized over a wide range of values, setting $\lambda_L = 1$ to avoid problems near zero ($\lambda_L = 1$ corresponds to $p^* = 0.459$) and $\lambda_H = 10000$ to avoid problems near infinity ($\lambda_H = 10000$ corresponds to $p^* = 0.001$). We set $\beta = 0.5$ so that edges for which we have no location data are equally likely to be present or absent in the graph; as a consequence, without location data, edge presence in the graph depends on expression data alone. We use simulated annealing as a heuristic search method to identify network structures with high scores because learning optimal Bayesian networks is known to be NP-hard.[14] The output of our DBN inference algorithm is the network structure with the highest BSM score among all those visited by the heuristic search during its execution.

### 4.1 Results using simulated data

We use simulated data from a synthetic cell cycle model to evaluate the accuracy of our algorithm and determine the relative utility of different quantities of available gene expression data. The synthetic cell cycle model involves 100 genes and a completely different regulatory network operates in each of the three phases of the cycle. The 100 genes include synthetic transcription factors, only some of which are involved in the cell cycle, and only some of which have simulated location data available. The target genes of the transcription factors are sometimes activated and sometimes repressed; some are under cell cycle control, but many are not. In addition, we include a number of additional genes whose expression is random and not regulated by genes in the model. The simulated gene expression data is generated using the (stochastic) Boolean Glass gene model.[17] Noisy $p$-values for the simulated location data associated with a subset of the regulators are generated with noise models of varying intensity.

We repeatedly conduct the following three experiments: score network structures with expression data alone, ignoring the log prior component $\log \mathcal{P}(S)$ in (3); score network structures with location data alone, using the prior component of the score as given by (7) and ignoring the log likelihood component $\log \mathcal{P}(D|S)$ in (3); and score network structures with both expression and location data. We use these experiments to evaluate the effects of location data with different noise characteristics, expression data of varying quantity, and different choices for $\beta$.

Each of our experiments is conducted on five independently-generated synthetic data sets and results are averaged over those five data sets. Most of the results are presented in the supplemental material, but Figure 2 offers a representative result. The vertical axis measures the (average) total number of errors: the sum of false positives and false negatives in the learned network; the total number of errors relative to the synthetic network in our experiment can range from 0 to 10000. As expected,

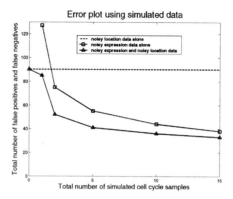

Figure 2. Total number of errors while learning a synthetic cell cycle network using (noisy simulated) expression and location data, separately and with both types of data together. The graph shows the effect of increasing the number of cell cycles worth of expression data, both with and without location data. The dashed horizontal line represents learning using location data alone.

the total number of errors drops sharply as the amount of available expression data increases. The figure demonstrates that our joint learning algorithm consistently reduces the total number of false positives and false negatives learned when compared to the error rate obtained using either expression or location data alone. Also, observe that the availability of location data means that we require typically only half as much expression data to achieve the same error rate as would be achieved with expression data alone, suggesting that the availability of location data can be used to compensate for small quantities of expression data.

### 4.2    Results using experimental data

We next apply our joint learning algorithm to uncover networks describing the regulation of transcription during the cell cycle in yeast. We use publicly available cell cycle gene expression data[18] and transcription factor binding location data.[19] The gene expression data consist of 69 time points collected over 8 cell cycles. Since these belong to different phases, the resultant number of time points in each phase is quite small. As a consequence, we choose to use only three states for the phase variable, by splitting the shortest phase $G_2$ in half and lumping the halves with the adjacent phases. Thus, the three states of our phase variable correspond roughly to $G_1$, $S + G_2$, and $G_2 + M$. To generate a phase label for each time point, we select characteristic genes known to be regulated during specific phases.[18] Guided by the expression of these characteristic genes, we can assign a phase label to each time point. This is done separately for each of the four synchronization protocols in the dataset (alpha, cdc15, cdc28, and elu).

We select a set of 25 genes, of which 10 are known transcription factors for

Table 1. Comparison of the highest scoring networks found in four different experiments with the gold standard network. As discussed in the text, the gold standard contains edges from only the 10 variables for which both location and expression data is available.

| Experiment | TP | TN | FP | FN |
|---|---|---|---|---|
| Expression data only | 7 | 181 | 20 | 32 |
| Location data only | 25 | 184 | 17 | 14 |
| Expression and location data (old prior) | 23 | 187 | 19 | 11 |
| Expression and location data (new prior) | 28 | 189 | 12 | 11 |

which we have available location data. The only important cell cycle transcription factor with location data missing from this set is FKH2; we are not able to use it in our analysis because expression data is missing for many of the time points. The remaining 15 genes in our set are selected on the basis of their known regulation by one or more of these 10 transcription factors. We apply our DBN inference algorithm on this set of 25 variables. Just as with the simulated data, we learn network structures using expression data alone, using location data alone, and jointly from both expression and location data. In the latter case, we evaluate both our old prior,[7] as before with a hard cutoff of $p = 0.001$, and our new informative prior.

As an evaluation criterion (which is more difficult in this context than in the synthetic network context), we create a "gold standard" network consisting of the set of edges that are known to exist from one of the 10 transcription factors with both expression and location data to any one of the other 24 genes in our set; we do not count edges from the other 15 genes when comparing with our gold standard since it would be difficult to determine whether recovered edges are true or false positives, and whether omitted edges are true or false negatives. The gold standard comes from a compiled list of evidence in the literature and from the Saccharomyces Genome Database (*http://www.yeastgenome.org*), but we have tried to ensure that it depends on neither the specific expression data nor the specific location data used in these experiments. Note also that the gold standard is likely not the true underlying regulatory network, but rather is the best we can do given the current understanding of the yeast cell cycle (a bronze standard?).

With these caveats in place, Table 1 shows the total number of positives and negatives that are true and false for the networks found in the four experiments, with respect to the gold standard network. We see that the location data by itself does noticeably better than the expression data, suggesting that this particular set of location data is quite insightful and/or that this particular set of expression data is quite limited in its quantity and quality. Despite the relatively poor performance of the expression data when considered in isolation, when we use our new informative prior to include evidence from the expression data along with the location data, the number of false positives and the number of false negatives are both reduced; in contrast, the old prior reduces the number of false negatives and increases the number of true negatives, but also increases the number of false positives and reduces the

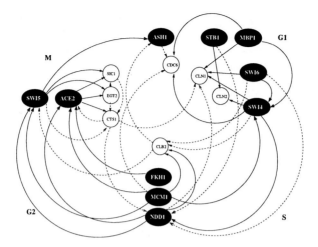

Figure 3. Partial regulatory network recovered using expression data from Spellman et al. and location data from Lee et al. Shaded elliptical nodes are transcription factors for which location data is available. Unshaded circular nodes are genes for which no location data is available. Solid edges represent interactions that have been independently verified in the literature. Dashed edges represent interactions that have not been verified; either the edge is incorrect or the evidence from the literature is inconclusive. Observe the cyclic regulation of transcription factors across phases of the cell cycle.

number of true positives. In contrast, the new prior uniformly outperforms all the other three methods.

From Table 1, we see that combining expression and location data with our new informative prior results in three fewer false negatives as compared to location data alone. These three are the binding of FKH1 to ACE2 ($p$-value= 0.0058), SWI4 to CLN2 ($p$-value= 0.005), and ACE2 to SIC1 ($p$-value= 0.0095). These edges are detected because while the evidence of the location data in isolation is just below threshold for inclusion, during the joint learning it is reinforced with evidence from expression data. Among the supposed false positives, we observe that both location and expression data provide evidence for the regulation of cyclin PCL2 by the transcription factor SWI6 although there is no known evidence of this interaction in the literature. Another interesting case is the regulation of transcription factor FKH1 by the transcription factor MBP1: although this interaction is detected by expression data alone, it is not detected when both location and expression data are used because the corresponding $p$-value of 0.93 is so high that the quantity of expression data is insufficient to overcome the location data evidence against inclusion of the edge.

Figure 3 shows part of the regulatory network recovered using our complete joint learning algorithm. The partial network consists of the 10 transcription factors with location and expression data, along with 7 of the other 15 genes selected at random; we do not show the full network to keep the figure as uncluttered as possible (for the full network, please see the supplemental material). The transcription factors are arranged according to the phase of the cell cycle in which they are maximally expressed. The figure shows that the important cell cycle transcription factors regulate each other in a cyclic fashion as has previously been observed.[19] The figure also shows that our model sometimes detects interactions between genes whose expression is correlated (e.g., CTS1, EGT2, and SIC1). Although these genes are not known to exhibit direct regulatory relationships, they are related in the sense that they are co-regulated by ACE2 and SWI5.

## 5 Conclusion and future work

In this paper, we demonstrate the benefits of recovering dynamic models of transcriptional regulatory networks by jointly learning from both gene expression data and transcription factor binding location data. Our method uses a new, factorable informative prior over network structures to incorporate location data into the learning process. Because location data provides direct evidence regarding the presence of an edge in a regulatory network, it fits well into the framework of our informative prior, and DBNs more generally. With this joint learning framework in place, we show that supplementing expression data with location data is useful both in increasing the accuracy of recovered networks, and in reducing the quantity of expression data needed to achieve an accuracy comparable to that of expression data alone. Since expression data is fairly expensive to generate, it is promising that the relative utility of the data can be further enhanced by combining it with other types of data. Different sources of data will have different noise characteristics and so may be able to reduce the overall error present in the learned network structure. We expect such joint learning techniques will become increasingly relevant in computational biology, especially as data of greater quality and diversity become available.

From a computational perspective, our method should scale well to networks of hundreds of interacting variables, as we have demonstrated here and elsewhere.[20] The primary limitation is not computational but statistical, and not so much with respect to the number of variables but with respect to the number of parents for each variable. As this number increases, larger and larger quantities of data are needed to learn an accurate DBN model.[4] On a related note, while nothing precludes us computationally from modeling a higher-order Markov process, we are constrained statistically by the limited quantity of available time-series expression data.

Note that although the interactions in our graphs can be oriented unambiguously (because time cannot flow backwards), that does not necessarily imply that the interactions are causal since we cannot account for cellular interactions that have not been measured. This can lead to latent variable problems in which we may learn spurious interactions between observed variables: a set of variables may appear to be correlated simply because we cannot observe their latent common cause. One of the main hopes of this line of research is that more direct causal information from alternative assays like transcription factor binding location data and protein-protein interaction data will ameliorate this problem if we can include them in the analysis framework in a principled way.

A number of directions remain in developing sophisticated joint learning methods for elucidating dynamic networks like the cell cycle. Most critically, we would like to incorporate a wider range of other sources of data like protein expression data, protein-protein interaction data, and DNA sequence data. Protein expression data can be added straightforwardly, but is not yet widely available. Nariai et al.[9] have developed a method for learning when to merge co-expressed regulators into complexes based on protein-protein interaction data that might be amenable to further generalization. Greater connection with the module approaches of Segal et al.[5] would also be fruitful.

Supplemental material is available from *http://www.cs.duke.edu/~amink/*.

## 6   Acknowledgments

The authors would like to thank the anonymous reviewers for their suggestions, which led to a number of improvements in the manuscript. AJH would like to gratefully acknowledge the support of the National Science Foundation under its CAREER award program.

## References

1. A. Hartemink, D. Gifford, T. Jaakkola, and R. Young. In *PSB*, p. 422–433, 2001.
2. N. Friedman, M. Linial, I. Nachman, and D. Pe'er. In *RECOMB*, ACM, 2000.
3. D. Husmeier. *Bioinformatics*, 19:2271–2282, 2003.
4. J. Yu, A. Smith, P. Wang, A. Hartemink, and E. Jarvis. *Bioinformatics*, to appear, 2004.
5. E. Segal, R. Yelensky, and D. Koller. *Bioinformatics*, 19:i273–i282, 2003.
6. Z. Bar-Joseph et al. *Nature Biotechnology*, 21(11):1337–1341, 2003.
7. A. Hartemink, D. Gifford, T. Jaakkola, and R. Young. In *PSB*, p. 437–449, 2002.
8. S. Imoto, T. Higuchi, T. Goto, and K. Tashiro. In *CSB*, IEEE, 2003.
9. N. Nariai, S. Kim, S. Imoto, and S. Miyano. In *PSB*, p. 384–395, 2004.
10. K. Murphy and S. Mian. University of California at Berkeley, 1999.
11. I. Ong, J. Glasner, and D. Page. *Bioinformatics*, 18:S241–S248, 2002.
12. A. Smith, J. Yu, T. Smulders, A. Hartemink, and E. Jarvis. In preparation, 2004.
13. N. Friedman, K. Murphy, and S. Russell. In *Proc. 14th UAI*, p. 139–147, 1998.
14. D. Heckerman. In M. I. Jordan, editor, *Learning in Graphical Models*, p. 301–354, Kluwer Academic Publishers, 1998.
15. T. Hughes et al. *Cell*, 102:109–126, 2000.
16. E. Segal, Y. Barash, I. Simon, N. Friedman, and D. Koller. In *RECOMB*, ACM, 2002.
17. R. Edwards and L. Glass. *Chaos*, 10:691–704, 2000.
18. P. Spellman et al. *Mol. Biol. Cell*, 9:3273–3297, 1998.
19. T. Lee et al. *Science*, 298:799–804, 2002.
20. A. Smith, E. Jarvis, and A. Hartemink. *Bioinformatics*, 18:S216–S224, 2002.

# GENOME-SCALE PROTEIN FUNCTION PREDICTION IN YEAST *SACCHAROMYCES CEREVISIAE* THROUGH INTEGRATING MULTIPLE SOURCES OF HIGH-THROUGHPUT DATA

YU CHEN[1,2 †] AND DONG XU[1,2 *]

[1] *UT-ORNL Graduate School of Genome Science and Technology, Oak Ridge, TN, USA.*

[2] *Digital Biology Laboratory, Computer Science Department, University of Missouri-Columbia, Columbia, MO, USA.*

As we are moving into the post genome-sequencing era, various high-throughput experimental techniques have been developed to characterize biological systems at the genome scale. Discovering new biological knowledge from high-throughput biological data is a major challenge for bioinformatics today. To address this challenge, we developed a Bayesian statistical method together with Boltzmann machine and simulated annealing for protein function prediction in the yeast *Saccharomyces cerevisiae* through integrating various high-throughput biological data, including protein binary interactions, protein complexes and microarray gene expression profiles. In our approach, we quantified the relationship between functional similarity and high-throughput data. Based on our method, 1802 out of 2280 unannotated proteins in the yeast were assigned functions systematically. The related computer package is available upon request.

## 1. Introduction

An immediate challenge of the post-genomic era is to assign biological functions to all the proteins encoded by the genome. For example, only one-third of all 6200 predicted genes in yeast *Saccharomyces cerevisiae* (Baker's yeast) were functionally characterized when the complete sequence of yeast genome became available[1]. At present, 4044 yeast genes have been annotated out of 6324 genes. Despite all the efforts, only 50-60% of genes have been annotated in most organisms with complete genomes. This leaves bioinformatics with the opportunity and challenge of predicting functions for unannotated proteins by developing effective and automated methods.

With ever-increasing flow of biological data generated by high-throughput methods, such as yeast two-hybrid systems[2], protein complexes identification by mass spectrometry[3,4] and microarray gene expression profiles[5], some computational approaches have been developed to use these data for gene function prediction. Cluster analysis of the gene-expression profiles is a common approach for predicting functions based on the assumption that genes with similar functions are likely to be co-expressed[6]. Using protein-protein

---

†Current address: Oncology Business Unit, Novartis Pharmaceuticals Corp. One Health Plaza, East Hanover, NJ 07936, USA.

* Corresponding author (xudong@missouri.edu).

interaction data to assign function for novel proteins is another approach. Proteins often interact with one another in an interaction network to achieve a common objective. It is therefore possible to infer the functions of proteins based on the functions of their interaction partners, also known as "guilt by association"[7]. Schwikowski et al.[7] applied a neighbor-counting method in predicting the function. They assigned function to an unknown protein based on the frequencies of its neighbors having certain functions. The method was improved by Hishigaki et al.[8] who used $\chi^2$ statistics. Both these approaches give equal significance to all the functions contributed by the protein neighbors in the interaction network. Other function prediction methods using high-throughput data include machine-learning approach[9] and Markov random fields[10,11]. MAGIC (Multisource Association of Genes by Integration of Clusters) approach to combine heterogeneous data for function assignment has been applied in yeast by Troyanskaya et al[12].

One major challenge for protein function prediction is that the errors (noises) in the high-throughput data have not been handled well and the rich information contained in high-throughput data has not been fully utilized given the complexity and the quality of high-throughput data[13]. A possible solution for this problem is Bayesian probabilistic model[14], which could lead to a coherent function prediction and reduce the effect of noise by combining information from diverse data sources within a common probabilistic framework and naturally weighs each information source according to the conditional probability relationship among information sources. Another major limitation of current function prediction methods based on "majority rule" assignment[7] is that the global properties of interaction network are underutilized since current methods often do not take into account the links among proteins of unknown functions. Vazquez et al.[15] recently proposed a global method to assign protein functions based on protein interaction network by minimizing the number of protein interactions among different functional categories.

To further overcome these limitations, we developed a computational framework for systematic protein function annotation at the genome scale. Our current study focuses on yeast *Saccharomyces cerevisiae*, where rich high throughput data are available. Compared with current methods, our method is distinctive in the following aspects: (1) Unannotated proteins can be assigned to various function categories of GO biological processes with probabilities. This is in contrast to many other prediction methods where proteins were predicted as yes or no without confidence assessment to a limited number of function categories (e.g., MIPS[16], which is less detailed than GO). (2) We quantitatively measured functional dependencies underlying each type of high-throughput data, including protein binary interactions, protein complexes, and microarray gene expression profiles) and coded them into "functional linkage graphs" (interaction network), where each node represents one protein and Bayesian probabilities were calculated to represent the function similarity using the weight of each edge between two proteins. (3) We developed a novel global function prediction method based on Boltzmann machine for function prediction with integration of functional linkage evidences from different types of high-throughput data. We may predict the function of an unannotated gene, even if none of its neighbors in the network has known function. Our method is robust for combining and propagating information systematically across the entire network based on the global optimization of the network configuration.

## 2. Data sources

All the high-throughput data were coded into an interaction network, which can be viewed as a weighted non-directed graph $Gp (D) = (Vp, Ep)$ with the vertex set $Vp = \{d_i \mid d_i \in D\}$ and the edge set $Ep = \{(d_i, d_j) \mid for\ d_i, d_j \in D\ and\ i \neq j\}$. Each vertex represents one protein and each edge represents one measured connection between the two linked proteins from any high-throughput data.

### *Protein-protein binary interaction data*
The protein-protein interaction data from high-throughput yeast two-hybrid interaction experiments were from Uetz et al.[17] and Ito et al.[18], together with 5075 unique interactions among 3567 proteins. We combined the yeast two-hybrid data with the protein-protein interaction data in the MIPS database (http://mips.gsf.de/proj/yeast/CYGD/db/). In total, 6516 unique binary interactions among 3989 proteins were used in this study.

### *Protein complexes*
The protein complex data were obtained from Gavin et al.[3] and Ho et al[4]. In the protein complexes, although it is unclear which proteins are in physical contact, the protein complex data contain rich information about functional relationship among involved proteins. For simplicity, we assigned binary interactions between any two proteins participating in a complex. Thus in general, if there are n proteins in a protein complex, we add n*(n-1)/2 binary interactions. This yields 49,313 edges to the interaction network.

### *Microarray gene expression data*
The gene-express profiles of microarray data were from Gasch et al.[19], which included 174 experimental conditions for all the genes in yeast. A Pearson correlation coefficient was calculated for each possible ORF pairs to quantify the correlation between the gene pairs.

## 3. Methods

### *3.1 Measurement of protein function similarity*
A particular gene product can be characterized with different types of function, including molecular function at the biochemical level (e.g. cyclase or kinase,

whose annotation is often more related to sequence similarity and protein structure) and the biological process at the cellular level (e.g. pyrimidine metabolism or signal transduction, which is often revealed in the high-throughput data of protein interaction and gene expression profiles). In our study, function annotation of protein is defined by GO (Gene Ontology) biological process[20]. The GO biological process ontology is available at http://www.geneontology.org. It has a hierarchical structure with multiple inheritance. After acquiring the biological process functional annotation for the known proteins along with their GO Identification (ID), we generated a numerical GO INDEX, which represents the hierarchical structure of the classification. The more detailed level of the GO INDEX, the more specific function a protein belongs to. The maximum level of GO INDEX is 12. The following shows an example of GO INDEX hierarchy, with the numbers on the left giving GO INDICES and the numbers in the brackets indicating GO IDs:

2          cellular process   (GO:0009987)
2-1        cell communication    (GO:0007154)
2-1-8      signal transduction      (GO:0007165)
2-1-8-1       cell surface receptor linked signal transduction   (GO:0007166)
2-1-8-1- 4         G-protein coupled receptor protein signaling pathway   (GO:0030454)
2-1-8-4- 4-12       signal transduction during conjugation with cellular fusion (GO:0000750)

In the SGD data (http://www.yeastgenome.org/), 4044 yeast proteins have been assigned one or more GO biological process IDs. We calculated protein function similarity by comparing the level of similarity that the two proteins share in terms of their GO INDICES. For example, if both ORF1 and ORF2 have annotated functions, assuming that ORF1 has a function represented by GO INDEX 2-1-8-1 and ORF2 has a function represented by GO INDEX 2-1-8. When compared with each other for the level of matching GO INDEX, they match with each other through 2-1-8, i.e., INDEX level 1 (2), INDEX level 2 (2-1) and INDEX level 3 (2-1-8).

### 3.2 Calculation of Bayesian probabilities

We calculated probabilities for two genes to share the same function based on different types of high-throughput data, i.e., microarray data, protein binary interaction data and protein complex data. With the assumption that $H = \{M, B, C\}$ denotes the interaction events in different types of high-throughput data, where $M$ represents two genes correlated in gene expression profiles with Pearson correlation coefficient $r$ in microarray data, $B$ represents a protein binary interaction and $C$ represents a protein complex interaction, the *posterior* probability that two proteins have the same function, $p(S|H)$, is computed using the Bayes' formulas:

$$p(S|H) = \frac{P(H|S)P(S)}{P(H)} \tag{1}$$

where $S$ represents the event that two genes/proteins have the same function at a given level of GO INDEX. The probability $p(S)$ is the relative frequency of genes/proteins whose functions are the same at the given level of GO INDEX by chance. $p(H|S)$ is the conditional (a priori) probability that two genes/proteins to have the event $H$ given that they have the same function at a given level of GO INDEX. The probability $p(H)$ is the frequency of $H$ in the entire data set, e.g., the frequency of gene expression correlated with coefficient $r$ over all gene pairs in yeast, which is calculated from the genome-wide gene expression profiles ($H = M$) or the relative frequency of two proteins having a known binary interaction over all possible pairs in yeast, which is estimated from the known protein interaction data set ($H = B$). The probabilities $p(H|S)$, and $p(S)$ are computed based on a set of proteins whose functions have been annotated in GO biological process.

To quantify the gene function relationship between the correlated gene expression pairs, we calculated the probabilities of such gene expression correlated pairs sharing the same function at each GO INDEX level, based on our early study[21]. Results show a higher probability of sharing the same function for broad functional categories (the high-order GO INDEX levels) or highly correlated genes in expression profiles (Figure 1A). Figure 1B shows the presence of information in highly correlated gene-expression pairs for their gene functional relationship in comparison to random pairs. In Figure 1, we only show the curves of GO INDEX 1, 2, 3 and 4. The other higher GO INDEX levels (from 5 to 10) have the same trend. Based on Figures 1, we decided to consider pairs with gene expression profile correlation coefficient $\geq 0.7$ for function predictions, as other pairs have little information for function prediction. The estimated probabilities of sharing the same function corresponding to gene pairs with $r \geq 0.7$ were smoothed by using a monotone regression function (the pool-adjacent-violators algorithm[22]) for function prediction of unannotated proteins.

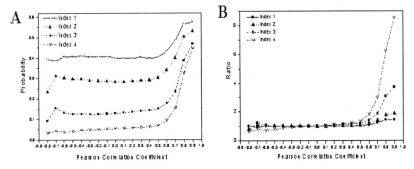

Figure 1. A: Probabilities of pairs sharing the same level of GO indices against Pearson correlation coefficient of microarray gene expression profiles. B: Normalized ratio for the percentage of gene pairs sharing the same level of GO indices ($p(S|M)$), against the percentage of random pairs sharing the same function ($p(S)$) versus Pearson correlation coefficient of microarray gene expression profiles.

The analysis result of the protein-protein interaction data is shown in Figure 2. The plots for protein binary interaction and complex interaction data, show a drop of probabilities of sharing the same function with an increase in the GO INDEX level, as seen in Figure 2A. A higher probability to share less specific, broader functional categories as represented by lower GO INDEX levels is observed. Comparison of our results with similar analysis on random pairs, shows a normalized ratio of protein-protein interaction pairs against the random pairs for sharing the same GO INDEX level (as seen in Figure 2B). Since the value is highly above 1, particularly for more specific function categories, there clearly exists a relationship between the protein-protein interaction data and function similarity. Such relationship can be utilized to make function predictions.

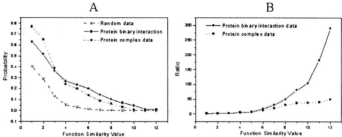

Figure 2. Functional relationship in yeast protein-protein interaction data. The horizontal axis in both plots shows the GO INDEX levels that two proteins share. (A) The probability of interacting proteins sharing the same levels of GO INDEX. (B) The normalized ratio of (A) compared with random pairs.

## 3.3 Protein function prediction

### 3.3.1 Local prediction

In the local prediction of an unannotated protein using its immediate neighbors in the network graph, we follow the idea of "guilt by association"[7], i.e., if an interaction partner of an unannotated protein $x$ has a known function, $x$ may share the same function, with a probability underlying the high-throughput data between $x$ and its partner. We identify the possible interactors for $x$ in each high-throughput data type (protein binary interaction, protein complex interaction and microarray gene expression with correlation coefficient $r \geq 0.7$). We assign functions to the unannotated proteins on the basis of common functions identified among the annotated interaction partners using the probabilities described in section 3.2. Furthermore, we assume that the information contents for protein function prediction from different sources of high-throughput data or different interaction partners are independent based on the early suggestion that the information from different high-throughput data are conditionally uncorrelated[23,24]. A protein can belong to one or more functional classes, depending upon its interaction partners and their functions. For example, protein $x$ is an unannotated protein with several interaction partners having known functions. With the assumption that $F_i$, $i = 1, 2, \ldots, n$, represents a collection of all the functions that the interaction partners have, a likelihood score function for protein $x$ to have function $F_i$, $G(F_i, x)$, is calculated as:

$$G(F_i, x) = 1 - (1 - P'(S_l|M)) * (1 - P'(S_l|B)) * (1 - P'(S_l|C)) \tag{2}$$

where $S_l$ represents the event that two proteins have the same function, $F_i$, whose GO INDEX has $l$ levels, $l = 1, 2, ..., 12$. $P'(S_l|M)$, $P'(S_l|B)$ and $P'(S_l|C)$ are the probabilities of interaction pairs to have the same function for gene expression correlation coefficient $\geq 0.7$ ($M$), protein binary interaction ($B$) and protein complex interaction ($C$), respectively. In each type of high-throughput data, one unannotated protein might have multiple interaction partners with function $F_i$. Suppose that there are $n_M$, $n_B$, and $n_C$ interaction partners with function $F_i$ in the three types of high throughput data, respectively. The combined probabilities $P'(S_l|M)$, $P'(S_l|B)$, and $P'(S_l|C)$ in equation (2) are calculated as:

$$P'(S_l|M) = 1 - \prod [1 - P_j(S_l|M)], \quad j = 1, 2, ....n_M . \tag{3}$$
$$P'(S_l|B) = 1 - \prod [1 - P_j(S_l|B)], \quad j = 1, 2, ....n_B . \tag{4}$$
$$P'(S_l|C) = 1 - \prod [1 - P_j(S_l|C)], \quad j = 1, 2, ....n_C . \tag{5}$$

$P_j(S_l|M)$, $P_j(S_l|B)$, and $P_j(S_l|C)$ were estimated probabilities retrieved from the probability curves calculated in section 3.2 for a single pair of genes/proteins. We also defined the likelihood score $G(F_i, x)$ as *Reliability Score* for each function, $F_i$. The final predictions are sorted based on the *Reliability Score* for each predicted GO INDEX. The *Reliability Score* represents the probability for the unannotated protein to have a function $F_i$, assuming that all the evidences from the high-throughput data are independent and only applicable to immediate neighbors in the network.

### 3.3.2 Global prediction

The major limitation of the local prediction method is that it only uses the information of immediate neighbors in a graph to predict a protein's function. In some cases, the uncharacterized proteins may not have any interacting partner with known function annotation and its function cannot be predicted based on the local prediction method. In addition, the global properties of the graph are underutilized since this analysis does not include the links among proteins of unknown function. In Figure 3 proteins 1, 2, 3 and 4 are annotated and proteins 5, 6, 7 and 8 have unknown functions. If we only use the local prediction method, the functions of proteins 3 and 4 can be predicted but the functions of proteins while 1 and 2 cannot since all the neighbors of proteins 1 and 2 are unannotated proteins. Moreover, the contributions of function assignment for protein 4 is not only from the neighbor proteins 7 and 8 whose functions are already known, but also from protein 1 when its functions is predicted through the following information propagation: proteins 5 and 6 → protein 3 → protein 2 → protein 1. Hence, the functional annotation of uncharacterized proteins should not only be decided by their direct neighbors but also controlled by the global configuration of the interaction network. Based on such global optimization strategies, we developed a new approach for predicting protein function. We used the Boltzmann machine to characterize the global stochastic behaviors of the network. A protein can be assigned to multiple functional classes, each with a certain probability.

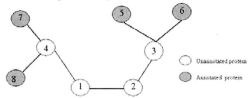

Figure 3: Illustration of protein function prediction from interaction network. Protein 1, 2, 3 and 4 are unannotated proteins. Protein 5, 6, 7 and 8 are annotated proteins with known functions.

In Boltzmann machine (BM), we consider a physical system with a set of states, $\alpha$, each of which has an energy, $H_\alpha$. In thermal equilibrium, given a temperature $T$, each of the possible states $\alpha$ occurs with probability:

$$P_\alpha = \frac{1}{R} e^{-H_\alpha / K_B T} \qquad (6)$$

where the normalizing factor $R = \sum_\alpha e^{-H_\alpha / K_B T}$ and $K_B$ is the Boltzmann's constant. This is called the Boltzmann-Gibbs distribution. It is derived based on the general assumptions about microscopic dynamics, and it can be applied to a stochastic network. In an undirected graphical model with binary-valued nodes, each node (protein) $i$ in the network has only one state value $Z$ (1 or 0). In our case, $Z = 1$ means that corresponding node (protein/gene) either has a known function, or it is ready for a function prediction. Now we consider the system going through a dynamic process from non-equilibrium to equilibrium, which corresponds the optimization process for the function prediction. For the state at time $t$ (optimization integration step $t$), node $i$ has the probability for $Z_{t,i}$ to be 1, $P(Z_{t,i} = 1 | Z_{t-1, j \neq i})$ and the probability is given as a sigmoid-function of the inputs from all the other nodes at time $t-1$:

$$P(Z_{t,i} = 1 | Z_{t-1, j \neq i}) = \frac{1}{1 + e^{-\beta \sum_{j \neq i} W_{ij} Z_{t-1, j \neq i}}} \qquad (7)$$

where $\beta$ is a parameter reversely proportional to the annealing temperature and $W_{ij}$ is the weight of the edge connecting proteins $i$ and $j$ in the interaction graph. $W_{ij}$ is calculated by combining the evidence from gene expression correlation coefficient $\geq 0.7$ ($M$), protein binary interaction ($B$) and protein complex interaction ($C$):

$$W_{ij} = \delta_j \sum_{k=1}^{12} G(F_k, i | j) = \delta_j \sum_{k=1}^{12} (1 - (1 - P(S_k | M))(1 - P(S_k | B))(1 - P(S_k | C))) \qquad (8)$$

where $S_k$ represents the event that two proteins $i$ and $j$ have the same function ($F_k$) at the GO INDEX level $k$, $k = 1, 2, \ldots, 12$. $G(F_k, i | j)$ is the reliability score for proteins $i$ and $j$ sharing the same function $F_k$. $P(S_k|M)$, $P(S_k|B)$ and $P(S_kC)$ were estimated probabilities retrieved from the probability curves calculated in section 3.2. $\delta_j$ is the modifying weight:

$$\delta_j = \begin{cases} 1 & \text{if } j \in \text{annotated proteins} \\ P(Z_{t-1,j} = 1) & \text{otherwise} \end{cases} \tag{9}$$

To achieve the global optimization, we conducted simulated annealing technique as the following process. First we set the initial state of all unannotated proteins (nodes) randomly to be 0 or 1. The state of any annotated protein is always 1. If an unannotated protein is assigned with the state 1, its function will be predicted based on its immediate neighbors with known functions using the local prediction method. Next starting with a high temperature, pick a node $i$ and compute its value $P_i$ according to equation (6), then update its state, till all the nodes in the network reach equilibrium. With gradually cooling down, the system might settle in a global optimization of network configuration if the sum of weights associated to all the unannotated proteins reaches the maximum value.

## 4. Results

We have implemented local and global methods to predict functions for unannotated proteins. We used *sensitivity* and *specificity* to measure the performance of our methods using ten-fold cross validation. We labeled all 4044 annotated proteins with known GO INDICES into folds 1 to 10. Each time, we pick one fold as the test dataset and the other nine folds as training data to calculate prior probabilities. We estimate the *sensitivity* to determine the success rate of the method and *specificity* to assess the confidence in the predictions. For a given set of proteins $K$, let $n_i$ be the number of the known functions for protein $P_i$. Let $m_i$ be the number of functions predicted for the protein $P_i$ by the method. Let $k_i$ be the number of predicted functions that are correct (the same as the known function). Thus *sensitivity* (SN) and *specificity* (SP) are defined as:

$$SN = \frac{\sum_1^K k_i}{\sum_1^K n_i} \tag{10} \qquad SP = \frac{\sum_1^K k_i}{\sum_1^K m_i} \tag{11}$$

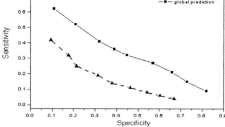

Figure 4: *Sensitivity-specificity* plot on the test set for the three prediction methods.

Figure 4 shows the *sensitivity* versus *specificity* of the methods with *Reliability score* cutoff from 0.1 to 0.9. It showed that the global prediction has a significantly better sensitivity-specificity plot than the local prediction. In our method, the highest specificity can only reach 70%. Some false positives generated in our method might be caused by the independence assumption of different sources of high-throughput data. Such assumption could be oversimplified due to biases inherent in data. For example, protein binary interactions are likely to be correlated in gene expression profiles. On the other hand, the predicted functions from our method could be true but they have not been determined by experiments yet, thus, they are not included in GO annotation.

| INDEX | Reliability Score | | | | | | | | |
|---|---|---|---|---|---|---|---|---|---|
| | ≥ 0.9 | ≥ 0.8 | ≥ 0.7 | ≥ 0.6 | ≥ 0.5 | ≥ 0.4 | ≥ 0.3 | ≥ 0.2 | ≥ 0.1 |
| 1 | 897 | 964 | 1045 | 1116 | 1185 | 1264 | 1331 | 1530 | 1707 |
| 2 | 847 | 922 | 978 | 1052 | 1130 | 1217 | 1315 | 1519 | 1707 |
| 3 | 710 | 801 | 883 | 955 | 1018 | 1102 | 1236 | 1491 | 1693 |
| 4 | 627 | 714 | 789 | 870 | 949 | 1052 | 1151 | 1433 | 1673 |
| 5 | 605 | 691 | 761 | 836 | 918 | 1016 | 1120 | 1405 | 1659 |
| 6 | 271 | 378 | 472 | 447 | 622 | 707 | 849 | 1128 | 1495 |
| 7 | 104 | 173 | 248 | 316 | 395 | 483 | 595 | 722 | 1159 |
| 8 | 14 | 31 | 48 | 68 | 103 | 147 | 194 | 299 | 680 |
| 9 | 0 | 1 | 2 | 3 | 4 | 4 | 11 | 20 | 105 |
| 10 | 0 | 0 | 0 | 0 | 0 | 0 | 0 | 0 | 6 |

Table 1. Number of unannotated genes with function predictions with respect to prediction confidence probabilities and index levels.

Using all the 4044 annotated proteins with known GO INDICES as the training set, we are able to assign functions to 1802 out of the 2280 unannotated proteins in yeast at different level of functions (different levels of GO INDICES). The detail prediction results can be queried at http://digbio.missouri.edu/~ychen/ProFunPred. The number of unannotated genes with function predictions with respect to the specificity and GO INDEX levels can be found in Table 1. Using our method, we not only assign general functional categories to unannotated protein but also assign very specific functions to some unannotated proteins. 104 unannotated proteins were assigned functions with reliability score >= 0.9 and GO INDEX level >= 7. Moreover, using our global prediction method we can assign functions for the proteins whose interacting partners are all unannotated proteins with unknown function. For example, all interacting partners of *YBR100W* have unknown functions. Using global function prediction, *YBR100W* was assigned function as "nucleic acid metabolism" (reliability score 0.8) and "response to DNA damage" (reliability score 0.8).

## 5. Discussion

Systematic and automated prediction of protein function using high-throughput data represents a major challenge in the post genomic era. To address this challenge, we developed a systematic method to assign function in an automated fashion using integrated computational analysis of yeast high-throughput data including binary interaction, protein complexes and gene expression microarray data, together with the GO biological process functional annotation. We applied Boltzmann machine for the global protein function annotation by combining and propagating information across the entire network. Our method is robust to obtain global optimization using simulated annealing. With six different sets of randomly selected starting points, we obtained exactly the same result as shown in Table 1.

Future work includes exploring better optimization methods and statistical models. To solve the optimization problem in Boltzmann machine, in contrast to the simulated annealing technique, a Bayesian learning of posterior distributions over parameters[25] provides a more elaborate and systematic estimation of maximum likelihood. In addition, supervised learning methods such as Conditional Random Fields[26] can also be alternative schemes to model this stochastic learning process. Furthermore, we will develop more elaborate model-based integrations to address the dependencies among different high-throughput data for protein function prediction.

## Acknowledgments

We would like to thank Drs. Jeffery Becker, Ying Xu, Loren Hausser, and Qiang Zhao for helpful discussions. This research is supported in part by the US Department of Energy's Genomes to Life program (http://www.doegenomestolife.org) under project, "Carbon Sequestration in *Synechococcus* Sp.: From Molecular Machines to Hierarchical Modeling" (www.genomes-to-life.org). This work is also partially funded by Nation Science Foundation (EIA-0325386).

## References

[1] Goffeau, A., Barrell, B. G., Bussey, H., Davis, R. W., Dujon, B., Feldmann, H., Galibert, et al *Science*. **546**:346-352, (1996).

[2] Chien, C., Bartel, P., Sternglanz, R., and Fields, S. *Proc. Natl. Acad. Sci. USA*. **88**:9578-9582, (1991).

[3] Gavin, A., Bosche, M., Krause, R., Grandi, P., Marzioch, M., Bauer, A., Schultz, J., Rick, J., Michon, A., Cruciat, C. *Nature*. **415**:141-147, (2002).

[4] Ho, Y., Gruhler, A., Heilbut, A., Bader, G.D., Moore, L., Adams, S., Millar, A., Taylor, P., Bennett, K., Boutilier, K., et al. *Nature*. **415**: 180-183, (2002).

[5] Brown, M., Grundy, W., Lin, D., Cristianini, N., Sugnet, C., Furey, T., Ares, M., and Haussler, D. *Proc. Natl. Acad. Sci. USA.* **97**:262-267, (2000)

[6] Pavlidis, P., and Weston, J. In Proceedings of the Fifth International Conference on Computational Molecular Biology (RECOMB2001), pp. 249 – 255, (2001)

[7] Schwikowski, B., Uetz, P., and Fields, S. *Nature Biotechnology.* **18**:1257-1261, (2000)

[8] Hishigaki, H., Nakai, K., Ono, T., Tanigami, A., and Takagi, T. *Yeast.* **18**: 523-531, (2001).

[9] Clare, A. and King, R. D. ECCB 2003 (also published as a journal supplement in Bioinformatics. **19**: ii42-ii49), (2003).

[10] Deng MH, Zhang K, Mehta S, Chen T, Sun FZ, The first IEEE Computer Society bioinformatics conference, CSB2002, 117-126, (2002).

[11] Letovsky, S. and Kasif, S. *Bioinformatics*, **19** Suppl 1:I197-I204, (2003).

[12] Troyanskaya, O., Dolinski, K., Owen, A., Altman, R., and Botstein, D. *Proc. Natl. Acad. Sci.* **100**:8348–8353, (2003)

[13] Chen, Y., and Xu, D. *Current Peptide and Protein Science.* **4**:159-181, (2003).

[14] Winkler, R. L. An introduction to Bayesian Inference and Decision, Holt, Rinehart and Winston Inc, (1972).

[15] Vazquez, A., Flammini, A., Maritan, A. and Vespignani, A. *Nat Biotechnol.* **21**:697-700, (2003).

[16] Mewes, H., Frishman, D., Guldener, U., Mannhaupt, G., Mayer, K., Mokrejs, M. *et al. Nucleic Acids Res.* **30**:31-34, (2002).

[17] Uetz, P., Giot, L., Cagney, G., Mansfield, T.A., Judson, R.S., Knight, J.R., Lockshon, D., Narayan, V., Srinivasan, M., Pochart, P., et al. *Nature.* **403**:623-627, (2000).

[18] Ito, T., Tashiro, K., Muta, S., Ozawa, R., Chiba, T., Nishizawa, M., Yamamoto, K., Kuhara, S., and Sakaki, Y. *Proc. Natl. Acad. Sci. USA.* **98**:4569-4574, (2001).

[19] Gasch, A. P., Spellman, P. T., Kao, C. M., Carmel-Harel, O., Eisen, M. B., Storz, G., Botstein, D. and Brown, P. O, *Mol Cell Biol,* **11**:4241-4257, (2000).

[20] The Gene Ontology Consortium. *Nature Genetics.* **25**:25-29, (2000).

[21] Joshi, T., Chen, Y., Becker, J. M., Alexandrov, N., Xu, D. *OMICS*, in press.

[22] Haerdle, W. Applied Nonparametirc Regression, Cambridge, UK. (1995).

[23] Jansen, R., Yu, H., Greenbaum, D., Kluger, Y., Krogan, N. J., Chung, S., Emili, A., et al. *Science.* **302**:449-453, (2003).

[24] Asthana S., King O. D., Gibbons F. D. and Roth F. P., *Genome Res.* **14**:1170-1175, (2004).

[25] Ackley D. H., Hinton G. E. and Sejnowski T. J., *Cognitive Science*, **9**:147-169, (1985).

[26] Lafferty J., McCallum A., and Pereira F., *International Conference on Machine Learning (ICML)* (2001).

# DISCOVERING TRANSCRIPTIONAL MODULES FROM MOTIF, ChIP-CHIP AND MICROARRAY DATA

TIJL DE BIE, PIETER MONSIEURS, KRISTOF ENGELEN, BART DE MOOR

*K.U.Leuven, ESAT-SCD, Kasteelpark Arenberg 10, 3001 Leuven, Belgium*

NELLO CRISTIANINI

*U.C.Davis Dept. of Statistics, 360 Kerr Hall, One Shields Ave, CA 95616, US*

KATHLEEN MARCHAL

*Centre of Microbial and Plant Genetics and ESAT-SCD, K.U. Leuven*
*Kasteelpark Arenberg 20, 3001 Leuven-Heverlee, Belgium*

We present a method for inference of transcriptional modules from heterogeneous data sources. It allows identifying the responsible set of regulators in combination with their corresponding DNA recognition sites (motifs) and target genes. Our approach distinguishes itself from previous work in literature because it fully exploits the knowledge of three independently acquired data sources: ChIP-chip data; motif information as obtained by phylogenetic shadowing; and gene expression profiles obtained using microarray experiments. Moreover, these three data sources are dealt with in a new and fully integrated manner. By avoiding approaches that take the different data sources into account sequentially or iteratively, the transparency of the method and the interpretability of the results are ensured. Using our method on biological data demonstrated the biological relevance of the inference.

## 1   Introduction

Nowadays, data representative of different cellular processes are being generated at large scale. Based on these *omics* data sources, the action of the regulatory network that underlies the organism's behavior can be observed.

Whereas until recently bioinformatics research was driven by the development of methods that deal with each of these data sources separately, the focus is now shifting drastically towards integrative approaches dealing with several data sources simultaneously. Indeed, technological and biological noise in the individual data sources is often so prohibitive and unavoidable that standard methods are bound to fail. Then only a combined use of *heterogeneous* and *independently acquired* information sources can help to solve the problem. Furthermore, these different points of view on the biological system allow gaining a holistic insight into the network studied. Therefore, the integration of heterogeneous data is an important, though non-trivial, challenge of current bioinformatics research.

In this study we focus on 3 types of *omics* data that give independent information on the composition of transcriptional modules, the basic building blocks of transcriptional networks in the cell: *ChIP-chip* data (*chromatin immuno-*

*precipitation on arrays*) provides information on the direct physical interaction between a regulator and the upstream regions of its target genes; *motif* information as obtained by phylogenetic shadowing describes the DNA recognition sites of these regulators; and *gene expression profiles* obtained using microarray experiments describe the expression behavior in the conditions tested. By integrating these three data sources, we aim at identifying the concerted action of regulators that elicit a characteristic expression profile in the conditions tested, the target genes of these regulators, and the DNA binding sites recognized by these regulators, thus fully specifying the relevant regulatory modules.

Previous successful approaches to integrative analyses in bioinformatics can be found in the class of *kernel methods* [7,19] and methods based on *graphical models* [4,5,13,14,15]. Still, to our knowledge, no successful attempts to solve the problem of module inference exploiting all 3 independently acquired ChIP-chip, motif and expression data have been made so far. Furthermore, most existing approaches that exploit the availability of heterogeneous data sources proceed in a *sequential* or an *iterative* way (see e.g. [8] for simultaneous detection of motifs and clustering of expression data, e.g. [2] for an iterative approach using ChIP-chip and expression data, and e.g. [10] for simultaneous motif detection and analysis of ChIP-chip data).

In this paper, we present an approach that is different in spirit from previous methods, taking the different data sources into account in a highly *concurrent* way. The performance of the algorithm was demonstrated using the Spellman dataset [16] as a benchmark.

## 2    Materials and Algorithms

### 2.1    Data Sources

As microarray benchmark set the Spellman dataset was used [16], which contains 77 experiments describing the dynamic changes of 6178 genes during the yeast cell cycle. The profiles were normalized (subtracting the mean of each profile and dividing by the standard deviation across the time points) and stored in a gene expression data matrix further denoted by $A$ with a row for each gene expression profile and a column for each condition.

Genome-wide location data performed by Lee et al. [9] were downloaded from http://web.wi.mit.edu/young/regulator_network. These contain data on the binding of 106 regulators to their respective target genes in rich medium. The ChIP-chip data matrix (further denoted by $R$) used in our study consists of *one minus* the p-values obtained from combined ratio's between immuno-precipitated and control DNA (see [9]). Thus, a large value (close to one) indicates that the regulator is probably present.

The motif data used in this study were obtained from a comparative genome analysis between distinct yeast species (phylogenetic shadowing) performed by Kellis et al. [6]. The authors describe the detection of 72 putative regulatory motifs in yeast. These motifs, available online as regular expressions, were transformed into the corresponding probabilistic representation (weight matrix): for each motif, the 20 *Saccharomyces cerevisiae* genes in which the motif was most reliably detected according to the scoring scheme of Kellis et al. [6] were selected. The intergenic sequences of these genes were subjected to motif detection based on Gibbs sampling [MotifSampler,18]. If the statistically overrepresented motif in this set of putatively co-expressed genes corresponded to the motif that was detected by the comparative motif search of [6] the motif model was retained. As such 53 of the 71 motifs could be converted into a weight matrix. This weight matrix was subsequently used to screen all intergenic sequences for the presence of the respective regulatory motifs using MotifLocator [11]. Absolute scores were normalized [11]. As the score distribution of the motif hits depends on the motif length and the degree of conservation of the motif, the distribution of the normalized scores differs between motifs. Therefore, normalized scores were converted into percentile values. This allows for an unbiased choice of the thresholds on the motif quality parameter in the algorithm. The matrix containing these percentile values is the motif data matrix $M$ that will be used in this work.

## 2.2 Module Construction Algorithm

The aim of the method is to find regulatory modules (which may be overlapping with each other), based on the gene expression, ChIP-chip, and Motif data matrices as specified above.

A module is fully specified by the set of genes it regulates (denoted by an index set $g$, pointing to the relevant set of rows of $R$, $M$ and $A$), in addition to the set of regulators (corresponding to the columns with indices in a set called $r$ in the ChIP-chip matrix $R$) and motifs (corresponding to the columns with indices $m$ in the Motif matrix $M$) that are responsible for the regulation of these genes. The goal of our method is to come up with regulatory modules specified in this way, by fully exploiting the heterogeneous data sources available.

We note that the principles behind the method developed here are based on ideas similar to those that laid the foundations for the Apriori algorithm, originally developed in the database community [1]. All implementations used in this paper have been done in matlab.

**Seed construction.** This is the main step of the algorithm, and allows the construction of a good guess (or *seed*) of the modules. The idealized goal of this step is to find a set of genes $g$, that have the same expression profile, and such that there exist sufficiently large sets of regulators $r$ and of motifs $m$ that are entirely present in all these genes. Since in practice it is not known exactly in which intergenic regions a certain motif occurs or where a regulator binds, we have to resort to the score matrices $R$ and $M$. Furthermore, the expression profiles $A$ of genes in a module will only be approximately equal, and possibly only in a set of conditions, so we relax this constraint to requiring a strong correlation instead of equality between them.

Formally, then the task to solve is:

---

Find all maximal gene sets $g$ for which there exist an $r$ of size $|r| \geq r_{min}$ and a set $m$ of size $|m| \geq m_{min}$, such that the following 3 constraints are satisfied:

    1. $R(i,j) > t_r$                  for all $i \in g$ and $j \in r$

    2. $M(i,j) > t_m$               for all $i \in g$ and $j \in m$

    3. $\mathrm{corr}(A(i,:), A(j,:)) > t_a$     for all $i,j \in g$

where $r_{min}$, $m_{min}$ and thresholds $t_r$, $t_m$ and $t_a$ are parameters of the method.

---

Here, a maximal set $g$ is defined as a set that cannot be extended with another gene without violating one or more of these constraints. In the following, we will use the term *valid set* for a gene set $g$ that satisfies these constraints.

Clearly it is computationally impossible to tackle this problem with a naive approach: the number of gene sets is exponentially large in the number of genes in the dataset, which is prohibitive even for the smallest genomes. However, it is trivial to verify that:

**Observation 1:** When a gene set does not satisfy the constraints, none of its supersets satisfy the constraints.

This means that we can build up the maximal sets incrementally, starting with valid sets of size one, and gradually expanding them. Concretely, the (already less naive) algorithm would then look like[1]:

---

[1] Notationally, we will use $L^i$ to denote the list containing all valid gene sets with $i$ genes. For an individual valid gene set we will use a bold face $g_k^i$, with a superscript $i$ to specify that it is an element of $L^i$ and thus contains $i$ genes, and with a subscript $k$ to distinguish it from the other gene sets in $L^i$. The $x$-th gene in this gene set is denoted as $g_k(x)$, for brevity without superscript.

- For all single genes, check if they satisfy constraints 1 and 2 (constraint 3 is trivially satisfied for singleton gene sets). Make a list $L^1$ of all singleton gene sets that contain such a valid gene.

- Set $i = 2$.

- While size($L^{i-1}$) $\neq 0$

  For $k=1$:size($L^{i-1}$), expand set $g_k^{i-1} = \{g_k(1), g_k(2), \ldots, g_k(i-1)\} \in L^{i-1}$ once for each gene $g$ that is not yet contained in $g_k^{i-1}$. Put the thus expanded sets $\{g_k(1), g_k(2), \ldots, g_k(i-1), g\}$ that satisfy the 3 constraints (to be verified in $R$, $M$ and $A$), in a list $L^i$.

  Set $i = i+1$.

Notice that following this strategy, a gene set can be constructed in different ways, by adding the genes to it in a different ordering (i.e. in different iterations $i$). This can be avoided by adding a gene to a gene set $g_k^{i-1}$ only whenever its row number $g$ is larger than that of all other genes already in $g_k^{i-1}$. Thus for every $g_k^i = \{g_k(1), g_k(2), \ldots, g_k(i)\} \in L^i$ we always have that $g_k(x) < g_k(y)$ for $x < y$.

Additionally, in this way we can easily keep the list $L^i$ of gene sets $g_k^i$ sorted as well, where the sorting is carried out first according to the first added gene and last according to the last added gene. More formally: $g_k^i$ preceeds $g_l^i$ in $L^i$ if and only if $g_k(\mathrm{argmin}_x(g_k(x) \neq g_l(x))) < g_l(\mathrm{argmin}_x(g_k(x) \neq g_l(x)))$ (this ordering of the list $L^i$ is indeed a total ordering relation.)

Still the number of expanded gene sets can be huge in every iteration: each of the gene sets $g_k^{i-1}$ in $L^{i-1}$ must be expanded by all genes $g > g_k(i-1)$, after which the validity has to be checked by looking at the matrices $R$, $M$ and $A$. This can still be too expensive. However, we can exploit the converse of Observation 1:

**Observation 2:** Whenever a gene set satisfies the constraints, all of its subsets satisfy the constraints.

Using this so-called *hereditary property* of the constraint set, in some cases we can conclude *a priori* —i.e. without checking in $R$, $M$ and $A$— if an extended gene set of size $i$ can possibly be valid or not: we simply have to check if all of its size $i-1$ subsets belong to $L^{i-1}$. Only if this is the case, we still have to access the data in $R$, $M$ and $A$; if it is not the case, we know without further investigation that the extended subset is invalid.

Specifically, assume that we expand the gene set $g_k^{i-1} = \{g_k(1), g_k(2), \ldots, g_k(i-2), g_k(i-1)\} \in L^{i-1}$ with $g$, leading to $\{g_k(1), g_k(2), \ldots, g_k(i-2), g_k(i-1), g\}$. Then, since for a valid size $i$ set each of its size $i-1$ subsets must be contained in $L^{i-1}$, also $\{g_k(1), g_k(2), \ldots, g_k(i-2), g\}$ must be contained in $L^{i-1}$. In other words: there has to exist a $g_l^{i-1} = \{g_l(1), g_l(2), \ldots, g_l(i-2), g_l(i-1)\} \in L^{i-1}$ for which $g_k(x) = g_l(x)$ for $x \leq i-2$, and $g = g_l(i-1)$. This can efficiently be ensured *constructively*, by exploiting the fact that the list $L^{i-1}$, and all $g_k^{i-1}$ themselves are sorted. Indeed, thanks to this, all gene sets $g_k^{i-1}$ that have the first $i-2$ genes in common occur consecutively in $L^{i-1}$. Therefore, to expand $g_k^{i-1}$ with an additional gene, we only have to screen the list $L^{i-1}$ starting at $g_{k+1}^{i-1}$ and move forward in $L^{i-1}$ for as long as the first $i-2$ genes are equal to $g_k(1)$, $g_k(2), \ldots$ and $g_k(i-2)$. For every gene set $g_l^{i-1}$ screened in this way, read the last gene $g_l(i-1)$ and append it to $g_k^{i-1}$, thus resulting in a candidate gene set of size $i$, potentially to be appended to $L^i$. To find out whether this candidate gene set is valid indeed, one still has to check the constraints explicitly. However, thus constructively exploiting the hereditary property, the number of queries to $R$, $M$ and $A$ is drastically reduced. Note that this strategy also ensures that $L^i$ is sorted automatically.

**Module validation.** In some cases the first step described above is not sufficient for adequate module inference. There are three reasons for this:

First, the seed construction method can be rather conservative in recruiting genes, since each of the genes in the module has to satisfy all 3 of the constraints. Therefore, in a second step, we calculate the mean of the expression profiles of the seed modules found in the first step, further called the *seed profile*. Then we can additionally recruit all genes with a high correlation with the seed profile to be incorporated in the module. In order to determine an optimal threshold value for this correlation, we compute the enrichment of each of the motifs and regulators in the genes that have an expression profile that achieves this threshold correlation with the seed profile. The logarithm of the p-value of the enrichment is then plotted as a function of this threshold (Figure (1)), and the threshold can be chosen such that this value is minimal.

Second, sometimes it is undesirable to a priori decide how many motifs and regulators we want in the module, or it may be difficult to choose the thresholds $t_r$, $t_m$ and $t_a$ (even though experiments show little dependence on these). Then one can first use the seed construction algorithm requiring only 1 regulator and motif, and with stringent thresholds, after which again the enrichment of all motifs and regulators can be plotted as a function of the correlation threshold with the mean profile of the seed module. For each such seed profile, the corresponding enrichment plot will visually hint at the number of motifs and regulators (namely the number of significantly enriched motifs and regulators).

Third, similarly, the enrichment plot allows excluding false positive motifs or regulators: when they are selected in step 1, but appear not to be enriched in the validation step, they are considered as a false positives and discarded.

To calculate the enrichment, we first calculate the mean score of the module for the particular motif or regulator. Note that the mean score of a module by random gene selection is approximately Gaussianly distributed (central limit theorem), with mean equal to the mean over all genes, and variance equal to the overall variance divided by the size of the module. Thus, we can calculate the enrichment as the logarithm of the p-value based on a Gaussian approximation.

Note that the p-values have been computed based on profiles that have been obtained from the data, such that they do not have a rigorous probabilistic interpretation here. Hence, we can only use them as explained above.

## 2.3   Calculating Overrepresentation of Functional Classes

Functional categories for each gene were obtained from MIPS [12]. Functional enrichment of the modules was calculated using the hypergeometric distribution [17], which assigns to each functional class a p-value.

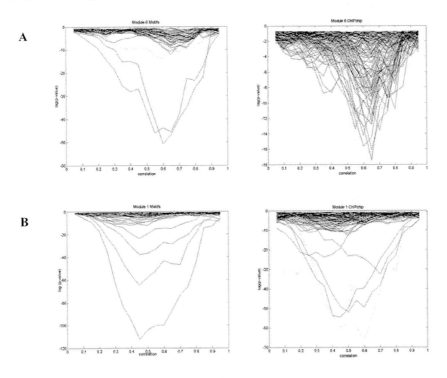

Figure 1: Two examples (panels A and B) of the module validation step for two seed profiles: on the left, the logarithms of the p-values (vertical axes) are plotted for all motifs as a function of the correlation threshold (horizontal axes). I.e. each line in the plot shows the p-values for the enrichment of its corresponding motif, as a function of the gene expression correlation threshold used. The right column shows similar plots for the regulators. Clearly panel A shows the results for a false positive prediction (module 6): the regulator (right figure) of the identified seed module turns out not to be significantly overrepresented in genes correlated with the seed profile. In panel B the results are displayed for the positive example described in the text.

## 3    Results

**Cell cycle related modules.** To test the reliability of our method, we used the well-studied Spellman dataset as benchmark. The analysis we performed using our two-step algorithm is illustrated by elaborating on the detection of the cell cycle related module 1. Using the seed detection step we searched for modules of genes having at least 1 common motif (1M) in their intergenic sequences and 1 common regulator (1R) showing a small p-value in the ChIP-chip data, and of which the expression profiles were mutually correlated with a minimal correlation of 0.7. This seed identification step then predicts several potential modules, and for each of them a seed profile can be calculated. For each of these modules we performed the module validation step.

Fig. 1A (right) shows how this validation step allows to detect that the regulator associated with this module is probably a false positive: when recruiting genes that correlate stronger than a certain threshold with the seed profile, it is not significantly enriched in the recruited gene set, no matter what the threshold is.

In Fig 1B, using the parameter settings of 1M/1R, we identified a potential seed module containing regulator 98 (Swi4) and motif M_11 (known as a Swi4 motif). Calculating the statistical overrepresentation of all motifs and regulators in genes correlated with the seed profile of this putative module showed that in this subset of genes indeed M_11 and Swi4 were overrepresented. The identified module seed thus is likely to be biologically relevant. These results also show that besides Swi4 and M_11, 3 additional motifs and regulators were overrepresented in subsets of genes correlated with the module seed profile, indicating the probable underestimation of the real module size. To verify whether these other regulators/motifs co-occur in the same subsets of genes and therefore comprise a larger module, we repeated the seed identification step using additional parameter settings (see Table 1 in the online supplement). From this result it appeared that we could recover a complete module consisting of the 3 overrepresented regulators (Mbp1, Swi4, Swi6) and 2 motifs (M_16, M_10) and that this module is present in genes displaying an expression profile that shows a correlation of at least 0.7 with the average seed profile. Checking the identities of the regulators and the motifs (regulators Mbp1, Swi4, Stb1 combined with the regulatory motifs Mbp1 (M_18, M_12) and Swi4 (M_11 and M67)) showed that we identified a previously extensively described regulatory module of the yeast cell cycle.

Besides this first module, 3 additional related cell cycle (Table 1) modules could be retrieved. Additional information on each of the separate modules can be found in the online supplement. Genes in the different modules showed peak expressions shifted in time relative to each other, as shown in Figure 1 of the online supplement. All of the predicted modules are conform the previously described knowledge on the cell cycle [2,3,9].

Table 1: Cell cycle related modules. Column 'R' contains the regulators, column 'M' the motifs, the column 'Functional Class: p-value' contains p-values for several functional classes, and the 'Seed Profile' column contains a plot with the expression profiles of the genes regulated by the module.

| | R | M | Functional Class: p-value | Seed Profile |
|---|---|---|---|---|
| Module 1 | Mbp1 Swi6 Swi4 Stb1 | M_18 (Mbp1) M_12 (Mbp1) M_11 (Swi4) M_67 (Swi4) | **10 CELL CYCLE AND DNA PROCESSING: 0** **10.03** cell cycle: 2.7e-5 **10.01** DNA processing: 1.3e-4 **42.04** cytoskeleton: 4.2e-3 | |
| Module 2 | Swi4 Mbp1 Swi6 FKH2 | M_18 (Mbp1) M_12 (Mbp1) M_11 (Swi4) M_8 (Mcm) | **40 CELL FATE: 5.2e-4** **40.01** cell growth / morphogenesis: 2.6e-3 **43 CELL TYPE DIFFERENTIATION: 5.2e-3** **43.01** f ungal/microorganismic cell type differentiation: 5.2e-3 **34.11** cellular sensing and response:5.3e-3 **01.05.01** C-compound and carbohydrate utilization: 6.8e-3 **10.03.04.03** chromosome condensation: 9.4e-3 | |
| Module 3 | NDD1 FKH2 Mcm1 | M_8 (Mcm) M_30 (Mcm) | **43 CELL TYPE DIFFERENTIATION: 3.6e-3** **43.01** fungal/microorganismic cell type differentiation: 3.6e-3 **10.03.03** cytokinesis (cell division) /septum formation: 4.8e-3 | |
| Module 4 | Swi5 (Ace2) | M_8 (Mcm) | **32.01** stress response: 3.2e-3 **10.03** cell cycle: 8.7e-3 | |

**Non cell cycle related modules.** Besides the modules primarily involved in cell cycle, other modules could be identified in the Spellman dataset (see Table 2). Module 5, consisting of Fhl1, Rap1 and Yap5, involved in the regulation of ribosomal proteins was previously also identified [9]. Note that it was identified from a profile that does not change significantly and consistently with the cell cycles. By our analysis we could pinpoint motif M_54 [6], as the regulatory motif correlated with this regulatory module. A second non cell cycle related module consisted of the genes regulated by the motifs M_7 and M_3 (identified as ESR1 and ESR2 [6]). For this module, related to transcription and ribosomal RNA processing only the motifs seemed informative (see module 6 in Table 2, and Figure 1A).

Table 2: Non cell cycle related modules.

| | R | M | Functional Class: p-value | Seed Profile |
|---|---|---|---|---|
| Module 5 | FKL1 Yap5 Rap1 | M_54 | 12 PROTEIN SYNTHESIS: 0 12.01 ribosome biogenesis: 0 | |
| Module 6 | / | M_3 (ESR1) M_7 (ESR2) | 11 TRANSCRIPTION: 0.000002 11.04 RNA processing: 0 11.04.01 rRNA processing: 0 | |

## 4    Discussion

We described a 2-step methodology combining ChIP-chip, motif and expression data to infer complete descriptions of transcriptional modules. The seed construction step predicts the putative modules consisting of regulators, their corresponding motifs and the elicited expression profile. The validation step filters false positive predictions and gives further insight into the module size.

The problem is attacked in a very direct way: the integration of the data sources is achieved in a one-shot-algorithm, and requires no iteration over the different data sources. While the running time was very reasonable for all experiments carried out for this paper, it heavily depends on the parameters. The more stringent they are set, the smaller the lists $L^i$ will be and the faster the algorithm will run. Further speed-ups are possible, but not needed for the experiments reported in this paper (for all parameter setting used, it remained below 1 hour for the slowest step, which is the seed identification, on an intel pentium 2GHz laptop with 512Mb RAM). Therefore we will not go into these here.

The Spellman dataset was used as a benchmark to test the performance of our method. Since this dataset and the yeast cell cycle have extensively been studied before [2,9], it is ideally suited for testing the reliability and biological relevance of the predictions. We were able to reconstruct 4 important modules known to be involved in cell cycle and also 2 non cell cycle related modules without using any prior biological knowledge or prior data reduction. These results indicate that predictions passing the module validation step are likely to be biologically relevant (no false positives present).

## 5   Conclusion

The 3 data types mutually agreeing with each other on the prediction of a module not only results in the most reliable predictions (as was the case for the cell cycle related modules), but also allows correlating a set of regulators with their corresponding regulatory motifs and elicited profiles in a very natural and direct way. On the other hand, because of the restricted number of experimental data yet available (ChIP-chip data not known for all regulators and tested in a limited set of conditions, expression data for specific conditions not available), and the questionable quality of the motif models, the presence of a signal in 1 data type can compensate for the lack of it in another data type, allowing still to retrieve the module.

While to our knowledge this is the first time these 3 independently acquired data sources are exploited in such a concurrent way for module identification, the approach is further extendible towards any number of information sources, and in principle towards the use of other data types. The only condition for an efficient method to exist is that the constraints the gene sets have to satisfy must be hereditary. This extension will be the subject of future work.

### Acknowledgments

KM is a post-doctoral researcher of the Fund for Scientific Research – Flanders (F.W.O–Vlaanderen). TDB is a research assistant with the Fund for Scientific Research – Flanders (F.W.O.–Vlaanderen). KE is a research assistant with the IWT.

This work is partially supported by: 1. IWT projects: GBOU-SQUAD-20160; 2. Research Council KULeuven: GOA Mefisto-666, GOA-Ambiorics, IDO genetic networks; 3. FWO projects: G.0115.01 (microarrays/oncology), G.0413.03 (inference in bioi), G.0388.03 (microarrays for clinical use), G.0229.03 (ontologies in bioi), and G.0241.04 (Functional Genomics); 4. IUAP V-22 (2002-2006).

Supplementary information:
www.esat.kuleuven.ac.be/~kmarchal/Supplementary_Info_PSB2005/SuppWebsiteYeastPSB.html

## References

1. R. Agrawal et al, *Proceedings of the 1993 ACM SIGMOD International Conference on Management of Data*, pp. 207-216 (1993)
2. Z. Bar-Joseph et al, *Nature Biotechnology*, **21**(11): 1337-1342 (2003)
3. M.C. Costanzo et al, *Nucleic Acids Research*, **28**(1): 73-76 (2000)
4. N. Friedman, *Science*, **303**(5659): 799-805 (2004)
5. A. Hartemink et al, *Proceedings of the Pacific Symposium on Biocomputing*, **7**: 437–449 (2002): 166-176 (2003)
6. M. Kellis et al, *Nature*, **423**(6937): 241-254 (2003)
7. G. Lanckriet et al, *Bioinformatics* Advance Access, bth294 (2004)
8. M. Lapidot and Y. Pilpel, *Nucleic Acids Research*, **31**(13): 3824-3828 (2003)
9. T.I. Lee et al, *Science*, **298**(5594): 799-804 (2002)
10. X.S. Liu et al, *Nature Biotechnology*, **20**(8): 835-839 (2002)
11. K. Marchal et al, *Genome Biology*, **5**(2): R9 (2004)
12. H.W. Mewes et al, *Nucleic Acids Research*, **32** Database issue D41-D44 (2004)
13. N. Narai et al, *Proceedings of the Pacific Symposium on Biocomputing*, **9**: 336-347 (2004)
14. E. Segal et al, *Bioinformatics*, **19** (Suppl 1): 264-272 (2003)
15. E. Segal et al, *Nature Genetics*, **34**(2)
16. P.T. Spellman et al, *Molecular Biology of the Cell*, **9**: 3273–3297 (1998)
17. S. Tavazoie et al, *Nature Genetics*, **22**(3):281-285 (1999)
18. G. Thijs et al, *Journal of Computational Biology*, **9**(2): 447-464 (2002)
19. J.P. Vert and M. Kanehisa, *Bioinformatics*, **19**: 238ii-234ii (2003)

# GENRATE: A GENERATIVE MODEL THAT FINDS AND SCORES NEW GENES AND EXONS IN GENOMIC MICROARRAY DATA [a]

BRENDAN J. FREY[1,2], QUAID D. MORRIS[1,2], WEN ZHANG[2],
NAVEED MOHAMMAD[2], TIMOTHY R. HUGHES[2]

[1] *Dept. of Electrical and Computer Engineering*
[2] *Banting & Best Dept. of Medical Research*
*University of Toronto, Toronto, ON, M5S 3G4, Canada*
*E-mail: frey@psi.toronto.edu*

Recently, researchers have made some progress in using microarrays to validate predicted exons in genome sequence and find new gene structures. However, current methods rely on separately making threshold-based decisions on intensity of expression, similarity of expression profiles, and arrangements of exons in the genome. We have taken a Bayesian approach and developed GenRate, a generative model that accounts for both genome-wide expression data taken from multiple conditions (*e.g.* tissues) and co-location and density of probes in DNA sequence data. GenRate balances probabilistic evidence derived from different sources and outputs scores (log-likelihoods) for each gene model, enabling the estimation of false-positive and false-negative rates. The model has a number of local minima that is exponential in the length of the DNA sequence data, so direct application of the EM learning algorithm produces poor results. We describe a novel way of parameterizing the model using examples from the data set, so that good solutions are found using an efficient algorithm. We apply GenRate to a subset of mouse genome-wide expression data that we have created, and discuss the statistical significance of the genes found by GenRate. Three of the highest-ranking gene structures found by GenRate, each containing thousands of bases from the genome, are confirmed using RT-PCR experiments.

## 1 Introduction

The use of DNA microarrays for the discovery of expressed elements in genomes is increasing with improvements in density, flexibility, and accessibility of the technology. Two general strategies have emerged. In the first, candidate elements (e.g. ORFs, genes, exons, RNAs) are identified computationally, and each is represented one or a few times on the array [10,12,4]. In the second, the entire genome sequence is "tiled"; for example, overlapping oligonucleotides encompassing both strands are printed on arrays, such that all possible expressed sequences are represented [12,6,11,13]. Both approaches, as well as independent analyses by other methods [8,4] have indicated that a substantially higher proportion of genomes are expressed than are currently annotated, underscoring the shortcomings of current sequence-based gene prediction algorithms and emphasizing the need for empirical analysis.

Microarrays do not inherently provide information regarding the length of the RNA or DNA molecules detected, nor do they inherently reveal whether

---

[a] This work was supported by a Premier's Research Excellence Award to Frey and a grant from the Canadian Institute for Health Research awarded to Hughes and Frey.

features designed to detect adjacent features on the chromosome are in fact detecting the same transcript. Co-expression (*i.e.* co-detection) of adjacent features can be taken as evidence supporting that the corresponding probes are indeed detecting the same molecular species. However, mRNAs, which account for the largest proportion of transcribed sequence in a genome, present a particular challenge in this paradigm. mRNAs are composed only of spliced exons, often separated in the genome (and in the primary transcript) by thousands to tens of thousands of bases of intronic sequence. Identifying the exons that comprise individual transcripts from genome- or exon-tiling data is not a trivial task, since falsely-predicted exons, overlapping features, transcript variants, and poor-quality measurements can confound assumptions based on simple correlation of magnitude or co-variation of expression.

We describe a generative model that jointly accounts for the stochastic nature of the arrangement of exons in genomic DNA and the noise properties in microarray data. The generative model, called GenRate, uses expression data taken from multiple conditions, accounts for co-location statistics of probes in DNA sequence data, and finds and scores gene structures. While the version of GenRate described here does not model expression variability introduced by alternative splicing, overlapping genes, and alternative transcription sites, in a future paper we will describe an extension, which does account for these effects.

## 2   Microarray Data

The microarray data are a subset of a full-genome data set to be described elsewhere[1]. Briefly, exons were predicted from Repeat-masked mouse draft genome sequence (Build 28) using five different exon-prediction programs. (While this data is based on putative exons, GenRate can be applied to any sequence-based expression data set, including genome tiling data.) A total of 63,041 non-overlapping exons were contained on chromosome 4. One 60-mer oligonucleotide probe for each exon was selected using conventional procedures, such that its binding free energy for the corresponding putative exon was as low as possible compared to its binding free energy with sequence elsewhere in the genome, taking into account other constraints on probe design. (For simplicity, we assume each probe has a unique position in the genome.) Arrays designs were submitted to Agilent Technologies (Palo Alto, California) for array production. Twelve diverse samples were hybridized to the arrays, each consisting of a pool of cDNA from poly-A selected mRNA from mouse tissues (37 tissues total were represented). The pools were designed to maximize the diversity of genes expressed between the pools, without diluting them beyond detection limits [2]. Scanned microarray images were quantitated with GenePix (Axon Instruments), complex noise structures (spatial trends, blobs, smudges) were removed from the images using our spatial detrending algorithm[3], and each set of 12 pool-specific images was calibrated using the VSN algorithm [15] (using a

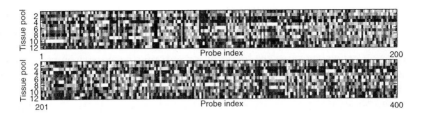

Figure 1: A small fraction of our data set, which consists of an expression measurement for each of 12 mouse tissue pools and 63,041 60-mer probes for putative exons arranged according to the order in Build 28 of the genome.

set of one hundred "housekeeping" genes represented on every slide). For each of the 63,041 probes, the 12 values were then normalized to have intensities ranging from 0 to 1.

Fig. 1 shows a portion of the data from chromosome 4. Because the probes are arranged according to their order in the genome, a consecutive sequence of similar expression profiles (columns) provides evidence of co-regulation of the corresponding putative exons, and thus provides evidence of a gene structure. For example, probes 4 to 14 have similar expression profiles (with high expression in the 10th tissue pool), which provides evidence of a gene. However, such visually obvious examples are relatively rare. More common examples that GenRate finds include complex gap patterns and noisy, albeit statistically significant patterns that are hard to identify visually.

## 3  Previous Work

Heuristics that group nearby putative exons using intensity of expression or co-regulation across experimental conditions can be used to approach this problem [12,6,13]. In [6], a dense activity map of RNA transcription is used to verify putative exons. A disadvantage of this approach is that it cannot detect weakly-expressed exons that have a large biological impact, due to high translational efficiency. In addition to detecting high levels of transcriptional activity, our approach finds correspondences in patterns of activation across multiple tissue pools, so even weakly-expressed exons that have tissue-dependent activity can be detected.

In [12], correlations between the expression patterns of nearby probes are used to merge probes into putative gene structures. A merge step takes place if the correlation exceeds 0.5, but not if the number of non-merged probes between the two candidate probes is greater than 5. Our approach differs from this approach in two ways. First, our algorithm doesn't make a sequence of threshold-based decisions, but instead uses distributions over gene lengths, gap lengths and probe similarity to jointly make decisions and compute maximum $a$

*posteriori* gene structures. So, for example, extraordinarily similar expression profiles may be merged even if there is a large gap between the corresponding exon. Second, since GenRate uses a principled generative probability model, decisions on gene structures are based on an automatic comparison of the likelihood of seeing the gene structure under the gene model and the background expression profile model. So, if the profiles for two probes are quite unusual compared to typical profiles in the data (*i.e.*, they have unusual patterns of tissue-specificity), the two probes may be merged into a gene structure even if their expression profiles are only weakly similar.

In the above previous work, it is not clear how to properly balance evidence provided by similarity of expression profiles with that provided by sequence features (*e.g.* gene length, intron lengths, intra-gene gap length). Recently researchers have successfully shown that complex probability models can be used effectively to combine different sources of information in genomics data (c.f. [14]). In a probability model, combining different sources of information is realized by a computationally efficient application of Bayes rule to combine sources of information in a probabilistic manner.

## 4 Generative Probability Model

GenRate can be applied to any genome-based expression data set, since it works on the assumption that the expression data is arranged in order on the genome. In our model, the probes are indexed by $i$ and the probes are ordered according to their locations in the genome. Denote the expression vector for probe $i$ by $\mathbf{x}_i$, which contains the levels of expression of probe $i$ across $K$ experimental conditions. In our data, there are $K = 12$ tissue pools. Since probe $i$ is selected from putative exon sequence data, it may in fact correspond to a false exon either between genes or within a gene. $e_i$ is a binary variable, where $e_i = 1$ indicates a true exon and $e_i = 0$ indicates a false exon. If probe $i$ is within a gene, the remaining length of the gene (in probes) including probe $i$ is $\ell_i$. $\ell_i = 0$ indicates that probe $i$ does not belong to the gene.

To model the relationships between the variables $\{\ell_i\}$ and $\{e_i\}$, we computed statistics using confirmed exons derived from four cDNA and EST databases: Refseq, Fantom II, Unigene, and Ensembl. The database sequences were mapped to Build 28 of the mouse chromosome using BLAT [9] and only unique mappings with greater than 95% coverage and greater than 90% identity were retained. Probes whose chromosomal location fell within the boundaries of a mapped exon were taken to be confirmed. The genes in these databases are obviously subject to selection bias, so statistics based on these genes will be biased, an effect we ignore for now.

We model the lengths of genes using a geometric distribution, with parameter $\lambda = 0.05$, which was estimated using cDNA genes. This approximation is accurate for lengths greater than 5. For shorter lengths, the accuracy of the prior is not critical, because the prior probability of starting a gene (see the next paragraph) dominates. Importantly, there is a significant computational advantage in using the memory-less geometric distribution. Using cDNA genes to select the length prior will introduce a bias, so other priors should be investigated, but in this paper we report results using the geometric prior.

The "control knob" that we use to vary the number of genes that GenRate finds is $\kappa$, the *a priori* probability of starting a gene at an arbitrarily chosen position. Combining the above distributions, and recalling that $\ell_i = 0$ indicates an inter-gene region, we have

$$P(\ell_i = 0|\ell_{i-1} = 0 \text{ or } 1) = 1 - \kappa$$
$$P(\ell_i|\ell_{i-1} = 0 \text{ or } 1) = \kappa \cdot 0.05 \exp(-0.05\ell_i), \quad \text{if } \ell_i > 0$$
$$P(\ell_i = n - 1|\ell_{i-1} = n) = 1, \quad \text{if } \ell_{i-1} > 1. \tag{1}$$

The expression "$\ell_{i-1} = 0$ or 1" occurs because a new gene may start immediately after the previous gene has finished.

From the data on confirmed genes, we found that within genes, each probe has a probability of 0.7 corresponding to a correct predicted exon. We assume that within genes, putative exons are true exons independently and with probability

$$P(e_i = 1|\ell_i > 0) = \epsilon, \tag{2}$$

where from the above data we estimated $\epsilon = 0.7$. Although we have not verified this assumption directly using the data, we find that the results obtained using this assumption give high sensitivity with high specificity (see below). Between genes, all putative exons are false, so $P(e_i = 1|\ell_i = 0) = 0$.

The similarity between the expression profiles belonging to the same gene is accounted for by a gene-specific prototype expression vector. While this model does not properly take into account effects introduced by alternative splicing, overlapping genes and alternative transcription sites, as shown in the experimental section, this model is sufficient for finding a large number of new exons and genes. In the gene model, each gene has a unique, hidden index variable and the prototype expression vector for gene $j$ is $\boldsymbol{\mu}_j$. We denote the index of the gene at probe $i$ by $c_i$. Different probes may have different sensitivities (for a variety of reasons, including free energy of binding), so we assume that each expression profile belonging to a gene is similar to a scaled version of the prototype. Since probe sensitivity is not tissue-specific, we use the same scaling factor for all $K$ tissues. Also, different probes will be offset by different amounts (*e.g.*, due to different average amounts of cross-hybridization), so we include a tissue-independent additive variable for each probe. Assuming the expression profile $\mathbf{x}_i$ for a true exon ($e_i = 1$) is equal to the corresponding prototype $\boldsymbol{\mu}_{c_i}$, plus isotropic Gaussian noise, we have

$$P(\mathbf{x}_i|e = 1, c_i, \{\boldsymbol{\mu}_j\}, a_i) =$$

$$\prod_{k=1}^{K} \frac{1}{\sqrt{2\pi a_{i3}^2}} \exp\left(-(x_{ik} - [a_{i1}\mu_{c_i,k} + a_{i2}])^2/2a_{i3}^2\right), \qquad (3)$$

where $a_{i1}$, $a_{i2}$ and $a_{i3}$ are the scale, offset and isotropic noise variance for probe $i$, collectively referred to as $a_i$. In a priori distribution $P(a_i)$ over these variables, the scale is assumed to be uniformly distributed in $[1/30, 30]$, which corresponds to a liberal assumption about the range of sensitivities of the probes. The offsets are assumed to be uniform in $[-0.5, 0.5]$ and the variance is assumed to be uniform in $[0, 1]$. These assumptions are naive and require further research, but we find they are sufficient for obtaining good results.

False exons are modelled using a background expression profile distribution,

$$P(\mathbf{x}_i|e_i = 0, c_i, a_i, \{\boldsymbol{\mu}_j\}) = P_0(\mathbf{x}_i) \qquad (4)$$

Since the background distribution doesn't depend on $c_i$, $a_i$ or $\{\boldsymbol{\mu}\}$, we also write it as $P(\mathbf{x}_i|e = 0)$. We obtained values for these probability densities by training a mixture of 100 Gaussians on the entire, unordered set of expression profiles, and including a component that is uniform over the range of expression profiles.

The structural relationships between the variables described above are indicated by the Bayesian network in Fig. 2. Often, when drawing Bayesian networks, the prototypes are considered as parameters and not shown. We include the prototypes in the Bayesian network to show that they induce long-range dependencies in the GenRate model. For example, if all of the prototypes are used to model gene structures in the first part of the chromosome, none will be left to model the remainder of the chromosome. So, during learning, prototypes must somehow be distributed in a fair fashion across the chromosome.

Combining the structure of the Bayesian network with the conditional distributions described above, we have a joint distribution,

$$P(\{\mathbf{x}_i\}, \{a_i\}, \{e_i\}, \{c_i\}, \{\ell_i\}, \{\boldsymbol{\mu}_j\}) = \qquad (5)$$

$$\prod_{i=1}^{N} P(\mathbf{x}_i|e_i, c_i, a_i, \{\boldsymbol{\mu}_j\})P(a_i)P(e_i|\ell_i)P(c_i|c_{i-1}, \ell_{i-1})P(\ell_i|\ell_{i-1}) \prod_{j=1}^{G} P(\boldsymbol{\mu}_j),$$

where in this expression $P(c_1|c_0, \ell_0)$ and $P(\ell_1|\ell_0)$ are equal to $P(\ell_1)$ and $P(c_1)$. Most of the components in the above model are described above. As for the gene indices, $c_i$, we assume that $c_i$ is ordered, starting at 1: $P(c_i = 1) = 1$.

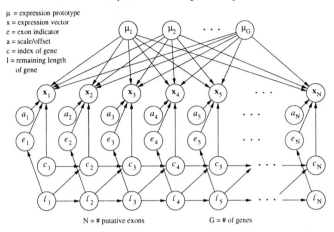

Figure 2: A Bayesian network showing the variables and parameters in GenRate.

Whenever a gene terminates, $c_i$ is incremented in anticipation of modelling the next gene, so $P(c_i = n | c_{i-1} = n, \ell_{i-1}) = 1$ if $\ell_{i-1} > 1$ and $P(c_i = n + 1 | c_{i-1} = n, \ell_{i-1}) = 1$ if $\ell_{i-1} = 1$. We assume the prototypes are distributed according to the background model: $P(\boldsymbol{\mu}_j) = P_0(\boldsymbol{\mu}_j)$.

## 5 Inference and Learning

Exact computation of the marginal probabilities or the MAP configuration in the above model is computationally intractable. A standard way of coping with this intractability in this form of model is to use the EM algorithm [16]. The EM algorithm fails spectacularly on this problem. This is not surprising, since the EM algorithm in long hidden Markov models is known to find poor local minima caused by suboptimal parsings of the long data sequence [19]. In the GenRate model, the EM algorithm gets stuck in local minima where prototypes are used to model weakly-evidenced gene patterns in one part of the chromosome, at the cost of not modelling strongly-evidenced gene patterns in another part of the chromosome.

To circumvent the problem of very poor local minima, we devised a novel scheme where the parameters ($\boldsymbol{\mu}$'s) are represented using examples from the data set. An additional advantage of this approach is that since learning consists of identifying nearby profiles as prototypes, learning can be performed in a single forward-backward pass (Viterbi pass).

The scheme is based on the fact that the model for each $\mathbf{x}_i$ is derived from nearby expression patterns, corresponding to nearby exons in the genomic

DNA. So, if $\mathbf{x}_i$ is an exon, there ought to be another $\mathbf{x}$ nearby that is a good representative of the profile for the gene. To accomplish this we replace each variables $c_i$ with a variable $r_i$, which indicates the distance, in indices, from $\mathbf{x}_i$ to the prototype $\mathbf{x}_j$ for the gene that $\mathbf{x}_i$ is in, $i.e.$ $r_i = j - i$. For example, $r_i = -1$ indicates that the profile immediately preceding $\mathbf{x}_i$ is the prototype for the gene to which $\mathbf{x}_i$ belongs. The new conditional distribution for $\mathbf{x}_i$ is

$$P_e(\mathbf{x}_i|e_i = 1, r_i, \mathbf{x}_{i+r_i}, a_i) =$$

$$\begin{cases} \prod_{k=1}^{K} \frac{1}{\sqrt{2\pi a_{i3}^2}} \exp\left(-(x_{ik} - [a_{i1}x_{i+r_i,k} + a_{i2}])^2/2a_{i3}^2\right), & \text{if } r_i \neq 0 \\ P(\mathbf{x}_i|e_i = 1, r_i, \mathbf{x}_{i+r_i}, a_i) = P(\mathbf{x}_i|e = 0), & \text{if } r_i = 0. \end{cases} \quad (6)$$

Here, $r_i$ acts as a switch to select the parent of $\mathbf{x}_i$. To ensure the $r_i$'s take on appropriate values, the conditional distribution for $r_i$ is given by $P(r_i = n - 1|r_{i-1} = n, \ell_{i-1}, \ell_i) = 1$ if $\ell_{i-1} > 1$ and $P(r_i|r_{i-1}, \ell_{i-1}, \ell_i) = \mathrm{Unif}(0, \ldots, \ell_i)$ if $\ell_{i-1} = 1$. This ensures that when a new gene starts, $r_i$ will be drawn randomly from within the length of the gene and that $r_i$ will decrement throughout the duration of the new gene. This model can be described using a factor graph [18].

The above model is a product of terms that has a Markov chain structure with tractable state complexity, so the forward-backward algorithm or Viterbi algorithm can be used for exact inference. Some readers may be concerned about the presence of the continuous variables $a_i$. However, these variables do not have parents, so they can be integrated or maximized for each configuration of the (discrete) variables in their Markov blankets ($r_i$ and $e_i$). We use the Viterbi algorithm to find the MAP configuration of the model. Our MATLAB implementation of GenRate takes approximately 3 minutes to process the 63,041 probes and 12 tissue pools in chromosome 4, with a restriction on the gene length of $W = 200$ probes.

The only free parameter in the model is $\kappa$, which sets the statistical significance of the genes found by GenRate. The score of each gene found by GenRate can be computed by taking the log-ratio of the probability under GenRate of the MAP path through the gene and the probability of the path corresponding to non-gene probes (intra-gene probes).

## 6  Experimental Results

Fig. 3 shows a snapshot of the GenRate view screen that contains interesting examples. After we set the probability of starting an exon at an arbitrary position ($\kappa$) to achieve a false positive rate of 1%, as described below, GenRate found 16,082 exons in chromosome 4, comprising 1,477 genes.

To determine how many of these predictions are new, we extracted confirmed genes derived from four cDNA and EST databases: Refseq, Fantom II,

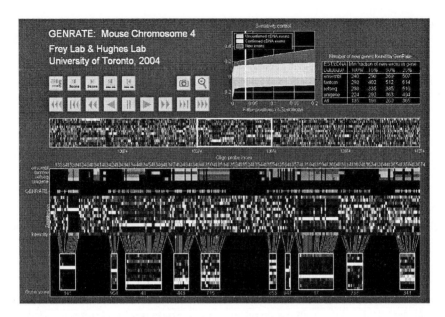

Figure 3: The GenRate MATLAB program shows the genomic expression data and predicted gene structures for a given false positive rate. Genes found by GenRate and genes in cDNA databases (Ensembl, Fantom II, RefSeq, Unigene), are identified by shaded blocks, each of which indicates that the corresponding exon is included in the gene. Each box at the bottom of the screen corresponds to a predicted gene and contains the normalized profiles for exons determined to be part of the gene. The corresponding raw profiles are connected to the box by lines. The rank of each gene is printed below the corresponding box.

Unigene, and Ensembl. The database sequences were mapped to Build 28 of the mouse chromosome using BLAT and only unique mappings with greater than 95% coverage and greater than 90% identity were retained. Probes whose chromosomal location fell within the boundaries of a mapped exon were taken to be confirmed.

The following table shows the number of new genes found by GenRate, relative to the 4 databases and the combined set of databases. Different levels of strictness are used to label each gene as new, ranging from 100% to 25% of the predicted exons as "unconfirmed" by these four cDNA databases. Note that 783 of the 1,477 genes found by GenRate have at least 50% exon overlap with the confirmed genes in all cDNA databases, but that 427 of the genes found by GenRate have no exon overlap with the confirmed genes in all cDNA databases. In the set of genes that are completely new, the minimum, median and maximum gene lengths (in number of exons) are 2, 15 and 66.

(a)                                    (b)

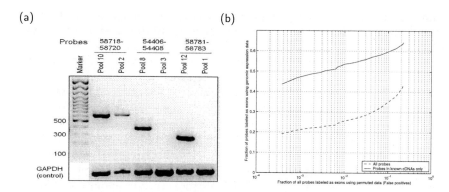

Figure 4: (a) RT-PCR results for three new genes identified by GenRate. The vertical axis corresponds to the weight of the RT-PCR product and the darkness of each band corresponds to the amount of product with that weight. (b) Fraction of predicted positives and positives confirmed by cDNA databases, versus fraction of positives predicted using randomly permuted data (false positives).

| | Number of new genes predicted by GenRate | | | |
|---|---|---|---|---|
| EST/cDNA | Minimum fraction of exons in each gene that are new | | | |
| Database | 100% | 75% | 50% | 25% |
| Ensembl | 1030 | 1157 | 1224 | 1284 |
| Fantom II | 606 | 776 | 902 | 1032 |
| RefSeq | 793 | 853 | 929 | 1032 |
| Unigene | 620 | 724 | 813 | 944 |
| All | 427 | 557 | 656 | 783 |

We are currently performing an extensive set of RT-PCR and Northern blotting experiments to verify the tissue-specific expression and exon structure of novel genes discovered by GenRate. Results on the first three genes tested (selected to have high scores and to overlap with no genes in the four cDNA databases) are shown in Fig. 4a. The two PCR primers for each predicted gene are from different exons separated by thousands of bases in the genome. For each predicted gene, we selected 1 tissue pool with high microarray expression, and 1 tissue pool with low expression. We included the ubiquitously-expressed gene GAPDH to ensure proper RT-PCR amplification. The RT-PCR results confirm the predicted genes and their tissue-specific expression.

An important motivation for approaching this problem using a probability model is that the model should be capable of balancing probabilistic evidence provided by the expression data and the genomic exon arrangements. For example, there are several expression profiles that occur frequently in the data (in particular, profiles where activity in a single tissue pool dominates). If two of these profiles are found adjacent to each other in the data, should they be labeled as a gene? Obviously not, since this event occurs with high probability, *even if the putative exons are arranged in random order.*

To test the statistical significance of the results obtained by GenRate, we constructed a new version of the chromosome 4 data set, where the columns (putative exons) are placed in random order. We then applied GenRate to the permuted data and compared the results to the results obtained on the original data, for varying levels of $\kappa$. Fig. 4b summarizes the results. The $x$-axis shows the fraction of the 63,041 probes that are labeled as exons *in the permuted data*. These can be viewed as false positives, since without knowing the order of the putative exons in the genome, we do not expect to be able to find gene structures with any statistical significance.

The plot shows two curves. The dashed curve is the fraction of all 63,041 probes that are labeled by GenRate as exons in the original (unpermuted) data. The solid curve is the fraction of exons from known genes (see above) that are labeled by GenRate as exons in the original (unpermuted) data. These curves demonstrate that GenRate is able to find predicted gene structures and gene structures compatible with cDNA databases with high statistical significance. For example, at a false positive rate of 1%, 25% of the probes in the original data are labeled as exons, and 54% of the probes in known cDNAs are correctly labeled as exons. This is a reasonable estimate of the proportion of genes that are expected to be expressed in the tissue pools represented in the data set.

## 7    Summary and Future Directions

GenRate is the first generative model that combines a model of genomic arrangement of putative exons with a model of expression patterns, for the purpose of exon and gene discovery. Applied to our microarray data set, GenRate identifies many new genes with a very low false-positive rate. Using RT-PCR to verify 3 new genes predicted by GenRate, we found that all 3 predicted exon sequences (each containing thousands of bases from the genome), are indeed transcriptionally active. We are developing an extension to GenRate, which accounts for alternative splicing, overlapping genes and alternative transcription sites, and applying it to our genome-wide data set of over 12,000,000 measurements. GenRate can be applied to any sequence-based data set, such as whole-genome tiling data.

## Bibliography

1. Frey BJ *et. al.* Full-genome exon profiling in *mus musculus.* In preparation.

2. Zhang W *et. al.* The functional landscape of mouse gene expression. Under review.

3. Shai O *et. al.* Spatial bias removal in microarray images. Univ. Toronto TR PSI-2003-21. 2003.

4. Hild M *et. al.* An integrated gene annotation and transcriptional profiling approach towards the full gene content of the Drosophila genome. *Genome Biol.* 2003;5(1):R3. Epub 2003 Dec 22.

5. Hughes TR *et. al.* Expression profiling using microarrays fabricated by an ink-jet oligonucleotide synthesizer. *Nat. Biotechnol.* 2001 Apr;19(4):342-7.

6. Kapranov P *et. al.* Large-scale transcriptional activity in chromosomes 21 and 22. *Science* 2002 May 3;296(5569):916-9.

7. Nuwaysir EF *et. al.* Gene expression analysis using oligonucleotide arrays produced by maskless photolithography. *Gen. Res.* 2002 Nov;12(11):1749-55.

8. FANTOM Consortium: RIKEN Genome Exploration Research Group Phase I & II Team. Analysis of the mouse transcriptome based on functional annotation of 60,770 full-length cDNAs. Okazaki Y *et. al. Nature.* 2002 Dec 5;420(6915):563-73.

9. Kent WJ. BLAT–the BLAST-like alignment tool. *Genome Res.* 2002 Apr;12(4):656-64.

10. Pen SG *et. al.* Mining the human genome using microarrays of open reading frames. *Nat. Genet.* 2000 Nov;26(3):315-8.

11. Rinn JL *et. al.* The transcriptional activity of human chromosome 22. *Genes Dev.* 2003 Feb 15;17(4):529-40.

12. Shoemaker DD *et. al.* Experimental annotation of the human genome using microarray technology. *Nature* 2001 Feb 15;409(6822):922-7.

13. Yamada K *et. al.* Empirical Analysis of Transcriptional Activity in the *Arabidopsis* Genome. *Science* 2003 October 31;302:842-6.

14. Segal E *et. al.* Genome-wide discovery of transcriptional modules from DNA sequence and gene expression. *Bioinformatics* 19:273-82, 2003.

15. Huber W *et. al.* Variance stabilization applied to microarray data calibration and to quantification of differential expression. *Bioinformatics* 18:S96-S104, 2002.

16. Dempster AP *et. al.* Maximum likelihood from incomplete data via the EM algorithm. *Proc. Royal Stat. Soc.* B-39:1–38, 1977.

17. Kschischang FR *et. al.* Factor graphs and the sum-product algorithm. *IEEE Trans. Infor. Theory* 47(2):498–519, February 2001.

18. Frey BJ. Extending factor graphs so as to unify directed and undirected graphical models. *Proc. UAI* 2003 August.

19. Ostendorf M. *IEEE Trans. Speech & Audio Proc.* 4:360, 1996.

# BAYESIAN JOINT PREDICTION OF ASSOCIATED TRANSCRIPTION FACTORS IN *BACILLUS SUBTILIS*

Y. MAKITA[1,2], M.J.L. DE HOON[1], N. OGASAWARA[3], S. MIYANO[1], AND K. NAKAI[1]

[1] *Human Genome Center, Institute of Medical Science, University of Tokyo 4-6-1 Shirokanedai, Minato-ku, Tokyo 108-8639, Japan*

[2] *School of Technology, Nagoya University Furocho Chikusa-ku Nagoya, Aichi 464-8603, Japan*

[3] *Graduate School of Biological Science, Nara Institute of Science and Technology, 8916-5 Takayama, Ikoma, Nara 630-0101, Japan*

Sigma factors, often in conjunction with other transcription factors, regulate gene expression in prokaryotes at the transcriptional level. Specific transcription factors tend to co-occur with specific sigma factors. To predict new members of the transcription factor regulon, we applied Bayes rule to combine the Bayesian probability of sigma factor prediction calculated from microarray data and the sigma factor binding sequence motif, the motif score of the transcription factor associated with the sigma factor, the empirically determined distance between the transcription start site to the *cis*-regulatory region, and the tendency for specific sigma factors and transcription factors to co-occur. By combining these information sources, we improve the accuracy of predicting regulation by transcription factors, and also confirm the sigma factor prediction. We applied our proposed method to all genes in *Bacillus subtilis* to find currently unknown gene regulations by transcription factors and sigma factors.

## 1. Introduction

In recent years, the genomes of more than one hundred bacteria have been sequenced and the respective coding regions have been found. Inferring the regulatory mechanism of those genes remains a difficult problem. For understanding the regulatory system on a genome-wide scale, gene expression data have been accumulated in microarray experiments for several organisms under various experimental conditions. Due to the complexity of the regulatory network and limits on the experimental accuracy, it is difficult to predict reliably which transcription factor (TF) regulates which genes.

One of the promising methods to predict regulation is supervised learning. However, it is powerful only if a sufficiently large training set is available, which is often not the case. Even in one of the best-studied bacteria, *B. subtilis*, only 20% of known TFs have more than 10 known binding sequences.[1] To address this problem, we consider combining other data under the biological context. In this paper, we focus on the joint prediction of sigma factors and associated TFs.

Sigma factors, which bind to the RNA polymerase complex, recognize specific DNA motifs that are located -35/-10 or -24/-12 basepairs from the transcription start site. For *B. subtilis*, 18 sigma factors are known. SigA is the primary sigma factor and regulates most genes, while secondary sigma factors activate specific groups of genes depending on cellular conditions. For example, the sigma factors SigE, SigF, SigG, SigK, and SigH are related to sporulation, while SigB is involved in stress response, and SigD regulates genes related to flagellar motion and chemotaxis. Similarly, other (non-sigma) TFs are involved in particular cellular processes. As a result, some combinations of sigma factors and TFs are often found to jointly regulate a gene, while other combinations do not occur often. As an extreme example, SigL, which belongs to the sigma54 family of enhancer-dependent sigma factors, can only direct transcription if one of the activating TFs AcoR, BkdR, LevR, RocR, or YplP is present.

Joint prediction of sigma factors and TFs is particulary important for SigA, which regulates about 90% of the *B. subtilis* genes. For differential regulation of these genes, additional TFs are therefore needed.

Previously, our group predicted which sigma factor regulates each gene in *B. subtilis* using 174 microarray data as well as experimentally known sigma factor binding motifs.[2] TF binding sites are typically located near the transcription start site, which can be found from the predicted sigma factor binding site. For example, in *Escherichia coli*, it is known that almost all activators have upstream binding sites near the transcription start site, whereas more than two third of repressors have at least one downstream binding site.[3]

Here, we aim to predict gene regulation by TFs by combining predicted sigma factor binding sites with the biological information of joint regulation by associated TFs, as well as the distribution of TF binding sites near the sigma factor binding site. Additionally, we consider TFs with more than one binding site for a specific gene, which can be used to improve the prediction accuracy.[4,5,6]

## 2. Method

To construct a suitable score function, we applied Bayesian statistics to combine the Bayesian probability of sigma factor prediction calculated from the microarray data and binding motif,[2] the Position Specific Score Matrix (PSSM) of the binding motif of the TF associated with the sigma factor, and the empirically determined distance between the transcription start site to the *cis*-regulatory region. We used the sigma factor predictions[2] to find the transcription start site and to determine which TFs may be expected to co-regulate the gene.

### 2.1. *Sigma factor prediction*

Previously, our group predicted gene regulation by sigma factors using the information of sigma factor binding motif and microarray data.[2] We extend this prediction to the full *B. subtilis* genome and to all sigma factors with known regulated genes, allowing genes to be regulated by more than one sigma factor. From this prediction, we find the Bayesian prior probability $P_{\text{prior}}(\sigma = \sigma^N)$ that a gene is regulated by $\sigma^N$, where $N \in \{A, B, D, E, F, G, H, K, L, W, X\}$.

### 2.2. *Combining sigma factors and transcription factors*

Specific TFs tend to occur with specific sigma factors, as shown in Table 1. In addition to four knowns gene, one more gene was predicted as an enhancer for SigL-regulated genes by our Pfam seach (PF00309)[7].

Table 1. Sigma factors and associated TFs in *B. subtilis*.

| Family | Sigma factor | Function | Cooperative transcription factors* |
|---|---|---|---|
| sigma70 | SigA | Housekeeping Early sporulation | AbrB(21)  AraR(3)  CcpA(40)  CcpC(3) ComA(6)  ComK(40)  CtsR(6)  DegU(15) DinR(6)  FNR(5)  Fur(21)  GlnR(4)  Hpr(6) PerR(7)  PucR(7)  PurR(11)  RocR(4) Spo0A(10)  TnrA(11)  Zur(3) |
| | SigE | Expressed in early mother cell | SpoIIID(4) |
| | SigH | Expressed in postexponential phase; competence and early sporulation | Spo0A(4) |
| | SigK | Expressed in late mother cell | GerE(13) SpoIIID(5) |
| sigma54 | SigL | Degradative enzymes | AcoR(1) BkdR(1) LevR(1) RocR(3) |

* The number in parentheses is the number of genes known to be regulated by each combination of sigma factor and TF. Genes whose sigma factor is unknown experimentally were assigned to the SigA regulon, which contains 90% of the *B.subtilis* genes[10].

From Table 1, we can estimate the probability that a gene is co-regulated by transcription factor $T_i$, given regulation by sigma factor $\sigma^N$:

$$P_{\text{prior}}(T = T_i|\sigma = \sigma^N) = \frac{\text{\# genes regulated by } T_i \text{ and } \sigma^N}{\text{\# genes regulated by } \sigma^N} \tag{1}$$

Some combinations of sigma factor and TF may exist that have not yet been found experimentally. To allow for this possibility, we add a pseudocount[8] $\frac{1}{k+1}\sqrt{\text{\# genes regulated by } T_i}$, where $k$ is the number of TFs under consideration, to the numerator, and $\sqrt{\text{\# genes regulated by } T_i}$ to the denominator. Note that $i$ runs from 0 to $k$, where 0 corresponds to a currently unknown transcription factor.

## 2.3. Motif search

The motif sequences can be described statistically by a position specific score matrix (PSSM) $W_{r,b}$ for each TF.[8] This matrix is the log-odds score of finding a nucleotide $b$ at position $r$ in the binding sequence motif of TF. The log-likelihood that a transcription factor $T_i$ binds a subsequence $S_i$ of the sequence $S$ upstream of a gene is then

$$M_i \equiv \ln \frac{P[S_i| \, T_i \text{ binds } S_i]}{P[S_i|\text{background}]} = \sum_{r=0}^{R-1} W_{r,S_i[r]} \tag{2}$$

where $R$ is the length of the motif. The PSSM was calculated from the known binding motifs of the genes in the regulon of each TF, as listed in the DBTBS database. For the matrix calculation based on $n$ known binding sites, we added $\sqrt{n}$ pseudocounts,[8] using a non-coding region background probability of 0.3185 for A and T, and 0.1815 for C and G.

## 2.4. Relative distance from transcription start site to TF binding site

Using the DBTBS data, we estimated the probability density distribution $f_{\text{dist}}(D_i)$ of the distance $D_i$ from the transcription start site to the binding site of transcription factor $T_i$, measured in base pairs, using a kernel density estimation based on Gaussian kernels.[9] Positive regulators tend to bind in front of the transcription start site, while negative regulators bind at or downstream of the transcription start site. About half of TFs we consider are dual purpose regulators, which regulate some genes positively and others negatively. Those dual TF binding sites are located over a wider range than single regulators.

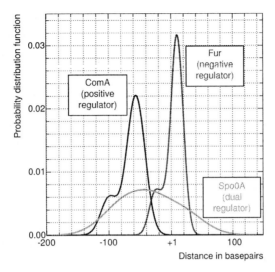

Figure 1. Distribution of the position of the TF binding site with the respect to the transcription start site.

As Figure 1 shows, the graph for positive regulators (ComA) and negative regulators (Fur) each have two peaks. The lower peaks correspond to TFs having two or more binding sites.

### 2.5. *Combining sigma factor and transcription factor prediction*

The joint probability that a gene is regulated by transcription factor $T_i$, $i \in 1..k$ and sigma factor $\sigma^N$ is denoted by $P(\sigma = \sigma^N, T = T_i)$. Here, $T_0$ corresponds to an unknown TF. For deriving the posterior joint probability, we combined the following three elements: the prior joint probability $P_{\text{prior}}(\sigma = \sigma^N, T = T_i)$, the maximum PSSM score in each promoter sequence $M_i$ calculated for $T_i$, and the distance $D_i$ between the transcription start site and the predicted TF binding site. $M_i$ and $D_i$ are calculated from the sequence region $S$ upstream of the gene. The Bayesian posterior probability that a gene is regulated by sigma factor $\sigma^N$ and transcription factor $T_i$, given the upstream sequence $S$, can be calculated as

$$
\begin{aligned}
&P\left(\sigma = \sigma^N, T = T_i | S\right) \\
&= \frac{P(S | \sigma = \sigma^N, T = T_i) P_{\text{prior}}(\sigma = \sigma^N, T = T_i)}{\sum_U \sum_j P(S | \sigma = \sigma^U, T = T_j) P_{\text{prior}}(\sigma = \sigma^U, T = T_j)},
\end{aligned} \tag{3}
$$

where in the denominator $U$ is summed over sigma factors A, B, D, E, F, G, H, K, L, X, and W. The prior probability $P_{\text{prior}}(\sigma = \sigma^N, T = T_i)$ is calculated as $P_{\text{prior}}(\sigma = \sigma^N)P_{\text{prior}}(T = T_i | \sigma = \sigma^N)$, as described above.

$P(S|\sigma = \sigma^N, T = T_i)$ is the conditional probability that an upstream sequence $S$ is generated, given that $\sigma^N$ and $T_i$ regulate the gene. The upstream sequence $S$ consists of the binding site $S_i$, described by the PSSM, and the remaining sequence $S \backslash S_i$. We can then decompose $P(S|\sigma = \sigma^N, T = T_i)$ into three parts:

$$P(S|\sigma = \sigma^N, T = T_i) = P(S_i | T = T_i) \cdot P(S \backslash S_i | \text{background}) \cdot f_{\text{dist}}(D_i). \quad (4)$$

The third factor is the probability that $S_i$ is generated at a distance $D_i$ from the transcription start site (Section 2.4). Here, the predicted position of the transcription start site depends on the sigma factor $\sigma^N$, as described previously.[2]

Dividing by the background probability yields

$$\frac{P(S|\sigma = \sigma^N, T = T_i)}{P(S|\text{background})} = \frac{P(S_i | T = T_i)}{P(S_i | \text{background})} f_{\text{dist}}(D_i) = e^{M_i} f_{\text{dist}}(D_i), \quad (5)$$

where $M_i$ is the maximum value of the PSSM score for transcription factor $T_i$ over the upstream region $S$. For an unknown transcription factor ($T = T_0$), however, this ratio is equal to unity.

Note that for $f_{\text{dist}}(D_i)$ uniform, this reduces to $e^{M_i}/D_{\max}$, where $D_{\max}$ is the size of the upstream region $S$ that we search. This then corresponds to the Bonferoni correction for multiple comparisons.

By combining these equations, we find the following expression for the posterior probability:

$$P[\sigma = \sigma^N, T = T_i | S] = \frac{\exp\left[\text{score}\left(\sigma^N, T_i\right)\right]}{\sum_U \sum_{j=0}^{k} \exp\left[\text{score}\left(\sigma^U, T_j\right)\right]}, \quad (6)$$

where we defined the score functions

$$\text{score}(\sigma^N, T_i) \equiv \ln P_{\text{prior}}(T = T_i | \sigma = \sigma^N) + \ln P_{\text{prior}}(\sigma = \sigma^N)$$
$$+ M_i + \ln f_{\text{dist}}(D_i), \quad (7)$$

while we drop the last two terms if $i = 0$. For genes that have more than two binding sites for the same transcription factor $T_i$, we add terms $(M_i + \ln f_{\text{dist}}(D_i))$ correspondingly.

## 2.6. *Example calculation*

We calculated the Bayesian posterior probability in Eq. (6) that the gene *rocA* is regulated by each sigma factor and by one of the TFs AcoR, BkdR, LevR, RocR, or on unknown TF. Table 2 shows that the (SigL, RocR) combination is by far the most likely. From biological experiments, *rocA* is known to be regulated by SigL and RocR, which serves as the transcriptional activator of arginine utilization operons.

## 3. Validation

### 3.1. *The sigma factor prediction aids in the TF prediction*

To verify the validity of combining the TF prediction with the sigma factor prediction, we examined the contribution of each term in Eq. (7). To assess the effect of using the sigma factor prediction for the TF prediction, we compare the two scores $M_i + \ln P(T = T_i | \sigma = \sigma^N)$ and $M_i$ (Table 3).

The negative dataset consists of genes regulated by sigma factors whose regulons do not contain any genes that are known to be regulated by the TF. The positive dataset are the genes known to be regulated by the TF. The specificity is given by $TP/(TP + FP)$ and the sensitivity is given by $TP/(TP + FN)$, where TP is true positive, FP is false positive, and FN is false negative.

Furthermore, the predicted sigma factor binding site $P_{\mathrm{prior}}(\sigma = \sigma^N)$ in Eq. (7) allows us to search for the TF motif nearby on the genome, as represented by the term in $f_{\mathrm{dist}}(D_i)$ in Eq. (7). We show the effect of including this term in Table 4.

As shown in these tables, both the sigma factor information and the transcription start sites greatly improve the specificity and the sensitivity of the TF prediction. The biological knowledge that specific sigma factors and TFs tend to co-occur is particularly informative, as shown in Table 3.

### 3.2. *The TF prediction aids in the sigma factor prediction*

We calculate the posterior probability that a gene is regulated by a specific sigma factor by summing Eq. (6) over $T_i$. As shown in Table 5, this posterior probability is more accurate than the prior probability in predicting sigma factors. While the prior probability already gives a very accurate prediction of sigma factor regulation, the accuracy of the posterior probability is even higher. We note that for unknown genes, the sigma factor prediction may be less accurate due to uncertainties in the operon structure.[2,11]

Table 2. Probability that *rocA* is regulated by various combinations of a sigma factor and TF.

| Sigma | TF | $M_i$ | $\ln(f_{\mathrm{dist}}(D_i))$ | $\ln(P_{\mathrm{prior}}(T = T_i | \sigma = \sigma^N))$ | $\ln(P_{\mathrm{prior}}(\sigma = \sigma^N))$ | Score | Probability |
|---|---|---|---|---|---|---|---|
| | AcoR | 6.41 | -6.45 | -5.54 | -2.04 | -7.62 | 0.000 |
| | BkdR | 4.53 | -4.98 | -5.54 | -2.04 | -8.03 | 0.000 |
| sigA | LevR | 3.21 | -4.93 | -5.54 | -2.04 | -9.3 | 0.000 |
| | RocR | 30.5 | -19.4 | -5.54 | -2.04 | 3.52 | 0.000 |
| | ND | - | - | -5.54 | -2.04 | -7.58 | 0.000 |
| | AcoR | 6.41 | -10.92 | -4.69 | -1.93 | -11.13 | 0.000 |
| | BkdR | 3.77 | -5.06 | -4.69 | -1.93 | -7.91 | 0.000 |
| sigB | LevR | 4.08 | -5.2 | -4.69 | -1.93 | -7.74 | 0.000 |
| | RocR | 30.5 | -13.74 | -4.69 | -1.93 | 10.14 | 0.000 |
| | ND | - | - | -4.69 | -1.93 | -6.62 | 0.000 |
| | AcoR | 6.41 | -8.23 | -3.63 | -6.55 | -12 | 0.000 |
| | BkdR | 4.53 | -7.01 | -3.63 | -6.55 | -12.66 | 0.000 |
| sigD | LevR | 3.48 | -5.42 | -3.63 | -6.55 | -12.12 | 0.000 |
| | RocR | 30.5 | -10.27 | -3.63 | -6.55 | 10.05 | 0.000 |
| | ND | - | - | -3.63 | -6.55 | -10.18 | 0.000 |
| | AcoR | 6.41 | -5.61 | -4.82 | -2.17 | -6.18 | 0.000 |
| | BkdR | 4.53 | -5.21 | -4.82 | -2.17 | -7.66 | 0.000 |
| sigE | LevR | 3.21 | -5.87 | -4.82 | -2.17 | -9.64 | 0.000 |
| | RocR | 30.5 | -22.45 | -4.82 | -2.17 | 1.06 | 0.000 |
| | ND | - | - | -4.82 | -2.17 | -6.99 | 0.000 |
| | AcoR | 6.41 | -10.78 | -3.57 | -5.14 | -13.08 | 0.000 |
| | BkdR | 3.77 | -5.08 | -3.57 | -5.14 | -10.03 | 0.000 |
| sigF | LevR | 4.08 | -5.24 | -3.57 | -5.14 | -9.87 | 0.000 |
| | RocR | 30.5 | -13.61 | -3.57 | -5.14 | 8.17 | 0.000 |
| | ND | - | - | -3.57 | -5.14 | -8.72 | 0.000 |
| | AcoR | 6.41 | -8.45 | -4.03 | -4.12 | -10.18 | 0.000 |
| | BkdR | 4.53 | -7.61 | -4.03 | -4.12 | -11.23 | 0.000 |
| sigG | LevR | 3.48 | -6.08 | -4.03 | -4.12 | -10.74 | 0.000 |
| | RocR | 30.5 | -10.3 | -4.03 | -4.12 | 12.06 | 0.003 |
| | ND | - | - | -4.03 | -4.12 | -8.14 | 0.000 |
| | AcoR | 0.76 | -5.84 | -4.59 | -4.14 | -13.82 | 0.000 |
| | BkdR | 4.53 | -5.5 | -4.59 | -4.14 | -9.71 | 0.000 |
| sigH | LevR | 3.48 | -4.66 | -4.59 | -4.14 | -9.92 | 0.000 |
| | RocR | 30.5 | -17.38 | -4.59 | -4.14 | 4.38 | 0.000 |
| | ND | - | - | -4.59 | -4.14 | -8.74 | 0.000 |
| | AcoR | 0 | -4.83 | -4.13 | -5.02 | -13.98 | 0.000 |
| | BkdR | 3.77 | -5.66 | -4.13 | -5.02 | -11.04 | 0.000 |
| sigK | LevR | 4.08 | -5.98 | -4.13 | -5.02 | -11.05 | 0.000 |
| | RocR | 30.5 | -12.98 | -4.13 | -5.02 | 8.37 | 0.000 |
| | ND | - | - | -4.13 | -5.02 | -9.15 | 0.000 |
| | AcoR | 6.41 | -10.46 | -1.74 | -0.57 | -6.36 | 0.000 |
| | BkdR | 4.53 | -6.02 | -1.74 | -0.57 | -3.8 | 0.000 |
| sigL | LevR | 3.48 | -5.06 | -1.74 | -0.57 | -3.89 | 0.000 |
| | RocR | 30.5 | -10.86 | -1.22 | -0.57 | 17.85 | 0.996 |
| | ND | - | - | -2.85 | -0.57 | -3.42 | 0.000 |
| | AcoR | 6.41 | -7.47 | -2.58 | -9.36 | -13 | 0.000 |
| | BkdR | 4.53 | -5.53 | -2.58 | -9.36 | -12.94 | 0.000 |
| sigX | LevR | 3.48 | -5.03 | -2.58 | -9.36 | -13.49 | 0.000 |
| | RocR | 30.5 | -12.53 | -2.58 | -9.36 | 6.03 | 0.000 |
| | ND | - | - | -2.58 | -9.36 | -11.94 | 0.000 |
| | AcoR | 6.41 | -8.51 | -3.99 | -7.82 | -13.91 | 0.000 |
| | BkdR | 4.53 | -7.74 | -3.99 | -7.82 | -15.01 | 0.000 |
| sigW | LevR | 2.96 | -5.71 | -3.99 | -7.82 | -14.55 | 0.000 |
| | RocR | 30.5 | -10.45 | -3.99 | -7.82 | 8.24 | 0.000 |
| | ND | - | - | -3.99 | -7.82 | -11.81 | 0.000 |

ND: TF unknown case.

Table 3.  The effect of sigma factor information on the TF prediction.

| TF | sigma | $M_i + \ln P(T = T_i \mid \sigma = \sigma^N)$ | | | | | $M_i$ | | | | |
|---|---|---|---|---|---|---|---|---|---|---|---|
| | | TP | FP | FN | SP | SN | TP | FP | FN | SP | SN |
| Spo0A | A,H | 8 | 0 | 0 | 100.0% | 100.0% | 5 | 3 | 3 | 62.5% | 62.5% |
| SpoIIID | E,K | 9 | 0 | 0 | 100.0% | 100.0% | 3 | 8 | 6 | 27.3% | 33.3% |
| GerE | K | 13 | 0 | 0 | 100.0% | 100.0% | 7 | 13 | 6 | 35.0% | 53.8% |
| SigL | L | 5 | 0 | 0 | 100.0% | 100.0% | 5 | 0 | 0 | 100.0% | 100.0% |
| Total | | 35 | 0 | 0 | 100.0% | 100.0% | 20 | 24 | 15 | 45.5% | 57.1% |

TP true positive, FP false positive, FN false negative, SP specificity, and SN sensitivity.

Table 4.  The effect of transcription start site information on TF prediction.

| TF | sigma | $M_i + \ln f_{\text{dist}}(D_i)$ | | | | | $M_i$ | | | | |
|---|---|---|---|---|---|---|---|---|---|---|---|
| | | TP | FP | FN | SP | SN | TP | FP | FN | SP | SN |
| Spo0A | A,H | 6 | 2 | 2 | 75.0% | 75.0% | 5 | 3 | 3 | 62.5% | 62.5% |
| SpoIIID | E,K | 5 | 6 | 4 | 45.5% | 55.6% | 3 | 8 | 6 | 27.3% | 33.3% |
| GerE | K | 7 | 2 | 6 | 77.8% | 53.8% | 7 | 13 | 6 | 35.0% | 53.8% |
| SigL | L | 5 | 0 | 0 | 100.0% | 100.0% | 5 | 0 | 0 | 100.0% | 100.0% |
| Total | | 23 | 10 | 12 | 69.7% | 65.7% | 20 | 24 | 15 | 45.5% | 57.1% |

## 4. Result

We applied our proposed method to jointly predict sigma factor and TFs for all genes in *B. subtilis* in order to find currently unknown gene regulations. Table 6 shows some predicted combinations for which a high posterior probability was found. For many proteins, the function is presently unknown. The sigma/TF prediction can suggest the cellular function of those proteins.

CcpA is one of the global repressor of the carbon catabolite repressors which bind to CRE site (TGWAANCGGNTNWCA)[10]. Our prediction shows that CcpA acts on some genes related to sugar metabolism (*sacP*, *fruR*, *yojA*) and dehydrogenase (*yrbE*), which is consistent with the known function of CcpA.

The sporulation genes, *spoIIP* and *spoIID* are known to be regulated by SigE. Both genes are required for complete dissolution of the asymmetric

Table 5.  The accuracy of the sigma factor prediction.

| sigma | prior | | | | | posterior | | | | |
|---|---|---|---|---|---|---|---|---|---|---|
| | TP | FP | FN | SP | SN | TP | FP | FN | SP | SN |
| SigE | 53 | 3 | 2 | 94.6% | 96.4% | 53 | 2 | 2 | 96.4% | 96.4% |
| SigH | 33 | 5 | 5 | 86.8% | 86.8% | 35 | 5 | 3 | 87.5% | 92.1% |
| SigK | 24 | 1 | 1 | 96.0% | 96.0% | 25 | 1 | 0 | 96.2% | 100.0% |
| SigL | 5 | 0 | 0 | 100.0% | 100.0% | 5 | 0 | 0 | 100.0% | 100.0% |
| Total | 115 | 9 | 8 | 92.7% | 93.5% | 118 | 8 | 5 | 93.7% | 95.9% |

Table 6.  Newly predicted gene regulations by TFs and sigma factors in *B. subtilis*.

| Sigma | RG | TF | posterior Prob. | | Function |
|-------|-----|------|-----------------|---|----------|
| SigA | *sacP* | CcpA | 0.997 | | PTS sucrose-specific enzyme IIBC component |
| | *yqgQ* | CcpA | 0.980 | * | unknown |
| | *yrzF* | CcpA | 0.976 | | unknown |
| | *yvfH* | CcpA | 0.972 | | unknown; similar to L-lactate permease |
| | *yvfK* | CcpA | 0.967 | | unknown; similar to maltose/maltodextrin-binding protein |
| | *yngI* | CcpA | 0.953 | | unknown; similar to long-chain acyl-CoA synthetase |
| | *yngI* | CcpA | 0.953 | | unknown; similar to long-chain acyl-CoA synthetase |
| | *ycsA* | CcpA | 0.947 | | unknown; similar to 3-isopropylmalate dehydrogenase |
| | *opuE* | CcpA | 0.916 | * | proline transporter |
| | *yrpD* | CcpA | 0.912 | | unknown; similar to unknown proteins from B. subtilis |
| | *ywqC* | CcpA | 0.904 | | unknown; similar to capsular polysaccharide biosynthesis |
| | *yvfI* | CcpA | 0.901 | | unknown; similar to transcriptional regulator (GntR family) |
| | *glcR* | ComK | 0.985 | | transcriptional repressor involved in the expression of the phosphotransferase system |
| | *aadK* | ComK | 0.971 | | aminoglycoside 6-adenylyltransferase |
| | *yufL* | ComK | 0.946 | | unknown; similar to two-component sensor histidine kinase [YufM] |
| | *yuiD* | ComK | 0.903 | | unknown; similar to unknown proteins |
| | *glmS* | CtsR | 0.968 | | L-glutamine-D-fructose-6-phosphate amidotransferase |
| | *yozM* | DinR | 0.949 | | unknown |
| | *ypoP* | Fur | 0.958 | | unknown; similar to transcriptional regulator (MarR family) |
| | *yodE* | TnrA | 0.938 | | unknown; similar to unknown proteins |
| SigE | *spoIIP* | SpoIIID | 0.961 | * | required for dissolution of the septal cell wall |
| | *spoIID* | SpoIIID | 0.960 | * | required for complete dissolution of the asymmetric septum |
| | *cwlD* | SpoIIID | 0.930 | * | N-acetylmuramoyl-L-alanine amidase (germination) |
| | *ylbJ* | SpoIIID | 0.910 | * | unknown; similar to unknown proteins |
| | *ytvA* | SpoIIID | 0.873 | | unknown; similar to protein kinase |
| | *yurH* | SpoIIID | 0.857 | | unknown; similar to N-carbamyl-L-amino acid amidohydrolase |
| | *greA* | SpoIIID | 0.849 | | transcription elongation factor |
| | *yugP* | SpoIIID | 0.827 | | unknown; similar to unknown proteins |
| | *yjkB* | SpoIIID | 0.813 | | unknown; similar to amino acid ABC transporter |
| | *ytxC* | SpoIIID | 0.754 | * | unknown; similar to unknown proteins |
| | *yqfZ* | SpoIIID | 0.745 | * | unknown; similar to unknown proteins |
| | *spoVE* | SpoIIID | 0.687 | * | required for spore cortex peptidoglycan synthesis |
| | *yugO* | SpoIIID | 0.671 | | unknown; similar to potassium channel protein |
| | *yqeW* | SpoIIID | 0.664 | | unknown; similar to Na+/Pi cotransporter |
| SigH | *yvyD* | Spo0A | 0.667 | * | general stress protein under dual control of sigB and sigH |
| SigK | *nucB* | GerE | 0.887 | | sporulation-specific extracellular nuclease |
| | *ytkC* | GerE | 0.851 | | unknown; similar to autolytic amidase |
| | *ywjE* | GerE | 0.820 | | unknown; similar to cardiolipin synthetase |
| | *ypgA* | GerE | 0.808 | | unknown; similar to unknown proteins |
| SigL | *yokK* | BkdR | 0.416 | | unknown |

* The sigma factor has been determined experimentally. In all cases shown in this table, the experimentally determined sigma factor agrees with the computational prediction. All predicted regulations by TFs shown in this table are currently unknown.

septum cell wall. We found the SpoIIID binding motif at +18 and +3 for *spoIIP* and at +24 for *spoIID*. From the location of the binding site, we infer that those genes might be negatively regulated. For the SigE-dependent asparagine synthetase gene *yisO*, we found three SpoIIID binding sites in the promoter region.

GerE is a transcriptional regulator required for the expression of late spore coat genes. It is predicted to regulate membrane phospholipid cardiolipin (*ywjE*) and permease (*yecA*). Since in addition it is known that GerE regulates N-acetylmuramoyl-L-alanine amidase, we expect the prediction for *ytkC*, which is similar to autolytic amidase, to be correct.

In *E. coli*, 17 operons are known to be regulated by SigL[12]. In *B. subtilis*, only six operons are known to be regulated by SigL. Whereas we may expect currently unknown SigL-regulated genes to exist in *B. subtilis*, our result suggests that there are few additional SigL regulated genes in the *B. subtilis* genome.

## 5. Discussion

Our result shows that the joint prediction of TFs is a powerful way both to confirm the sigma prediction and to predict new members of the TF regulon. As the joint prediction of sigma factors and TFs is a supervised learning method, it can make better use of known biological facts than unsupervised methods. This method can also detect genes regulated by two or more different sigma factors. For example, spoIVCB is initially transcribed under the direction of SigE acting in conjunction with SpoIIID. Later in sporulation, SigK-mediated transcription of spoIVCB is repressed by GerE. In our method, we can calculate the probability that spoIVCB is regulated by SigK with GerE and by SigE with SpoIIID separately. This method can also be applied to other organisms such as *E. coli*, cyanobacteria and yeast, for which some regulatory relations are known.

## Acknowledgments

We thank Seiya Imoto for his kind advice on the statistical analysis. This research was supported by Grant-in-Aid for Scientific Research on Priority Areas and JSPS Fellow ship of the Ministry of Education, Science, Sports and Culture.

# References

1. Y. Makita, M. Nakao, N. Ogasawara, and K. Nakai. DBTBS: Database of transcriptional regulation in *Bacillus subtilis* and its contribution to comparative genomics. *Nucleic Acids Res.*, 1:32 Database issue:D75-7, 2004. http://dbtbs.hgc.jp.
2. M.J.L. de Hoon, Y. Makita, S. Imoto, K. Kobayashi, N. Ogasawara, K. Nakai and S. Miyano, Predicting gene regulation by sigma factors in *Bacillus subtilis* from genome-wide data. *Bioinformatics*, 20 Suppl 1:I102-I108, 2004.
3. M.M. Babu and S.A. Teichmann, Functional determinants of transcription factors in *Escherichia coli*: protein families and binding sites. *TRENDS in Genetics*, 19(2):75-79, 2003.
4. M.L. Bulyk, A.M. McGuire, N. Masuda, and G.M. Church. A motif co-occurrence approach for genome-wide prediction of transcription-factor-binding sites in *Escherichia coli*. *Genome Res.*, 14(2):201-8, 2004
5. S. Sinha, M. Tompa. Discovery of novel transcription factor binding sites by statistical overrepresentation. *Nucleic Acids Res.* 30(24):5549-60, 2002.
6. E. Segal and S. Sharan. A Discriminative Model for Identifying Sparial *cis*-Regulatory Modules. In Proc. 8th Inter. Conf. on Research in Computational Molecular Biology (RECOMB), 2004.
7. A. Bateman, L. Coin, R. Durbin, R.D. Finn, V. Hollich, S. Griffiths-Jones, A. Khanna, M. Marshall, S. Moxon, E.L.L. Sonnhammer, D.J. Stud holme, C. Yeats, and S.R. Eddy. The Pfam Protein Families Database. *Nucleic Acids Res.*, 1:32 Database Issue:D138-141 2004.
8. R. Durbin, S. Eddy, A. Krogh, G. Mitchison. Biological Sequence Analysis: Probabilistic Models of Proteins and Nucleic Acids, Cambridge University Press, Cambridge, UK. 1998.
9. B.W. Silverman. Density Estimation for Statistics and Data Analysis. Chapman and Hill, London, 1986.
10. A.L. Sonenshein, J.A. Hoch, and R. Losick. *Bacillus subtilis* and its closest relatives: From genes to cells. ASM Press, Washington, DC, 2001.
11. M.J.L. de Hoon, S. Imoto, K. Kobayashi, N. Ogasawara, and S. Miyano. Predicting the operon structure of *Bacillus subtilis* using operon length, intergene distance, and gene expression information. PSB 2004:276-87.
12. L. Reitzer and B.L. Schneider. Metabolic context and possible physiological themes of sigma(54)-dependent genes in *Escherichia coli*. *Microbiol Mol Biol Rev.*, 65(3):422-44, 2001.

# MODULEFINDER: A TOOL FOR COMPUTATIONAL DISCOVERY OF *CIS* REGULATORY MODULES

ANTHONY A. PHILIPPAKIS[†,1,3,4], FANGXUE SHERRY HE[†,1,3],
MARTHA L. BULYK[*,1,2,3,4]

[1]*Division of Genetics, Department of Medicine,* [2]*Department of Pathology,*
[3]*Harvard/MIT Division of Health Sciences & Technology (HST), and*
[4]*Harvard University Biophysics Program*
*Brigham & Women's Hospital and Harvard Medical School*
*Boston, MA 02115*
*Email: {aphilippakis, mlbulyk}@receptor.med.harvard.edu, sherryhe@mit.edu*

Regulation of gene expression occurs largely through the binding of sequence-specific transcription factors (TFs) to genomic binding sites (BSs). We present a rigorous scoring scheme, implemented as a *C* program termed "ModuleFinder", that evaluates the likelihood that a given genomic region is a *cis* regulatory module (CRM) for an input set of TFs according to its degree of: (1) homotypic site clustering; (2) heterotypic site clustering; and (3) evolutionary conservation across multiple genomes. Importantly, ModuleFinder obtains all parameters needed to appropriately weight the relative contributions of these sequence features directly from the input sequences and TFBS motifs, and does not need to first be trained. Using two previously described collections of experimentally verified CRMs in mammals and in fly as validation datasets, we show that ModuleFinder is able to identify CRMs with great sensitivity and specificity.

## 1. Introduction

Recent technological advances have enabled both the sequencing of a large number of genomes and the generation of expansive gene expression datasets. Still, little is known about how these gene expression patterns are precisely regulated through the binding of sequence-specific transcription factors (TFs) to their DNA binding sites (BSs). Of particular interest is the organization of TF binding sites (TFBSs) into *cis* regulatory modules (CRMs) that coordinate the complex spatio-temporal patterns of gene expression, and to use that information to identify the CRMs themselves. Mapping TFs to their target CRMs, however, is significantly complicated in higher eukaryotic genomes by the large proportion of non-protein-coding sequence. Since a typical TFBS can be as short as ~5 base pairs (bp), matches to its motif occur frequently by chance alone, with many of these occurrences presumably not acting to modulate gene expression. Therefore, a central challenge that must be overcome is distinguishing functional TFBSs from spurious motif matches.

---

[†] These authors contributed equally.
[*] Corresponding author.

To date, three indicators have been used to identify functional TFBSs. First, functional BSs for some TFs tend to occur in clusters, with multiple BSs occurring in close proximity (homotypic clustering). Second, searching for clusters containing BSs for 2 or more TFs that are believed to co-regulate can enrich for likely CRMs (heterotypic clustering). Finally, functional TFBSs are frequently conserved across evolutionarily divergent organisms[1]. Cross-species sequence conservation in particular has enormous potential for filtering sequence space, as many genomes have recently been sequenced, and many more are slated to be sequenced (http://www.genome.gov/10002154). The discriminatory power of phylogenetic footprinting for identifying *cis* regulatory elements is therefore expected to continue to increase through the use of more genomes[2-6]. In order to appropriately incorporate information on conservation across multiple genomes, however, a measure of TFBS conservation is required that weights each alignment genome according to its evolutionary distance not only from the query genome, but also relative to the other alignment genomes. For example, given a candidate TFBS in the human genome, observing conservation in chicken should be weighted more heavily than conservation in mouse, as mouse is evolutionarily closer to human. Moreover, if the candidate site were also conserved in rat, then this additional conservation should be weighted only slightly, given the evolutionary proximity of mouse and rat.

While numerous groups have developed approaches for the prediction of CRMs, none is optimized for practical applications. Specifically, many approaches[7-9] have been based on binary scoring schemes, wherein all regions containing a threshold number of occurrences for a given combination of TFBSs are returned. These approaches suffer from the limitation that they do not prioritize among the predictions, an important feature for experimentalists as only a limited number of candidate CRMs can feasibly be validated. Additionally, the threshold value determined in any given biological system is unlikely to be generalizeable from one set of TFs and CRM type to another; thus, the appropriate discriminatory criterion must be re-discovered with each application. Alternatively, among existing continuous scoring schemes, many require large training sets[10,11]. Such approaches cannot be applied to a system in which there are only a handful of known examples, as is frequently the case in practical applications. Finally, among approaches that employ continuous scoring schemes and do not require training[12-15], most do not systematically integrate BS clustering and conservation. We are aware of only one other approach that combines all three indicators[16], but it is computationally rather slow and requires the user to specify a single sequence window size for the search. Since CRMs are known to vary greatly in size, a scoring scheme is needed that evaluates clustering and conservation over windows of varying sizes[15].

We have developed a statistically rigorous scoring scheme that for any given genomic region integrates into a single score the degree of: (1) homotypic clustering; (2) heterotypic clustering; and (3) evolutionary conservation across multiple genomes. Similar to programs such as BLAST[17], our score is an objective measure of the statistical significance of the observed degree of clustering and conservation that is independent of the genome and TFBSs under consideration. Thus, the scoring scheme obtains all parameters needed to appropriately weight the relative contribution of each input alignment and TFBS motif directly from the sequences and motifs themselves, and so does not need to first be trained. We have implemented this scoring scheme as a $C$ program called "ModuleFinder," that is algorithmically efficient and has an intuitive interface. Using two previously described collections of experimentally verified CRMs (mammalian skeletal muscle[18] and *D. melanogaster* segmentation genes[7]), we show that ModuleFinder is able to identify CRMs with ~95% sensitivity and ~95% specificity.

## 2. Methods

Methods that evaluate the overall degree of conservation for a given region have been successful in identifying *cis* regulatory elements in metazoan genomes[2,6]; they do not, however, necessarily identify the CRMs through which a given set of TFs exert their regulatory roles (i.e., the TFs' "target" CRMs). Since our ultimate goal is to identify candidate CRMs that are bound by a given set of TFs, we have developed a scoring scheme that specifically considers the conservation of a particular set of TFBSs comprising a given transcriptional regulatory model. For this, we developed a novel statistical framework that builds on earlier work. Blanchette *et al.* stated the substring parsimony problem and presented a rigorous and efficient algorithmic procedure for solving it[19]; this model was applied to the identification of candidate DNA motifs. Moses *et al.* used mixture models to evaluate conservation within a tree, and applied it to the identification of candidate DNA motifs from sets of co-expressed genes[20]; this was similar to an approach given by Prakash *et al.*[21] Here we present a related approach for identifying candidate CRMs from input TFBS motifs.

### 2.1. Scoring Scheme

We define a *word* to be a short sequence on the DNA alphabet {A,C,G,T}, and a *motif* to be a collection of words all of the same length. ModuleFinder takes as input a collection of arbitrarily many motifs $\{m_1...m_m\}$, where each motif $m_i$ is composed of arbitrarily many words of length $l_i$. It also takes as input a set of sequences $G = \{g_1,...g_n\}$ corresponding to genomic regions that are to be searched for instances of these motifs, as well as two sets of genomic sequences, $A = \{a_1,...,a_n\}$ and $B = \{b_1,...,b_n\}$, extracted from evolutionarily divergent organisms and then aligned to the sequences of $G$. Here, we primarily illustrate the scoring scheme for the case of two alignment genomes, but include comments on the extension to fewer or more alignments. For any $g_j$, let $g_{j,k}$ denote the base at the $k$th position and $(g_{j,k}...g_{j,k+l})$ denote the subsequence of length $l$ beginning at position $k$. If there is a match to a given motif $m_i$ at position $k$ of sequence $g_j$, we define it to be *conserved in A* (respectively, $B$), if it is true that the subsequence $(a_{j,k}...a_{j,k+l})$ (respectively, $(b_{j,k}...b_{j,k+l})$) is also a word in motif $m_i$. Note that we are not assuming that $g_{j,k}...g_{j,k+l} = a_{j,k}...a_{j,k+l}$, but merely that they are both words in $m_i$.

Our basic approach is to scan each sequence in $G$ with a series of nested windows (i.e., overlapping windows of differing sizes). In each window we count the number of occurrences of each motif and the number of these that are conserved in $A$ and $B$. We then evaluate the likelihood of observing this number of matches and conserved matches under the appropriate null hypothesis, and return those windows that are statistically significant. Specifically, let $X = (X_1,..., X_m)$ be the vector whose components indicate the number of occurrences for each motif individually in a given window, and let $Y = (Y_1,..., Y_m)$ and $Z = (Z_1,..., Z_m)$ be the corresponding vectors indicating that $Y_i$ and $Z_i$ out of $X_i$ occurrences are conserved in $A$ and $B$, respectively. The window score is obtained by finding the probability of observing $(X,Y,Z)$. This quantity will vary according to the likelihood of conservation in $A$ and/or $B$, the motif frequency, and the window width. Thus, this probability can be represented by:

$$P_{\Gamma,\alpha,w}(X,Y,Z) \tag{1}$$

where $\Gamma$ parameterizes conservation likelihood, $\alpha$ parameterizes motif frequencies, and $w$ is the window width. Observe that:

$$P_{\Gamma,\alpha,w}(X,Y,Z) = P_\Gamma(Y,Z\,|\,X)P_{\alpha,w}(X) \tag{2}$$

where the relevant parameters can be split between terms in the Markov decomposition, as $P_{\alpha,w}(X)$ is unaffected by conservation likelihood, and $P_\Gamma(Y,Z\,|\,X)$ is unaffected by motif frequency and window size.

For a single motif $m_i$, the term $P_{\alpha_i,w}(X_i)$ of Eq. (2) is the likelihood of observing $X_i$ occurrences under the null hypothesis that the motif matches are distributed at random. This has been proved to be well-approximated by a Poisson distribution, provided the motif occurs infrequently and the words comprising it do not exhibit extensive self-overlap.[22] Thus, $P_{\alpha_i,w}(X_i) = e^{-\lambda_i}(\lambda_i^{X_i} / X_i!)$, where $\lambda_i = \alpha_i * w$. The value of $\alpha_i$ will itself be determined by both the words comprising $m_i$, as well as genomic word frequencies. To obtain it, we estimate the frequency of each word in $m_i$ by a seventh order Markov approximation based on genomic word frequencies, and then sum these frequencies for all words in the motif.

For multiple motifs, the joint probability is given by assuming independence:

$$P_{\alpha,w}(X_1,...X_m) = \prod_{i=1}^{m} \left(P_{\alpha_i,w}(X_i)\right)$$

This is a simplifying assumption to make the computation tractable; the error in this approximation has, however, been proved to be bounded[22].

The computation of the second term of Eq. (2), $P_\Gamma(Y,Z\,|\,X)$, is complicated by two factors. First, the score must reflect not only the evolutionary distances of $A$ and $B$ to $G$, but also the distances of $A$ and $B$ to each other. Thus, $\Gamma$ must re-parameterize $P_\Gamma(Y,Z\,|\,X)$ so that it becomes smaller as $A$ and $B$ grow more distant from $G$, and as the correlation between $A$ and $B$ decreases. Second, the quantity $P_\Gamma(Y,Z\,|\,X)$ will depend not only on the phylogeny of $A$, $B$ and $G$, but also on the degeneracy of the motifs $m_i$. Since we have defined a given motif match to be conserved in $A$ or $B$ if there is a motif occurrence (but not necessarily an exact word match) at the same position in these aligned sequences, a more degenerate motif has a greater likelihood of being conserved.

We account for these difficulties as follows. Define $\Gamma^1_{A,B}$ to be the covariance matrix representing the relative proportions of $A$ and $B$ that can be aligned against $G$; thus, $\Gamma^1_{0,0}$ gives the proportion of sequence in $G$ for which neither $A$ nor $B$ could be aligned, $\Gamma^1_{1,0}$ and $\Gamma^1_{0,1}$ give the proportion for which either $A$ or $B$ (but not both) could be aligned, and $\Gamma^1_{1,1}$ gives the proportion for which both $A$ and $B$ could be aligned. Similarly, for each motif $m_i$, define $\Gamma^{i,2}_{A,B}$ to be the covariance matrix representing the relative likelihoods of exact conservation of $l_i$ positions (i.e., $(g_{j,k}...g_{j,k+l}) = (a_{j,k}...a_{j,k+l})$) in $A$ and/or $B$. Here, we have observed non-independence of exact conservation likelihood between adjacent positions, so we model it as a first order Markov chain.

Conservation of a completely degenerate motif is parameterized by $\Gamma^1_{A,B}$, and conservation of a motif composed of a single word is parameterized by $\Gamma^{i,2}_{A,B}$. The parameterization of a generic motif is between these extremes; for this, let $P_{i,j,k}$ be the matrix giving the frequency of nucleotide $j \in \{A,C,G,T\}$ at position $k$ $\in \{1,...,l_i\}$ in motif $m_i$, and let $E_i$ be the average entropy of the motif:

$$E_i = -\frac{1}{2l_i} \sum_{k=1}^{l_i} \sum_{j \in \{A,C,G,T\}} P_{i,j,k} \log_2 P_{i,j,k}$$

Hence, $E_i=1$ for a completely degenerate motif, $E_i=0$ for a motif composed of a single word, and $E_i$ increases monotonically and smoothly between these extremes as the motif degeneracy increases. Therefore, we take our parameterization of $\Gamma_i$ for $m_i$ to be a weighted average of $\Gamma^1_{A,B}$ and $\Gamma^{i,2}_{A,B}$:

$$\Gamma^i = E_i\Gamma^{i,1}_{A,B} + (1-E_i)\Gamma^{i,2}_{A,B}$$

We then use $\Gamma_i$ to compute $P_{\Gamma^i}(Y_i, Z_i\,|\,X_i)$. In a sequence window containing $X_i$ matches to motif $m_i$, let $a_i$ be the number that are not conserved in either $A$ or $B$, let $b_i$ and $c_i$ be the number conserved in either $A$ or $B$ (but not both), and let $d_i$ be the number that are conserved in both $A$ and $B$. The following equations hold:

$$a_i + b_i + c_i + d_i = X_i \qquad (3\text{-}5)$$

$$b_i + d_i = Y_i \qquad c_i + d_i = Z_i$$

$P(Y_i, Z_i|X_i)$ is therefore given by the following multinomial:

$$P_{\Gamma^i}(Y_i, Z_i \mid X_i) = \sum \left( \frac{X_i!}{a_i! b_i! c_i! d_i!} \right) \left( (\Gamma_{0,0}^i)^a \cdot (\Gamma_{1,0}^i)^b \cdot (\Gamma_{1,0}^i)^c \cdot (\Gamma_{1,1}^i)^d \right) \tag{6}$$

where the summation is performed over all values of $a_i$, $b_i$, $c_i$ and $d_i$ satisfying Eqs. (3)-(5). To achieve computational efficiency, we make use of the following 1-dimensional parameterization, where $X_i$, $Y_i$ and $Z_i$ remain fixed as $d_i$ is varied:

$$a_i = X_i - Y_i - Z_i + d_i \tag{7-9}$$
$$b_i = Y_i - d_i \qquad c_i = Z_i - d_i$$

Thus, the summation of Eq. (6) can be performed by simply taking each value of $d_i$ in the range $0 \le d_i \le \min(Y_i, Z_i)$.

If one desires to only input one genome, it is sufficient to set $A=B$. The relevant parameters then simplify, and the preceding multinomial distribution collapses to a binomial distribution with parameter $\gamma_i = \Gamma_{1,1}^i$ :

$$P_{\gamma_i}(Y_i \mid X_i) = \binom{X_i}{Y_i} \gamma_i^{Y_i} (1 - \gamma_i)^{X_i - Y_i}$$

This parameterization can also be easily generalized to more than 2 alignment genomes by replacing the matrix $\Gamma^i$ with an appropriate tensor.

This derived value of $P_{\Gamma,\alpha,w}(X,Y,Z)$ alone is insufficient for determining statistical significance, since a measurement of distance into the appropriate tail of the distribution is also required. Therefore, we perform a summation of $P_{\Gamma,\alpha,w}(X,Y,Z)$ extending from the observed value of $(X,Y,Z)$ and including all values of $(X,Y,Z)$ with an increased degree of clustering and conservation (we use log values to simplify the numerical analysis):

$$S_{\Gamma,\alpha,w}(X,Y,Z) = \log_{10}\left( \sum_{X=X}^{\infty} \sum_{Y=Y}^{\tilde{X}} \sum_{Z=Z}^{\tilde{X}} \prod_{i=1}^{m} \left( P_{\Gamma^i,\alpha_i,w}(\tilde{X}_i, \tilde{Y}_i, \tilde{Z}_i) \right) \right) \tag{10}$$

$$= \sum_{i=1}^{m} \log_{10}\left( \sum_{\tilde{X}_i=X_i}^{\infty} \sum_{\tilde{Y}_i=Y_i}^{\tilde{X}_i} \sum_{\tilde{Z}_i=Z_i}^{\tilde{X}_i} P_{\Gamma^i,\alpha_i,w}(\tilde{X}_i, \tilde{Y}_i, \tilde{Z}_i) \right) = \sum_{i=1}^{m} S_{\Gamma^i,\alpha_i,w}(X_i,Y_i,Z_i)$$

Therefore, the output score $S_{\Gamma,\alpha,w}(X,Y,Z)$ for a given window is the linear sum of scores for the input motifs, $S_{\Gamma_i,\alpha_i,w}(X_i,Y_i,Z_i)$, where each such term has been automatically weighted so that more degenerate motifs contribute less. Observe also that $S_{\Gamma_i,\alpha_i,w}(X_i,Y_i,Z_i)=0$ if and only if $X_i=0$, and that $S_{\Gamma_i,\alpha_i,w}(X_i,Y_i,Z_i)$ increases monotonically with increasing values of $(X_i,Y_i,Z_i)$, as desired.

## 2.2. Implementation and Availability

ModuleFinder has been implemented in $C$. To minimize runtime, we pre-process each sequence of $G$ with suffix arrays[23] for efficient searching; additionally, as the algorithm proceeds, a look-up table is kept that contains a list of scores for all observed window sizes $w$ and motif matches $(X, Y, Z)$. ModuleFinder can scan ~120 Mb/hr using window sizes of 300-700 bp with an increment size of 50 bp and one alignment genome on a Pentium 4 computer. The compiled code, along with README files and appropriately formatted genomes and alignments for human, mouse, rat, fly, worm and yeast based on the latest UCSC assemblies[24] are available for download at our website (http://the_brain.bwh.harvard.edu).

Two additional features were included for improved practical applicability. First, it is known that TFs frequently bind to DNA as homo- and hetero-dimers. We have added to ModuleFinder the ability to take pairs of TFBSs as input, along with minimum and maximum spacer lengths between sites. The score of the dimer is computed by evaluating the probability of each component motif as in Eq. (1), then taking the product of these probabilities and summing them over all input spacings. Second, ModuleFinder allows a certain amount of 'wiggle room' to compensate for the potential existence of local misalignments. Specifically, given an input value $r$, a motif match $(g_{j,k}....g_{j,k+l})$ is considered conserved in $A$ if there is any subsequence of $(a_{j,k-r}...a_{j,k+r+l})$ that is a word in $m_i$. Although this does increase the likelihood of conservation, the effect is miniscule for small values of $r$ ($1 \leq r \leq 5$) and has frequently identified potentially conserved sites that would have been missed otherwise.

## 3. Results

### 3.1 Validation of ModuleFinder on human skeletal muscle CRMs

In order to evaluate ModuleFinder, we used a set of positive control regions previously compiled by Wasserman et al.[18] This test dataset comprises 27[a] skeletal muscle CRMs that have been demonstrated to direct transcription in skeletal muscle or a suitable cell-culture model system[18]. Each region contains a validated BS for at least one of the following 5 TFs: the Myf family (total of 39 TFBSs in the positive control set), Mef2 (26 TFBSs), SRF (20 TFBSs), Tef (12 TFBSs) and Sp1 (13 TFBSs). Of these 27 regions, 23 are located within 5 kb upstream of translational Start, and 2 within introns. As negative controls, 1000 regions of size 200 bp were randomly selected to positionally match the positive control regions: 852 (=(23/27)*1000) regions were within 5 kb of translational Start for a randomly chosen RefGene[24] gene, and the remaining 148 were within introns. This matching of chromosomal locations was performed as ModuleFinder accounts for local word frequencies, which vary throughout the genome; in particular, promoter regions are known to be GC-rich.

We ran ModuleFinder on the positive and negative control regions with window sizes of 100-200 bp (increment size = 10 bp), using human sequence alone,

---

[a] The original collection gave 28 genes, but we removed the gene *Rb1* as there were no confirmed TFBSs for the listed TFs.

human/mouse/rat (H/M/R) alignments and human/mouse/chicken alignments (H/M/C) obtained from UCSC Genome Browser (hg16, mm3, rn3, galGal2)[24]. Currently, two alternative strategies for representing TFBSs have been used by various groups in computational searches for CRMs: exact word matches to known BSs[9,15], and position weight matrices (PWMs)[7,10-13], which allow for extrapolation to additional BSs. To determine which of these representations had greater discriminatory power, we performed our searches both ways, using a PWM threshold value of 1 standard deviation (SD) below the motif average[25]. We used a "jack-knife" strategy[11] for these searches, whereby the BSs for each CRM were excluded from the construction of the PWM used to search that CRM, and similarly the exact word matches from each CRM were excluded in the search of that CRM. In addition, since *in vitro* binding experiments had been performed for Mef2[26] and SRF[27], we also added those BSs to both searches.

| | Human Alone | | Human/Mouse/Rat | | Human/Mouse/Chicken | |
|---|---|---|---|---|---|---|
| | Exact | PWM | Exact | PWM | Exact | PWM |
| Sens. | 88.9% | 92.6% | 92.6% | 96.3% | 92.6% | 92.6% |
| Spec. | 90.1% | 89.2% | 88.8% | 94.4% | 87.4% | 94.4% |
| *p*-val | $1.19 \times 10^{-8}$ | $6.5 \times 10^{-10}$ | $2.5 \times 10^{-10}$ | $1.4 \times 10^{-10}$ | $4.5 \times 10^{-10}$ | $7.1 \times 10^{-10}$ |

**Table 1.** ModuleFinder was run on a human skeletal muscle dataset using both exact word matches and PWMs. The searches were done using no alignments, mouse/rat alignments, and mouse/chicken alignments. For each search, sensitivity ("Sens."), specificity ("Spec."), and a t-test on the means ("*p*-val") were computed, as compared to matched random regions.

The results of these evaluations are shown in **Table 1**. Here, we have reported those values for sensitivity and specificity which maximally discriminate between the positive and negative control sets (i.e., using the threshold score such that the difference between the sensitivity and specificity is minimized). Since there was great variability in score among the positive control regions (see **Figure 1**; i.e., the top positive control region received a score of -11.23 and the worst positive control region scored only -1.22 (positive controls: mean = -4.69, SD = 2.24; negative controls: mean = -0.42, SD = 0.81)), we also performed a t-test on the positive and negative control region means, in order to measure the effectiveness of ModuleFinder on regions falling far from the threshold score.

On this dataset, ModuleFinder achieved a maximum sensitivity of 96.3% and specificity of 94.4% on the H/M/R PWM searches. Moreover, the PWM approach consistently gave better discrimination than exact word matches. Much of this improved discrimination, however, is an artifact of the jack-knife procedure, which has a stronger effect on exact match searches. Here, using the complete set of BSs (i.e., without the jack-knife), exact word matching achieves 100% sensitivity and 95.1% specificity (we removed degenerate flanking sequences for all searches with exact words). In addition, these results indicate that the H/M/R searches reliably outperformed the H/M/C searches. There are two possible explanations for this: 1) the chicken genome is not yet complete, and the appropriate alignment regions may not have been sequenced yet; 2) the underlying mechanisms of transcriptional regulation are not actually conserved in an organism as distant as chicken. Neither of these hypotheses can be ruled out until the completion of the chicken genome.

Since ModuleFinder was specifically developed to integrate homotypic clustering, heterotypic clustering, and conservation, we wanted to determine which of these features were most contributory to discriminatory power. In order to assess this, we ran ModuleFinder on the positive and negative control regions

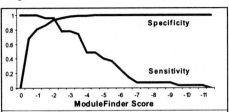

**Figure 1:** Sensitivity and specificity of ModuleFinder on skeletal muscle test regions, versus randomly selected control regions.

using no alignments, one alignment (each of mouse, rat and chicken), and two alignments (H/M/R and H/M/C). These searches were repeated with each TF individually, as well as with all 5 TFs together. In **Figure 2**, we show the negative logarithm of the *p*-values obtained from t-tests on the positive versus negative control regions for each of these searches. Here the mouse and rat alignments improved discriminatory power, but little was gained by using both genomes, because of their evolutionary proximity. Somewhat surprisingly, using chicken actually reduced discrimination relative to human alone. This was unexpected, as it implies that our negative controls are more likely to be conserved than these 27 regions. However, this effect could be an artifact of the small size of the positive controls and gaps in the chicken genome (only 13/27 positive controls had any alignable chicken sequence).

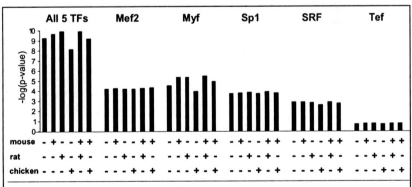

**Figure 2:** Negative log of p-values obtained from t-test on means between positive and negative controls. ModuleFinder was run with various combinations of mouse, rat, and chicken alignments (indicated by +/-), using all 5 TFs together, and each TF alone.

Finally, at least four other algorithms have used overlapping subsets of this dataset as positive controls[11-13,28], achieving sensitivities between 59% and 66%, and specificities between 95.3% and 97.1% (see **Table 2**). Thus, ModuleFinder appears to have comparable specificity but greater sensitivity. However, note the following caveats for this comparison. First, because ModuleFinder uses evolutionary conservation as a central component and because few vertebrate genomes have been sequenced, we limited our searches to the subset of the original compilation for which human/rodent alignments were available[18]. The other algorithms tested on this dataset did not consider conservation, and thus used the original, larger compilation that included CRMs obtained from diverse organisms including chicken, hamster, rabbit, pig and cow[11]. Frith et al.[12,13] trimmed this larger set[11] to a subset of 27 regions, but their subset overlapped with ours by only 15 genes. Second, each group used a different set of negative controls. The original paper by Wasserman et al.[11] used a set of negative control regions similar to our set; it was composed of 200 bp regions selected from the Eukaryotic Promoter Database. Comet and Cister were each tested on 300 bp regions that were selected to overlap well-characterized transcriptional Starts[12,13]. Finally, MSCAN[28] measured specificity by looking at the "hit rate" in contiguous stretches of the *Fugu* genome.

| Algorithm | Sensitivity | Specificity |
|---|---|---|
| Logistic Regression[11] | 60% | 96% |
| Cister[12] | 59% | 97.1% |
| COMET[13] | 59% | 95.3% |
| MSCAN[28] | 66% | NA |
| ModuleFinder | 96% | 94% |

**Table 2.** Relative performance of ModuleFinder: Sensitivities and specificities, as reported by groups using overlapping subsets of the skeletal muscle dataset. Logistic regression, Cister, Comet and ModuleFinder specificities refer to 200-300bp portions of the human genome; the MSCAN specificity was ascertained using large stretches of *Fugu* sequence.

### 3.2 Other validations of ModuleFinder

In addition to the mammalian skeletal muscle set, we have also tested ModuleFinder on a *D. melanogaster* dataset that comprises 20 transcriptional enhancers from 9 genes known to be co-regulated during anterior-posterior segmentation of fly embryos[7]. Using the *D. melanogaster/D. pseudoobscura* alignments and a protocol similar to that described in Section 2.1, ModuleFinder was able to discriminate this collection of CRMs from randomly chosen noncoding regions with 95% sensitivity and 95% specificity (Philippakis et al., manuscript in preparation). In addition to these *in silico* confirmations, we have also successfully applied ModuleFinder to predict CRMs in three biological systems: (1) development of the fly pericardium (Michaud et al., manuscript in preparation), (2) development of fly muscle founder cells (Philippakis et al., manuscript in preparation), and (3) mammalian myogenesis (Warner et al., manuscript in preparation). For mammalian myogenesis, we applied the same 5

TFs and their BSs as described above; indeed much of the work presented here was done for the explicit purpose of selecting optimized BSs and sequence alignments before attempting to predict novel mammalian CRMs.

## 4. Discussion and Future Directions

We have presented a statistically rigorous approach for scoring windows of genomic sequence according to their likelihood of containing BSs for a collection of input TFs. The approach systematically integrates homotypic clustering, heterotypic clustering and evolutionary conservation across multiple genomes into a single, objective scoring scheme that does not require training. Additionally, our algorithm, implemented as a *C* program called "ModuleFinder," is publicly available for download, along with pre-processed genomes and alignments for yeast, worm, fly, mouse, rat, and human, at our lab website (http://the_brain.bwh.harvard.edu). The current version of ModuleFinder considers up to two alignment genomes as input, and we are currently expanding it to accept arbitrarily many genomes.

We have tested ModuleFinder on a set of human skeletal muscle CRMs using a variety of genome alignments and TFBSs, and have achieved a maximum sensitivity and specificity of 96% and 94%. On this dataset, improved sensitivity and specificity were achieved by using mouse and rat alignments in the searches, whereas chicken alignments actually decreased sensitivity and specificity. Furthermore, PWMs resulted in improved sensitivity and specificity over exact TFBS matches. Preliminary results indicate that ModuleFinder can successfully predict novel CRMs in human myoblasts (Warner *et al.*, manuscript in preparation). In addition, on a *D. melanogaster* segmentation gene dataset with *D. pseudoobscura* as the alignment genome, ModuleFinder achieved sensitivity and specificity of 95% and 95%. We have also predicted and experimentally validated several novel CRMs in the developing fly mesoderm (Philippakis *et al.*, manuscript in preparation). We expect that in the future we and others will use ModuleFinder to further refine transcriptional regulatory models for CRMs in particular biological systems and thus discover how the associated TFBSs are organized to confer specific gene expression patterns.

## 5. Acknowledgments

The authors thank Bertrand Huber and Jason Warner for helpful discussion, and Mike Berger, Mark Umbarger, and Roman Yelensky for critical reading of the manuscript. This work was funded in part by a PhRMA Foundation Informatics Research Starter Grant (M.L.B.), a William F. Milton Fund Award (M.L.B.), and NIH/NHGRI R01 HG02966-01 (M.L.B.). A.A.P. was supported in part by a National Defense Science and Engineering Graduate Fellowship from the Department of Defense, and an Athinoula Martinos Fellowship from HST.

### References

1. Bulyk, M.L. *Genome Biol.* **5**, 201 (2003).
2. Boffelli, D. et al. *Science* **299**, 1391-4 (2003).
3. Cliften, P. et al. *Science* **301**, 71-6 (2003).
4. Kellis, M., Patterson, N., Endrizzi, M., Birren, B. & Lander, E.S. *Nature* **423**, 241-54 (2003).
5. McGuire, A.M., Hughes, J.D. & Church, G.M. *Genome Res.* **10**, 744-57 (2000).
6. Thomas, J.W. et al. *Nature* **424**, 788-93 (2003).
7. Berman, B.P. et al. *Proc. Natl. Acad. Sci. USA* **99**, 757-62 (2002).
8. Halfon, M.S., Grad, Y., Church, G.M. & Michelson, A.M. *Genome Res.* **12**, 1019-28 (2002).
9. Markstein, M., Markstein, P., Markstein, V. & Levine, M.S. *Proc. Natl. Acad. Sci. USA* **99**, 763-8 (2002).
10. Krivan, W. & Wasserman, W.W. *Genome Res.* **11**, 1559-66 (2001).
11. Wasserman, W.W. & Fickett, J.W. *J. Mol. Biol.* **278**, 167-81 (1998).
12. Frith, M.C., Hansen, U. & Weng, Z. *Bioinformatics* **17**, 878-89 (2001).
13. Frith, M.C., Spouge, J.L., Hansen, U. & Weng, Z. *Nucleic Acids Res.* **30**, 3214-24 (2002).
14. Rajewsky, N., Vergassola, M., Gaul, U. & Siggia, E.D. *BMC Bioinformatics* **3**, 30 (2002).
15. Rebeiz, M., Reeves, N.L. & Posakony, J.W. *Proc. Natl. Acad. Sci. USA* **99**, 9888-93 (2002).
16. Sinha, S., van Nimwegen, E. & Siggia, E.D. *Bioinformatics* **19 Suppl 1**, i292-301 (2003).
17. Altschul, S.F., Gish, W., Miller, W., Myers, E.W. & Lipman, D.J. *J. Mol. Biol.* **215**, 403-10 (1990).
18. Wasserman, W.W., Palumbo, M., Thompson, W., Fickett, J.W. & Lawrence, C.E. *Nat. Genet.* **26**, 225-8 (2000).
19. Blanchette, M., Schwikowski, B. & Tompa, M. *J. Comput. Biol.* **9**, 211-23 (2002).
20. Moses, A.M., Chiang, D.Y. & Eisen, M.B. *Pac. Symp. Biocomput.*, 324-35 (2004).
21. Prakash, A., Blanchette, M., Sinha, S. & Tompa, M. *Pac. Symp. Biocomput.*, 348-59 (2004).
22. Reinert, G., Schbath, S. & Waterman, M.S. *J. Comput. Biol.* **7**, 1-46 (2000).
23. Irving, R.W. & Love, L. *Technical Report no. TR-2001082 of the Computing Science Department of Glasgow University* (2001).
24. http://www.genome.ucsc.edu.
25. Stormo, G.D. *Bioinformatics* **16**, 16-23 (2000).
26. Andres, V., Cervera, M. & Mahdavi, V. *J. Biol. Chem.* **270**, 23246-9 (1995).
27. http://www.cognia.com.
28. Johansson, O., Alkema, W., Wasserman, W.W. & Lagergren, J. *Bioinformatics* **19 Suppl 1**, i169-76 (2003).

# RANDOM FOREST SIMILARITY FOR PROTEIN-PROTEIN INTERACTION PREDICTION FROM MULTIPLE SOURCES

YANJUN QI

*School of Computer Science, Carnegie Mellon University, Pittsburgh, PA 15213, USA*

JUDITH KLEIN-SEETHARAMAN

*Department of Pharmacology, University of Pittsburgh School of Medicine, Pittsburgh, PA 15213, USA*

ZIV BAR-JOSEPH

*School of Computer Science, Carnegie Mellon University, Pittsburgh, PA 15213, USA*

One of the most important, but often ignored, parts of any clustering and classification algorithm is the computation of the similarity matrix. This is especially important when integrating high throughput biological data sources because of the high noise rates and the many missing values. In this paper we present a new method to compute such similarities for the task of classifying pairs of proteins as interacting or not. Our method uses direct and indirect information about interaction pairs to constructs a random forest (a collection of decision tress) from a training set. The resulting forest is used to determine the similarity between protein pairs and this similarity is used by a classification algorithm (a modified kNN) to classify protein pairs. Testing the algorithm on yeast data indicates that it is able to improve coverage to 20% of interacting pairs with a false positive rate of 50%. These results compare favorably with all previously suggested methods for this task indicating the importance of robust similarity estimates.

## 1 Background

Protein-protein interactions play key role in many biological systems. These involve complex formation and various pathways which are used to carry out biological processes. Correctly identifying the set of interacting proteins can shed new light on the functional role of various proteins within the complex environment of the cell.

High throughput methods including two-hybrid screens [3,4] and mass spectrometry [5,6] have been used in an attempt to characterize the large set of interacting proteins in yeast. While some promising results were generated from these large scale experiments, these data sets are often incomplete and exhibit high false positive and false negative rates [1]. In addition to the high throughput experimental datasets that specifically look for protein interaction, other datasets provide indirect information about interaction pairs. For example, it has been shown that many interacting pairs are co-expressed [1] and that proteins in the same complex are in some cases bound by the same transcription factor(s) [18]. Sequence data was also be used to infer such interactions (for example by relying on domain-domain interactions [14]). Each of these datasets provides partial information about the interacting pairs and thus, in order to reliably infer protein-protein interactions one needs to integrate evidence from these different biological sources. Deriving such an accurate and complete set of interactions from these data sources is an important computational and biological problem.

When combining these disparate datasets one needs to solve a number of computational problems. First, the data is noisy and contains many missing values (for example, for two-hybrid (Y2H) system, the interactions involving membrane proteins may be undetectable and the system also suffers from high false positive rate due to factors like fortuitous activation of reporter genes [3,4]). In addition, some of these data sources are categorical (for example, synthetic lethal [1]) while others are continuous (for example, mRNA co-expression). Finally, there is the issue of weighting the different data sources. Some should have more weight than others (for example, intuitively the direct information should be weighted higher than the indirect information).

In this paper we present a method that overcomes the above problems by using random forest [19] to compute similarity between protein interaction pairs. We construct a set of decision trees such that each tree contains a random subset of the attributes. Next, protein pairs are propagated down the trees and a similarity matrix based on leaf occupancy is calculated for all pairs. Decision trees and the randomization strategy within random forest can handle categorical data and can automatically weight the different data sources based on their ability to distinguish between interacting and non interacting pairs. Because the trees are generated from random subsets of the possible attributes, missing values can be handled by filled in by an iterative algorithm. Finally, a weighted k nearest neighbor algorithm, where distances are based on the computed similarity, is used to classify pairs as interacting or not.

Our method was tested on yeast. We used several direct and indirect data sources and compared our results to previously suggested algorithms and to many other classification algorithms. As we show in Results, the method described above outperformed all other methods.

## 2    Related work

von Mering [1] is one of the first to discuss the problem of accurately inferring protein interactions from high throughput data sources. In that paper they have relied on the intersection of four direct experiments to identify interacting pairs.

While this method resulted in a low false positive rate, the coverage was also very low. Less than 3% of interacting pairs were recovered using this method compared to a reference set.

In order to improve coverage, Jansen et al. [11] combined direct and indirect data sources using naïve Bayes and a fully-connected Bayesian network. Unlike our method naïve Bayes assumes conditional independence between the attributes, which is clearly a problem for these datasets (for example, co-expression and co-binding are clearly correlated attributes). Indeed, as we show in section 4 (results), our method outperforms naïve Bayes for this task.

Gilchrist et al. [13] proposed a Bayesian method to integrate information from direct high-throughput experiment [5, 6]. However, while their method works well it requires a large number of direct measurements for each protein pair. Such data is not available for most pairs. In contrast, by integrating indirect information our method can correctly detect interacting pairs if only limited direct information is available.

Lan et al. [12] constructed a decision tree to predict co-complexed protein pairs by integrating direct and indirect genomic and proteomic data. While our method also relies on decision trees, it constructs many such trees and not one. This is important when missing values are an issue. Consider two (dependent) attributes A and B that are both useful for the classification task. Assume A is slightly better than B. In that case A can be chosen as the root node of the tree and since B is highly correlated with A, it will not be used at a high level in the tree. Now, if a protein pair lacks a measurement for A but has a value for B it might not be classified correctly based on the low weight assigned to B. In contrast, when using the random forest approach, B will be selected by many trees as a high level split (see Methods). This allows our method to deal more effectively with these noisy datasets.

## 3    Method

We use multiple high throughput datasets to construct a $d$-dimensional vector $X_i$ for every pair of proteins. Each item in this vector summarizes one of these datasets (for example, are these two proteins bound by the same transcription factor? what is their expression correlation? see [21] for a complete list of attributes in each vector). Given these vectors the task of protein interaction prediction can be presented as a binary classification problem. That is, given $X_i$ does the $i$th pair interact ($Y_i=1$) or not ($Y_i=-1$).

Note that there are a number of unique characteristic to this classification task. First, the number of non interacting pairs is much larger than the number of

534

interacting pairs (as much as ~600 to 1 [section 4.2]) and so false negatives may be more costly than false positives. Second, there is no negative training dataset available for this task. Third, the features are of different types. Some of the entries in $X_i$ are categorical while others are numerical. Finally, many entries in each of these vectors are missing.

In order to overcome these difficulties we divide the classification task into two parts (see Figure 1). We first compute a similarity measure between genes pairs (overcoming noise, missing values and the different types of data) and then use this similarity to classify protein pairs taking into account the skewed distribution of positives and negatives. Below we discuss in detail each of these two parts.

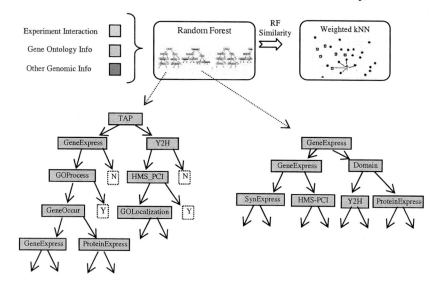

Figure 1. Classification process. To generate the random forest we select for each tree a subset of training data. Next, for every node in these trees a random subset of the attributes is chosen and the attribute achieving the best division is selected. Once trees are grown, all protein pairs (remaining training and test sets) are propagated down the trees and similarity is computed based on leaf occupancy (see text). Using the computed similarity a weighted KNN algorithm is used to rank pairs by the resulting interaction score.

## 3.1    Random forest (RF) and RF similarity

We use random forest [19] to determine the similarity of protein pairs. Random forest (RF) was initially introduced as a classification algorithm, though it can also be used to compute similarities. RF constructs a large set of independent decision trees (see below). Results from these trees are combined for the classification or similarity calculation task.

**Decision tree:** A decision tree is a binary tree with nodes corresponding to attributes in the input vectors. Tree nodes are used to determine how to propagate a given attribute set down the tree. Nodes can either be threshold nodes or categorical nodes. Decision trees also contain terminal (or leaf) nodes that are labeled as -1 or 1. In order to classify a protein pair as interacting or not, this pair is propagated down the tree and decision is made based on the terminal node that is reached. Decision trees are grown using a training set. At each node the algorithm searches for an attribute that best separates all instances in that node. If the attribute perfectly classifies all instances so that all instances in one of the two descendent nodes have the same label then this node becomes a terminal node with the appropriate label. Otherwise, the above process is repeated until all instances are at terminal nodes.

**Random forest:** Random forest [19] uses a collection of independent decision trees instead of one tree. Denote by $\Theta$ the set of possible attributes (or variables on which nodes can be split) and by $h(x,\Theta)$ a tree grown using $\Theta$ to classify a vector $x$. Using these notations a random forest $f$ is defined as:

$$f = \{h(x,\Theta_k)\}, k = 1,2,...,K \tag{1}$$

Where $\Theta_k \subseteq \Theta$. That is, a random forest is a collection of trees, where each tree is grown using a subset of the possible attributes. For the $k$th tree $\Theta_k$ is randomly selected, and is independent of the past random vectors $\Theta_1,...,\Theta_{k-1}$. In order to classify, using each of the trees 'votes' for one of the classes and the most popular class is assigned to input $\mathbf{x}$.

One of the main reasons random forests perform better than a single decision tree is their ability to utilize redundant features and the independence of the different classifiers (trees) used. This is also important in our case since if a pair has values for one redundant feature but not the other, we can still use this feature for the similarity calculation process.

Specifically we have grown the random forest in the following way: Each tree is grown on a bootstrap sample of the training set (this helps in avoiding overfitting). A number $m \ll M$ (M is the total number of attributes) is specified, and for each node in the tree the split is chosen from $m$ variables that are selected at random out of the total M attributes. Once the trees are grown, they can be used to estimate missing values and to compute similarities as follows.

**Handling missing values:** Random forest can be used to estimate missing values. For *training* data missing values are first assigned the median of all values in the same class (or the most frequent value for categorical data). Next, the data vectors are run on the forest, and missing data is re-estimated based on pairs that share the same terminal leaves with this pair. This process can be iterated until these estimates converge.

For *test* data, we first replicate the attribute vector and then apply a similar procedure to each of the two replicas. Initially, missing values in the first and second replicas are set to the mean values of the positive and negative classes, respectively. Next, these two replicas are propagated down the trees and the values for each are re-estimated based on neighboring pairs. This process is iterated and the final values are determined from the class that receives the most number of votes in the different trees.

**Random Forest Similarity:** For a given forest $f$, we compute the similarity between two pairs of proteins pairs $X_1$ and $X_2$ in the following way. For each of the two pairs we first propagate their values down all trees within $f$. Next, the terminal node position for each pair in each of the trees is recorded. Let $\mathbf{Z}_1 = (Z_{11}, ..., Z_{1K})$ be these tree node positions for $X_1$ and similarly define $\mathbf{Z}_2$. Then the similarity between pair $X_1$ and $X_2$ is set to: ($I$ is the indicator function.)

$$S(X_1, X_2) = \frac{1}{K} \sum_{i=1}^{K} I(Z_{1i} == Z_{2i})$$ (2)

As we discuss in Results, we partition our training set to two parts. The first is used to generate the random forest. The second is used to train the kNN algorithm. In order to compute similarities efficiently, the following algorithm is used. Given a random forest with $K$ trees and up to $N$ terminal nodes in each tree we first generate a $N*K$ vector $V$ where each entry in $V$ contains a linked list of the kNN training set pairs that reside in that node. Given a new test pair we first propagate it down all trees (in $O(N*K)$ time) and for each of the terminal nodes it arrives at we find the corresponding set of training pairs from $V$. For each such pair we increase their similarity count by one. Thus, for a given test pair it takes only $O(|S_{train}|+N*K)$ to compute its similarity to all the training points, where $S_{train}$ is the training set and $|S|$ represents the number of elements in $S$.

### 3.2 Classifying protein pairs

We use a weighted version of the k-Nearest Neighbor (kNN) algorithm to classify pairs as interacting or not. While we have tried a number of classifiers for this data (see Results) the main advantage of kNN for this task is its ability to classify based on both, similarity and dissimilarity (as opposed to similarity alone). As can be seen in Figure 2, while non interacting pairs are similar to each other, the main distinguishing feature of interacting pairs is their distance from (or dissimilarity with) non interacting pairs. Due to the highly skewed distribution of interacting and non interacting pairs, it is likely that the closest pair to an interacting pair will be a non interacting pair (though their similarity might not be high). Decision trees (or RF) may use these to incorrectly classify an interacting pair as non interacting.

However, kNN can take into account the magnitude of the similarity, and if it is too weak can still classify the pair as an interacting pair.

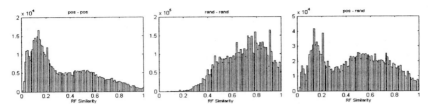

Figure 2. The pairwise RF similarity. Three histograms of pairwise similarities between all positive pairs (left) all random pairs (center) and between all positive and all random pairs. Note that while the random set is fairly tight, the positive set exhibits a greater diversity and is also far (on average) from most random samples. kNN can utilize this fact by relying on the actual distance to the closest neighbors (see text for details).

Specifically, given a set of training examples $(X_i, Y_i)$, and a query $X_q$, we calculate the interaction score for $q$ using the weighted mean of its neighbor's $Y_i$ values, where the weight depends on the similarity of each of training pairs to $q$:

$$f(q) = \sum_{p=1}^{k} S(X_q, X_{neighbor\,(p)}) * Y_{neighbor\,(p)} \qquad (3)$$

Here $S(X_i, X_q)$ is the similarity between $i$ and $q$ as computed by RF and $Y \in \{1, -1\}$. The *test* set can then be ranked by the $f(q)$ interaction scores. A cutoff $t$ can be derived using a training set to threshold this ranking list such that $q$ is classified as interacting if $f(q) > t$. In particular, $t$ does not have to be equal to 0. In fact, in our case $t < 0$ meaning that even though this pair is (on average) closer to non interacting pairs, since it is not close enough, it is classified as an interacting pair.

## 4    Results

We first discuss the biological features we used for the attributes vectors. Next we present results for applying our classification method for determining protein interaction pairs in yeast.

### 4.1    Attribute set

As mentioned in the introduction, there are many high throughput biological data sources related to protein-protein interaction. The method described in this paper is general and can be used with any type of biological data. Thus, while we have tried to use as much data sources as we could, when a new data source (such as protein expression arrays) becomes available, the method discussed in this paper can take advantage of that data as well. For the results presented in this section we used a total of 15 attributes for each protein pair (see website [21]). Overall, these data sources can be divided into three categories: Direct experimental data sets (two-hybrid screens and mass spectrometry), indirect high throughput data sets (gene expression, protein-DNA binding etc.) and sequence based data sources (domain information, gene fusion etc.). In addition to combining these data sources, our method can also indicate which of the different data sources is better for predicting protein interactions as discussed below.

## 4.2    Reference set

We need a reference set to train/test the algorithm. For the positive set (or the interacting pairs) ~4000 yeast protein pairs are derived from the database of interacting proteins (DIP [17]). This set is composed of interacting protein pairs which have been experimentally validated, and thus can serve as a reliable positive set. Unlike positive interactions, it is rare to find a confirmed report on non interacting yeast pairs. Here we follow [12] which have used a random set of protein pairs as their negative set instead. This selection is justified because of the small fraction of interacting pairs in the total set of potential protein pairs. It is estimated Supplementary [21] that only ~1 in 600 possible protein pairs actually interact [10, 13] and thus, over 99.8% of our random data is indeed non interacting which is probably better than the accuracy of most training negative data. Actually this extremely unbalanced class distribution of our reference set motives the weighted kNN ranking step in our algorithm.

## 4.3    Important attributes

Biologically, it is of particular interest to identify the attributes and data sources that contribute the most to our ability to classify protein pairs. Such an analysis can help uncover relationships between different data sources which are not directly apparent. In addition, it can help identify what data sources should be generated for determining interaction in other species (for example, in humans). One way to determine such a set using random forest is to score attributes based on the levels of nodes that use them to split the data. Since each node splits the data using the best available attribute, attributes used in higher levels in the tree contribute more than those used in lower levels.

As a rough estimate for the contribution of each attribute we have counted the percentage of nodes that use this attribute in the top four levels of all trees in our trained random forest model. Of the 15 features we used, gene co-expressed had the highest score with 18% of top nodes using it to split the input data. Next came three features: protein co-expression, domain-domain interaction and GO co-process, each with ~11% of the nodes. These were followed by TAP mass spectrometry data (8%) GO co-localization (6%), Y2H screens (4%) and HMS-PCI (4%). (see

Supplementary [21] complete list). Interestingly, indirect information played a very important role in the decision process though it is likely that this results from the fact that direct experiments cover less than 30% of all protein pairs. However, mass spectrometry data are clearly more significant than Y2H data, consistent with the notion that mass spectrometric identification of protein-protein interaction is less prone to artifacts than Y2H experiments. It is particularly encouraging that co-expression and GO features contribute such strong components to the prediction, clearly supporting the notion that a large amount of indirect data that measures biologically relevant information is helpful in predicting interaction partners.

### 4.4    Performance comparison

We use precision vs. recall curve to perform the comparisons completely.
**Precision**: Among the pairs identified as interacting by the classifier, what is the fraction (or percentage) that is truly interacting?
**Recall:** For the known interaction pairs, what is the percentage that is identified?

Let $d$ be the number pairs identified as interacting by the classifier, $T$ be the number of pairs labeled as interacting, and $c$ be the number of pairs correctly identified as interacting, then precision and recall are defined as:

$$\text{Precision} = c/d \quad \text{Recall} = c/T \qquad (4)$$

In other words, precision is the accuracy of our predictor whereas recall is the coverage of the classifier. Note that even 50% precision can useful. For example, biologists studying a specific protein can extract a list of potential interacting partners computationally first and carry out further experiments knowing that on average 50% of their experiments will identify true interacting pair. This is a much better ratio than if the set of potential pairs was randomly selected.

For our algorithm, in each cross validation run, we divided our training set into two equal parts. The first was used to construct the random forest and the second was used by kNN algorithm. Thus, our algorithm uses the same amount of training data as the other algorithms we compare to (see below).

In order to generate a precision / recall curve we use different thresholds as discussed above. For the other classification methods, we generate the curve in a similar manner. For instance, for the naïve Bayes classifier, we can use the naïve Bayes prediction probability of a test point to arrive at a ranked list.

Figure 3 shows a comparison between our method and a number of other classification algorithms. The figure on the left compares our method with a weighted kNN that uses Euclidean distance instead of the random forest similarity, with the naïve Bayes method and with a single decision tree. In general, for a wide range of high precision values our method outperforms the other methods. It is especially interesting to compare our method with the kNN method using Euclidian distance. As can be seen, using the robust similarity estimates from the random forest results greatly improves the classification results. We have also compared our algorithm to a classification method that only uses the resulting random forest (based on popular vote) to classify protein pairs and to a number of other popular classifiers including Support Vector Machine (SVM), logistic regression and Adaboost (right figure). In all cases our algorithm performed better for precision values that are higher than 0.32. Specifically, holding precision fixed at 0.5 (or 50%) our algorithm achieved a recall rate of 20% while logistic regression achieved 14% recall, random forest and SVM achieved 11% and Adaboost had a 7% recall rate. Finally, we note that while the methods we have used to compare our algorithm with were inspired by previous work (such as single decision tree [12] and naïve Bayes [11]), we have used a slightly different feature set and a different training set compared to each of these two papers. Thus, the results reported for these methods here are different from the ones reported in these papers. See website [21] for details about the implementation of the other classification methods.

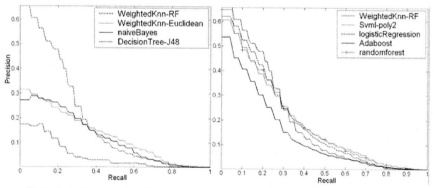

Figure 3. Precision vs. Recall curves. Left: Comparison of weighted kNN using random forest similarity, weighted kNN using Euclidean distance, naïve Bayes and a single decision tree (J48); Right: Comparison of weighted kNN using random forest similarity, logistic regression, support vector machine, Adaboost and random forest classifier.

*4.5   Predicting interactions for the yeast pheromone response pathway*

To analyze the results for their utility in the design of new experiments, we compared the predictions of our method to their labels for one specific pathway, the yeast pheromone pathway. The yeast mating factors MATα/a bind to their cognate membrane receptors, Ste2/3, members of the G protein coupled receptor family.

Subsequent binding and activation of the G protein induces a MAP kinase signaling pathway via the G protein $\beta\gamma$ subunit. We selected 25 proteins that are known to participate in this pathway. We applied our algorithm (using a different training set) to classify the 300 (25*24/2) potential interacting pairs. Our algorithm classified 44 of these pairs as interacting. 31 of these pairs (70.45%) were known to interact while only 2 (4.55%) are verified to be wrong predictions. The remaining 11 pairs (25%) are new predictions that are reasonable and would functionally make sense. They form two clusters: The first involves the possible interaction between the STE5 anchor protein and the receptors. The receptor would then make additional interactions due to STE5s' anchoring function. The second cluster is a possible interaction between the most downstream components of the signaling cascade including BEM1, BNI1 and FUS1, mediating cell fusion (see website [21] for details). These new biological hypothesis can be used to design future lab experiments.

## 5    Discussion and future work

In this paper we presented a method for predicting protein-protein interactions by integrating diverse high-throughput biological datasets. Our method works in two steps. First, a similarity measure is computed between protein pairs. Then a classification algorithm uses the computed similarities to classify pairs as interacting or not. We have applied our algorithm to the task of classifying protein pairs in yeast. As we have shown, our algorithm outperforms previous methods suggested for this task and can also derive meaningful biological results for known pathways.

In this paper we have used random forest to learn a similarity function between protein pairs. Recently, a number of methods have been suggested for learning distance matrices [22]. We would like to test some of these methods and see if they can improve our classification accuracy. Specifically, it will be challenging to apply these methods to datasets with missing values, as is the case here.

Interestingly, many of the features determined to be important using our method are indirect measurements. This opens the possibility of extending this method to determine interacting pairs in organisms where little high throughput direct information is available, such as humans.

**Acknowledgements** This research work is supported by National Science Foundation Information Technology Research grant number 0225656.

**References**

1. Von Mering C, et al., Comparative assessment of large-scale data sets of protein-protein interactions. *Nature* 417:399-403, 2002
2. Bader GD, Hogue CWV. Analyzing yeast protein-protein interaction data obtained from different sources. *Nature Biotechnology* 20:991-997, 2003
3. Uetz P, et al., A comprehensive analysis of protein-protein interactions in Saccharomyces cerevisiae. *Nature.* 403(6770):623-7, 2000
4. Ito T, et al., A comprehensive two-hybrid analysis to explore the yeast protein interactome., *Proc Natl Acad Sci,* 10;98(8):4569-74, 2001
5. Gavin AC, et al., Functional organization of the yeast proteome by systematic analysis of protein complexes. *Nature.* 415(6868):141-7, 2002
6. Ho Y, et al., Systematic identification of protein complexes in Saccharomyces cerevisiae by mass spectrometry. *Nature* 415(6868), 2002
7. Enright AJ, et al., Protein interaction maps for complete genomes based on gene fusion events. *Nature.* 402(6757):86-90, 1999
8. Huh WK, et al, Global analysis of protein localization in budding yeast. *Nature.* 425(6959):686-91, 2003
9. Ghaemmaghami S, et al., Global analysis of protein expression in yeast. *Nature.* 425(6959):737-41, 2003
10. Tong A.H.Y. et al. Global Mapping of the Yeast Genetic Interaction Network. *Science.* 303: 808-813, 2004
11. R Jansen, et al., A Bayesian networks approach for predicting protein-protein interactions from genomic data, *Science* 302: 449-53, 2003
12. Lan V. Zhang, et al., Predicting co-complexed protein pairs using genomic and proteomic data integration, *BMC Bioinformatics.* 5 (1): 38, 2004
13. Gilchrist MA, et al. A statistical framework for combining and interpreting proteomic datasets. *Bioinformatics.* 20(5):689-700, 2004
14. Deng M, et al., Inferring domain-domain interactions from protein-protein interactions. *Genome Res.* 12(10):1540-8, 2002
15. Lee et al., Transcriptional Regulatory Networks in Saccharomyces cerevisiae, *Science* 298:799-804, 2002
16. Gene Ontology: tool for the unification of biology. The Gene Ontology Consortium (2000), *Nature Genet.* 25: 25-29, Dec. 2003
17. Xenarios I, et al., DIP: The Database of Interacting Proteins: 2001 update, *Nucleic Acids Res.* 29(1):239-41, 2001
18. Bar-Joseph Z, et al., Computational discovery of gene modules and regulatory networks, *Nat Biotechnol.* (11):1337-42, 2003
19. Breiman, L.Random Forests, *Machine Learning,* 45, 5-32, 2001
20. Elion, E.A. Ste5: a meeting place for MAP kinases and their associates, *Trends Cell Biol.* 5, 322-7, 1995
21. Supporting and supplementary information for this paper is available at http://www.cs.cmu.edu/~qyj/psb05_PPI.html
22. E.P. Xing, et al. Distance Metric Learning, with application to Clustering with side-information, *NIPS,* 2002